Significance of Tests for Petroleum Products

8th Edition

Salvatore J. Rand, Editor

ASTM Stock Number MNL1-8TH

INTERNATIONAL
Standards Worldwide
ASTM International
100 Barr Harbor Drive
PO Box C700
West Conshohocken, PA 19428-2959

Printed in U.S.A.

Library of Congress Cataloging-in-Publication Data

Significance of tests for petroleum products. — 8th ed. / [edited by] Salvatore J. Rand.
p. cm.
ISBN 978-0-8031-7001-8
1. Petroleum—Testing. 2. Petroleum products—Testing. I. Rand, Salvatore J., 1933–

TP691.M36 2009
665.5′38—dc22 2010003804

Printed in Newburyport, MA,
May, 2010

Foreword

THIS PUBLICATION, Significance of Tests for Petroleum Products: 8th Edition, was sponsored by ASTM committee D02 on Petroleum Products and Lubricants. The editor is Salvatore J. Rand, Consultant, North Fort Myers, Florida. This is the 8th edition of Manual 1 in the ASTM International manual series.

Dedication

To Mary, Cathy, Jeanne, Joseph, and John

Acknowledgments

This manual was brought to fruition by the combined efforts of many individuals. I would like to convey my sincerest appreciation to all of them, particularly the publication staff of ASTM International, especially Kathy Dernoga and Monica Siperko, who have given us much behind-the-scenes guidance and assistance from the outset of this venture. I also wish to thank Christine Urso of the American Institute of Physics, who was responsible for this logistically challenging project of handling the 24 chapters and 37 authors involved in this publication. In addition, I wish to convey my accolades to the authors, who are all experts in their particular fields and who bring a broad spectrum of topics and interests to this manual. They have devoted considerable time, energy, and resources to support this endeavor. I am also grateful to the 46 experts who reviewed the various chapters, who through their perusal of the chapters and their suggestions permitted good manuscripts to be made better. Finally, I would like to extend my appreciation to the industrial and governmental employers of all those involved in this publication. They ultimately make it possible for us to produce manuals such as this for the benefit of those who use petroleum standards worldwide.

Contents

1

Introduction

Salvatore J. Rand[1]

TECHNOLOGY IS UNDERGOING ACCELERATING change. No longer do people have to wait months or even years for analytical methods to be submitted to ASTM International, tested, and voted for approval. The response of the various committees of ASTM International to new developments in the industrial and petroleum industries, and to unexpected occurrences in the field, is both swift and focused. It is because of this unprecedented and exponential increase in new testing methods that *Manual 1* is being revised only 6 years after its prior publication.

Committee D02 on Petroleum Products and Lubricants has assumed the responsibility of revising *Manual on Significance of Tests for Petroleum Products (ASTM Manual Series: MNL 1)*, although other national and international standards organizations contribute significantly to the development of standard test methods for petroleum products. These organizations include the Energy Institute (EI), formerly known as the Institute of Petroleum in the United Kingdom, the Deutsches Institut für Normung (DIN) in Germany, the Association Française de Normalisation (AFNOR) in France, the Japanese Industrial Standards (JIS) in Japan, the CEN (European Committee for Standardization), and the International Organization for Standardization (ISO). Selected test methods from these organizations have been cross-referenced where relevant with ASTM International standards in selected chapters in this publication. There are discussions presently in progress to harmonize many standard test methods so they are technically equivalent to one another.

The chapters in this manual are not intended to be research papers or exhaustive treatises of a particular field. The purpose of the discussions herein is to answer two questions: What are the relevant tests that are done on various petroleum products and why do we perform these particular tests? All tests are designed to measure properties of a product such that the "quality" of that product may be described. I consider a workable definition of a quality product to be "that which meets agreed-on specifications." It is not necessary that the quality of a product be judged by its high purity, although it may very well be, but merely that it meets specifications previously agreed on among buyers, sellers, regulators, transferors, etc. The various chapters in this publication discuss individual or classes of petroleum products and describe the standardized testing that must be done on those products to assure all parties involved that they are dealing with quality products.

Since publication of the previous edition of the manual, not only has the number available but also the type of some petroleum products undergone dramatic changes. The result is that most products have had changes incorporated into their methods of test, and that these new procedures have been standardized and accepted into specifications as required. The generic petroleum products discussed in this eighth edition of *Manual 1* are similar to those products described in the chapters of the previous edition. All chapters with one exception have been updated to reflect new specification and testing standards, where applicable. Chapter 21, "Methods for the Environmental Testing of Petroleum Products," has been reprinted in its entirety from the previous edition because the test procedures and protocols have been essentially unchanged and the discussion of toxicity and biodegradation of petroleum products is relevant to today's products. In the discussion of some of the various petroleum products, selected sections of chapters have been retained from the seventh edition for the sake of completeness and to provide more complete background information. The authors of the chapters in the seventh edition have been credited in the footnotes to the appropriate chapters where necessary.

This edition has been enlarged by the inclusion of three new chapters to more fully reflect today's new products and new testing procedures, while the original 21 chapters contained in the seventh edition have been retained and updated. One new chapter, "Biodiesel," has been added in response to the worldwide interest in developing renewable fuels. In addition to oxygenates, which are generally blended for gasoline engines, specifications for diesel fuel are being changed to incorporate materials of biological origin for the purpose of sustainability of fuels products. Government regulators are mandating the use of biodiesel fuels ("biodiesels") and are presently in discussions with petroleum companies and engine manufacturers to ensure conformance with published timetables for the use of these fuels. Committee D02 has responded with the development of specifications and new test methods, as described in this new chapter.

Another new chapter is entitled "Synthetic Liquid Fuels." Again, due to the worldwide interest in diminishing dependence on traditional petroleum fuels, research in alternative fuels is being conducted by many organizations including petroleum companies. Specifications and test methods for synthetic fuels are continually being developed by Committee D02 to define the characteristics of these new materials, and these are discussed in the new chapter.

The various petroleum products, including crude oils, have always been tested to determine the qualitative and quantitative nature of inorganic substances contained therein. This is discussed in the new chapter "Determination of Inorganic Species in Petroleum Products and Lubricants." The techniques used are many and varied, the product and the nature and concentration of the inorganic species. In

[1] *Consultant, North Fort Myers, FL.*

addition, a number of unexpected problems have recently arisen in the field regarding inorganic materials affecting the performance of petroleum fuels. One such problem is the deposition of silicon dioxide on gasoline engine parts due to the contamination of gasoline with very small quantities of silicon. Another problem is the inactivation of silver alloy–sensing units in fuel tanks with the use of some low-sulfur gasoline fuels. Still another concern is the deposition of sulfate-containing materials in fuel metering systems and on fuel dispenser filters when certain ethanol batches are blended with gasoline. These problems require methods that measure inorganic contaminants at extremely low levels using new techniques, all of which are under development in Committee D02.

Many of the test procedures described in this manual are newer correlative methods, which represent the way of the future due to their simplicity, objectivity, economy, and, in many cases, portability. Quality assurance methods must be integrated into analytical procedures and protocols, so that we can demonstrate that these methods provide accuracy and precision equal to or better than the referee methods they supersede. A major thrust in analytical chemistry at the present is the development of methods that count individual molecules. While we have not yet achieved this level of sensitivity in the testing of petroleum products, when these new tools do arrive, and they will, we will be able to determine the concentration of an analyte in a petroleum product with 100% accuracy.

The chapters that follow show that the technology associated with the testing of petroleum products is advancing at an increasingly rapid rate. They also demonstrate that ASTM International continues to be the foremost standardization organization in the world.

2

Automotive Spark-Ignition Engine Fuel

Lewis M. Gibbs,[1] Ben R. Bonazza,[2] and Robert L. Furey[3]

AUTOMOTIVE SPARK-IGNITION ENGINE FUEL consists of gasoline or gasoline-oxygenate blends used in internal combustion spark-ignition engines, as opposed to engine fuels used in diesel or compression-ignition engines. These spark-ignition engine fuels are used primarily in passenger car and highway truck service. They are also used in off-highway utility trucks, farm machinery, two- and four-stroke cycle marine engines, and other spark-ignition engines used in a variety of service applications.

ASTM D4814, Specification for Automotive Spark-Ignition Engine Fuel, defines *gasoline* as a volatile mixture of liquid hydrocarbons, containing small amounts of additives. A *gasoline-oxygenate blend* is defined as a fuel consisting primarily of gasoline, along with a substantial amount of one or more oxygenates. An *oxygenate* is an oxygen-containing, ashless organic compound, such as an alcohol or ether, which can be used as a fuel or fuel supplement. Ethanol is the predominant oxygenate in use today. Spark-ignition engine fuel includes both gasolines and gasoline-oxygenate blends.

Gasoline is a complex mixture of relatively volatile hydrocarbons that vary widely in their physical and chemical properties. It is a blend of many hydrocarbons derived from the fractional distillation of crude petroleum and from complex refinery processes that increase either the amount or the quality of gasoline.

The hundreds of individual hydrocarbons in gasoline typically range from those having just four carbon atoms (designated C_4, composed of butanes and butenes) to those having as many as 11 carbon atoms (designated C_{11}, such as methylnaphthalene). The types of hydrocarbons in gasoline are paraffins, isoparaffins, naphthenes, olefins, and aromatics. The properties of commercial gasolines are predominantly influenced by the refinery practices that are used and partially influenced by the nature of the crude oils from which they are produced. Finished gasolines have a boiling range from about 30 to 225°C (86 to 437°F) in a standard distillation test.

Gasoline may be blended, or may be required to be blended, with oxygenates to improve the octane rating, extend the fuel supply, reduce vehicle exhaust emissions, or comply with regulatory requirements. The oxygenated components of spark-ignition engine fuel include aliphatic ethers, such as methyl *tert*-butyl ether (MTBE), and alcohols such as ethanol. The ethers are allowed by U.S. Environmental Protection Agency (EPA) regulations to be used in concentrations where they provide not more than 2.7 mass percent oxygen in the final fuel blend. Because of concerns over ground water contamination, MTBE is banned in many states and is no longer widely used in the United States. Ethanol and certain other alcohols may provide not more than 3.7 mass percent oxygen in the fuel. Legal restrictions exist on the use of methanol in gasoline, and it is not currently intentionally added to any gasolines marketed in the United States. These restrictions will be discussed later. The federal Renewable Fuel Standard (RFS) established under the Energy Independence and Security Act of 2007 requires a national minimum volume usage requirement of ethanol that increases annually until 2022. In addition, a number of states or portions of states mandate that spark-ignition engine fuel contain 10 volume percent ethanol blended with gasoline.

Spark-ignition engine fuels are blended to satisfy diverse automotive requirements. In addition, the fuels are exposed to a variety of mechanical, physical, and chemical environments. Therefore, the properties of the fuel must be balanced to give satisfactory engine performance over an extremely wide range of operating conditions. The prevailing standards for fuel represent compromises among the numerous quality, environmental, and performance requirements. Antiknock rating, distillation characteristics, vapor pressure, sulfur content, oxidation stability, corrosion protection, and other properties are balanced to provide satisfactory vehicle performance. In most gasolines, additives are used to provide or enhance specific performance features.

In recent years, there has been an ever-growing body of governmental regulations to address concerns about the environment. Initially, most of the regulations were aimed at the automobile and have resulted in technologies that have significantly reduced vehicle emissions. Regulations have also been aimed at compositional changes to the fuels that result in reduced vehicle emissions. The first major change in fuel composition was the introduction of unleaded gasoline in the early 1970s, followed by the phase-down of lead levels in leaded gasoline (1979–1986). Most passenger cars and light-duty trucks beginning with the 1975 model year have required unleaded fuel.

In 1989, the U.S. EPA implemented gasoline volatility regulations. Reductions in fuel vapor pressure limits during the summer were implemented under these regulations, followed by further reductions in 1992.

Beginning in 1987, several states required the addition of oxygenates to gasoline during the winter months in certain geographic areas to reduce vehicle carbon monoxide emissions. The added oxygenates are especially effective in reducing carbon monoxide during a cold start with older vehicles. When a vehicle is started cold, the catalyst is inactive and the computer is not controlling the air-fuel ratio in closed-loop mode. Added oxygen in the fuel leans the vehicle's fuel mixture, lowering carbon monoxide emissions.

The Clean Air Act Amendments of 1990 required additional compositional changes to automotive spark-ignition

[1] *Chevron Products Company, Richmond, CA.*
[2] *TI Automotive (retired), Lapeer, MI.*
[3] *Furey & Associates, LLC, Rochester Hills, MI.*

engine fuels. In November 1992, 39 areas failing to meet the federal standard for carbon monoxide were required to implement oxygenated fuel programs similar to those mentioned previously. There are also provisions in the act that address ozone nonattainment. Beginning in 1995, the use of a cleaner-burning "reformulated gasoline" was required in the nine worst ozone nonattainment areas. Other ozone nonattainment areas have the option of participating in the program. Federal reformulated gasoline is a gasoline-oxygenate blend certified to meet the specifications and emission reduction requirements established by the Clean Air Act Amendments of 1990; therefore, it would be more correctly referred to as federal reformulated spark-ignition engine fuel. Federal and state regulations frequently use the term "gasoline" to cover both gasoline and gasoline-oxygenate blends. (See ASTM Committee D02 on Petroleum Products and Lubricants Research Report D02: 1347, Research Report on Reformulated Spark-Ignition Engine Fuel for reformulated gasoline requirements and test methods.)

This chapter summarizes the significance of the more important physical and chemical characteristics of automotive spark-ignition engine fuel and describes pertinent test methods for defining or evaluating these properties. Information on governmental requirements is also provided. This discussion applies only to those fuels that can be used in engines designed for spark-ignition engine fuel. It does not include fuels that are primarily oxygenates, such as M85, a blend of 85 volume percent methanol and 15 volume percent gasoline, or E85, a blend of 85 volume percent ethanol and 15 volume percent gasoline, which are for use in flexible fuel vehicles. These fuels and the oxygenates commonly used in gasoline are discussed in detail in Chapter 3. [See ASTM D5797, Specification for Fuel Methanol (M70-M85) for Automotive Spark-Ignition Engines, or ASTM D5798, Specification for Fuel Ethanol (Ed75-Ed85) for Automotive Spark-Ignition Engines.]

GRADES OF SPARK-IGNITION ENGINE FUEL

Until 1970, with the exception of one brand of premium-grade fuel marketed on the East Coast and southern areas of the United States, all grades of automotive fuel contained lead alkyl compounds to increase the antiknock rating. The Antiknock Index [the average of the Research Octane Number (RON) and the Motor Octane Number (MON)] of the leaded premium-grade fuel pool increased steadily from about 82 at the end of World War II to about 96 in 1968. During the same time, the Antiknock Index of the leaded regular grade followed a parallel trend from about 77 to 90. Leaded fuel began to be phased out during the 1970s, and in 1996 all lead was banned from highway fuel.

In 1971, U.S. passenger car manufacturers began a transition to engines that would operate satisfactorily on fuels with lower octane ratings, namely, a minimum RON of 91. This octane level was chosen because unleaded fuels are needed to prolong the effectiveness of automotive emission catalyst systems and because unleaded fuels of 91 RON could be produced in the required quantities using refinery processing equipment then available. In 1970, fuel marketers introduced unleaded and low-lead fuels of this octane level to supplement the conventional leaded fuels already available.

Beginning in July 1974, the U.S. EPA mandated that most service stations have available a grade of unleaded fuel defined as having a lead content not exceeding 0.013 gram of lead/liter (g Pb/L) [0.05 gram of lead/U.S. gallon (0.05 g Pb/gal)] and a RON of at least 91. (This was changed to a minimum Antiknock Index of 87 in 1983, and the requirement was dropped in 1991.) Starting in the 1975 model year, most spark-ignition engine–powered automobiles and light-duty trucks required the use of unleaded fuel. With this requirement, low-lead fuels [0.13 g Pb/L (0.5 g Pb/gal)] disappeared. In addition, leaded premium began to be superseded by unleaded premium in the late 1970s and early 1980s. In the mid-1980s, an unleaded midgrade fuel became widely available, and many fuel marketers now offer three grades of unleaded fuel: regular, midgrade, and premium. Lead usage in motor fuels was banned entirely in California effective in 1992 and was banned from all U.S. reformulated fuels in 1995 and from all U.S. motor fuels in 1996. Leaded fuel can still be produced for off-road use and for use as a racing fuel.

ANTIKNOCK RATING

The definitions and test methods for antiknock rating for automotive spark-ignition engine fuels are set forth in Appendix X1 in ASTM D4814, Specification for Automotive Spark-Ignition Engine Fuel. Antiknock rating and volatility are perhaps the two most important characteristics of spark-ignition engine fuel. If the antiknock rating of the fuel is lower than that required by the engine, knock occurs. Knock is a high-pitch, metallic rapping noise. Fuel with an antiknock rating higher than that required for knock-free operation generally does not improve performance. However, vehicles equipped with knock sensors may show a performance improvement as the antiknock rating of the fuel is increased, provided that the antiknock rating of the fuel is lower than that required by the engine. Conversely, reductions in fuel antiknock rating may cause a loss in vehicle performance. The loss of power and the damage to an automotive engine due to knocking are generally not significant until the knock intensity becomes severe and prolonged.

Knock depends on complex physical and chemical phenomena highly interrelated with engine design and operating conditions. It has not been possible to characterize completely the antiknock performance of spark-ignition engine fuel with any single measurement. The antiknock performance of a fuel is related intimately to the engine in which it is used and the engine operating conditions. Furthermore, this relationship varies from one engine design to another and may even be different among engines of the same design, due to normal production variations.

The antiknock rating of a spark-ignition engine fuel is measured in single-cylinder laboratory engines. Two methods have been standardized and are presented in ASTM D2699/IP 237, Test Method for Research Octane Number of Spark-Ignition Engine Fuel, and ASTM D2700/IP 236, Test Method for Motor Octane Number of Spark-Ignition Engine Fuel. Another method used for quality control in fuel blending is given in ASTM D2885/IP 360, Test Method for Research and Motor Method Octane Ratings Using On-Line Analyzers.

These single-cylinder engine test procedures use a variable-compression-ratio engine. The Motor method operates at a higher speed and inlet mixture temperature than the Research method. The procedures relate the knocking characteristics of a test fuel to standard fuels, which are blends of two pure hydrocarbons: 2,2,4-trimethylpentane ("isooctane") and *n*-heptane. These blends are called *primary reference fuels*. By definition, the octane number of isooctane is 100, and the octane number of *n*-heptane is 0. At octane levels below 100, the octane number of a given fuel is the percentage by volume of isooctane in a blend with *n*-heptane that knocks with the same intensity at the same compression ratio as the fuel when compared by one of the standardized engine test methods. The octane number of a fuel greater than 100 is based on the volume of tetraethyl lead that must be added to isooctane to produce knock with the same intensity as the fuel. The volume of tetraethyl lead in isooctane is converted to octane numbers greater than 100 by use of tables included in the Research and Motor methods.

The octane number of a given blend of either isooctane and *n*-heptane or tetraethyl lead in isooctane is, by definition, the same for the Research and Motor methods. However, the RON and MON will rarely be the same for commercial fuels. Therefore, when considering the octane number of a given fuel, it is necessary to know the engine test method. RON is, in general, the better indicator of antiknock rating for engines operating at full throttle and low engine speed. MON is the better indicator at full throttle, high engine speed, and part throttle, low and high engine speed. The difference between RON and MON is called "sensitivity." According to recent surveys of U.S. commercial fuels, the average sensitivity is about 9 units for unleaded regular grade and about 10 units for unleaded premium grade.

For most automotive engines and operating conditions, the antiknock performance of a fuel will be between its RON and MON. The exact relationship is dependent on the vehicle and operating conditions. Antiknock Index [the average of RON and MON, that is, (R + M)/2] is a currently accepted method of relating RON and MON to actual road antiknock performance in vehicles. U.S. Federal Trade Commission (FTC) regulations require a label on each service station dispensing pump showing the minimum (R + M)/2 value of the fuel dispensed. For fuels sold in the United States, regular grade is typically 87 (R + M)/2 (often slightly lower at high altitudes), midgrade is typically about 89, and premium is typically 91 or higher. Other grades also exist. The terms used to describe the various grades (e.g., regular, midgrade, super, premium, etc.) vary among fuel marketers and location. With the FTC regulation, a consumer can match the (R + M)/2 value specified in the owner's manual with the value on the pump. Because octane quality is a marketing issue, ASTM does not specify a minimum Antiknock Index in ASTM D4814.

VOLATILITY

The volatility characteristics of a spark-ignition engine fuel are of prime importance to the driveability of vehicles under all conditions encountered in normal service. The large variations in operating conditions and wide ranges of atmospheric temperatures and pressures impose many limitations on a fuel if it is to give satisfactory vehicle performance. Fuels that vaporize too readily in pumps, fuel lines, carburetors, or fuel injectors will cause decreased fuel flow to the engine, resulting in hard starting, rough engine operation, or stoppage (vapor lock). Under certain atmospheric conditions, fuels that vaporize too readily can also cause ice formation in the throat of a carburetor, resulting in rough idle and stalling. This problem occurs primarily in older cars. Conversely, fuels that do not vaporize readily enough may cause hard starting and poor warm-up driveability and acceleration. These low-volatility fuels may also cause an unequal distribution of fuel to the individual cylinders.

The volatility of automotive spark-ignition engine fuel must be carefully "balanced" to provide the optimum compromise among performance features that depend on the vaporization behavior. Superior performance in one respect may give serious trouble in another. Therefore, volatility characteristics of automotive fuel must be adjusted for seasonal variations in atmospheric temperatures and geographic variations in altitude. Four common volatility properties are described later. The effect of these volatility parameters on the performance of the vehicle is also presented.

Vapor Pressure

One of the most common measures of fuel volatility is the vapor pressure at 37.8°C (100°F) measured in a chamber having a 4:1 ratio of air to liquid fuel. ASTM D323, Test Method for Vapor Pressure of Petroleum Products (Reid Method), can be used for hydrocarbon-only gasolines and gasoline-ether blends but not for gasoline-alcohol blends because traces of water in the apparatus can extract the alcohol from the blend and lead to incorrect results. Therefore, this method is no longer listed as an acceptable test method for spark-ignition engine fuels in ASTM D4814.

To avoid the alcohol–water interaction problem in Test Method D323, a similar method using the same apparatus and procedure, but maintaining dry conditions, has been developed–ASTM D4953, Test Method for Vapor Pressure of Gasoline and Gasoline-Oxygenate Blends (Dry Method). The results are reported as Dry Vapor Pressure Equivalent (DVPE) rather than Reid Vapor Pressure (RVP), which is only determined using Test Method D323. For hydrocarbon-only gasolines, there is no statistically significant difference in the results obtained by Test Methods D323 and D4953. Advances in instrumentation have led to the development of three other methods that can be used for both gasolines and gasoline-oxygenate blends. They are ASTM D5190, Test Method for Vapor Pressure of Petroleum Products (Automatic Method), ASTM D5191, Test Method for Vapor Pressure of Petroleum Products (Mini Method), and ASTM D5482, Test Method for Vapor Pressure of Petroleum Products (Mini Method-Atmospheric). The precision (repeatability and reproducibility) of these three methods is much better than that for D4953. Another method, ASTM D6378, Test Method for Determination of Vapor Pressure (VP_X) of Petroleum Products, Hydrocarbons, and Hydrocarbon-Oxygenate Mixtures (Triple Expansion Method), is reported to not require air saturation and cooling of the sample before testing. This test method reports results as VP_X. An equation is provided in the test method to convert the results to DVPE to determine compliance with Specification D4814 maximum limits.

The U.S. EPA and the California Air Resources Board use the D5191 test method. However, each uses a slightly different equation than that used by ASTM to calculate vapor pressure from the instrument's total pressure reading. The equation used depends on the brand of the instrument.

Distillation

The tendency of a fuel to vaporize is also characterized by determining a series of temperatures at which various percentages of the fuel have evaporated, as described in ASTM D86, Test Method for Distillation of Petroleum Products at Atmospheric Pressure. A plot of the results is commonly called the *distillation curve*. The 10, 50, and 90 volume percent evaporated temperatures are often used to characterize the volatility of spark-ignition engine fuel. The fuel also can be characterized by reporting the percentage evaporated at specified temperatures (e.g., E100). Two gas chromatography test methods that can be used to determine the distillation characteristics are ASTM D3710, Test Method for Boiling Range Distribution of Gasoline and Gasoline Fractions by Gas Chromatography, and D7096, Test Method for Determination of the Boiling Range Distribution of Gasoline by Wide-Bore Capillary Gas Chromatography. Two distillation test methods requiring considerably smaller sample sizes than Test Method D86 are D7344, Test Method for Distillation of Petroleum Products at Atmospheric Pressure (Mini Method), and D7345, Test Method for Distillation of Petroleum Products at Atmospheric Pressure (Micro Distillation Method). To improve the correlation of reported results with those of Test Method D86, bias corrections are provided.

Driveability Index

While each area of the distillation curve is important, the combination of the various points that describe the whole curve must be taken into account to describe adequately vehicle driveability. The ASTM Driveability Task Force, using data from the Coordinating Research Council (CRC) and others, has developed a correlation between various distillation points and vehicle cold-start and warm-up driveability. This correlation is called Driveability Index (DI) and is defined as: $DI = 1.5 \times T_{10} + 3.0 \times T_{50} + 1.0 \times T_{90} + 1.33°C\ (2.4°F) \times$ Ethanol Volume percent, where T_{10}, T_{50}, and T_{90} are the temperatures at the 10, 50, and 90 % evaporated points of a Test Method D86 distillation, respectively; 1.33 is the coefficient for the volume percent ethanol present when the distillation results are determined in degrees Celsius; and 2.4 is the coefficient when distillation results are determined in degrees Fahrenheit. The ethanol correction term is required because the reduction in the T_{50} resulting from the addition of ethanol does not provide as much improvement in driveability as would such a reduction by a hydrocarbon.

Vapor-Liquid Ratio

The vaporization tendency of spark-ignition engine fuel can also be expressed in terms of vapor-to-liquid ratio (V/L) at temperatures approximating those found in critical parts of the fuel system. The standard test method is ASTM D5188, Test Method for Vapor-Liquid Ratio Temperature Determination of Fuels (Evacuated Chamber Method). This method is applicable to samples for which the determined temperature is between 36 and 80°C and the V/L is between 8:1 and 75:1.

The fuel temperature at a V/L of approximately 20 ($T_{V/L=20}$) was shown to be indicative of the tendency of a fuel to cause vapor lock, as evidenced by loss of power during full-throttle accelerations. V/L–temperature relationships were originally developed for vehicles equipped with carburetors and suction-type fuel pumps. The applicability of such relationships to late-model vehicles equipped with fuel injection and pressurized fuel systems has been shown by CRC test programs. Appendix X2 of ASTM D4814 includes a computer method and a linear equation method that can be used for estimating V/L of spark-ignition engine fuels from vapor pressure and distillation test results. However, until recently these estimation methods were not applicable to gasoline-oxygenate blends. ASTM D4814 in Appendix X2 now provides equations for correcting the estimated values applicable to ethanol blends.

Volatility and Performance

In general terms, the following relationships between volatility and performance apply:

1. High vapor pressures and low 10 % evaporated temperatures are both conducive to ease of cold starting. However, under hot operating conditions, they are also conducive to vapor lock and increased vapor formation in fuel tanks, carburetors, and fuel injectors. The amount of vapor formed in fuel tanks and carburetors, which must be contained by the evaporative emissions control system, is related to the vapor pressure and distillation temperatures. Thus, a proper balance of vapor pressure and 10 % evaporated temperature must be maintained and seasonally adjusted for good overall performance.
2. Although vapor pressure is a factor in the amount of vapor formed under vapor locking conditions, vapor pressure alone is not a good index. A better index for measuring the vapor locking performance of spark-ignition engine fuels is the temperature at which the V/L is 20 at atmospheric pressure. The lower the temperature at which V/L = 20, the greater is the tendency to cause vapor lock and hot-fuel–handling driveability problems. Vapor lock is much less of a problem for fuel-injected cars, which have pressurized fuel systems. However, driveability symptoms are similar to carbureted vehicles; a too-volatile fuel in fuel-injected cars can cause hard starting and rough idling, and in the extreme, the car will not start.
3. The distillation temperature at which 50 % of the fuel has evaporated is a broad indicator of warm-up and acceleration performance under cold-starting conditions. The lower the 50 % evaporated temperature, the better is the performance. (This statement is not always valid for gasoline-oxygenate blends, especially those containing alcohol.) The temperatures for 10 and 90 % evaporated are also indicators of warm-up performance under cold-starting conditions, but to a lesser degree than the 50 % evaporated temperature. Lowering the 50 % evaporated point, within limits, also has been shown to reduce exhaust hydrocarbon emissions.
4. The temperatures for 90 % evaporated and the final boiling point, or end point, indicate the amount of relatively high-boiling components in gasoline. A high 90 % evaporated temperature, because it is usually associated with higher density and high-octane number components, may contribute to improved fuel economy and resistance to knock. If the 90 % evaporated temperature and the end point are too high, they can cause poor mixture distribution in the intake manifold and combustion chambers, increased hydrocarbon emissions, excessive combustion chamber deposits, and dilution of the crankcase oil.
5. DI represents the entire distillation curve. Lower values of DI mean greater volatility, which equates to better cold-start and warm-up driveability until some minimum level is reached where no further improvement is observed. If the DI is too high, vehicle cold-start and warm-up driveability can be adversely affected. Maximum DI for each volatility class is limited by ASTM D4814 and other specifications developed by motor vehicle manufacturers and by fuel suppliers. A DI specification limit allows a refiner more flexibility in blending spark-ignition engine fuel that provides proper cold-start and warm-up driveability, compared to tight restrictions on individual distillation points. As ambient temperature is reduced, fuels with lower DI

are required. The impact of oxygenates on DI and driveability is not well established. Some testing has shown that at the same DI level, poorer driveability occurs with oxygenated fuels. Other data have not shown this effect. The oxygenate effect may depend on the ambient temperature, type of oxygenate, and DI level of the fuel. The DI equation now contains a correction for ethanol blends. The CRC continues to investigate this issue.

ASTM D4814, Specification for Automotive Spark-Ignition Engine Fuel, includes a table of six volatility classes for vapor pressure, distillation temperatures, and DI, and a separate vapor lock protection table of six volatility classes for $T_{V/L=20}$. A combination of limits from these two tables defines the fuel volatility requirements for each month and geographic area of the United States. The specification also accounts for the EPA regulations on vapor pressure and state implementation plan (SIP) vapor pressure limits approved by the EPA. These volatility characteristics have been established on the basis of broad experience and cooperation between fuel suppliers and manufacturers and users of automotive vehicles and equipment. Spark-ignition engine fuels meeting this specification have usually provided satisfactory performance in typical passenger car service. However, certain equipment or operating conditions may require or permit variations from these limits. Modern vehicles, designed to exacting tolerances for good emission control, fuel economy, and driveability, may require more restrictive limits.

OTHER PROPERTIES

In addition to providing acceptable antiknock performance and volatility characteristics, automotive spark-ignition engine fuels must also provide for satisfactory engine and fuel system cleanliness and durability. The following properties have a direct bearing on the overall performance of a fuel.

Workmanship and Contamination

A finished fuel is expected to be visually free of undissolved water, sediment, and suspended matter. It should be clear and bright when observed at 21°C (70°F). It should also be free of any adulterant or contaminant that may render the fuel unacceptable for its commonly used applications. Physical contamination may occur during refining or distribution of the fuel. Control of such contamination is a matter requiring constant vigilance by refiners, distributors, and marketers. Solid and liquid contamination can lead to restriction of fuel metering orifices, corrosion, fuel line freezing, gel formation, filter plugging, and fuel pump wear. ASTM D2709, Test Method for Water and Sediment in Distillate Fuels by Centrifuge, or ASTM D2276/IP 216, Test Method for Particulate Contaminant in Aviation Fuel by Line Sampling, can be used to determine the presence of contaminants. Appendix X6 of ASTM D4814 contains a recommendation that all fuel dispensers be equipped with filters of 10-micron (micrometer) or less nominal pore size at point of delivery to the customer.

Petroleum products pick up microbes during refining, distribution, and storage. Most growth takes place where fuel and water meet. Therefore, it is most important to minimize water in storage tanks. Microbial contamination in fuel was not much of a problem until lead was removed. Appendix X5 of ASTM D4814 discusses microbial contamination and references ASTM D6469, Guide for Microbial Contamination in Fuels and Fuel Systems.

Lead Content

Constraints imposed by emission control regulations and health concerns have led to the exclusive availability of unleaded fuels for street and highway use. Leaded fuel is still allowed for nonroad use, such as for farm equipment and for racing. The lead content of unleaded fuel is limited to a maximum of 0.013 g Pb/L (0.05 g Pb/gal), but typical lead contents in U.S. unleaded fuels are 0.001 g Pb/L or less. Although the EPA regulations prohibit the deliberate addition of lead to unleaded fuels, some contamination by small amounts of lead can occur in the distribution system. Such occurrences are rare, because leaded fuel has been eliminated from the market.

The following methods are suitable for determining the concentration of lead in spark-ignition engine fuel:

FOR LEADED FUEL

ASTM D3341, Test Method for Lead in Gasoline–Iodine Monochloride Method

ASTM D5059/IP 228, Test Methods for Lead in Gasoline by X-Ray Spectroscopy

FOR UNLEADED FUEL

ASTM D3237, Test Method for Lead in Gasoline by Atomic Absorption Spectroscopy

ASTM D3348, Test Method for Rapid Field Test for Trace Lead in Unleaded Gasoline (Colorimetric Method)

ASTM D5059/IP 228, Test Methods for Lead in Gasoline by X-Ray Spectroscopy

Phosphorus Content

In the past, phosphorus compounds were sometimes added to leaded fuels as combustion chamber deposit modifiers. However, because phosphorus adversely affects exhaust emission control system components, particularly the catalytic converter, today EPA regulations limit its concentration in unleaded fuel to a maximum of 0.0013 g P/L (0.005 g P/gal). Furthermore, phosphorus may not be intentionally added to unleaded fuel in any concentration. The concentration of phosphorus can be determined by ASTM D3231, Test Method for Phosphorus in Gasoline.

Manganese Content

In the 1970s, methylcyclopentadienyl manganese tricarbonyl (MMT) was added to some unleaded fuels for octane improvement. However, the use of MMT was banned in 1977 in California. In October 1978, the EPA banned its use in unleaded fuel throughout the United States because it increased vehicle hydrocarbon emissions in various test programs, including the 63-vehicle CRC program in 1977. In 1995, after much testing and court action, MMT was granted a waiver by the EPA for use at a maximum concentration of 0.008 g Mn/L (0.031 g Mn/gal). According to the EPA's website, "the Agency determined that MMT added at 1/32 gpg Mn will not cause or contribute to regulated emissions failures of vehicles." Nevertheless, the use of MMT remains controversial. The EPA's website notes the agency's uncertainty about the health risks of using MMT. The manganese content of spark-ignition engine fuel can be determined by ASTM D3831, Test Method for Manganese in Gasoline by Atomic Absorption Spectroscopy.

Sulfur Content

Crude petroleum contains sulfur compounds, most of which are removed during refining. The maximum amount of sulfur as specified in ASTM D4814 is 0.0080 mass percent, which is the federal per-gallon maximum limit. The federal regulations also have a refinery annual average maximum limit of 0.0030 mass percent. There are a few exceptions for qualified refineries that expire by the end of 2010.

Sulfur oxides formed during combustion may be converted to acids that promote rusting and corrosion of engine parts and exhaust systems. Sulfur oxides formed in the exhaust are undesirable atmospheric pollutants. However, the contribution of automotive exhaust to total sulfur oxide emissions is negligible. Sulfur also reduces the effectiveness of exhaust gas catalytic converters. In 1996, California set an average maximum limit of 0.0030 mass percent and then at the end of 2003 lowered it to 0.0015 mass percent. More details on sulfur requirements are presented later in this chapter.

The sulfur content of spark-ignition engine fuel can be determined by the following methods:

- ASTM D1266/IP 107, Test Method for Sulfur in Petroleum Products (Lamp Method)
- ASTM D2622, Test Method for Sulfur in Petroleum Products by Wavelength Dispersive X-Ray Fluorescence Spectrometry
- ASTM D3120, Test Method for Trace Quantities of Sulfur in Light Liquid Hydrocarbons by Oxidative Microcoulometry
- ASTM D4045, Test Method for Sulfur in Petroleum Products by Hydrogenolysis and Rateometric Colorimetry
- ASTM D4294, Test Method for Sulfur in Petroleum and Petroleum Products by Energy-Dispersive X-Ray Fluorescence Spectrometry
- ASTM D5453, Test Method for Determination of Total Sulfur in Light Hydrocarbons, Spark Ignition Engine Fuel and Engine Oil by Ultraviolet Fluorescence
- ASTM D6334, Test Method for Sulfur in Gasoline by Wavelength Dispersive X-Ray Fluorescence
- ASTM D6445, Test Method for Sulfur in Gasoline by Energy-Dispersive X-Ray Fluorescence Spectrometry
- ASTM D6920, Test Method for Total Sulfur in Naphthas, Distillates, Reformulated Gasolines, Diesels, Biodiesels, and Motor Fuel by Oxidative Combustion and Electrochemical Detection
- ASTM D7039, Test Method for Sulfur in Gasoline and Diesel Fuel by Monochromatic Wavelength Dispersive X-Ray Fluorescence Spectrometry

It is important to review the sulfur content determination minimum and maximum levels before selecting a test method to ensure it is applicable for the test sample of interest.

The presence of free sulfur or reactive sulfur compounds can be detected by ASTM D130/IP 154, Test Method for Detection of Copper Corrosion from Petroleum Products by the Copper Strip Tarnish Test, or by ASTM D4952, Test Method for Qualitative Analysis for Active Sulfur Species in Fuels and Solvents (Doctor Test). Sulfur in the form of mercaptans can be determined by ASTM D3227/IP 342, Test Method for (Thiol Mercaptan) Sulfur in Gasoline, Kerosene, Aviation Turbine, and Distillate Fuels (Potentiometric Method).

Gum and Stability

During storage, fuels can oxidize slowly in the presence of air and form undesirable oxidation products such as peroxides and/or gum. These products are usually soluble in the fuel, but the gum may appear as a sticky residue on evaporation. These residues can deposit on carburetor surfaces, fuel injectors, and intake manifolds, valves, stems, guides, and ports. ASTM D4814 limits the solvent-washed gum content of spark-ignition engine fuel to a maximum of 5 mg/100 mL. ASTM D381/IP 131, Test Method for Gum Content in Fuels by Jet Evaporation, is used to determine gum content.

Many fuels are deliberately blended with nonvolatile oils or additives or both, which remain as residues in the evaporation step of the gum test. A heptane-washing step is, therefore, a necessary part of the procedure to remove such materials, so that the solvent washed gum may be determined. The unwashed gum content (determined before the heptanes-washing step) indicates the presence of nonvolatile oils or additives. ASTM Test Method D381/IP 131 also is used to determine the unwashed gum content. There is no specification limit for unwashed gum content in ASTM D4814.

Automotive fuels usually have a very low gum content when manufactured but may oxidize to form gum during extended storage. ASTM D525/IP 40, Test Method for Oxidation Stability of Gasoline (Induction Period Method), is a test to indicate the tendency of a fuel to resist oxidation and gum formation. It should be recognized, however, that the method's correlation with actual field service may vary markedly under different storage conditions and with different fuel blends. Most automotive fuels contain special additives (antioxidants) to prevent oxidation and gum formation. Some fuels also contain metal deactivators for this purpose. Commercial fuels available in service stations move rather rapidly from refinery production to vehicle usage and are not designed for extended storage. Fuels purchased for severe bulk storage conditions or for prolonged storage in vehicle fuel systems generally have additional amounts of antioxidant and metal deactivator added.

Although not designed for automotive fuel, ASTM D873, Test Method for Oxidation Stability of Aviation Fuels (Potential Residue Method), is sometimes used to evaluate the stability of fuel under severe conditions, and like ASTM D525, it can indicate the tendency of the fuel to oxidize. No correlation has been established between the results of this test and actual automotive service, but the comparative rankings of fuels tested by D873 are often useful.

Peroxides are undesirable in automotive fuel because they can attack fuel system elastomers and copper commutators in fuel pumps. Peroxides can participate in an autocatalytic reaction to form more peroxides, thus accelerating the deterioration of fuel system components. Also, peroxides reduce the octane rating of the fuel. Hydroperoxides and reactive peroxides can be determined by ASTM D3703, Test Method for Peroxide Number of Aviation Turbine Fuels, or by ASTM D6447, Test Method for Hydroperoxide Number of Aviation Turbine Fuels by Voltammetric Analysis.

Density and Relative Density

ASTM D4814 does not set limits on the density of spark-ignition engine fuels, because the density is fixed by the other chemical and physical properties of the fuel. Density relates to the volumetric energy content of the fuel–the more dense the fuel, the higher is the volumetric energy content. Density is important, also, because fuel is often bought and sold with reference to a specific temperature, usually 15.6°C (60°F). Because the fuel is usually not at the specified temperature, volume correction factors based on the change in density with temperature are used to correct the volume to that temperature. Volume correction factors for oxygenates differ somewhat from those for hydrocarbons, and work is in progress to determine precise correction factors for gasoline-oxygenate blends.

Rather than using absolute density (in units of kg/m^3, for example), relative density is often used. Relative density, or specific gravity, is the ratio of the mass of a given volume of fuel at a given temperature to the mass of an equal volume of water at the same temperature. Most automotive fuels have relative densities between 0.70 and 0.78 at 15.6°C (60°F).

The American Petroleum Institute (API) gravity is often used as a measure of a fuel's relative density, although this practice is now discouraged with the move toward the use of SI units. API gravity is based on an arbitrary hydrometer scale and is related to specific gravity at 15.6°C (60°F) as follows:

$$\text{API Gravity, Deg.} = \frac{141.5}{\text{sp gr}(15.6/15.6^\circ\text{C})} - 131.5 \qquad (1)$$

Fuel density is determined by ASTM D1298/IP 160, Test Method for Density, Relative Density (Specific Gravity), or API Gravity of Crude Petroleum and Liquid Petroleum Products by Hydrometer Method, or by ASTM D4052/IP 365, Test Method for Density and Relative Density of Liquids by Digital Density Meter.

Rust and Corrosion

Filter plugging and engine wear problems are reduced by minimizing rust and corrosion in fuel distribution and vehicle fuel systems. Modifications of ASTM D665/IP 135, Test Method for Rust-Preventing Characteristics of Inhibited Mineral Oil in the Presence of Water, are sometimes used to measure rust protection of fuels.

Silver Corrosion

Reactive trace sulfur compounds (elemental sulfur, hydrogen sulfide, and mercaptans) present in fuel under some circumstances can corrode or tarnish silver alloy fuel gage in-tank sending units, causing them to fail. To minimize the failure of the silver fuel gage sending units, the fuel must pass the silver corrosion test method described in Annex A1 in ASTM Specification D4814. The test method uses the ASTM Test Method D130 test apparatus except a silver coupon replaces the normal copper one. ASTM is working to develop a new silver corrosion test method. ASTM D4814 limits the silver corrosion rating to a maximum of "1."

Hydrocarbon Composition

The three major types of hydrocarbons in gasoline are the saturates (paraffins, isoparaffins, naphthenes), olefins, and aromatics. They are identified by ASTM D1319/IP 156, Test Method for Hydrocarbon Types in Liquid Petroleum Products by Fluorescent Indicator Adsorption. This method ignores oxygenates in the fuel and only measures the percentages of saturates, olefins, and aromatics in the hydrocarbon portion of the fuel. Therefore, the results must be corrected for any oxygenates that are present. ASTM D6293, Test Method for Oxygenates and Paraffin, Olefin, Naphthene, Aromatic (O-PONA) Hydrocarbon Types in Low-Olefin Spark Ignition Engine Fuel by Gas Chromatography, is another method. A more detailed compositional analysis can be determined using one of the following methods: ASTM D6729, Test Method for Determination of Individual Components in Spark Ignition Engine Fuels by 100 Metre Capillary High Resolution Gas Chromatography, ASTM D6730, Test Method for Determination of Individual Components in Spark Ignition Engine Fuels by 100 Metre Capillary (with Precolumn) High Resolution Gas Chromatography, or ASTM D6733, Test Method for Determination of Individual Components in Spark Ignition Engine Fuels by 50 Metre Capillary High Resolution Gas Chromatography.

The amount of benzene can be determined by ASTM D4053, Test Method for Benzene in Motor and Aviation Gasoline by Infrared Spectroscopy. The amounts of benzene and other aromatics can be determined by ASTM D3606, Test Method for Benzene and Toluene in Finished Motor and Aviation Gasoline by Gas Chromatography, although there are interferences from methanol and ethanol. ASTM D5580, Test Method for the Determination of Benzene, Toluene, Ethylbenzene, p/m-Xylene, o-Xylene, C_9 and Heavier Aromatics, and Total Aromatics in Finished Gasoline by Gas Chromatography, and ASTM D5769, Test Method for Determination of Benzene, Toluene, and Total Aromatics in Finished Gasoline by Gas Chromatography/Mass Spectrometry, can also be used. Another method for the determination of aromatics is ASTM D5986, Test Method for Determination of Oxygenates, Benzene, Toluene, C_8-C_{12} Aromatics and Total Aromatics in Finished Gasoline by Gas Chromatography/Fourier Transform Infrared Spectroscopy. The benzene content of reformulated gasoline is limited to 1 volume percent by legislation, because benzene is considered toxic and a carcinogen. Beginning in 2011 under the Mobile Source Air Toxics (MSAT) Rule, benzene will be controlled for all gasoline at a refinery maximum average of 0.62 volume percent with a credit and trading program.

The total olefin content of automotive fuel can be determined by ASTM D6296, Test Method for Total Olefins in Spark-Ignition Engine Fuels by Multi-dimensional Gas Chromatography, or by ASTM D6550, Test Method for Determination of Olefin Content of Gasolines by Supercritical-Fluid Chromatography. The latter method has recently been designated by the California Air Resources Board as their standard test method for olefins.

Oxygenates

Oxygenates are discussed in detail later in this chapter, and additional information on oxygenates is presented in Chapter 3. Nevertheless, it is appropriate to mention here that alcohols or ethers are often added to gasoline to improve octane rating, extend the fuel supply, or reduce vehicle emissions. Certain governmental regulations require such addition, as will be discussed. Consequently, it is often necessary to determine the oxygenate content or the oxygen content of spark-ignition engine fuels. ASTM D4815, Test Method for Determination of MTBE, ETBE, TAME, DIPE, *tertiary*-Amyl Alcohol and C_1 to C_4 Alcohols in Gasoline by Gas Chromatography, can be used to determine the identity and concentrations of low-molecular-weight aliphatic alcohols and ethers. Alternative methods for determining the amounts of oxygenates are ASTM D5599, Test Method for Determination of Oxygenates in Gasoline by Gas Chromatography and Oxygen Selective Flame Ionization Detection, and ASTM D5845, Test Method for Determination of MTBE, ETBE, TAME, DIPE, Methanol, Ethanol and *tert*-Butanol in Gasoline by Infrared Spectroscopy. Appendix X4 in Specification D4814 describes a procedure for calculating the oxygen content of the fuel from the oxygenate content.

Additives

Fuel additives are used to provide or enhance various performance features related to the satisfactory operation of engines, as well as to minimize fuel handling and storage problems. These chemicals complement refinery processing in attaining the desired level of product quality. The most commonly used additives are listed in Table 1. With few

TABLE 1—Commercial Spark-Ignition Engine Fuel Additives

Class	Function	Additive Type
Oxidation inhibitors (antioxidants)	Minimize oxidation and gum formation	Aromatic amines and hindered phenols
Corrosion inhibitors	Inhibit ferrous corrosion in pipelines, storage tanks, and vehicle fuel systems	Carboxylic acids and carboxylates
Silver corrosion inhibitors	Inhibit corrosion of silver fuel gage sender units	Substituted thiadiazole
Metal deactivators	Inhibit oxidation and gum formation catalyzed by ions of copper and other metals	Chelating agent
Carburetor/injector detergents	Prevent and remove deposits in carburetors and port fuel injectors	Amines, amides, and amine carboxylates
Deposit control additives	Remove and prevent deposits throughout fuel injectors, carburetors, intake ports and valves, and intake manifold	Polybutene amines and polyether amines
Demulsifiers	Minimize emulsion formation by improving water separation	Polyglycol derivatives
Anti-icing additives	Minimize engine stalling and starting problems by preventing ice formation in the carburetor and fuel system	Surfactants, alcohols, and glycols
Antiknock compounds	Improve octane quality of automotive fuel	Lead alkyls and methylcyclopentadienyl manganese tricarbonyl
Dyes, markers	Identification of automotive fuel	Oil-soluble solid and liquid dyes, organic fluorescent compounds

Note. Some materials are multifunctional or multipurpose additives, performing more than one function. Source: SAE J312-Automotive Gasolines, Society of Automotive Engineers, Inc.

exceptions, standardized test methods are not available to determine the identity and concentration of specific additives. As mentioned previously, standard test methods are available for determining lead, manganese, and oxygenate content.

U.S. LEGAL REQUIREMENTS FOR SPARK-IGNITION ENGINE FUEL

Fuel Composition

The U.S. EPA has established vehicle exhaust and evaporative emissions standards as part of the U.S. effort to attain acceptable ambient air quality. To meet these EPA vehicle requirements, extensive modifications have been made to automotive engines and emissions systems. Because some fuel components can harm the effectiveness of vehicle emissions control systems, the EPA also exercises control over automotive fuels. EPA regulations on availability of unleaded automotive fuels, and on limits of lead, phosphorus, and manganese contents in the fuel, have been mentioned.

In addition, the Clean Air Act Amendments of 1977 prohibit the introduction into U.S. commerce, or increases in the concentration of, any fuel or fuel additive for use in 1975 and later light-duty motor vehicles, which is not "substantially similar" to the fuel or fuel additives used in the emissions certification of such vehicles.

The EPA considers fuels to be "substantially similar" if the following criteria are met:

1. The fuel must contain carbon, hydrogen, and oxygen, nitrogen, and/or sulfur, exclusively, in the form of some combination of the following:
   a. Hydrocarbons
   b. Aliphatic ethers
   c. Aliphatic alcohols other than methanol
      (i) Up to 0.3 % methanol by volume
      (ii) Up to 2.75 % methanol by volume with an equal volume of butanol or higher-molecular-weight alcohol
   d. A fuel additive at a concentration of no more than 0.25 % by weight, which contributes no more than 15 ppm sulfur by weight to the fuel.
2. The fuel must contain no more than 2.0 % oxygen by weight, except fuels containing aliphatic ethers and/or alcohols (excluding methanol) and must contain no more than 2.7 % oxygen by weight. (*Note.* As mentioned previously, ethanol and certain other alcohols have received waivers allowing as much as 3.7 % oxygen in the fuel.)
3. The fuel must possess, at the time of manufacture, all of the physical and chemical characteristics of an unleaded gasoline, as specified by ASTM Standard D4814-88, for at least one of the Seasonal and Geographical Volatility Classes specified in the standard. (*Note.* The EPA's February 11, 1991, notice specified the 1988 version of D4814.)
4. The fuel additive must contain only carbon, hydrogen, and any one or all of the following elements: oxygen, nitrogen, and/or sulfur.

Fuels or fuel additives that are not "substantially similar" may only be used if a waiver of this prohibition is obtained from the EPA. Manufacturers of fuels and fuel additives must apply for such a waiver and must establish to the satisfaction of the EPA that the fuel or additive does not cause or contribute to a failure of any emission control device or system over the useful life of the vehicle for which it was certified. Under prior law, if the EPA Administrator had not acted to grant or deny the waiver within 180 days after its filing, the waiver was treated as granted. The waiver process has been changed to now require the EPA to act within 270 days. The EPA has granted several waivers for gasoline-oxygenate blends. The reader is referred to the EPA for the latest information on waivers and the conditions under which they may be used.

Any fuel or fuel additive that had a waiver as of May 27, 1994, has to have had a supplemental registration with additional toxics data by November 27, 1994, to continue marketing the material. These registered products are subjected to a three-tier toxicological testing program. A new fuel or additive that was not registered as of May 27, 1994, will not be registered until all Tier 1 and Tier 2 information has been supplied. At present, no methanol-containing fuel additive has obtained a supplemental registration, and therefore, the addition of methanol to gasoline is prohibited.

Volatility

Concerns over increased evaporative emissions prompted the EPA to promulgate regulations that, beginning in 1989, reduced fuel vapor pressure. Spark-ignition engine fuels sold between June 1 and September 15 of each year were limited to maximum vapor pressures of 9.0, 9.5, or 10.5 psi, depending on the month and the region of the country. (Vapor pressure restrictions applied to fuels in the distribution system as early as May 1.) In 1992, the EPA implemented Phase II of the volatility controls, which limited fuels sold between June 1 and September 15 to a maximum vapor pressure of 9.0 psi. The regulations are more restrictive in ozone nonattainment areas in the southern and western areas of the United States, where fuels sold during certain months of the control period are limited to a maximum vapor pressure of 7.8 psi. The EPA permits conventional (i.e., not reformulated) fuels containing between 9 and 10 volume percent ethanol to have a vapor pressure 1.0 psi higher than the maximum limit for other fuels.

California was the first state to control spark-ignition engine fuel vapor pressure and, in 1971, mandated a maximum vapor pressure limit of 9.0 psi. By 1992, the maximum vapor pressure limit was lowered to 7.8 psi. In 1996, it was further lowered to 7.0 psi maximum. A number of other states have set maximum limits on vapor pressure in certain areas as part of their SIPs. The EPA vapor pressure limits and the EPA-approved SIP limits are an integral part of ASTM D4814.

Sulfur Regulations

California's Phase 2 reformulated gasoline specification limited the maximum sulfur content of fuel to 30 ppm average, with an 80 ppm cap. On December 31, 2003, new Phase 3 specifications lowered the sulfur maximum to 15 ppm average and the cap limits to 60 ppm. The cap limits were further reduced to 30 ppm on December 31, 2005.

Federal Tier 2 regulations required that in 2004, refiners meet an annual corporate average sulfur level of 120 ppm, with a cap of 300 ppm. In 2005, the required refinery average was 30 ppm, with a corporate average of 90 ppm and a cap of 300 ppm. Both of the average standards can be met with the use of credits generated by other refiners who reduce sulfur levels early. Beginning in 2006, refiners were required to meet a final 30 ppm average with a cap of 80 ppm. Fuel produced for sale in parts of the western United States must comply with a 150-ppm refinery average and a 300-ppm cap through 2006 but are required to meet the 30-ppm average/80-ppm cap by 2007. Refiners demonstrating a severe economic hardship may apply for an extension of up to two years. The regulations provide for some special sulfur limit exemptions for small refineries relating to the early introduction of ultralow sulfur diesel fuel, but these all expire at the end of 2010. The regulations include an averaging program. Some states include fuel sulfur limits in their SIPs.

Sampling, Containers, and Sample Handling

Correct sampling procedures are critical to obtain a sample representative of the lot intended to be tested. ASTM D4057, Practice for Manual Sampling of Petroleum and Petroleum Products, provides several procedures for manual sampling. ASTM D4177, Practice for Automatic Sampling of Petroleum and Petroleum Products, provides automatic sampling procedures. For volatility determinations of a sample, ASTM D5842, Practice for Sampling and Handling of Fuels for Volatility Measurement, contains special precautions for sampling and handling techniques to maintain sample integrity. ASTM D4306, Practice for Aviation Fuel Sample Containers for Tests Affected by Trace Contamination, should be used to select appropriate containers, especially for tests sensitive to trace contamination. Also ASTM D5854, Practice for Mixing and Handling of Liquid Samples of Petroleum and Petroleum Products, provides procedures for container selection and sample mixing and handling. For octane number determination, protection from light is important. Collect and store fuel samples in an opaque container, such as a dark brown glass bottle, metal can, or minimally reactive plastic container, to minimize exposure to UV emissions from sources such as sunlight or fluorescent lamps.

Oxygenated Fuel Programs and Reformulated Spark-Ignition Engine Fuel

In January 1987, Colorado became the first state to mandate the use of oxygenated fuels in certain areas during the winter months to reduce vehicle carbon monoxide (CO) emissions. By 1991, areas in Arizona, Nevada, New Mexico, and Texas had also implemented oxygenated-fuels programs.

The 1990 amendments to the Clean Air Act require the use of oxygenated fuels in 39 CO nonattainment areas during the winter months, effective November 1992. The program had to be implemented by the states using one of the following options. If averaging is allowed, the average fuel oxygen content must be at least 2.7 mass percent, with a minimum oxygen content of 2.0 mass percent in each gallon of fuel. Without averaging, the minimum oxygen content of each fuel must be 2.7 mass percent. (This is equivalent to about 7.3 volume percent ethanol or 15 volume percent MTBE.) The first control period was November 1, 1992, through January or February 1993, depending on the area. Subsequent control periods can be longer in some areas. Over time a number of states have come into conformance with CO regulations, and only about eight states still require wintertime ethanol requirements.

Beginning in 1995, the nine areas with the worst ozone levels, designated as extreme or severe, were required to sell reformulated spark-ignition engine fuel. Later four additional areas were added, but two are still pending implementation. Areas with less severe ozone levels were permitted to participate in ("opt-in" to) the program. Initially, about 37 other ozone nonattainment areas opted into participating in the program. Since then, about 17 have chosen to opt-out of the program. The reformulated fuel program is directed toward reducing ground level ozone and toxics concentrations.

The Clean Air Act Amendments of 1990 set specific guidelines for reformulated spark-ignition engine fuel for 1995 through 1997. Fuels sold in the control areas were required to meet the specifications of what is called the "Simple Model." Limits were established for vapor pressure (June 1 through September 15) and benzene content, deposit control additives were required in all fuels, and the use of heavy-metal additives was prohibited. A minimum oxygen content of 2.0 mass percent was required all year (averaged). The sulfur and olefin contents and the 90 % evaporated temperature were not allowed to exceed 125 % of the average values of the refiner's 1990 fuels. The use of the "Simple Model" expired December 31, 1997.

Effective January 1, 1998, a "Complex Model" had to be used for determining conformance to standards for reformulated fuel blends. Fuel properties in the "Complex Model" included vapor pressure, oxygen content, aromatics content, benzene content, olefins content, sulfur content, E200 and E300 (distillation properties), and the particular oxygenate used. The benzene limit, the ban on heavy metals, the minimum oxygen content, and the requirement for a deposit control additive remained the same as under the "Simple Model." As a result of the adoption of the RFS, the minimum oxygen requirement for reformulated fuel was eliminated effective in 2006.

The Clean Air Act Amendments of 1990 also contain an antidumping provision. In the production of reformulated spark-ignition engine fuel, a refiner cannot "dump" into its "conventional" fuel pool those polluting components removed from the refiner's reformulated fuel. These requirements apply to all fuel produced, imported, and consumed in the United States and its territories.

In 1992, California instituted its Phase 1 fuel regulations, which were followed in 1996 by its Phase 2 reformulated fuel regulations. The Phase 2 specifications controlled vapor pressure, sulfur content, benzene content, aromatics content, olefins content, 50 % evaporated point, and 90 % evaporated point. These same variables were used in California's "Predictive Model," which is similar to the federal "Complex Model," but with different equations. Beginning December 31, 2003, California required fuel to meet a Phase 3 reformulated fuel regulation.

An excellent source of information on reformulated fuels (federal and California) and their associated requirements can be found in the ASTM Committee D02 on Petroleum Products and Lubricants Research Report D02: 1347, Research Report on Reformulated Spark-Ignition Engine Fuel for current federal and state future reformulated fuel (cleaner burning fuels) requirements and approved test methods.

Renewable Fuel Standard

In 2007, the EPA finalized regulations for the RFS, which was authorized by the Energy Policy Act of 2005. The RFS establishes a minimum requirement for the volume of renewable fuels blended into automotive spark-ignition and diesel fuels. The national minimum volume requirement started at 4.0 billion gallons per year of renewable fuel in 2006 and increases to 7.5 billion gallons per year in 2012. Each producer and importer of fuel in the United States is obligated to demonstrate compliance with this requirement based on the pro rata share of fuel it produces or imports. With the passage of the Energy Independence and Security Act (EISA) of 2007, the amount of renewable fuels required was increased to 15.2 billion gallons per year in 2012 and ends with a requirement of 36.0 billion gallons per year by 2022. The proportional requirement for cellulosic biofuel in the act begins in 2010 and scales up to 16.0 billion gallons per year by 2022.

Deposit Control Additive Requirements

California in 1992 and the EPA in 1995 required the use of deposit control additives to minimize the formation of fuel injector and intake valve deposits. Both California and the EPA required that additives be certified in specified test fuels in vehicle tests. The fuel injector test procedure is ASTM D5598, Test Method for Evaluating Unleaded Automotive Spark-Ignition Engine Fuel for Electronic Port Fuel Injector Fouling, and the intake valve deposit test procedure is ASTM D5500, Test Method for Vehicle Evaluation of Unleaded Automotive Spark-Ignition Engine Fuel for Intake Valve Deposit Formation. ASTM developed more recent, nonvehicle versions of these tests for consideration by the EPA. These are ASTM D6201, Test Method for Dynamometer Evaluation of Unleaded Spark-Ignition Engine Fuel for Intake Valve Deposit Formation, and ASTM D6421, Test Method for Evaluating Automotive Spark-Ignition Engine Fuel for Electronic Port Fuel Injector Fouling by Bench Procedure.

GASOLINE-OXYGENATE BLENDS

Blends of gasoline with oxygenates are common in the U.S. marketplace and, in fact, are required in certain areas, as discussed previously. These blends consist primarily of gasoline with substantial amounts of oxygenates, which are oxygen-containing, ashless, organic compounds such as alcohols and ethers. The most common oxygenate in the United States is ethanol. MTBE was widely used but has been phased out in many states because of concern over ground water pollution. It is still used in some European countries as an octane trimming agent. Other ethers, such as ethyl *tert*-butyl ether (ETBE), *tert*-amyl methyl ether (TAME), and diisopropyl ether (DIPE), are receiving some attention, but have not yet achieved widespread use. Like MTBE, these ethers also are banned in some states. Methanol/*tert*-butyl alcohol mixtures were blended with gasoline on a very limited scale in the early 1980s but cannot be used now until they have a supplemental toxics registration. When methanol was used as a blending component, it had to be accompanied by a cosolvent (a higher-molecular-weight alcohol) to help prevent phase separation of the methanol and gasoline in the presence of trace amounts of water. EPA waiver provisions also required corrosion inhibitors in gasoline-methanol blends.

ASTM D4806, Specification for Denatured Fuel Ethanol for Blending with Gasolines for Use as Automotive Spark-Ignition Engine Fuel, describes a fuel-grade ethanol that is suitable for blending with gasoline. ASTM D5983, Specification for Methyl Tertiary-Butyl Ether (MTBE) for Downstream Blending with Automotive Spark-Ignition Fuel, provides limits for blending MTBE in gasoline.

Sampling of Gasoline-Oxygenate Blends

Sampling of blends can be conducted according to the procedures discussed earlier; however, water displacement must not be used, because of potential problems associated with the interaction of water with oxygenates contained in some spark-ignition engine fuels.

Test Methods for Gasoline-Oxygenate Blends

Some of the test methods originally developed for gasoline can be used for gasoline-oxygenate blends, while certain other test methods for gasoline are not suitable for blends. To avoid the necessity of determining in advance whether a fuel contains oxygenates, ASTM D4814 now specifies test methods that can be used for both gasolines and gasoline-oxygenate blends. This has been made possible by experience with some test methods, modification of existing test methods, and the development of new ones. Gasoline-ethanol blends are not included in the scopes of many test methods, and the precision statements do not apply. ASTM is working to modify the scopes and develop precision statements for the test methods specified in ASTM D4814 to cover gasoline-ethanol blends. Additional test methods and limits need to be developed to protect against incompatibility with elastomers and plastics, corrosion of metals, and other factors that may affect vehicle performance and durability.

In general, the test methods discussed previously for determining distillation temperatures, lead content, sulfur content, copper corrosion, solvent-washed gum, and oxidation stability can be used for both gasolines and gasoline-oxygenate blends. In some cases, standard solutions with which to calibrate the instrument must be prepared in the same type of fuel blend as the sample to be analyzed.

Some of the test methods for vapor pressure and vapor/liquid ratio are sensitive to the presence of oxygenates in the fuel, and approved procedures were discussed earlier in this chapter.

Water Tolerance

The term "water tolerance" is used to indicate the ability of a gasoline-alcohol blend to dissolve water without phase separation. Gasoline and water are almost entirely immiscible and will readily separate into two phases. Gasoline-alcohol blends will dissolve some water but will also separate into two phases when contacted with more water than they can dissolve. This water can be absorbed from ambient air or can occur as liquid water in the bottom of tanks in the storage, distribution, and vehicle fuel system. When gasoline-alcohol blends are exposed to a greater amount of water than they can dissolve, about 0.1 to 0.7 mass percent water, they separate into a lower alcohol-rich aqueous phase and an upper alcohol-poor hydrocarbon phase. The aqueous phase can be corrosive to metals, and the engine cannot operate on it. Because the fuel pump is at the bottom of an automotive fuel tank, the aqueous phase will be sent to the engine if the fuel separates. Therefore, this type of phase separation is undesirable. Separation occurs more readily with the lower-molecular-weight alcohols and at lower alcohol concentrations. With ethanol, the 10 volume percent levels used in the United States are easily handled; however, the 5 volume percent levels used in Europe are much more sensitive to separation. Several years of experience in California with 5.7 volume percent ethanol has shown no phase separation problems using ethanol meeting a 1.0 volume percent maximum water content limit.

Phase separation can usually be avoided if the fuels are sufficiently water free initially and care is taken during distribution to prevent contact with water. Formation of a haze must be carefully distinguished from separation into two distinct phases with a more or less distinct boundary. Haze formation is not grounds for rejection. Actual separation into two distinct phases is the criterion for failure. The test method originally developed to measure the water tolerance of ethanol blends was determined in an interlaboratory study to not be sufficiently accurate and was withdrawn. The limits were removed from the specification section of ASTM D4814 and placed in Appendix X8 for reference. The need for a water tolerance test is still thought to be important, and a water tolerance specification would be included in ASTM D4814 if a suitable test can be developed.

Compatibility with Plastics and Elastomers

Plastics and elastomers used in current automotive fuel systems such as gaskets, O-rings, diaphragms, filters, seals, etc., may be affected in time by exposure to motor fuels. These effects include dimensional changes, embrittlement, softening, delamination, increase in permeability, loss of plasticizers, and disintegration. Certain gasoline-oxygenate blends can aggravate these effects.

The effects depend on the type and amount of the oxygenates in the blend, the aromatics content of the gasoline, the generic polymer and specific composition of the elastomeric compound, the temperature and duration of contact, and whether the exposure is to liquid or vapor.

Currently, there are no generally accepted tests that correlate with field experience to allow estimates of tolerance of specific plastics or elastomers to oxygenates.

Metal Corrosion

Corrosion of metals on prolonged contact with gasolines alone can be a problem, but it is generally more severe with gasoline-alcohol blends. When gasoline-alcohol blends are contacted by water, the aqueous phase that separates is particularly aggressive in its attack on fuel system metals. The tern (lead-tin alloy) coating on fuel tanks and aluminum, magnesium, and zinc castings and steel components such as fuel senders, fuel lines, pump housings, and injectors are susceptible.

A number of test procedures, other than long-term vehicle tests, have been used or proposed to evaluate the corrosive effects of fuels on metals. The tests range from static soaking of metal coupons to operation of a complete automotive fuel system. None of these tests has yet achieved the status of an ASTM standard.

Applicable ASTM Specifications

ASTM	Title
D4806	Specification for Denatured Fuel Ethanol for Blending with Gasolines for Use as Automotive Spark-Ignition Engine Fuel
D4814	Specification for Automotive Spark-Ignition Engine Fuel
D5797	Specification for Fuel Methanol (M70-M85) for Automotive Spark-Ignition Engines
D5798	Specification for Fuel Ethanol (Ed75-Ed85) for Automotive Spark-Ignition Engines
D5983	Specification for Methyl Tertiary-Butyl Ether (MTBE) for Downstream Blending with Automotive Spark-Ignition Engine Fuel
D02:1347	Committee D02 Research Report on Reformulated Spark-Ignition Engine Fuel

Applicable ASTM/IP Test Methods

Before using any test method, the scope shall be reviewed to make sure the test method is applicable to the product being tested and that the specified measurement range covers the area of interest.

ASTM	IP	Title
D86		Test Method for Distillation of Petroleum Products at Atmospheric Pressure
D130	154	Test Method for Detection of Copper Corrosion from Petroleum Products by the Copper Strip Tarnish Test
D323		Test Method for Vapor Pressure of Petroleum Products (Reid Method)
D381		Test Method for Gum Content in Fuels by Jet Evaporation
D525	40	Test Method for Oxidation Stability of Gasoline (Induction Period Method)
D665	135	Test Method for Rust-Preventing Characteristics of Inhibited Mineral Oil in the Presence of Water
D873	138	Test Method for Oxidation Stability of Aviation Fuels (Potential Residue Method)
D1266	107	Test Method for Sulfur in Petroleum Products (Lamp Method)
D1298	160	Test Method for Density, Relative Density (Specific Gravity), or API Gravity of Crude Petroleum and Liquid Petroleum Products by Hydrometer Method
D1319	156	Test Method for Hydrocarbon Types in Liquid Petroleum Products by Fluorescent Indicator Adsorption
D2276	216	Test Method for Particulate Contaminant in Aviation Fuel by Line Sampling
D2622		Test Method for Sulfur in Petroleum Products by Wavelength Dispersive X-Ray Fluorescence Spectrometry
D2699	237	Test Method for Research Octane Number of Spark-Ignition Engine Fuel
D2700	236	Test Method for Motor Octane Number of Spark-Ignition Engine Fuel
D2709		Test Method for Water and Sediment in Distillate Fuels by Centrifuge
D2885	360	Test Method for Research and Motor Method Octane Ratings Using On-Line Analyzers
D3120		Test Method for Trace Quantities of Sulfur in Light Liquid Petroleum Hydrocarbons by Oxidative Microcoulometry
D3227	342	Test Method for (Thiol Mercaptan) Sulfur in Gasoline, Kerosene, Aviation Turbine, and Distillate Fuels (Potentiometric Method)
D3231		Test Method for Phosphorus in Gasoline
D3237		Test Method for Lead in Gasoline by Atomic Absorption Spectroscopy
D3341		Test Method for Lead in Gasoline–Iodine Monochloride Method
D3348		Test Method for Rapid Field Test for Trace Lead in Unleaded Gasoline (Colorimetric Method)
D3606		Test Method for Determination of Benzene and Toluene in Finished Motor and Aviation Gasoline by Gas Chromatography
D3703		Test Method for Peroxide Number of Aviation Turbine Fuels
D3710		Test Method for Boiling Range Distribution of Gasoline and Gasoline Fractions by Gas Chromatography
D3831		Test Method for Manganese in Gasoline by Atomic Absorption Spectroscopy
D4045		Test Method for Sulfur in Petroleum Products by Hydrogenolysis and Rateometric Colorimetry
D4052	365	Test Method for Density and Relative Density of Liquids by Digital Density Meter
D4053		Test Method for Benzene in Motor and Aviation Gasoline by Infrared Spectroscopy
D4057		Practice for Manual Sampling of Petroleum and Petroleum Products
D4177		Practice for Automatic Sampling of Petroleum and Petroleum Products
D4294		Test Method for Sulfur in Petroleum and Petroleum Products by Energy-Dispersive X-Ray Fluorescence Spectrometry
D4306		Practice for Aviation Fuel Sample Containers for Tests Affected by Trace Contamination
D4815		Test Method for Determination of MTBE, ETBE, TAME, DIPE, *tertiary*-Amyl Alcohol and C_1 to C_4 Alcohols in Gasoline by Gas Chromatography
D4952		Test Method for Qualitative Analysis for Active Sulfur Species in Fuels and Solvents (Doctor Test)
D4953		Test Method for Vapor Pressure of Gasoline and Gasoline-Oxygenate Blends (Dry Method)
D5059	228	Test Methods for Lead in Gasoline by X-Ray Spectroscopy
D5188		Test Method for Vapor-Liquid Ratio Temperature Determination of Fuels (Evacuated Chamber Method)
D5190		Test Method for Vapor Pressure of Petroleum Products (Automatic Method)
D5191		Test Method for Vapor Pressure of Petroleum Products (Mini Method)
D5453		Test Method for Determination of Total Sulfur in Light Hydrocarbons, Spark Ignition Engine Fuel and Engine Oil by Ultraviolet Fluorescence

ASTM	IP	Title
D5482		Test Method for Vapor Pressure of Petroleum Products (Mini Method-Atmospheric)
D5500		Test Method for Vehicle Evaluation of Unleaded Automotive Spark-Ignition Engine Fuel for Intake Valve Deposit Formation
D5580		Test Method for Determination of Benzene, Toluene, Ethylbenzene, p/m-Xylene, o-Xylene, C_9 and Heavier Aromatics, and Total Aromatics in Finished Gasoline by Gas Chromatography
D5598		Test Method for Evaluating Unleaded Automotive Spark-Ignition Engine Fuel for Electronic Port Fuel Injector Fouling
D5599		Test Method for Determination of Oxygenates in Gasoline by Gas Chromatography and Oxygen Selective Flame Ionization Detection
D5769		Test Method for Determination of Benzene, Toluene, and Total Aromatics in Finished Gasoline by Gas Chromatography/Mass Spectrometry
D5842		Practice for Sampling and Handling of Fuels for Volatility Measurement
D5845		Test Method for Determination of MTBE, ETBE, TAME, DIPE, Methanol, Ethanol and *tert*-Butanol in Gasoline by Infrared Spectroscopy
D5854		Practice for Mixing and Handling of Liquid Samples of Petroleum and Petroleum Products
D5986		Test Method for Determination of Oxygenates, Benzene, Toluene, C_8-C_{12} Aromatics and Total Aromatics in Finished Gasoline by Gas Chromatography/Fourier Transform Infrared Spectroscopy
D6201		Test Method for Dynamometer Evaluation of Unleaded Spark-Ignition Engine Fuel for Intake Valve Deposit Formation
D6293		Test Method for Oxygenates and Paraffin, Olefin, Naphthene, Aromatic (O-PONA) Hydrocarbon Types in Low-Olefin Spark Ignition Engine Fuel by Gas Chromatography
D6296		Test Method for Total Olefins in Spark-Ignition Engine Fuels by Multi-dimensional Gas Chromatography
D6334		Test Method for Sulfur in Gasoline by Wavelength Dispersive X-Ray Fluorescence
D6378		Test Method for Determination of Vapor Pressure (VP_X) of Petroleum Products, Hydrocarbons, and Hydrocarbon-Oxygenate Mixtures (Triple Expansion Method)
D6421		Test Method for Evaluating Automotive Spark-Ignition Engine Fuel for Electronic Port Fuel Injector Fouling by Bench Procedure
D6445		Test Method for Sulfur in Gasoline by Energy-Dispersive X-Ray Fluorescence Spectrometry
D6447		Test Method for Hydroperoxide Number of Aviation Turbine Fuels by Voltammetric Analysis
D6469		Guide for Microbial Contamination in Fuels and Fuel Systems
D6550		Test Method for Determination of Olefin Content of Gasolines by Supercritical-Fluid Chromatography
D6729		Test Method for Determination of Individual Components in Spark Ignition Engine Fuels by 100 Metre Capillary High Resolution Gas Chromatography
D6730		Test Method for Determination of Individual Components in Spark Ignition Engine Fuels by 100 Metre Capillary (with Precolumn) High Resolution Gas Chromatography
D6733		Test Method for Determination of Individual Components in Spark Ignition Engine Fuels by 50 Metre Capillary High Resolution Gas Chromatography
D6920		Test Method for Total Sulfur in Naphthas, Distillates, Reformulated Gasolines, Diesels, Biodiesels, and Motor Fuel by Oxidative Combustion and Electrochemical Detection
D7039		Test Method for Sulfur in Gasoline and Diesel Fuel by Monochromatic Wavelength Dispersive X-Ray Fluorescence Spectrometry
D7096		Test Method for Determination of the Boiling Range Distribution of Gasoline by Wide-Bore Capillary Gas Chromatography
D7344		Test Method for Distillation of Petroleum Products at Atmospheric Pressure (Mini Method)
D7345		Test Method for Distillation of Petroleum Products at Atmospheric Pressure (Micro Distillation Method)

3

Fuel Oxygenates

Marilyn J. Herman[1] and Lewis M. Gibbs[2]

FUEL OXYGENATES ARE WIDELY USED IN THE United States. In the late 1970s and early 1980s, as lead antiknocks were removed from motor gasoline, gasoline producers used oxygenates to offset the loss in octane number from the removal of lead. In the 1990s, oxygenates were required by the government as an emission reduction control strategy. More recently, the United States has required the use of renewable fuels in order to help reduce U.S. dependence on foreign sources of oil. In December 2007, the President signed into law the Energy Independence and Security Act of 2007 (P.L. 110-140). The Energy Independence and Security Act of 2007 (EISA) significantly expands and increases the Renewable Fuels Standard established under the Energy Policy Act of 2005 requiring the use of 9.0 billion gallons of renewable fuel in 2008, increasing to 36 billion gallons by 2022. In 2022, 21 billion gallons of the total renewable fuels requirement must be obtained from cellulosic ethanol and other advanced biofuels.

Under the Clean Air Act, oxygenates have been used as an emission control strategy to reduce carbon monoxide (CO) in wintertime oxygenated fuel programs and as a required component in federal reformulated gasoline programs to help reduce ozone. The Clean Air Act Amendments (CAA) of 1990 require states with areas exceeding the national ambient air quality standard for carbon monoxide to implement programs requiring the sale of oxygenated gasoline containing a minimum of 2.7 weight percent oxygen during the winter months.

The Clean Air Act Amendments also require the use of reformulated gasoline (RFG) in those areas of the United States with the most severe ozone pollution. Under the Clean Air Act Amendments and the Energy Policy Act of 1992, Congress enacted legislation requiring the use of alternative fuels and alternative fuel vehicles. Fuels containing high concentrations of ethanol or methanol, where oxygenate is the primary component of the blend, qualify as alternative fuels. E85, a blend of 85 volume percent ethanol and 15 volume percent hydrocarbons, and M85, a blend of 85 volume percent methanol and 15 volume percent hydrocarbons, may be used in specially designed vehicles to comply with state and local alternative fuel programs.

An *oxygenate* is defined under ASTM specifications as an oxygen-containing, ashless, organic compound, such as an alcohol or ether, which can be used as a fuel or fuel supplement. A *gasoline-oxygenate blend* is defined as a fuel consisting primarily of gasoline along with a substantial amount (more than 0.35 mass percent oxygen, or more than 0.15 mass percent oxygen if methanol is the only oxygenate) of one or more oxygenates.

While there are several oxygenates that can be used to meet federal oxygen requirements in gasoline, ethanol is currently the primary oxygenate used to comply with Clean Air Act requirements. While methyl *tertiary*-butyl ether (MTBE) had been used to meet Clean Air Act requirements, state legislation banning the use of MTBE in gasoline has virtually eliminated its use in the United States. Other oxygenates, such as methanol, *tertiary*-amyl methyl ether (TAME), ethyl *tertiary*-butyl ether (ETBE), and diisopropyl ether (DIPE) have been used in much smaller quantities. In the early 1980s, methanol/*tertiary*-butyl alcohol mixtures were blended with gasoline on a limited scale. When methanol is used as a blending component, it must be accompanied by a co-solvent (a higher molecular weight alcohol) to help prevent phase separation of the methanol and gasoline in the presence of trace amounts of water.

Oxygenated fuels are subject to a number of federal regulations. The U.S. Environmental Protection Agency regulates the allowable use of oxygenates in unleaded gasoline and is responsible for promulgating regulations and enforcing the Renewable Fuels Standard program. The Alcohol and Tobacco Tax and Trade Bureau (TTB) of the Department of Treasury regulates the composition of alcohol used for fuel. The Internal Revenue Service (IRS) regulates the characteristics of fuels qualifying for special tax treatment.

This chapter focuses on ethanol and other oxygenates for use as blending components in fuel or for use as high ethanol content fuels in spark-ignition engines. This chapter summarizes the significance of the more important physical and chemical characteristics of these oxygenates and the pertinent test methods for determining these properties. Information on government regulations and tax incentives for oxygenated fuels is provided. ASTM specifications for oxygenates and other biofuels discussed are:

- ASTM D4806, Specification for Denatured Fuel Ethanol for Blending with Gasolines for Use as Automotive Spark-Ignition Engine Fuel, covers a denatured fuel ethanol suitable for blending up to 10 volume percent with gasoline.
- ASTM D5798, Specification for Fuel Ethanol (Ed75-Ed85) for Automotive Spark-Ignition Engines, covers a fuel blend, nominally 75 to 85 volume percent denatured fuel ethanol and 25 to 15 additional volume percent hydrocarbons for use in ground vehicles with automotive spark-ignition engines.
- ASTM D5797, Specification for Fuel Methanol M70-M85 for Automotive Spark-Ignition Engine Fuels, covers a fuel blend, nominally 70 to 85 volume percent methanol and 30 to 15 volume percent hydrocarbons for use in ground vehicles with automotive spark-ignition engines.

[1] *Herman and Associates, Washington, DC.*
[2] *Chevron Products Company, Richmond, CA.*

- ASTM D5983, Specification for Methyl *Tertiary*-Butyl Ether (MTBE) for Downstream Blending for Use in Automotive Spark-Ignition Engine Fuel, covers requirements for fuel grade methyl *tertiary*-butyl ether utilized in commerce, terminal blending, or downstream blending with fuels for spark-ignition engines.
- ASTM D6751, Specification for Biodiesel Fuel Blend Stock (B100) for Distillate Fuels, covers low-sulfur biodiesel (B100) for use as a blend component with diesel fuel oils as defined by ASTM D975, Specification for Diesel Fuel Oils.
- ASTM D7467, Standard Specification for Diesel Fuel Oil, Biodiesel Blend (B6 to B20), covers fuel blend grades of 6 to 20 volume percent biodiesel with the remainder being a light middle or middle distillate diesel fuel, collectively designated as B6 to B20.
- ASTM D975, Standard Specification for Diesel Fuel Oils, covers biodiesel blends containing up to 5 volume percent biodiesel fuel.
- ASTM D396, Standard Specification for Fuel Oils, covers biodiesel blends containing up to 5 volume percent biodiesel fuel.

Gasoline and gasoline-oxygenate blends are subject to the limits and test methods contained in ASTM D4814, Standard Specification for Automotive Spark-Ignition Engine Fuel. This chapter does not address the physical and chemical characteristics of finished gasoline-oxygenate blends. The properties and significance of gasoline-oxygenate blends are discussed in the chapter, "Automotive Spark-Ignition Engine Fuel." Diesel fuel and diesel fuel-biodiesel fuel blends up to 5 volume percent biodiesel fuel are subject to the limits and test methods contained in ASTM D975, Standard Specification for Diesel Fuel Oils, and ASTM D7467, Standard Specification for Diesel Fuel Oil, Biodiesel Blend (B6 to B20). The properties and significance for these fuels are discussed in the chapter, "Fuels for Land and Marine Diesel Engines for Non-Aviation Gas Turbines." Fuel oils and fuel oil-biodiesel fuel blends up to 5 volume percent biodiesel fuel are subject to the limits and test methods contained in ASTM D396 Standard Specification for Fuel Oils. The properties and significance for these fuels are discussed in the chapter, "Burner, Heating, and Lighting Fuels."

GOVERNMENT REGULATIONS

Oxygenated Fuels and Reformulated Gasoline

Section 211(m) of the Clean Air Act Amendments of 1990 requires states with wintertime carbon monoxide (CO) nonattainment areas having design values of 9.5 parts per million (ppm) CO or more based on 1988 and 1989 data to submit revisions to their State Implementation Plans (SIPs) to establish a wintertime oxygenated gasoline program. The Act requires that any gasoline sold or dispensed to the ultimate consumer in a carbon monoxide nonattainment area during the regulated time period must contain not less than 2.7 weight percent oxygen. A number of areas have been redesignated into attainment for carbon monoxide and are no longer required to have a wintertime oxygenated gasoline program. At the present time, ten areas have wintertime oxygenated fuels programs.

The Clean Air Act also requires the use of reformulated gasoline (RFG) in certain areas in order to reduce vehicle emissions of toxic and ozone-forming compounds. Section 211(k)(l) of the Clean Air Act, as amended, prohibits the sale of conventional gasoline (gasoline that has not been certified as reformulated) in the nine largest metropolitan areas with the most severe summertime ozone levels, as well as other ozone nonattainment areas that opt in to the program. Under the Clean Air Act, RFG was originally required in the following nine areas of the country with the highest levels of ozone: Baltimore, Chicago, Hartford, Houston, Los Angeles, Milwaukee, New York, Philadelphia, and San Diego. Any area reclassified as a severe ozone nonattainment area is required to have reformulated gasoline. Subsequent to passage of the Clean Air Act, additional areas in the county were reclassified as severe ozone nonattainment areas and were thus required to be a covered area under the federal RFG program. Other areas with less severe air pollution problems were allowed to opt into the reformulated gasoline program.

The Clean Air Act established a two-phase program for the implementation of RFG. Federal Phase I RFG requirements began January 1, 1995, and were in effect until December 31, 1999. Phase II performance standards began January 1, 2000. Under Phase I, the EPA required reformulated gasoline to achieve a 15 percent reduction in volatile organic emissions and toxic air pollutants. During Phase II, the EPA requires a 5.5 % reduction in NOx, as well as further reductions in volatile organic emissions and toxic air pollutants. General requirements (under the 1990 Clean Air Act Amendments as amended by the Energy Policy Act of 2005) for federal reformulated gasoline are a maximum 1.0 volume percent benzene content, a limit on heavy metals, and not causing an increase in emissions of oxides of nitrogen.

Additional information on federal and state reformulated gasoline requirements and test methods is provided in ASTM Committee D02 Research Report, D02:1347, Research Report on Reformulated Spark-Ignition Engine Fuel.

Renewable Fuels Standard

On August 8, 2005, the President signed into law the Energy Policy Act of 2005 (P.L. 109-58). This legislation made significant revisions to the federal RFG program, and established a Renewable Fuels Standard (RFS) mandating the use of 4 billion gallons of renewable fuels in the U.S. starting in 2006, increasing to 7.5 billion gallons by the year 2012. Other key provisions of the Energy Policy Act of 2005 (EPACT 2005) included elimination of the minimum 2.0 weight percent oxygen requirement in RFG, establishment of a credits trading program, consolidation of VOC Control Regions, establishment of small refiner provisions, modifications to the mobile source air toxics program and baselines, commingling of compliant RFG fuels, and other fuel related provisions.

In response to EPACT 2005, the EPA enacted a rule to eliminate the minimum oxygen content requirement for RFG both nationally and in California. The rule eliminating the oxygen content requirement for Federal RFG areas in California became effective April 24, 2006. The rule eliminating the oxygen requirement for all other RFG areas became effective May 5, 2006.

For 2006, EPA adopted the default renewable fuels standard set forth in EPACT 2005. The final rule establishing the complete RFS program for 2007 and beyond became effective September 1, 2007. The rule established annual renewable fuel standards through 2012, defined the responsibilities

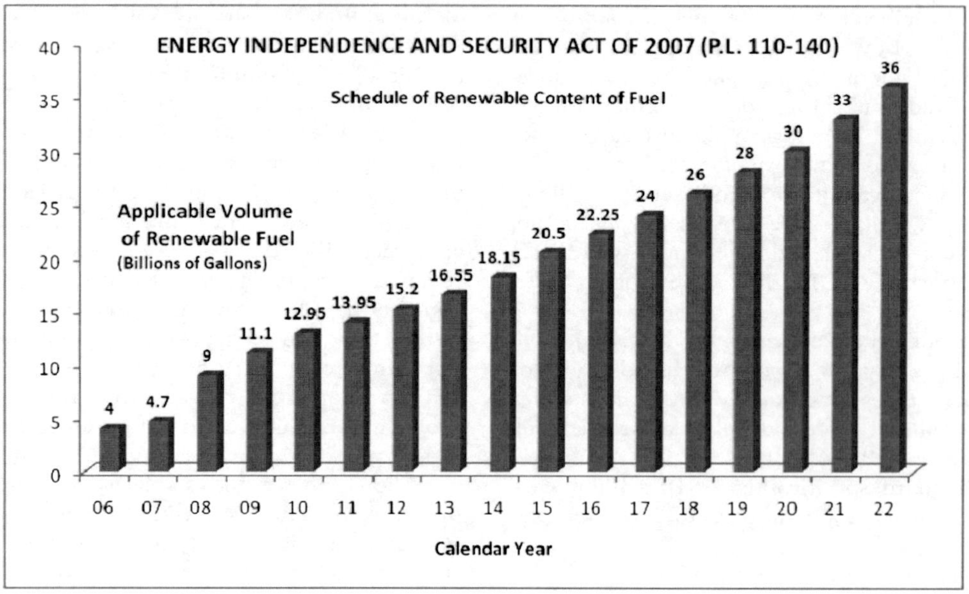

Fig. 1

of refiners and other fuel producers and importers, established a credit trading system and set forth recordkeeping and reporting requirements. Under the RFS regulations, any party that produces or imports gasoline for use in the U.S. is considered an obligated party and is required to meet the annual renewable fuels standard through the purchase of renewable identification numbers (RINs). Qualifying small refiners and small refineries are exempt from meeting the renewable fuel requirements through 2010. Gasoline producers located in Alaska, and noncontiguous U.S. territories are exempt from the RFS program indefinitely, but may opt into the program. Hawaii opted into the program on January 1, 2008.

On December 19, 2007, the President signed into law "the Energy Independence and Security Act of 2007" (P.L. 110-140). This legislation significantly expanded and increased the Renewable Fuels Standard established under the Energy Policy Act of 2005. Section 202 of the Energy Independence and Security Act of 2007 (EISA) requires the use of 9 billion gallons of renewable fuel in 2008, increasing to 36 billion gallons by 2022. EISA also establishes targets for cellulosic ethanol and other advanced biofuels. By 2022, 21 billion gallons of the total renewable fuels requirement must be obtained from cellulosic ethanol and other advanced biofuels. Figure 1 shows the applicable volumes of total renewable fuels required under the EISA of 2007.

The EISA of 2007 establishes annual standards for four categories of renewable fuel: cellulosic biofuel, biomass-based diesel, total advanced biofuel, and total renewable fuel. To qualify under any of these categories, a renewable fuel must meet a certain lifecycle greenhouse gas emission threshold, unless the fuel is produced in a facility that had commenced construction prior to enactment of the legislation. The U.S. Environmental Protection Agency is charged with issuing regulations to implement the Renewable Fuel Standard provisions of EISA 2007. The new regulations will address how the four standards will be set and how obligated parties will comply with the four standards in addition to other detailed provisions of the legislation.

State Biofuel Mandates

In order to encourage the use of renewable fuels, a number of states have enacted laws mandating the use of ethanol, biodiesel, and/or other biofuels. Certain states require that a designated percentage of the gasoline and/or diesel pool be comprised of renewable fuel. Other states have adopted a content requirement mandating that each gallon of gasoline or diesel fuel contain a certain percentage of biofuel. At the present time, twelve states and one county have adopted legislation mandating the use of ethanol, biodiesel, and/ or other renewable fuels. Those states are Florida, Hawaii, Iowa, Louisiana, Massachusetts, Minnesota, Missouri, Montana, New Mexico, Oregon, Pennsylvania, and Washington as well as the county of Portland, OR.

Clean Fleets and Alternative Fuels Program

The Clean Air Act Amendments of 1990 created a Clean Fleets Program to introduce clean fuel vehicles nationwide. In model-year 1996, automobile manufacturers were required to produce 150,000 clean-fueled cars and light trucks per year under a California pilot program. For model years 1999 and thereafter, manufacturers must produce 300,000 clean fuel vehicles each year.

Beginning in model year 1998, 22 cities classified as having serious, severe, and extreme ozone nonattainment areas plus Denver, Colorado, for carbon monoxide nonattainment purchased clean fuel vehicles for their fleets. Marginal and moderate ozone nonattainment areas are not required to participate, but may elect to do so.

The Energy Policy Act of 1992 defines alternative fuels as natural gas, propane, and blends of alcohol with gasoline or other fuels containing 85 volume percent or more alcohol, hydrogen, fuels derived from biomass, and liquid fuels derived from coal and electricity. Vehicles can be flexible-fuel or dual-fuel, but must use alternative fuels within the nonattainment areas.

Under the legislation, designated federal, state, and fuel provider fleets are required to replace their gasoline-powered vehicles with alternative fuel vehicles over time. Many fleets are choosing flexible fuel vehicles (FFVs) that can operate on 85 volume percent ethanol (E85), gasoline, or any combination of gasoline and alcohol in the same tank.

Private-sector companies that produce alternative fuels, such as natural gas companies or electric utilities, are required to introduce alternative fuel vehicles into their fleets as follows: 30 % in model-year 1996, 50 % in model-year 1997, 70 % in model-year 1998, 90 % in model-year 1999 and thereafter. The minimum federal fleet requirements for light-duty alternative fuel vehicles are as follows: 5,000 in fiscal year 1993, 7,500 in fiscal year 1994, 10,000 in fiscal year 1995, 25 % in fiscal year 1996, 33 % in fiscal year 1997, 50 % in fiscal year 1998, and 75 % in fiscal year 1999 and thereafter.

State governments are required to purchase alternative fuel vehicles in the following amounts: 10 % in model-year 1996, 15 % in model-year 1997, 25 % in model-year 1998, 50 % in model-year 1999, and 75 % in model-year 2000 and thereafter.

Regulations Governing Oxygenated Fuels

The use of oxygenates in blends with unleaded gasoline is governed by Section 211 (f) of the Clean Air Act and EPA Fuel and Fuel Additive Registration regulations at 40 CFR 79. EPA waivers granted under Section 211(f), the "substantially similar" Interpretive Rule, and compliance with EPA registration requirements govern the allowable amounts of oxygenates that may be added to unleaded gasoline.

Section 211(f)(1)(A) of the Clean Air Act prohibits fuel or fuel additive manufacturers from introducing into commerce, or increasing the concentration in use, of any fuel or fuel additive for general use in light-duty motor vehicles that is not substantially similar to any fuel or fuel additive utilized in the certification of any 1975 or subsequent model year vehicle or engine under Section 206 of the act. EPA treats a fuel or fuel additive as "substantially similar" if the following criteria are met:

1. The fuel must contain carbon, hydrogen, and oxygen, nitrogen, and/or sulfur, exclusively in the form of some of the following:
    a. Hydrocarbons;
    b. Aliphatic ethers;
    c. Aliphatic alcohols other than methanol;
        (i) Up to 0.3 volume percent methanol;
        (ii) Up to 2.75 volume percent methanol with an equal volume of butanol, or higher molecular weight alcohol;
2. A fuel additive at a concentration of no more than 0.25 weight percent that contributes no more than 15 ppm sulfur by weight to the fuel. The fuel must contain no more than 2.0 weight percent oxygen, except fuels containing aliphatic ethers and/or alcohols (excluding methanol) must contain no more than 2.7 weight percent oxygen.
3. The fuel must possess, at the time of manufacture, all of the physical characteristics of an unleaded gasoline as specified in ASTM Standard D4814-88 for at least one of the seasonal and geographic volatility classes specified in the standard.
4. The fuel additive must contain only carbon, hydrogen, and any one or all of the following elements: oxygen, nitrogen, and/or sulfur.

For those fuels or fuel additives that are not "substantially similar," the manufacturer may apply for a waiver of the prohibitions as provided in Section 211 (f) (4) of the Clean Air Act. The applicant must establish that the fuel or fuel additive will not cause or contribute to a failure of any emission control device or system (over the useful life of any vehicle in which such device or system is used). Prior to December 2007, if the EPA administrator did not act to grant or deny an application within 180 days of receipt of the waiver application, the waiver was deemed as granted. Under the EISA of 2007, the EPA now is required to grant or deny an application for a waiver within 270 days of its filing.

However, in order to be marketed, fuels receiving a waiver or fuels permitted under the "Substantially Similar" rule must also comply with EPA fuels and fuel additive registration requirements.

Under 40 CFR 79, "Registration of Fuels and Fuel Additives," any manufacturer of motor vehicle gasoline or diesel fuel, or an additive for use in gasoline or diesel fuel, must register with the Environmental Protection Agency prior to the proposed introduction into commerce of the fuel or fuel additive. On May 27, 1994, under Section 211(b) of the Clean Air Act, the EPA promulgated a rule adding health effects information and testing requirements to the agency's existing registration program for motor vehicle fuels and fuel additives.

For fuels/fuel additives registered before May 27, 1994, Tier 1 data and evidence of a suitable contractual arrangement for completion of Tier 2 requirements were required to be submitted to EPA by May 27, 1997. Fuels/fuel additives not registered as of May 27, 1994, are considered either "registrable" or "new." "Registrable" fuels/fuel additives are compositionally similar to currently registered products in general use and may be marketed upon EPA's receipt of the basic registration data. "New" fuels/fuel additives must complete all testing requirements before registration and introduction into commerce.

With respect to methanol, before the establishment of the health effects testing requirement in 1994, methanol co-solvent combinations were allowed under several waivers and/or the Substantially Similar Interpretive Rule at levels up to 3.5 weight percent oxygen. The health effects testing regulations established separate testing categories for each oxygenate used at a concentration of 1.5 weight percent oxygen or greater.

Testing is under way for MTBE, ethanol, ETBE, *Tertiary* butyl alcohol (TBA), DIPE, and TAME. Any other oxygenate or combination of oxygenates previously allowed by the Interpretive Rule or waiver, such as a methanol co-solvent combination, is now limited to the baseline category, which must contain less than 1.5 weight % oxygen. Methanol blends, which contribute greater than 1.5 weight % total oxygen to the fuel, may not be marketed unless a health effects testing program is conducted. Because the current potential for methanol blends is limited, it is unlikely that any fuel manufacturer would perform the health effects testing necessary for higher usage.

Table 1, "EPA Waivers and Substantially Similar Levels for Oxygenated Fuels," summarizes oxygenated fuels granted a waiver from EPA or permitted under the EPA "Substantially Similar" interpretive rule.

Alcohol and Tobacco Tax and Trade Bureau

The Alcohol and Tobacco Tax and Trade Bureau (TTB) is the government agency responsible for regulating distilled spirits

TABLE 1—EPA Waivers and Substantially Similar Levels for Oxygenated Fuels

Oxygenate	Max. Limit (volume percent)	Action	Other Restrictions
Ethanol	10 %	Gasohol Waiver, 1978 (Gas Plus)	Ethanol must be anhydrous (200 proof).
Ethanol	Blends of less than 10 %	Interpretive Rule, 1982 (Ashland Oil)	Ethanol must be anhydrous.
Ethanol Additive (Synco 76)	Proprietary stabilizer mixed with anhydrous ethanol and denatured with methyl isobutyl ketone. Additive must include 67 % hexanol, 4 % pentanol, 2 % octanol, and 27 % araffinates derived from the coal liquefaction process	Waiver granted to Synco 76 Fuel Corporation, 1982	Must be used in 1:20 ratio with ethanol (1/4 gal stabilizer to 5 gal ethanol added to 45 gal of finished unleaded gasoline. Must meet ASTM volatility requirements for time of year and location.
Ethanol/MTBE	Up to 2 % MTBE	EPA Rulings, 1986 and 1988	MTBE allowed only as a result of commingling during storage or transport, and not purposefully added.
ARCONOL	0 to 7 % TBA	Waiver granted to Arco Corporation, 1979	Must meet ASTM volatility limits.
Methanol/*Tertiary* Butyl Alcohol (TBA)[A]	2.75 % methanol, 2.75 % TBA, 5.5 % total alcohol, 2.0 % max. oxygen (w)	Waiver granted to Sun Oil Company, 1979	Must meet ASTM volatility limits. Amount of methanol to cosolvent cannot exceed a ratio of 1:1.
Methanol without Cosolvents[A]	Cannot exceed 0.3 % straight methanol (i.e., without cosolvents)	Substantially Similar Rule, 1981, modified in 1991	Must meet ASTM volatility limits for one of ASTM volatility classes.
Methanol with Cosolvents[A]	2.75 % with equal butanol or other higher molecular weight alcohols	Substantially Similar Rule 1981, modified in 1991	Must meet ASTM volatility limits for one of ASTM volatility classes.
OXINOL[A]	4.75 % methanol, 4.75 % gasoline grade TBA (GTBA), max. 3.5 % oxygen (w)	Waiver granted to Arco Corporation, 1981	Not to exceed 1:1 ratio methanol/GTBA. Must meet ASTM volatility limits.
Methanol/Cosolvents[A] (Sun Waiver)	5 % methanol, 2.5 % cosolvent alcohols. Max. 3.7 % oxygen (w)	Waiver granted to Sun Refining and Marketing, 1985	Must meet ASTM volatility limits.
Methanol/Cosolvents[A] (Dupont Waiver)	5 % methanol, 2.5 % cosolvent alcohols having a carbon number of 4 or less (i.e., ethanol, propanol, butanol, and/or GTBA)	Waiver granted to E. I. DuPont de Nemours and Company, 1985 modified in 1986 and 1987	Must contain one of three specified corrosion inhibitors, must meet ASTM volatility limits.
OCTAMIX[A]	5 % methanol, 2.5 % cosolvent alcohols with a carbon number of 8 or less. Pentanols, hexanols, heptanols, and octanols or mixtures are limited to a maximum of 40 % (w) Heptanols and octanols are limited to 5 % max. (w)	Waiver granted to Texas Methanol Corporation, 1988	Corrosion inhibitor required. Must meet ASTM D439-85a, plus maximum temperature for phase separation and alcohol purity.
MTBE	7 % MTBE	Waiver granted to Arco Corporation, 1979	Must meet ASTM volatility limits.
MTBE	15 %	Waiver granted to Sun Refining and Marketing, 1988	Must meet ASTM volatility limits.
Ethers and Aliphatic Alcohols	2.7 % oxygen (w)	Substantially Similar, Interpretive Rule, 1991	Must meet ASTM volatility limits.

[A]Methanol blends, which contribute greater than 1.5 weight % oxygen to a fuel, may not be marketed unless a health effects testing program is conducted.

and alcohol fuel plants (AFPs). The TTB requires that all ethanol must be denatured in accordance with two specified formulae in order to render it unfit for human consumption. For fuel quality purposes, ASTM limits the allowable denaturants approved by the TTB for fuel ethanol. [Additional information on denaturants is provided under "Denatured Fuel Ethanol for Blending with Gasoline."]

Internal Revenue Service

Effective January 1, 2009, under the Food, Conservation, and Energy Act of 2008 (P.L.110-246) (2008 Farm Bill), Congress reduced the maximum allowable denaturant content for purposes of claiming the tax credit from 5 volume percent to 2 volume percent of the volume of alcohol. The IRS of the Department of Treasury is responsible for developing regulations to implement the denaturant provisions of the 2008 Farm Bill. In order to provide temporary "safe harbor" and study the issue further, IRS issued a temporary rule indicating that IRS will not challenge a claim to a credit or payment with respect to the volume of denaturant in alcohol fuel mixtures as long as the added denaturants do not exceed 2.5 % of the volume of alcohol. Final regulations are under development.

FEDERAL TAX INCENTIVES FOR RENEWABLE FUELS

In order to encourage the use of renewable energy and alternative fuels, the federal government has provided a number of tax incentives for biofuels, including ethanol, biodiesel, renewable diesel, and cellulosic biofuel. Some of the tax incentives provided by Congress for biofuels include:

- An excise tax credit from the federal excise tax on gasoline or income tax credits for ethanol not derived from petroleum, natural gas, or coal (including peat);
- An excise tax credit from the federal excise tax on diesel fuel or income tax credits for each gallon of biodiesel and renewable diesel used by the taxpayer in producing a qualified biodiesel mixture for sale or use in a trade or business;
- An excise tax credit for producers of cellulosic biofuel;
- An excise tax exemption from the federal excise tax on special motor fuels for fuel containing at least 85% methanol, ethanol, or other alcohol produced from natural gas; and
- Income tax credits for blenders of alcohol mixtures, users of straight alcohol fuel, and small ethanol producers.

Present law provides a tax credit for alcohol used as fuel. The credit is allowed to those who blend the alcohol and gasoline mixture for sale or use in their trade or business. "Alcohol fuel mixtures" containing ethanol are provided a $0.45 cents per gallon excise tax credit for each gallon of ethanol used in the mixture. Under the 2008 Farm Bill, "H.R. 6124 Food, Conservation, and Energy Act of 2008" (P.L. 110-246), effective January 1, 2009, the ethanol tax incentive was reduced from $0.51 cents to $0.45 cents per gallon for qualifying mixtures. Under the excise tax credit system, gasoline refiners and marketers are required to pay the full rate of tax (18.4 cents per gallon) on the total gasoline-ethanol mixture (including the ethanol portion), but are able to claim a $0.45 per gallon tax credit or refund for each gallon of ethanol used in the mixture. The credit is paid on the amount of alcohol added to the fuel mixture. Under present law, the current excise tax credit for ethanol is in effect until December 31, 2010.

The Food, Conservation and Energy Act of 2008, H.R. 2419, includes an income tax credit for producers of cellulosic alcohol and other cellulosic biofuels. The credit is $1.01 per gallon. If the cellulosic biofuel is ethanol, this amount is reduced by the amount of credit available for alcohol fuels.

Present law provides an excise tax credit for qualifying biodiesel mixtures. The credit or payment amount is $1.00 per gallon. A credit is available for each gallon of biodiesel used by the taxpayer in producing a qualified biodiesel mixture for sale or use in a trade or business. A qualified biodiesel mixture is a mixture of biodiesel and diesel fuel that (1) is sold by the taxpayer producing the mixture to any person for use as a fuel, or (2) is used as a fuel by the taxpayer producing the mixture. The IRS has determined that a renewable diesel mixture is treated as a biodiesel mixture and thus qualifies for the tax credit allowable for biodiesel. Under present law, the tax credit for biodiesel mixtures is in effect until December 31, 2009.

FUEL ETHANOL

Ethanol has been used in gasoline blends (known as "gasohol") in the U.S. for many years. Under the EPA waiver for gasohol, a maximum 10 volume percent ethanol may be used. While in 1978 ethanol was used as an octane enhancer and gasoline extender, market penetration of gasoline-ethanol blends has significantly expanded in response to government programs mandating increased use of renewable fuels. During 2008, the U.S. ethanol industry produced 9 billion gallons of ethanol. Ethanol blended gasoline now represents approximately 70 % of the U.S. motor fuel market.

While the majority of fuel ethanol marketed in the U.S. is used as a blending component in gasoline in concentrations up to 10 volume percent, ethanol is also being used as fuel, in concentrations as high as 85 volume percent ethanol in specially designed flexible fuel vehicles. Ethanol is an approved alternative fuel as defined in the Energy Policy Act. In certain locations throughout the country, ethanol fuels are being used in fleets, urban buses and heavy-duty engines.

Under current regulations, EPA limits the maximum amount of ethanol that can be added to unleaded gasoline at 10 volume percent. However, due to increased market penetration of ethanol in response to government mandates, it is likely that in the near future a "blend wall" will occur – where by all U.S. gasoline will contain the maximum amount of ethanol, i.e., 10 volume percent, currently allowed under EPA regulations. In order to surpass this "blend wall" and enable greater use of renewable fuels, the U.S. government and industry are conducting a major research program to collect the necessary data to determine whether higher level ethanol blends in amounts higher than 10 volume % can be used safely in conventional vehicles, small engine, and marine applications without jeopardizing emission requirements and durability.

ASTM has adopted two specifications governing the properties and limits of fuel ethanol:

- ASTM D4806, Specification for Denatured Fuel Ethanol for Blending with Gasolines for Use as Automotive Spark-Ignition Engine Fuel, covers a denatured fuel-grade ethanol that is suitable for blending up to 10 volume percent with gasoline.
- ASTM D5798, Specification for Fuel Ethanol (Ed75-Ed85) for Automotive Spark-Ignition Engines, covers a fuel blend, nominally 75 to 85 volume percent denatured fuel

ethanol and 25 to 15 additional volume percent hydrocarbons for use in flexible fuel and dedicated E85 fuel vehicles with automotive spark-ignition engines.

Following is a discussion of ASTM specifications and test methods for fuel ethanol. [*Note*: Blends of ethanol and gasoline are also governed by the limits and test methods of ASTM D4814, Specification for Automotive Spark-Ignition Fuel. See the chapter "Automotive Spark-Ignition Engine Fuel" for a discussion of these limits and test methods.]

Test Methods for Gasoline-Oxygenate Blends

Many of the original test methods used to measure the properties of gasoline-oxygenate blends were developed for hydrocarbons, and are not necessarily applicable for gasoline-oxygenate blends. A new program is under way within ASTM to review existing test methods and determine their applicability for biofuels. As part of this program, ASTM is working to modify the scopes and develop precision statements for the test methods specified in D4814, D4806, and D5798. In certain cases, modifications have already been made to the scopes of various test methods. In other instances, where no standardized test method currently exists, (such as a direct test for denaturant content in ethanol), new test methods may need to be developed.

DENATURED FUEL ETHANOL FOR BLENDING WITH GASOLINE

ASTM D4806, Standard Specification for Denatured Fuel Ethanol for Blending with Gasolines for Use as Automotive Spark-Ignition Engine Fuel, establishes limits and test methods for denatured fuel ethanol. This specification covers nominally anhydrous denatured fuel ethanol intended to be blended with unleaded or leaded gasolines at 1 to 10 volume percent for use as an automotive spark-ignition engine fuel.

The specified property limits in ASTM D4806 limits are not designed for blending E85. Denatured fuel ethanol for blending E85 must conform with the limits and test methods in D5798, Specification for Fuel Ethanol (Ed75-Ed85) for Automotive Spark-Ignition Engines.

Denaturants

The TTB, formerly the Bureau of Alcohol, Tobacco, and Firearms, under the U.S. Department of Treasury, requires that certain materials must be added to ethanol under a formula approved by the TTB in order to make the alcohol unfit for beverage or internal human medicinal use and therefore not subject to alcohol beverage tax. These materials are called denaturants.

Under ASTM D4806, the only denaturants allowed to be used for fuel ethanol shall be natural gasoline, gasoline components, or unleaded gasoline at a minimum concentration of two parts by volume per 100 parts by volume of fuel ethanol. The denatured fuel ethanol covered by ASTM D4806 may contain between 1.96 and 5.0 volume percent denaturant. As discussed earlier, the IRS is developing new regulations to implement a federal law requiring that for tax purposes, the maximum allowable denaturant is 2 volume percent of the total alcohol.

The denaturant content is determined by the ratio of metered denaturant and ethanol volumes at the time of denaturing. Approved ASTM analytical methods do not exist to determine that the amount of denaturant added during the denaturing process or contained in the denatured fuel ethanol are within the appropriate limits allowed by the TTB.

One denatured formula specifically designed for fuel use by the TTB is Formula CDA-20. It requires that for every 100 gal of ethanol of not less than 195 proof, a total of 2.0 gal of denaturant be added. Another fuel alcohol rendered unfit for beverage use and manufactured at an alcohol fuel plant requires the addition of 2 gal or more of materials listed by the TTB director to each 100 gal of ethanol. ASTM does not allow certain formulas permitted by federal denaturant regulations because they can be harmful to automotive engines. ASTM D4806 does not allow the use of hydrocarbons with an end boiling point higher than 225°C (437°F) as determined by ASTM D86, Test Method for Distillation of Petroleum Products at Atmospheric Pressure. ASTM D4806 prohibits such denaturants because they can adversely affect fuel stability, automotive engines, and fuel systems. Prohibited denaturants are methanol that does not meet ASTM D1152, Specification for Methanol (Methyl Alcohol), pyrroles, turpentine, ketones, and tars (high-molecular weight pyrolysis products of fossil or nonfossil vegetable matter).

The California Air Resources Board (ARB) has adopted additional restrictions on denaturants. ARB regulations limit the amounts of benzene, aromatics, and olefins present. ASTM D5580, Standard Test Method for Determination of Benzene, Toluene, Ethylbenzene, p/m-Xylene, o-Xylene, C9 and Heavier Aromatics, and Total Aromatics in Finished Gasoline by Gas Chromatography, is used to determine the benzene and aromatics contents of the denaturant. ASTM D6550, Standard Test Method for Determination of Olefin Content of Gasolines by Supercritical-Fluid Chromatography, is used to determine the olefins content of the denaturant.

Water Content

The water content of denatured fuel ethanol must be limited when blended with gasoline. Blends of fuel ethanol and gasoline have a limited solvency for water. This solvency will vary with the ethanol content, the temperature of the blend, and the aromatics content of the base gasoline. A fuel made by blending 10 volume percent fuel ethanol with a gasoline containing 14 volume percent aromatics and 0.6 mass percent dissolved water (about 0.5 volume percent) will separate into a lower alcohol-rich aqueous phase and an upper hydrocarbon phase if cooled to about 7°C (45°F). As normal spark-ignition engines will not run on the aqueous phase material, such a separation is likely to cause serious operating problems. Because some degree of water contamination is practically unavoidable in transport and handling, and because gasoline-ethanol blends are hygroscopic, the water content of the denatured fuel ethanol must be limited when blended with gasoline to reduce the risk of phase separation.

ASTM E203, Test Method for Water Using Volumetric Karl Fischer Titration, is generally the only consistently reliable procedure for the determination of water in denatured fuel ethanol. ASTM E203 includes modifications required to run the test in the presence of alcohols. Because the addition of denaturants will normally affect specific gravity, specific gravity methods such as ASTM D891, Test Methods for Specific Gravity, Apparent, of Liquid Industrial Chemicals, and ASTM D3505, Test Method for Density or Relative Density of Pure Liquid Chemicals, are generally not suitable for determining the water content of denatured fuel ethanol.

Solvent Washed Gum

Solvent washed gum is important because it can contribute to deposits on the surfaces of carburetors, fuel injectors, and intake manifolds, ports, valves, and valve guides. Solvent washed gum consists of fuel-insoluble gum. The fuel-insoluble portion can clog fuel filters. Both types of gum can be deposited on surfaces when the fuel evaporates.

ASTM D381, Test Method for Gum Content in Fuels by Jet Evaporation, is used to determine solvent washed gum. This test method is used to detect the presence of high-boiling, heptane-insoluble impurities and measures the amount of residue remaining after the fuel evaporates and after a heptane wash is performed. However, the precision statements for ASTM D381 were developed using only data on hydrocarbons and may not be applicable to denatured fuel ethanol.

pHe

pHe is a measure of the acid strength of alcohol fuels. The pHe of ethanol is important to reduce the risk of fuel injector failure and engine cylinder wear. When the pHe of ethanol used as a fuel for automotive spark-ignition engines is below 6.5, fuel pumps can malfunction as a result of film forming between the brushes and commutator, fuel injectors can fail from corrosive wear, and excessive engine cylinder wear can occur. When the pHe is above 9.0, fuel pump plastic parts can fail. The adverse effects are less when ethanol is used at concentrations of 10 volume percent or less in gasoline.

ASTM D6423, Test Method for Determination of pHe of Ethanol, Denatured Fuel Ethanol, and Fuel Ethanol (Ed75-Ed85), is used to determine the pHe levels of fuel ethanol. The test method is applicable to fuels containing nominally 70 volume % ethanol or higher, as described in ASTM D4806, Specification for Denatured Fuel Ethanol for Blending With Gasolines for Use as Automotive Spark-Ignition Engine Fuel, and ASTM D5798, Specification for Fuel Ethanol (Ed75-Ed85) for Automotive Spark-Ignition Engines. The pHe value will depend somewhat on the fuel blend, the stirring rate, and the time the electrode is in the fuel.

Chloride Ion Content

Low concentrations of chloride ions are corrosive to many metals. ASTM D7319, Test Method for Determination of Total and Potential Sulfate and Inorganic Chloride in Fuel Ethanol by Direct Injection Suppressed Ion Chromatography, or ASTM D7328, Test Method for Determination of Total and Potential Inorganic Sulfate and Total Inorganic Chloride in Fuel Ethanol by Ion Chromatography Using Aqueous Sample Injection, is used to determine inorganic chloride content in ethanol.

Sulfate Content

The presence of small amounts of inorganic sulfates in denatured fuel ethanol under the right conditions can contribute to turbine meter deposits and the premature plugging of fuel dispensing pump filters in the fuel distribution system. The sulfates also have been shown to cause fuel injector sticking, resulting in engine misfiring and poor driveability in automobiles. ASTM D7318, Test Method for Total Inorganic Sulfate in Ethanol by Potentiometric Titration, ASTM D7319, Test Method for Determination of Total and Potential Sulfate and Inorganic Chloride in Fuel Ethanol by Direct Injection Suppressed Ion Chromatography, or ASTM D7328, Test Method for Determination of Total and Potential Inorganic Sulfate and Total Inorganic Chloride in Fuel Ethanol by Ion Chromatography Using Aqueous Sample Injection, is applicable to determine sulfate content in ethanol.

Copper Content

Copper is an active catalyst for the low-temperature oxidation of hydrocarbons. Experimental work has shown that copper concentrations higher than 0.012 mg/kg in commercial gasoline may significantly increase the rate of gum formation. ASTM D1688, Test Methods for Copper in Water, Test Method A, has been modified for determining the copper content of denatured fuel ethanol. The modification of Test Method A (atomic absorption, direct) consists of mixing reagent grade ethanol (which may be denatured according to the TTB Formula 3A or 30) in place of water as the solvent or diluent for the preparation of reagents and standard solutions. Because a violent reaction may occur between the acid and the ethanol, use water, as specified, in the acid solution part of the procedure to prepare the stock copper solution. Use ethanol for the rinse and final dilution only. The precision of this modified method has not been determined, but the precision is expected to be similar to the precision of Test Method A in ASTM D1688.

Acidity

Denatured fuel ethanol may contain additives such as corrosion inhibitors and detergents that may affect the titratable acidity (acidity as acetic acid) of the finished fuel ethanol. Very dilute aqueous solutions of low-molecular weight organic acids such as acetic acid (CH_3COOH) are highly corrosive to many metals. It is necessary to keep such acids at a very low level. ASTM D1613, Test Method for Acidity in Volatile Solvents and Chemical Intermediates Used in Paint, Varnish, Lacquer, and Related Products, is used to determine the acidity of denatured fuel ethanol.

Appearance

Denatured fuel ethanol is required to be visibly free of suspended or precipitated contaminants (clear and bright). Turbidity or evidence of precipitation normally indicates major contamination. This shall be determined at indoor ambient temperature unless otherwise agreed upon between the supplier and the purchaser.

Ethanol Purity

The presence of even small quantities of some organic oxygen compounds other than ethanol may adversely affect the properties of gasoline-ethanol blends. ASTM D5501, Test Method for the Determination of Ethanol Content of Denatured Fuel Ethanol by Gas Chromatography, determines the ethanol and methanol contents of denatured fuel ethanol.

FUEL ETHANOL: ED75-ED85

ASTM D5798, Specification for Fuel Ethanol (Ed75-Ed85) for Automotive Spark-Ignition Engines, covers a fuel blend, nominally 75 to 85 volume percent denatured fuel ethanol and 25 to 15 additional volume percent hydrocarbons for use in ground vehicles with automotive spark-ignition engines designed to use it, which are designated as dedicated E85 vehicles or as flexible fuel vehicles. *Fuel ethanol (Ed75-Ed85)* is defined as a blend of ethanol and hydrocarbons of which the measured ethanol portion is nominally 70 to 79 volume percent.

Denatured fuel ethanol for blending E85 is required to meet the limits and test methods of D5798, Specification for Fuel Ethanol (Ed75-Ed85) for Automotive Spark-Ignition Engines. Since the specified property limits in ASTM D4806 limits are not designed for blending E85, blenders of E85 need to ensure that the final E85 blend conforms with the requirements of ASTM D5798.

While the performance requirements of ASTM D5798 are based on the best technical information available, these requirements are still under development. Certain performance limits in ASTM D5798 are likely to change in the future as improvements in vehicle technology occur and greater field experience is gained from field use of fuel ethanol vehicles.

The ethanol content of fuel ethanol (Ed75-Ed85) is a critical parameter since it affects the capability of the fuel metering system of dedicated Ed75-Ed85 vehicles to establish the proper air/fuel ratio for optimum vehicle operation. This is much less of a concern for flexible fuel vehicles than for dedicated Ed75-Ed85 vehicles. Ethanol content may also affect the lubricating properties of the fuel, water tolerance, and the ability to meet cold and cool area volatility requirements. The inclusion of impurities, some denaturants, and contaminants, can adversely affect the properties and performance of fuel ethanol (Ed75-Ed85). The quantities of some of these materials are controlled by specified property limits. The limits on water, higher molecular weight alcohols, and methanol and on types of denaturants as well as minimums on the amount of ethanol and hydrocarbons limit, but do not prevent, the presence of trace materials. ASTM D5501, Test Method for the Determination of Ethanol Content of Denatured Fuel Ethanol by Gas Chromatography, is used to determine ethanol and methanol contents.

Vapor Pressure

Denatured fuel ethanol has a low vapor pressure, and the addition of volatile hydrocarbons to make fuel ethanol (Ed75-Ed85) is required for adequate cold startability. The addition of hydrocarbons that are too volatile can contribute to hot fuel handling problems. Higher vapor pressures are required at colder ambient temperatures, while lower volatility fuels are less prone to hot fuel-handling problems at higher (summertime) ambient temperatures. Excessive vapor pressure contributes to evaporative emissions.

Lower and upper limits on vapor pressure for the three volatility classes are used to define the acceptable range of volatile components to ensure adequate vehicle performance. Vapor pressure is varied for seasonal and climatic changes by providing three vapor pressure classes for fuel ethanol (Ed75-Ed85). Class 1 encompasses geographic areas with six-hour tenth percentile minimum ambient temperature greater than 5°C (41 °F). Class 2 encompasses geographic areas with six-hour tenth percentile minimum ambient temperature greater than −5°C (23°F) but less than +5°C (41°F). Class 3 encompasses geographic areas with six-hour tenth percentile minimum ambient temperature less than or equal to −5°C (23°F). ASTM D4953, Test Method for Vapor pressure of Gasoline and Gasoline-Oxygenate Blends (Dry Method), ASTM D5190, Test Method for Vapor Pressure of Petroleum Products (Automatic Method), or ASTM D5191, Test Method for Vapor Pressure of Petroleum Products (Mini Method), shall be used to determine the vapor pressure of Ed75-Ed85.

Hydrocarbons

Hydrocarbons are deliberately added to provide improved cold startability and warm-up driveability. For cold ambient conditions, increasing the hydrocarbon content improves cold startability. The addition of hydrocarbons also contributes to flame visibility (luminous flame), nonexplosive air-fuel mixtures in storage tanks (rich mixture vapor space), and denaturation (malodorant and taste deterrent). The hydrocarbon portion of the fuel must be unleaded. While the composition of the hydrocarbons added to fuel ethanol is not controlled, the hydrocarbons should be stable, noncorrosive, and be in the boiling range of spark-ignition engine fuel as specified in Specification D4814, Standard Specification for Automotive-Spark Ignition Engine Fuel. It should have sufficient vapor pressure to meet the requirements of ASTM D5798. Approved ASTM analytical methods do not exist to determine the amount of hydrocarbon added during the blending process.

Acidity

Ed75-Ed85 as well as denatured fuel ethanol may contain additives such as corrosion inhibitors and detergents that could affect the titratable acidity (acidity as acetic acid) of the fuel. [See acidity discussion under "Denatured Fuel Ethanol for Blending with Gasoline" for additional information.] ASTM D1613, Test Method for Acidity in Volatile Solvents and Chemical Intermediates Used in Paint, Varnish, Lacquer, and Related Products, is used to determine the acidity of denatured fuel ethanol.

pHe

The pHe of Ed75-Ed85 is important to reduce the risk of fuel injector failure and engine cylinder wear. [See pHe discussion under "Denatured Fuel Ethanol for Blending with Gasoline" for additional information.] The adverse effects are believed to be greater when ethanol is used at higher concentrations than in a 10 volume percent blend with gasoline. ASTM D6423, Test Method for Determination of pHe of Ethanol, Denatured Fuel Ethanol, and Fuel Ethanol (Ed75-Ed85), is used to determine the pHe levels of fuel ethanol. The test method is applicable to fuels containing nominally 70 volume percent ethanol or higher, as described in ASTM D4806, Specification for Denatured Fuel Ethanol for Blending With Gasolines for Use as Automotive Spark-Ignition Engine Fuel, and ASTM D5798, Specification for Fuel Ethanol (Ed75-Ed85) for Automotive Spark-Ignition Engines.

Gum Content, Solvent Washed and Unwashed

Solvent washed gum can contribute to deposits on the surface of carburetors, fuel injectors, and intake manifolds, ports, valves, and valve guides. The impact of solvent washed gum on engines that can operate on fuel ethanol (Ed75-Ed85) has not been fully established but is based on limited experience gained with M70-M85 fuels in field tests and from historic gasoline limits. Performance effects depend on where the deposits form and the amount of deposit. The test for solvent washed gum content measures the amount of residue after the evaporation of the fuel and following a heptane wash. The heptane wash removes the heptane-soluble, nonvolatile material, such as additives, carrier oils used with the additives, and diesel fuel.

Unwashed gum content consists of fuel-insoluble and fuel-soluble gum. The fuel-insoluble portion can clog fuel filters. Both can be deposited on surfaces when the fuel evaporates. The difference between the unwashed and solvent washed gum content values can be used to assess the presence and amount of nonvolatile material in the fuel. Additional analytical testing is required to determine if the material is additive, carrier oil, diesel fuel, or other.

The unwashed gum content limit is intended to limit high-boiling contaminants, like diesel fuel, that can affect engine performance, yet allow the use of appropriate levels of deposit control additives with carrier fluids in fuel ethanol (Ed75-Ed85). ASTM D381, Test Method for Gum Content in Fuels by Jet Evaporation, is used to determine unwashed and solvent washed gum. Because the precision statements for ASTM D381 were developed using only data on hydrocarbons, they may not be applicable to fuel ethanol (Ed75-Ed85).

Ionic Chloride

Ionic (inorganic) chloride is corrosive to many metals, and it is desirable to minimize ionic chlorine compounds in fuel ethanol (Ed75-Ed85). An inorganic chloride limit of a maximum 1 mg/kg has been found to be adequate in protecting fuel system components. ASTM D7319, Test Method for Determination of Total and Potential Sulfate and Inorganic Chloride in Fuel Ethanol by Direct Injection Suppressed Ion Chromatography or ASTM D7328, Test Method for Determination of Total and Potential Inorganic Sulfate and Total Inorganic Chloride in Fuel Ethanol by Ion Chromatography Using Aqueous Sample Injection, determines inorganic chloride content in Ed75-Ed85.

Sulfur

The limit on sulfur content is included to protect against engine wear, deterioration of engine oil, corrosion of exhaust system parts, and exhaust catalyst deactivation. Sulfur content can be determined using ASTM D1266, Test Method for Sulfur in Petroleum Products (Lamp Method), ASTM D2622, Test Method for Sulfur in Petroleum Products by Wavelength Dispersive X-Ray Fluorescence Spectrometry, ASTM D3120, Test Method for Trace Quantities of Sulfur in Light Liquid Petroleum Hydrocarbons by Oxidative Microcoulometry, or ASTM D5453, Test Method for Determination of Total Sulfur in Light Hydrocarbons, Motor Fuels and Oil by Ultraviolet Fluorescence. With ASTM D2622, prepare the calibration standards using ethanol (reagent grade) as the solvent to prevent errors caused by large differences in carbon/hydrogen ratios.

Lead

Most vehicles equipped to operate on fuel ethanol (Ed75-Ed85) are equipped with exhaust catalysts that control emissions of aldehydes (formaldehyde and acetaldehyde) as well as regulated emissions. Lead compounds deactivate the catalyst and are limited to trace amounts. ASTM D5059, Test Methods for Lead in Gasoline by X-Ray Spectroscopy, is used to determine lead content.

Phosphorus

Phosphorus deactivates exhaust catalysts and is limited by federal regulations to trace amounts. ASTM D3231, Test Method for Phosphorus in Gasoline, is used to determine phosphorus levels.

Water

The solubility of hydrocarbons in fuel ethanol (Ed75-Ed85) decreases with lowering temperature and increasing water content. Separation of the hydrocarbon from the fuel will adversely affect cold starting, driveability, and denaturing. Water may affect the calibration of some types of composition sensors of flexible-fuel vehicles. Water also reduces the energy content of the fuel and thus adversely affects fuel economy and power. Because some degree of water contamination is unavoidable in transport and handling, and because fuel ethanol (Ed75-Ed85) is miscible with water, the water content of fuel ethanol (Ed75-Ed85) is limited to reduce the potential for problems. ASTM E203, Test Method for Water Using Karl Fischer Titration or E1064, Test Method for Water in Organic Liquids by Coulometric Karl Fischer Titration, is a suitable test method for determining water content.

Copper

Copper is an active catalyst for low-temperature oxidation of hydrocarbons. Experimental work has shown that copper concentrations higher than 0.012 mg/kg in commercial gasolines may significantly increase the rate of gum formation. ASTM D1688, Test Methods for Copper in Water, is used to determine copper content.

FUEL METHANOL: M70-M85

ASTM D5797, Standard Specification for Fuel Methanol M70-M85 for Automotive Spark-Ignition Engines, covers a fuel blend, nominally 70 to 85 volume percent methanol and 30 to 15 volume percent hydrocarbons for use in ground vehicles with automotive spark-ignition engines. *Fuel methanol (M70-M85)* is defined as a blend of methanol and hydrocarbons of which the methanol portion is nominally 70 to 85 volume percent.

Methanol

The methanol content of M70-M85 is a crucial parameter, as it affects the capability of the fuel metering system of the vehicle to establish the proper air/fuel ratio for optimum vehicle operation. This is much less of a concern for flexible-fuel vehicles (FFVs) than for dedicated M70-M85 vehicles. Methanol content affects the lubrication properties of the fuel and affects the water tolerance of the M70-M85. The inclusion of impurities and contaminants, except for deliberately added hydrocarbons or additives, can impact adversely on the properties and performance of fuel methanol (M70-M85) as an automotive spark-ignition engine fuel. The quantities of some of these materials are limited by specified property limits. Trace amounts of unspecified materials including higher alcohols, methyl formate, acetone, and dimethyl ether can be present. The maximum limit on water, the maximum limit on higher alcohols, and minimum and maximum limits on hydrocarbon/aliphatic ether content control the amount of some impurities and contaminants.

Test Method for Determination of Methanol in Fuel Methanol (M70-M85) for Automotive Spark-Ignition Engines, which appears in Annex A1 of ASTM D5797, provides a procedure for measuring methanol content by gas chromatography for fuels containing 70 to 95 volume percent methanol. However, the precision of this test method may not be adequate.

Hydrocarbons

Hydrocarbons are deliberately added to provide improved cold startability and warm-up driveability. The addition of hydrocarbons also contributes to flame visibility (luminous flame), nonexplosive air/fuel mixtures in storage tanks (rich mixture vapor space), and denaturation (malodorant and taste deterrent). The hydrocarbon portion of the fuel must be unleaded. While the composition of the hydrocarbons added to the fuel methanol is not controlled, the hydrocarbons should be stable, noncorrosive, and be in the boiling range of spark-ignition engine fuel provided in ASTM D4814, Specification for Automotive Spark-Ignition Engine Fuel.

ASTM D4815, Test Method for Determination of MTBE, ETBE, TAME, DIPE, *tertiary*-Amyl Alcohol and C_1 to C_4 Alcohols in Gasoline by Gas Chromatography, with modifications, may be used to determine higher alcohols, MTBE, and other ethers. Water may also be determined if the gas chromatograph is equipped with a thermal conductivity detector. As an alternative, ASTM E203, Test Method for Water Using Karl Fischer Titration, can be used for measurement of water. The concentration of methanol, other alcohols, and water can be added, and the sum subtracted from 100 to provide an estimate of the percent of hydrocarbons/aliphatic ethers. The precision of such a technique is not known. An alternative test method, Test Method for Determination of Hydrocarbon/Aliphatic Ether Content of Fuel Methanol (M70-M85) for Spark-Ignition Engines, is under development and appears in Annex A2 of ASTM D5797. Its reported precision is poor.

Vapor Pressure

Vapor pressure is varied for seasonal and climatic changes by providing three vapor pressure classes for M70-M85. The addition of volatile hydrocarbons improves cold startability. The addition of too many volatile hydrocarbons can cause hot fuel-handling problems. When blending with gasoline during the wintertime, higher hydrocarbon content may be necessary to obtain required volatility. Higher vapor pressures are required in the wintertime for cold starting, and lower vapor pressures are needed in the summertime to prevent hot fuel handling problems. Excessive vapor pressure for a given ambient condition can contribute to evaporative emissions. Lower and upper limits on vapor pressure for three volatility classes are used to define the acceptable range of the volatile components to ensure proper vehicle performance. Three vapor pressure classes of fuel are provided to satisfy vehicle performance requirements under different climatic conditions. The schedule for seasonal and geographical distribution indicates the appropriate vapor pressure class for each month in all areas of the United States based on altitude and expected air temperatures.

ASTM D4953, Test Method for Vapor Pressure of Gasoline and Gasoline-Oxygenate Blends (Dry Method), ASTM D5190, Test Method for Vapor Pressure of Petroleum Products (Automatic Method), or ASTM D5191, Test Method for Vapor Pressure of Petroleum Products (Mini Method), shall be used to determine the vapor pressure of M70-M85.

Luminosity

When pure methanol burns, it produces a blue, smokeless, nonluminous flame that is nearly invisible in daylight. Thus, it is difficult to know when a fire exists and to fight such a fire. A desirable property for M70-M85 fuel is that it maintains a clearly visible flame throughout the duration of a burn. It would be hazardous for the visible flame to disappear before the fire was extinguished. To make a methanol flame visible, materials such as aromatic hydrocarbons are added to methanol. In general, it has been established that unleaded gasoline having greater than 30 volume percent aromatics content when used as the hydrocarbon portion of M70-M85 will result in an M70-M85 fuel that will meet a requirement of a clearly visible flame throughout most of a burn. However, luminosity performance is dependent on the types of aromatics present in the hydrocarbon portion.

Appendix X2 of ASTM D5797, Test Method for Luminosity of Fuel Methanol (M70-M85) for Automotive Spark-Ignition Engines, covers a procedure to determine if a fuel methanol (M70-M85) composition produces a luminous flame throughout the duration of a burn by comparing its luminosity performance under controlled conditions to that of ethanol. This test method is not adequate for use in its present form and is provided for information only.

Acidity

Very dilute aqueous solutions of low-molecular-weight organic acids such as formic acid are highly corrosive to many metals. It is necessary to keep such acids at a very low level. ASTM D1613, Test Method for Acidity in Volatile Solvents and Chemical Intermediates Used in Paint, Varnish, Lacquer, and Related Products, shall be used to determine acidity.

Gum Content, Solvent Washed and Unwashed

The test for solvent washed gum content measures the amount of residue after evaporation of the fuel and following a heptane wash. The heptane wash removes the heptane-soluble, nonvolatile material such as additives, carrier oils used with additives, and diesel fuels. Unwashed gum consists of fuel-insoluble gum and fuel-soluble gum. The fuel-insoluble portion can clog fuel filters. Both can be deposited on surfaces when the fuel evaporates.

Solvent washed gum content can contribute to deposits on the surfaces of carburetors, fuel injectors, and intake manifolds, ports, valves, and valve guides. The impact of solvent-washed gum on engines operating on fuel methanol (M70-M85) has not been fully established. Performance effects depend on where the deposits form and the amount of deposit.

The difference between the unwashed and solvent washed gum content values can be used to assess the presence and amount of nonvolatile soluble material in the fuel. Additional analytical testing is required to determine if the material is an additive, carrier fluid, diesel fuel, or other. The unwashed gum content limit is intended to limit high-boiling contaminants, like diesel fuel, that can affect engine performance, yet allow the proper dosage of deposit-control additives with carrier oils normally added to the hydrocarbon portion of the fuel methanol (M70-M85).

ASTM D381, Test Method for Gum Content in Fuels by Jet Evaporation, is used for determining unwashed and solvent washed gum. However, because the precision statements for ASTM D381 were developed using only data on hydrocarbons, this test method may not be applicable to fuel methanol (M70-M85).

Ionic Chloride
Ionic (inorganic) chloride is corrosive to many metals, and it is desirable to minimize ionic chlorine compounds in fuel methanol (M70-M85). An inorganic chloride limit of a maximum 1 mg/kg has been found to be inadequate in protecting some fuel system components. ASTM D7319, Test Method for Determination of Total and Potential Sulfate and Inorganic Chloride in Fuel Ethanol by Direct Injection Suppressed Ion Chromatography or ASTM D7328, Test Method for Determination of Total and Potential Inorganic Sulfate and Total Inorganic Chloride in Fuel Ethanol by Ion Chromatography Using Aqueous Sample Injection, is used to determine inorganic chloride content in M75-M85.

Lead
Most vehicles equipped to operate on fuel methanol (M70-M85) are equipped with exhaust catalysts that control emissions of aldehydes (formaldehyde and acetaldehyde) as well as regulated emissions. Lead compounds deactivate the catalysts and are limited to trace amounts to prevent this problem. ASTM D5059, Test Methods for Lead in Gasoline by X-Ray Spectroscopy, is used to determine lead content. However, when using this test method, prepare the calibration standards using methanol (reagent grade) as the solvent to prevent errors caused by large differences in carbon-hydrogen ratios.

Phosphorus
Phosphorus deactivates exhaust catalysts and is limited to trace amounts. ASTM D3231, Test Method for Phosphorus in Gasoline is used to determine the presence of phosphorus.

Water
The solubility of hydrocarbons in fuel methanol decreases with lowering temperature and increasing water content. Separation of the hydrocarbon from the fuel will adversely affect cold starting and driveability, luminosity, and taste-deterrence. Water may affect the calibration of some types of composition sensors of flexible-fuel vehicles. Water also reduces the energy content of the fuel and thus adversely affects fuel economy and power. Because some degree of water contamination is practically unavoidable in transport and handling, and because fuel methanol is miscible with water, the water content of fuel methanol is limited to reduce the potential for problems. ASTM E203, Test Method for Water Using Karl Fischer Titration, is a suitable test method for determining water content of fuel methanol (M70-M85).

Sulfur
The limit on sulfur content is included to protect against engine wear, deterioration of engine oil, corrosion of exhaust system parts, and exhaust catalyst deactivation. Sulfur content can be determined using ASTM D1266, Test Method for Sulfur in Petroleum Products (Lamp Method), ASTM D2622, Test Method for Sulfur in Petroleum Products by Wavelength Dispersive X-Ray Fluorescence Spectrometry, ASTM D3120, Test Method for Trace Quantities of Sulfur in Light Liquid Petroleum Hydrocarbons by Oxidative Microcoulometry, or ASTM D5453, Test Method for Determination of Total Sulfur in Light Hydrocarbons, Motor Fuels and Oil by Ultraviolet Fluorescence. For ASTM D2622, the calibration standards should be prepared using methanol (reagent grade) as the solvent to prevent errors caused by large differences in carbon/hydrogen ratios.

METHYL *TERTIARY*-BUTYL ETHER FOR BLENDING WITH GASOLINE
ASTM D5983, Specification for Methyl *Tertiary*-Butyl Ether (MTBE) for Downstream Blending for Use in Automotive Spark-Ignition Engine Fuel, covers requirements for fuel grade MTBE utilized in commerce, terminal blending, or downstream blending with fuels for spark-ignition engines.

MTBE may be used as a blending component for automotive spark-ignition engine fuel to meet the oxygenate requirements of clean air programs or improve the antiknock quality of certain types of fuels. EPA regulations govern the allowable amounts of MTBE and other oxygenates that may be added to unleaded gasoline. MTBE is also subject to various state regulations that may ban or restrict the use of MTBE in gasoline.

Appearance
Suspended materials, sediments, or contaminants in the MTBE, which cause a cloudy or colored appearance, may adversely affect the performance of the finished fuel blend in automotive spark-ignition engines. In addition, a cloudy or colored appearance may indicate excessive water or contamination by materials not directly measured under this specification. Appearance should be clear and bright. ASTM D4176, Test Method for Free Water and Particulate Contamination in Distillate Fuels (Visual Inspection Procedures), Procedure 1, shall be used for determining appearance.

Methyl Tertiary-*Butyl Ether Purity*
The MTBE minimum purity level limits the quantities of contaminants. A minimum MTBE content of 95.0 mass percent has been established. ASTM Test Method D5441, Test Method for Analysis of Methyl *Tertiary*-Butyl Ether (MTBE) by Gas Chromatography, is used to measure MTBE content.

Sulfur
Sulfur and sulfur-containing compounds contribute to engine wear, deterioration of engine oil, exhaust catalyst deactivation, and corrosion of exhaust system parts in spark-ignition engine systems. The limit on sulfur is included to ensure that the finished blend of fuel is not detrimental to these systems. ASTM D4045, Test Method for Sulfur in Petroleum Products by Hydrogenolysis and Rateometric Colorimetry, may be used to determine sulfur content. However, the sample may require dilution with a sulfur-free diluent.

Solvent Washed Gum Content
The test for solvent washed gum content measures the amount of residue after evaporation of the fuel and following a heptane wash. The heptane wash removes the heptane-soluble, nonvolatile material such as additives, carrier oils used with additives, and diesel fuels. Solvent washed gum consists of fuel-insoluble gum. The fuel-insoluble portion can clog fuel filters. Both can be deposited on surfaces when the fuel evaporates. The solvent washed gum content test may also indicate contamination of the MTBE during shipping and storage. The limit is included to ensure that finished blends of gasoline do not contain excess solvent washed gum and handling contamination is minimized.

Solvent washed gum can contribute to deposits on the surfaces of carburetors, fuel injectors, and intake manifolds, ports, valves, and valve guides. The impact of solvent washed gum on malfunctions of modern engines is not well established, and the current limit has been assumed from the historic gasoline limit rather than from any recent correlative work. Performance effects depend on where the deposits form and the amount of deposit. ASTM D381, Test Method for Gum Content in Fuels by Jet Evaporation, is used to determine solvent washed gum. However, because the precision statements for this test method were developed using only data on hydrocarbons, this test method may not be applicable to MTBE.

Copper Strip Corrosion
The limit for copper strip corrosion is included to ensure that the MTBE does not contribute to copper corrosion. Fuels must pass the copper strip corrosion test to minimize corrosion in fuel systems due to sulfur compounds in the fuel. ASTM D130, Test Method for Detection of Copper Corrosion from Petroleum Products by the Copper Strip Tarnish Test, is used to measure copper corrosion.

Methanol Content
Methanol content in MTBE is limited to a maximum 0.5 mass percent. Methanol is one of the reactants in the production of MTBE and is a potential contaminant. Methanol contributes to vapor pressure increase and poorer water tolerance of finished fuel blends. Also, the methanol content of unleaded fuel is limited by EPA regulations. ASTM D5441, Test Method for Analysis of Methyl *Tertiary*-Butyl Ether (MTBE) by Gas Chromatography, may be used to measure the mass percent of methanol.

Water Content
Blends of MTBE and hydrocarbon gasoline have a limited solvency for water. This solvency will vary with the chemical composition, temperature, and MTBE content of the fuel. Excess water (which may be soluble in the MTBE) may not be soluble in the gasoline-MTBE blend and could result in a hazy fuel that does not meet the *clear and bright* requirement of Specification D4814. The water content of MTBE used for blending with hydrocarbon gasoline is limited to reduce the risk of haze formation. ASTM E203, Test Method for Water Using Volumetric Karl Fischer Titration, or ASTM E1064, Test Method for Water in Organic Liquids by Coulometric Karl Fischer Titration, is used to determine the water content of MTBE.

Vapor Pressure
The vapor pressure of a finished fuel blend must be high enough to ensure ease of engine starting. Excessive vapor pressure, however, may contribute to vapor lock or high evaporative emissions and running losses. The vapor pressure of MTBE is controlled to prevent adversely affecting the vapor pressure of the finished blend. The EPA regulates the summertime vapor pressure of finished MTBE fuel blends. In addition, vapor pressure exceeding the limits may indicate contamination by a light hydrocarbon. ASTM D4953, Test Method for Vapor Pressure of Gasoline and Gasoline-Oxygenate Blends (Dry Method), is used to determine the vapor pressure of MTBE.

BIODIESEL FUEL
ASTM D6751, Specification for Biodiesel Fuel Blend Stock (B100) for Distillate Fuels, covers low-sulfur biodiesel (B100), for use as a blend component with diesel fuel oils as defined by ASTM D975, Specification for Diesel Fuel Oils, and by ASTM D7467, Standard Specification for Diesel Fuel Oil.

Biodiesel, designated B100, is defined as a fuel comprised of mono-alkyl esters of long chain fatty acids derived from vegetable oils or animal fats. *Diesel fuel* is defined as a light or middle petroleum distillate fuel. A *biodiesel blend* is defined as a blend of biodiesel fuel with petroleum-based diesel fuel.

Biodiesel is typically produced by a reaction of a vegetable oil or animal fat with an alcohol such as methanol or ethanol in the presence of a catalyst to yield mono-alkyl esters and glycerin, which is removed. Biodiesel derives approximately 10 % of its mass from the reacted alcohol. The alcohol used in the reaction may or may not come from renewable resources. Biodiesel has been generally blended in the United States in concentrations of 5 volume percent (B5) and 20 volume percent biodiesel (B20).

Following is a discussion of the major physical and chemical properties of biodiesel.

Flash Point
The flash point for biodiesel is used as the mechanism to limit the level of unreacted alcohol remaining in the finished fuel. The flash point is also of importance in connection with legal requirements and safety precautions involved in fuel handling and storage and are normally specified to meet insurance and fire regulations. Typical values are over 160°C. The limit for biodiesel flash point has been set at 130°C minimum to ensure an actual value of 100°C minimum.

ASTM D93, Test Methods for Flash Point by Pensky-Martens Closed Cup Tester, can be used except where other methods are prescribed by law. ASTM D3828, Test Methods for Flash Point by Small Scale Closed Tester, or ASTM D6450, Standard Test Method for Flash Point by Continuously Closed Cup (CCCFP) Tester, can also be used. However, the precision and bias of ASTM D3828 and ASTM D6450 with biodiesel is not known and is currently under investigation. ASTM D93 shall be the referee method.

Viscosity
Minimum viscosity levels are important to protect against power loss due to fuel injection pump and injector leakage. Maximum viscosity levels are limited by engine design and size and by the characteristics of the fuel injection system. The upper limit of 6.0 mm^2/s at 40°C for biodiesel viscosity is higher than the maximum allowable viscosity in Specification D975 Grade 2-D and 2-D low sulfur (4.1 mm/s at 40°C). Blending biodiesel with diesel fuel close to its upper limit could result in a viscosity level exceeding ASTM D975, Specification for Diesel Fuel Oils. ASTM D445, Test Method for Kinematic Viscosity of Transparent and Opaque Liquids (and the Calculation of Dynamic Viscosity), is used to determine viscosity.

Sulfated Ash
Ash-forming materials may be present in biodiesel as abrasive solids, soluble metallic soaps, and unremoved catalysts. Abrasive solids and unremoved catalysts can contribute to fuel injector, fuel pump, piston, and ring wear, as well as engine deposits. Soluble metallic soaps have little effect on wear but may contribute to filter plugging and engine

deposits. Use ASTM D874, Test Method for Sulfated Ash from Lubricating Oils and Additives, for determining sulfated ash.

Sulfur

The effect of sulfur content on engine cylinder wear and deposits appears to vary considerably in importance and depends largely on operating conditions. Fuel sulfur can also affect emissions control systems performance, and various limits on sulfur have been imposed for environmental reasons. Most B100 contains less than 5 ppm sulfur. Some biodiesel produced from used cooking oils has been found to contain slightly higher levels of sulfur (15 to 30 ppm).

ASTM D5453, Test Method for Determination of Total Sulfur in Light Hydrocarbons, Spark Ignition Engine Fuel, Diesel Engine Fuel and Engine Oils by Ultraviolet Fluorescence, can be used for determining sulfur content. Other test methods may also be suitable for determining up to 0.05 % sulfur in biodiesel fuels such as ASTM D1266, Test Method for Sulfur in Petroleum Products (Lamp Method), ASTM D2622, Test Method for Sulfur in Petroleum Products by Wavelength Dispersive X-Ray Fluorescence Spectrometry, ASTM D3120, Test Method for Trace Quantities of Sulfur in Light Liquid Petroleum Hydrocarbons by Oxidative Microcoulometry, and ASTM D4294, Test Method for Sulfur in Petroleum Products by Energy-Dispersive X-Ray Fluorescence Spectroscopy. However, these test methods may provide falsely high results, and their precision and bias with biodiesel is not known. ASTM D5453 shall be the referee test method.

Copper Strip Corrosion

This test serves as a measure of possible difficulties with copper and brass or bronze parts of the fuel system. The presence of acids or sulfur-containing compounds can tarnish the copper strip, thus indicating the possibility for corrosion. ASTM D130, Test Method for Detection of Copper Corrosion from Petroleum Products by the Copper Strip Tarnish Test, 3-h test at 50°C, shall be used for determining copper strip corrosion.

Cetane Number

Cetane number is a measure of the ignition quality of the fuel and influences white smoke and combustion roughness in some engines. The cetane number requirements depend on engine design, size, nature of speed and load variations, and on starting and atmospheric conditions.

ASTM D613, Test Method for Cetane Number of Diesel Fuel Oil, shall be used for determining cetane number. ASTM D6890, Test Method for Determination of Ignition Delay and Derived Cetane Number (DCN) of Diesel Fuel Oils by Combustion in a Constant Volume Chamber may also be used. ASTM D976, Test Methods for Calculated Cetane Index of Distillate Fuels, and ASTM D4737, Test Method for Calculated Cetane Index by Four Variable Equation, should not be used to calculate the cetane number of biodiesel or biodiesel blends, since these test methods may yield falsely low results.

Cloud Point

Cloud point is of importance since it defines the temperature at which a cloud or haze of crystals appears in the fuel under prescribed test conditions. Biodiesel generally has a higher cloud point than petroleum-based diesel. To ensure trouble-free operation in cold climates, the cloud point of biodiesel and its impact on cold flow properties of the final blend should be taken into account.

ASTM D2500, Test Method for Cloud Point of Petroleum Oils, can be used for determining cloud point. ASTM D3117, Test Method for Wax Appearance Point of Distillate Fuels, may also be used because the two test methods are closely related. ASTM D5773, Test Method for Cloud Point of Petroleum Products (Constant Cooling Rate Method), may also be used. ASTM D2500 shall be the referee test method. However, the precision and bias of these test methods for biodiesel are not known and are currently under investigation.

Cold Soak Filterability

Some substances that are soluble or appear to be soluble in biodiesel at room temperature will, upon cooling at temperatures above the cloud point or standing at room temperature for extended periods, come out of solution. These substances can cause filter plugging. ASTM D7501 Standard Test Method for Determination of Fuel Filter Blocking Potential of Biodiesel (B100) Blend Stock by Cold Soak Filtration Test (CSFT), provides an accelerated means of assessing the propensity for these substances to plug filters.

Carbon Residue

Carbon residue gives a measure of the carbon-depositing tendencies of a fuel oil. While not directly correlating with engine deposits, this property is considered an approximation. Although biodiesel is in the distillate boiling range, most biodiesel boils at approximately a constant temperature and it is difficult to leave a 10 % residual upon distillation.

ASTM D4530, Test Method for Determination of Carbon Residue (Micro Method), can be used for determining carbon residue. The sample is first distilled to remove 90 volume percent. The remaining bottoms are subjected to the test procedure. The results are reported as the percentage carbon residue on 10 % distillation residue. ASTM D189, Test Method for Conradson Carbon Residue of Petroleum Products, or ASTM D524, Test Method for Ramsbottom Carbon Residue of Petroleum Products, may also be used. ASTM D4530 shall be the referee method.

Acid Number

Acid number is used to determine the level of free fatty acids or processing acids that may be present in biodiesel. Biodiesel with a high acid number has been shown to increase fuel system deposits and may increase the likelihood for corrosion. Acid number measures a different phenomenon for biodiesel than petroleum-based diesel. The acid number for biodiesel measures free fatty acids or degradation by-products not found in petroleum-based diesel. Increased recycle temperatures in new fuel system designs may accelerate fuel degradation that could result in high acid values and increased filter plugging potential. ASTM D664, Test Method for Acid Number of Petroleum Products by Potentiometric Titration, can be used for determining acid number. ASTM D3242, Test Method for Acidity in Aviation Turbine Fuel, or ASTM D974, Test Method for Acid and Base Number by Color-Indicator Titration, may also be used. ASTM D664 shall be used as the referee test method.

Free Glycerin

Free glycerin is a measure of the amount of glycerin remaining in the biodiesel after processing. High levels of free glycerin can cause injector deposits, as well as clogged fuel systems, and result in a buildup of free glycerin in the bottom of storage and fuel systems. ASTM D6584, Test Method for Determination of Free and Total Glycerin in B100 Biodiesel Methyl Esters by Gas Chromatography, is used to determine free glycerin.

Total Glycerin

Total glycerin is the sum of the free glycerin and the glycerin portion of any unreacted or partially reacted oil or fat. Low levels of total glycerin ensure that high conversion of the oil or fat into its mono-alkyl esters has taken place. High levels of mono-, di-, and triglycerides can cause injector deposits and may adversely affect cold weather operation and filter plugging. ASTM D6584, Test Method for Determination of Free and Total Glycerin in B100 Biodiesel Methyl Esters by Gas Chromatography, determines total glycerin.

Calcium and Magnesium

Calcium and magnesium may be present in biodiesel as abrasive solids or soluble metallic soaps. Abrasive solids can contribute to injector, fuel pump, piston, and ring wear, as well as to engine deposits. Soluble metallic soaps have little effect on wear, but they may contribute to filter plugging and engine deposits. High levels of calcium and magnesium compounds may also be collected in exhaust particulate removal devices, are not typically removed during passive or active regeneration, and can create increased back pressure and reduced time to service maintenance. There is no standard ASTM test method available, but Test Method EU14538, Fat and Oil Derivatives–Fatty Acid Methyl Esters (FAME)–Determination of Ca, K, Mg and Na Content by Optical Emission Spectral Analysis with Inductively Coupled Plasma (ICP OES) can be used.

Water and Sediment

Contamination by water and particulates can adversely affect the performance of fuel filters and fuel injectors. ASTM D2709, Test Method for Water and Sediment in Middle Distillate Fuels by Centrifuge, is the preferred test method. ASTM D1796, Test Method for Water and Sediment in Fuel Oils by the Centrifuge Method (Laboratory Procedure), may also be used.

Phosphorus Content

Phosphorus levels must be kept low, since the presence of phosphorus can damage catalytic converters used in emission control systems. Complying with a phosphorus limit of 10 ppm maximum should not be a problem, since most biodiesel produced in the United States has a phosphorus content below 1 ppm. Biodiesel from other sources may contain higher levels of phosphorus. ASTM D4951, Test Method for Determination of Additive Elements in Lubricating Oils by Inductively Coupled Plasma Atomic Emission Spectrometry, shall be used for measuring phosphorus.

Sodium and Potassium Combined Content

Sodium and potassium may be present in biodiesel as abrasive solids or soluble metallic soaps. Abrasive solids can contribute to injector, fuel pump, piston and ring wear, and also to engine deposits. Soluble metallic soaps have little effect on wear, but they may contribute to filter plugging and engine deposits. High levels of sodium or potassium compounds may also be collected in exhaust particulate removal devices, are not typically removed during passive or active regeneration, and they can create increased back pressure and reduced period to service maintenance. Sodium and potassium, combined can be determined using EN14538 Fat and Oil Derivatives–Fatty Acid Methyl Ester (FAME)–Determination of Ca, K, Mg and Na Content by Optical Emission Spectral Analysis with Inductively Coupled Plasma (ICP OES). Test Method UOP 391 Trace Metals in Petroleum Products or Organics by AAA may also be used. Test Method EN 14538 shall be the referee test method.

Reduced Pressure Distillation

Biodiesel exhibits a boiling point rather than a distillation curve. The fatty acid chains in the raw oils and fats from which biodiesel is produced are mainly comprised of straight chain hydrocarbons with 16 to 18 carbons that have similar boiling temperatures. The atmospheric boiling point of biodiesel generally ranges from 330 to 357°C. The reduced pressure distillation limit of 360°C is not problematic and was added to ensure the fuel has not been adulterated with high boiling contaminants. ASTM D1160, Test Method for Distillation of Petroleum Products at Reduced Pressure, shall be used for determining reduced pressure distillation.

Oxidation Stability

Products of oxidation in biodiesel can take the form of various acids or polymers, which, if in high enough concentration, can cause fuel system deposits and lead to filter clogging and fuel system malfunctions. Additives designed to retard the formation of acids and polymers can significantly improve the oxidation stability performance of biodiesel. There is no ASTM test method to determine this property. ASTM Specification D6751 specifies the use of EN14112, the Rancimate Test, Fat and Oil Derivatives–Fatty Acid Methyl Esters (FAME)–Determination of Oxidation Stability (Accelerated oxidation test).

DENSITY AND RELATIVE DENSITY

None of the ASTM specifications set limits on the density of oxygenates because the density is fixed by the other chemical and physical properties of the materials. Density relates to the volumetric energy content of the fuel–the denser the fuel, the higher the volumetric energy content although the oxygen present reduces the energy content. Density is important because oxygenates are often bought and sold with the volume corrected to a specific temperature, usually 15.6°C (60°F). Volume correction factors for oxygenates differ somewhat from those for hydrocarbons, and work is in progress to determine precise correction factors for gasoline-oxygenate blends.

Oxygenate density is determined by ASTM D4052/IP 365, Test Method for Density and Relative Density of Liquids by Digital Density Meter.

SAMPLING, CONTAINERS, AND SAMPLE HANDLING

Using the correct sampling procedures are critical for all fuels and fuel components. ASTM D4057, Practice for Manual Sampling of Petroleum and Petroleum Products, provides several procedures for manual sampling. ASTM D4177, Practice for Automatic Sampling of Petroleum and Petroleum Products provides automatic sampling procedures. For volatility determinations of a sample, ASTM D5842, Practice for Sampling and Handling of Fuels for Volatility Measurement, contains special precautions for sampling and handling techniques to maintain sample integrity. ASTM D4306, Practice for Aviation Fuel

Sample Containers for Tests Affected by Trace Contamination, should be used to select appropriate containers especially for tests sensitive to trace contamination. Also ASTM D5854, Practice for Mixing and Handling of Liquid Samples of Petroleum and Petroleum Products, provides procedures for container selection and sample mixing and handling. For octane or cetane number determination, protection from light is important. Collect and store sample fuels in an opaque container, such as a dark brown glass bottle, metal can, or minimally reactive plastic container to minimize exposure to UV emissions from sources such as sunlight or fluorescent lamps. For sampling of oxygenated materials water displacement must not be used, because of potential problems associated with the interaction of water with oxygenates.

Applicable ASTM Specifications

ASTM	Title
D396	Specification for Fuel Oils
D975	Specification for Diesel Fuel Oils
D1152	Specification for Methanol (Methyl Alcohol)
D1193	Specification for Reagent Water
D4806	Specification for Denatured Fuel Ethanol for Blending with Gasolines for Use as Automotive Spark-Ignition Engine Fuel
D4814	Specification for Automotive Spark-Ignition Engine Fuel
D5797	Specification for Fuel Methanol (M70-M85) for Automotive Spark-Ignition Engines
D5798	Specification for Fuel Ethanol (Ed75-Ed85) for Automotive Spark-Ignition Engines
D5983	Specification for Methyl Tertiary-Butyl Ether (MTBE) for Downstream Blending with Automotive Spark-Ignition Fuel
D6751	Specification for Biodiesel Fuel Blend Stock (B100) for Distillate Fuels
D7467	Specification for Diesel Fuel Oil, Biodiesel Blend (B6 to B20)
D02:1347	Committee D02 Research Report on Reformulated Spark-Ignition Engine Fuel

Applicable ASTM/IP Test Methods

Before using any test method, the Scope shall be reviewed to make sure the test method is applicable to the product being tested and that the specified measurement range covers the area of interest.

ASTM	IP	Title
D86	154	Test Method for Distillation of Petroleum Products at Atmospheric Pressure
D93	13	Test Methods for Flash Point by Pensky-Martens Closed Cup Tester
D130	131	Test Method for Detection of Copper Corrosion from Petroleum Products by the Copper Strip Tarnish Test
D189	71	Test Method for Conradson Carbon Residue of Petroleum Products
D381		Test Method for Gum Content in Fuels by Jet Evaporation
D445	14	Test Method for Kinematic Viscosity of Transparent and Opaque Liquids (and the Calculation of Dynamic Viscosity)
D524		Test Method for Ramsbottom Carbon Residue of Petroleum Products
D613	163	Test Method for Cetane Number of Diesel Fuel Oil
D664		Test Method for Acid Number of Petroleum Products by Potentiometric Titration
D874	139	Test Method for Sulfated Ash from Lubricating Oils and Additives
D891		Test Methods for Specific Gravity, Apparent, of Liquid Industrial Chemicals
D974		Test Method for Acid and Base Number by Color-Indicator Titration
D976	107	Test Methods for Calculated Cetane Index of Distillate Fuels
D1160	154	Test Method for Distillation of Petroleum Products at Reduced Pressure
D1266	13	Test Method for Sulfur in Petroleum Products (Lamp Method)
D1613	131	Test Method for Acidity in Volatile Solvents and Chemical Intermediates Used in Paint, Varnish, Lacquer, and Related Products
D1688		Test Methods for Copper in Water
D1796		Test Method for Water and Sediment in Fuel Oils by the Centrifuge Method (Laboratory Procedure)
D2622		Test Method for Sulfur in Petroleum Products by Wavelength Dispersive X-ray Fluorescence Spectrometry
D2709		Test Method for Water and Sediment in Middle Distillate Fuels by Centrifuge
D3117		Test Method for Wax Appearance Point of Distillate Fuels
D3120		Test Method for Trace Quantities of Sulfur in Light Liquid Petroleum Hydrocarbons by Oxidative Microcoulometry

ASTM	IP	Title
D3242		Test Method for Acidity in Aviation Turbine Fuel
D3505		Test Method for Density or Relative Density of Pure Liquid Chemicals
D3828		Test Methods for Flash Point by Small Scale Closed Tester
D4045		Test Method for Sulfur in Petroleum Products by Hydrogenolysis and Rateometric Colorimetry
D4052	84	Standard Test Method for Density and Relative Density of Liquids by Digital Density Meter
D4057		Practice for Manual Sampling of Petroleum and Petroleum Products
D4176		Test Method for Free Water and Particulate Contamination in Distillate Fuels (Visual Inspection Procedures)
D4177		Practice for Automatic Sampling of Petroleum and Petroleum Products
D4294		Test Method for Sulfur in Petroleum Products by Energy-Dispersive X-Ray Fluorescence Spectroscopy
D4306		Standard Practice for Aviation Fuel Sample Containers for Tests Affected by Trace Contamination
D4530		Test Method for Determination of Carbon Residue (Micro Method)
D4737		Test Method for Calculated Cetane Index by Four Variable Equation
D4815		Test Method for Determination of MTBE, ETBE, TAME, DIPE, tertiary-Amyl Alcohol and d to C_4 Alcohols in Gasoline by Gas Chromatography
D4951		Test Method for Determination of Additive Elements in Lubricating Oils by Inductively Coupled Plasma Atomic Emission Spectrometry
D4953		Test Method Test Method for Vapor Pressure of Gasoline and Gasoline-Oxygenate Blends (Dry Method)
D5059		Test Methods for Lead in Gasoline by X-Ray Spectroscopy
D5190		Test Method for Vapor Pressure of Petroleum Products (Automatic Method)
D5191		Test Method for Vapor Pressure of Petroleum Products (Mini Method)
D5441		Test Method for Analysis of Methyl tert-Butyl Ether (MTBE) by Gas Chromatography
D5453		Test Method for Determination of Total Sulfur in Light Hydrocarbons, Spark Ignition Engine Fuel, Diesel Engine Fuel, and Engine Oils by Ultraviolet Fluorescence
D5501		Test Method for the Determination of Ethanol Content of Denatured Fuel Ethanol by Gas Chromatography
D5580		Standard Test Method for Determination of Benzene, Toluene, Ethylbenzene, p/m-Xylene, o-Xylene, C_9 and Heavier Aromatics, and Total Aromatics in Finished Gasoline by Gas Chromatography
D5773		Test Method for Cloud Point of Petroleum Products (Constant Cooling Rate Method)
D5854		Standard Practice for Mixing and Handling of Liquid Samples of Petroleum and Petroleum Products—Designation: Manual of Petroleum Measurement Standards
D6423		Test Method for Determination of pHe of Ethanol, Denatured Fuel Ethanol, and Fuel Ethanol (Ed75-Ed85)
D6450		Test Method for Flash Point by Continuously Closed Cup (CCCFP) Tester
D6550		Standard Test Method for Determination of Olefin Content of Gasolines by Supercritical-Fluid Chromatography
D6584		Test Method for Determination of Free and Total Glycerin in B100 Biodiesel Methyl Esters by Gas Chromatography
D6890		Test Method for Determination of Ignition Delay and Derived Cetane Number (DCN) of Diesel Fuel Oils by Combustion in a Constant Volume Chamber
D7318		Test Method for Total Inorganic Sulfate in Ethanol by Potentiometric Titration
D7319		Test Method for Determination of Total and Potential Sulfate and Inorganic Chloride in Fuel Ethanol by Direct Injection Suppressed Ion Chromatography
D7328		Test Method for Determination of Total and Potential Inorganic Sulfate and Total Inorganic Chloride in Fuel Ethanol by Ion Chromatography Using Aqueous Sample Injection
D7501		Standard Test Method for Determination of Fuel Filter Blocking Potential of Biodiesel (B100) Blend Stock by Cold Soak Filtration Test (CSFT)
E203		Test Method for Water Using Volumetric Karl Fischer Titration
E1064		Test Method for Water in Organic Liquids by Coulometric Karl Fischer Titration

Applicable EN Test Methods	
EN14112	The Rancimat Test, Fat and Oil Derivatives—Fatty Acid Methyl Ester (FAME)—Determination of Oxidation Stability (Accelerated Oxidation Test).
EN14538	Fat and oil derivatives—Fatty Acid Methyl Ester (FAME)—Determination of Ca, K, Mg and Na Content by Optical Emission Spectral Analysis with Inductively Coupled Plasma (ICP OES)

4

Fuels for Land and Marine Diesel Engines and for Nonaviation Gas Turbines

Steven R. Westbrook[1] and Richard T. LeCren[2]

THE DIESEL ENGINE IS NOW FULLY ESTABLISHED in a variety of applications on land and in marine use. On land, it serves to power trains, buses, trucks, and automobiles and to run construction, petroleum drilling, agricultural, and other off-road equipment. At sea, it serves both to provide main propulsion power and to run auxiliaries.

Gas turbine engines also serve in a wide range of applications. Over half of the larger industrial gas turbines are in electric-generation use. Other uses include gas pipeline transmission, co-generation systems, and transportation. Cruise ships use gas turbines in newer ships to reduce vibration and stack emissions. In the military, gas turbines power a number of combatant ships both as main propulsion units and as the power source for auxiliary uses. Gas turbines are also used to power some military ground vehicles such as main battle tanks.

The quality criteria and methods for testing fuels for land and marine diesel engines and for nonaviation gas turbines are sufficiently similar to address in a common chapter. Obviously, certain criteria and tests will apply to one or the other rather than both. For example, cetane number, which is a critical property for automotive diesel fuels, is of limited significance for gas turbine fuels.

DIESEL ENGINE

The diesel engine is a high-compression, self-ignition engine. The heat of the high compression ignites the fuel; no spark plug is used. The diesel cycle consists of charging the combustion chamber with air, compressing the air, injecting the fuel, which ignites spontaneously, expanding the burned gases, and expelling the products of combustion.

Diesel engines may be designed to operate on a four- or two-stroke cycle. Each type has advantages and disadvantages, so the choice depends on the application. The four-stroke cycle has better volumetric efficiency, good combustion characteristics, and positive exhaust gas scavenging. The principal advantage of the two-stroke cycle is compactness in relation to power output.

Diesel engines vary greatly in size, power output, and operating speeds. While some small units develop only a few brake horsepower, at the other extreme there are engines having cylinder diameters of greater than 1,000 mm, developing several thousand horsepower per cylinder. Operating speeds are almost as diverse as size and power output. They range from below 100 revolutions per minute (rpm) for some larger engines to 4,000 rpm and above for those used in automotive and other vehicle prime mover applications. The entire range of diesel engines can be divided into three broad classification groups indicated in Table 1.

It should not be surprising that diesel engines also vary extensively in their requirements for fuel. Selection of the proper fuel is not a simple procedure but depends on many variables. Among the most important considerations are:

1. Engine size and design
2. Operating speed and load changes
3. Frequency of speed and load changes
4. Maintenance considerations
5. Atmospheric conditions
6. Fuel price and availability
7. Fuel handling system

Each of the foregoing factors plays a part in dictating the fuel to be chosen for a diesel engine. The relative influence of each factor is determined by the specific application and installation involved.

NONAVIATION GAS TURBINE

The simple-cycle, gas-turbine engine operates on the Brayton or joule cycle consisting of adiabatic compression, constant pressure heating, and adiabatic expansion. By adding a heat exchanger to transfer heat from the turbine exhaust gas to the combustor inlet air, thermal efficiency can be increased. Air is compressed in the compressor through axial or centrifugal stages or both and directed toward the combustion chamber. Here, part of the air mixes with vaporized or atomized fuel and supports combustion. The remainder of the air passes around the flame, cooling the metal surfaces, and combining with those gases that are rapidly expanding from combustion. The resulting gas stream is then expanded through one or more turbine wheels, which drive the compressor and provide the output power. In a simple-cycle gas turbine, the gas is then exhausted; in a regenerative engine, the exhaust gas is directed through a heat exchanger to heat the combustor inlet air.

Gas turbine engines cover a range from 50 to nearly 100,000 hp. The selection of a gas-turbine fuel oil for use in a given gas turbine requires consideration of the following factors: availability of the fuel, design of the gas turbine and fuel handling system, and maintenance and operating requirements.

COMBUSTION PROCESS

Diesel Combustion

The following discussion of diesel combustion is only a general overview. A more detailed discussion can be found elsewhere [1]. The fuel used in all diesel engines—precombustion-chamber,

[1] *Southwest Research Institute, San Antonio, TX.*
[2] *Consultant Engineer, Gas Turbine Combustion & Fuels, Retired.*

TABLE 1—Range of Diesel Engines

Classifications	Speed Range	Conditions	Typical Applications[A]
Low speed	Below 300 rpm	Sustained heavy load, constant speed	Marine main propulsion; electric power generation
Medium speed	300 to 1,000 rpm	Fairly high load and relatively constant speed	Marine auxiliaries: stationary power generators; pumping units
High speed	Above 1,000 rpm	Frequent and wide variation in load and speed	Road transport vehicles: diesel locomotives; construction equipment

[A] There are many other typical applications that are not listed.

direct-injection, two-cycle, four-cycle–passes through the following processes:

1. Storage, pumping, and handling
2. Filtering
3. Heating (if necessary)
4. Atomization and mixing with air
5. Combustion
6. Power extraction
7. Heat exchange and exhaust

Additionally, marine diesel fuels containing residual oil components require centrifuging during the fuel-handling process.

The fuel properties are significant in all these processes and particularly influence combustion and resultant energy extraction. In any combustion process, there are at least three basic requirements:

1. Formation of a mixture of fuel and air
2. Ignition of the fuel/air mixture
3. Completion of combustion of the fuel/air mixture

In the diesel engine, these requirements are met as indicated diagrammatically in Fig. 1. Figure 2 shows a typical pressure–versus–crank angle diagram for a diesel engine combustion chamber. Prior to the injection of the fuel, air alone is compressed and raised to a high temperature during the compression stroke. The final compression pressure and resultant air temperature will vary with compression ratio, speed, and engine design. Shortly before the end of compression, at a point controlled by the fuel injection timing system, one or more jets of fuel are introduced into the main combustion chamber or the precombustion chamber.

Ignition does not occur immediately on injection. The fuel droplets must first be vaporized by heat from the compressed air. The fuel/air mixture finally reaches a temperature at which self-ignition occurs, and the flame begins to spread.

Engine design, fuel and air inlet temperatures, degree of atomization of the fuel, and fuel composition control the duration of the delay period between injection and ignition. This delay period is known commonly as "ignition delay."

Injection of fuel is continued during the ignition delay. The ignition delay period must be short in order to avoid "diesel knock," which is caused by very rapid burning or detonation of relatively large amounts of fuel gathered in the cylinder before combustion begins. Once the flame has spread completely, the only fuel in the cylinder is that being injected into the burning mixture. This fuel burns almost instantaneously. The final part of the combustion cycle is the completion of burning after injection has ceased.

Pressure charging is used frequently as a means for increasing the amount of air delivered to an engine without increasing its size. A compressor, either directly coupled to the crankshaft (a supercharger) or driven by a turbine using the heat energy in the exhaust gases (a turbocharger), is used to increase the amount of combustion air available. Consequently, the engine is able to burn a greater quantity of fuel. The amount of fuel is ultimately limited by the thermal and mechanical stresses that can be tolerated by engine components.

Nonaviation Gas Turbine Combustion

In a gas turbine engine, fuel is burned continuously at peak cycle pressure to heat the air to moderate temperatures. The combustor is essentially a direct-fired air heater in which fuel is burned with less than one third of the available air, and the combustion products are mixed with the remaining air to cool them to the maximum temperature allowed by metallurgical considerations in the turbine section.

Although gas-turbine combustors vary widely in design arrangement, all perform the same basic functions in much the same way. The three main methods of fuel injection are as follows:

PRESSURE ATOMIZING

Most combustors use an atomizer in which fuel is forced under high pressure through an orifice. This breaks up the liquid fuel into small droplets, which greatly increases the surface-to-volume ratio. The droplets are then introduced to the primary zone of the combustor where they vaporize and burn; the two processes occur simultaneously.

AIR ATOMIZING

Most combustors today use some form of air atomization. In this approach, the pressure drop across the combustor is used to generate a jet or stream of air. The fuel is injected into this stream. The actual arrangement varies in different combustors. It is particularly useful in low emission combustors where some degree of premixing before combustion takes place is required. The fuel pressure required is somewhat less than that required by pressure atomizing systems.

PREVAPORIZING (VAPORIZING)

This method nearly always relies on direct heating of the fuel. Fuel, together with a proportion of the primary air, a mixture too rich to burn, is supplied to the vaporizing tubes located in the primary zone of the combustor. The heat released in the primary zone vaporizes the fuel, and the

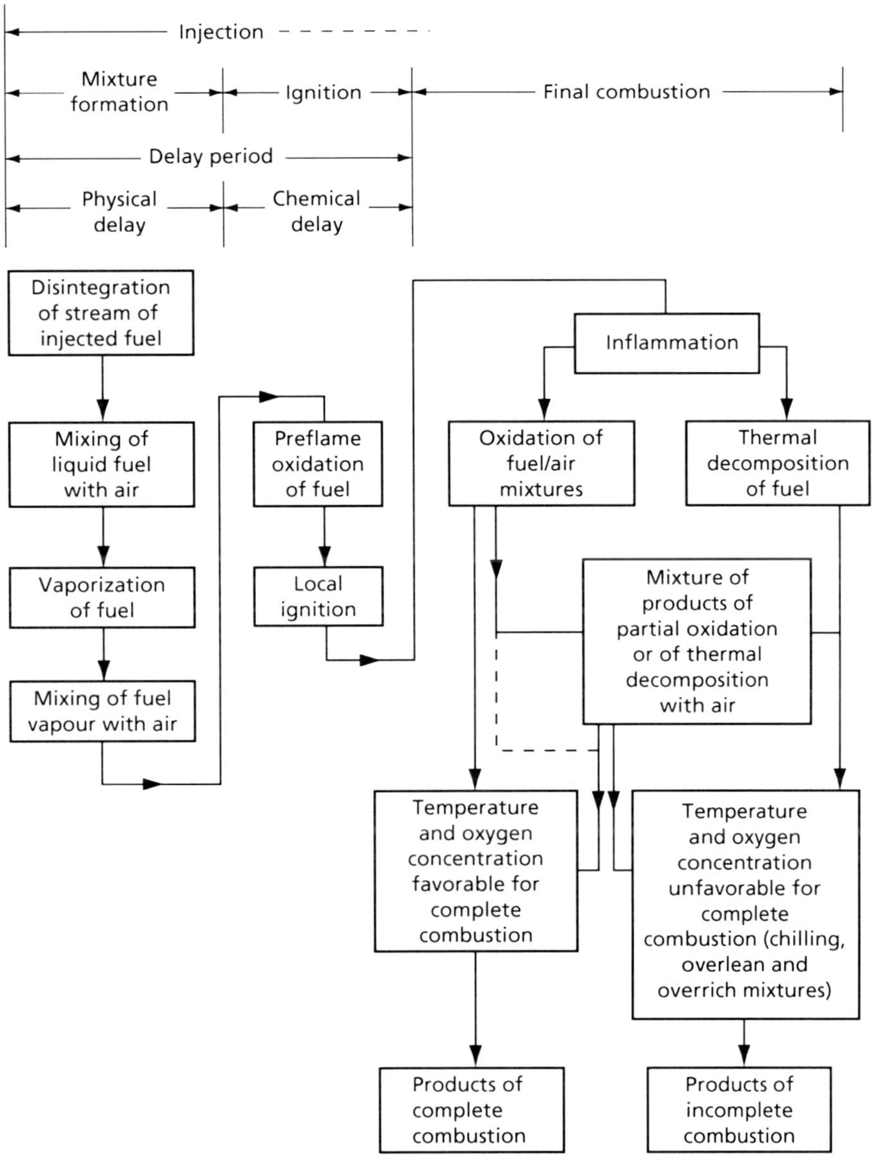

Fig. 1—Outline of the combustion process of the diesel engine (From M. A. Eliot, July 1949, "Combustion of Diesel Fuel," *SAE Trans.,* Vol. 3, No. 3, p. 492).

resulting rich mixture is discharged into the primary zone in an upstream direction. The remainder of the primary air is admitted into the primary zone and burning of the mixture occurs.

Combustor inlet air temperature, which depends on engine pressure ratio and load, varies from about 121 to 454°C (250 to 850°F) in various nonregenerative engines. With regeneration, combustor inlet temperatures may be 371 to 649°C (700 to 1,200°F). Combustor output temperatures range from 649 to 1,038°C (1,200 to 1,900°F) or above. Combustor pressures can range from 310 kPa (45 psia) to as high as 21.1 kg/cm (300 psia).

GENERAL FUEL CHARACTERISTICS AND SPECIFICATIONS

The basic fuel requirements for land and marine diesel engines and for nonaviation gas turbines are:

- Satisfactory ignition and combustion under the conditions existing in the combustion chamber
- Suitability for handling by the injection equipment
- Convenient handling at all stages from the refinery to the engine fuel tank without suffering degradation and without harming any surface which it may normally contact

Diesel and nonaviation gas-turbine fuels were originally straight-run products obtained from the distillation of crude oil. Today, through various refinery processes, these fuels may contain varying amounts of selected cracked distillates. This permits an increase in the volume of available fuel at minimum cost. The boiling range of middle distillate fuels is approximately 150 to 400°C (300 to 755°F). The relative merits of the fuel types to be considered will depend on the refining practices used, the nature of crude oils from which they are produced, and the additive package (if any) used. Additionally, up to 5 volume percent biodiesel is also allowed in diesel fuel under specification D975.

The broad definition of fuels for land and marine diesel engines and for nonaviation gas turbines covers many possible combinations of volatility, ignition quality, viscosity, gravity, stability, and other properties. Various classifications or

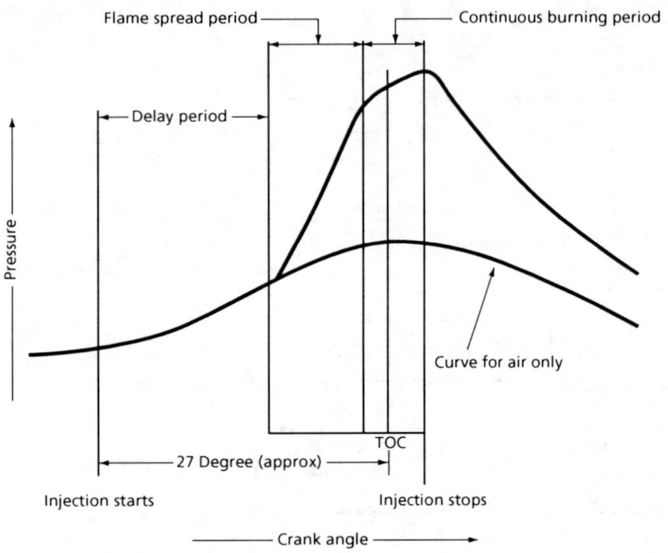

Fig. 2—Pressure variations in a diesel engine cylinder during combustion.

specifications are used in different countries to characterize these fuels and thereby to establish a framework for definition and reference. The ASTM Specification for Diesel Fuel Oils (D975) is one of the most widely used specifications for diesel fuels. The ASTM Specification for Gas Turbine Fuel Oils (D2880) has some similar requirements, as a comparison of Tables 2 and 3, the detailed chemical and physical property requirements for the two classes of fuels, will show.

Blended marine diesel fuels are a relative newcomer to the range of available fuels. The development and use of marine diesel engines for deep-draft vessel propulsion became so pervasive during the 1970s that diesel power generation now predominates in waterborne commerce. Steamships have been displaced from their major role in the world's fleets.

As diesel power came to prominence, equipment manufacturers developed engine designs and fuel handling processes to use the new blended fuels more efficiently. Petroleum refiners, marine fuel suppliers, and vessel operators quickly found that existing specifications and test methods were not completely adequate for their needs. Ad hoc quality and test procedures proliferated. Fuel quality disputes frequently resulted in long, expensive, and inconclusive litigation. The need for a general standard for marine fuels became clear. To fill the void, the British Standards Institution published BSMA 100, the first generally accepted marine fuel specification, in 1982. This document was superseded in 1987 by the publication of ISO 8217 by the International Standards Organization. ISO 8217 represents a cooperative effort by national standards organizations such as BSI and ASTM to develop an international consensus on minimum marine fuel requirements. ASTM adopted Standard Specification D2069 as an ASTM equivalent to ISO 8217; however, it has since been withdrawn because ISO 8217 is used and accepted throughout the world as the specification for commercial marine fuels.

TABLE 2—Detailed Requirements for Diesel Fuel Oils[A]

Property	ASTM Test Method[B]	Grade						
		No. 1-D S15	No. 1-D S500[C]	No. 1-D S5000[D]	No. 2-D S15[E]	No. 2-D S500[C,E]	No. 2-D S5000[D,E]	No. 4-D[D]
Flash Point, °C, min.	D93	38	38	38	52[E]	52[E]	52[E]	55
Water and Sediment, percent volume, max	D2709	0.05	0.05	0.05	0.05	0.05	0.05	...
	D1796	...	...	...	...	...	...	0.50
Distillation Temperature, °C 90 %, percent volume recovered	D86							
min		...	...	...	282[E]	282[E]	282[E]	...
max		288	288	288	338	338	338	...
Kinematic Viscosity, mm²/s at 40°C	D445							
min		1.3	1.3	1.3	1.9[E]	1.9[E]	1.9[E]	5.5
max	...	2.4	2.4	2.4	4.1	4.1	4.1	24.0
Ash percent mass, max	D482	0.01	0.01	0.01	0.01	0.01	0.01	0.10
Sulfur, ppm (μg/g)[F] max	D5453	15	...	...	15	...	...	...
percent mass, max	D2622[G]	...	0.05	...	...	0.05	...	...
percent mass, max	D129	...	...	0.50	...	...	0.50	2.00
Copper strip corrosion rating, max (3 h at a minimum control temperature of 50°C)	D130	No. 3	No. 3	No. 3	No. 3	No. 3	No. 3	...
Cetane number, min[H]	D613	40[I]	40[I]	40[I]	40[I]	40[I]	40[I]	30[I]

TABLE 2—Detailed Requirements for Diesel Fuel Oils[A] (Continued)

Property	ASTM Test Method[B]	Grade						
		No. 1-D S15	No. 1-D S500[C]	No. 1-D S5000[D]	No. 2-D S15[E]	No. 2-D S500[C,E]	No. 2-D S5000[D,E]	No. 4-D[D]
One of the following properties must be met:								
(1) Cetane index, min.	D976–80[G]	40	40	...	40	40	...	...
(2) Aromaticity, percent volume, max	D1319[G]	35	35	...	35	35	...	...
Operability Requirements								
Cloud point, °C, max or	D2500	[J]	[J]	[J]	[J]	[J]	[J]	...
LTFT/CFPP, °C, max	D4539/ D6371							
Ramsbottom carbon residue on 10 % distillation residue, percent mass, max	D524	0.15	0.15	0.15	0.35	0.35	0.35	...
Lubricity, HFRR @ 60°C, micron, max	D6079	520	520	520	520	520	520	...
Conductivity, pS/m or Conductivity Units (C.U.), min	D2624/ D4308	25[K]	25[K]	25[K]	25[K]	25[K]	25[K]	...

[A] To meet special operating conditions, modifications of individual limiting requirements may be agreed on between purchaser, seller, and manufacturer.
[B] The test methods indicated are the approved referee methods. Other acceptable methods are indicated in 5.1.
[C] Under United States regulations, if Grades No. 1–D S500 or No. 2–D S500 are sold for tax exempt purposes then, at or beyond terminal storage tanks, they are required by 26 CFR Part 48 to contain the dye Solvent Red 164 at a concentration spectrally equivalent to 3.9 lb per thousand barrels of the solid dye standard Solvent Red 26, or the tax must be collected.
[D] Under United States regulations, Grades No.1–D S5000, No. 2–D S5000, and No. 4–D are required by 40 CFR Part 80 to contain a sufficient amount of the dye Solvent Red 164 so its presence is visually apparent. At or beyond terminal storage tanks, they are required by 26 CFR Part 48 to contain the dye Solvent Red 164 at a concentration spectrally equivalent to 3.9 lb per thousand barrels of the solid dye standard Solvent Red 26.
[E] When a cloud point less than −12°C is specified, as can occur during cold months, it is permitted and normal blending practice to combine Grades No. 1 and No. 2 to meet the low temperature requirements. In that case, the minimum flash point shall be 38°C, the minimum viscosity at 40°C shall be 1.7 mm^2/s, and the minimum 90% recovered temperature shall be waived.
[F] Other sulfur limits can apply in selected areas in the United States and in other countries.
[G] These test methods are specified in 40 CFR Part 80.
[H] Where cetane number by Test Method D 613 is not available, Test Method D 4737 can be used as an approximation.
[I] Low ambient temperatures as well as engine operation at high altitudes may require the use of fuels with higher cetane ratings.
[J] It is unrealistic to specify low temperature properties that will ensure satisfactory operation at all ambient conditions. In general, cloud point (or wax appearance point) Low Temperature Flow Test, and Cold Filter Plugging Point Test may be used as an estimate of operating temperature limits for Grades No. 1–D S500; No. 2–D S500; and No. 1–D S5000 and No. 2–D S5000 diesel fuel oils. However, satisfactory operation below the cloud point (or wax appearance point) may be achieved depending on equipment design, operating conditions, and the use of flow-improver additives as described in X5.1.2. Appropriate low temperature operability properties should be agreed on between the fuel supplier and purchaser for the intended use and expected ambient temperatures. Test Methods D 4539 and D 6371 may be especially useful to estimate vehicle low temperature operability limits when flow improvers are used. Due to fuel delivery system, engine design, and test method differences, low temperature operability tests may not provide the same degree of protection in various vehicle operating classes. Tenth percentile minimum air temperatures for U.S. locations are provided in Appendix X5 as a means of estimating expected regional temperatures. The tenth percentile minimum air temperatures can be used to estimate expected regional target temperatures for use with Test Methods D 2500, D 4539, and D 6371. Refer to X5.1.3 for further general guidance on test application.
[K] The electrical conductivity of the diesel fuel is measured at the time and temperature of the fuel at delivery. The 25 pS/m minimum conductivity requirement applies at all instances of high velocity transfer (7 m/s) but sometimes lower velocities, see 8.1 for detailed requirements) into mobile transport (for example, tanker trucks, rail cars, and barges).

In 1985, ASTM published *Marine Fuels, STP 878*. This STP provides a review of research and developments completed, in progress, or planned up to 1985. The published documents yield an understanding of the problems and complexities of marine fuel production. ASTM D975, ASTM D2880, and ISO 8217 do not provide complete quality specifications for the fuels defined in those documents. In general, they present minimum requirements necessary to ensure reasonable operation under the conditions described.

Diesel Fuels

When first published, D975 contained specifications for "three grades of diesel fuel oils suitable for various types of diesel engines." The grades were No. 1-D, No. 2-D, and No. 4-D. Table 4 provides a description, of each grade, from the 1948 standard and the 2008 standard. Note that the description of Grade 2-D in the 1948 version did not include automotive applications, reflecting the fact that diesel engines were for heavy-duty use and large trucks. In the 2008 version of the standard, the grades are distinguished by the maximum allowable sulfur in each grade. The Sxxx designation was adopted to distinguish grades by sulfur rather than using words such as "Low Sulfur," as previously, because the word descriptions were thought to be imprecise. S5000 grades correspond to the so-called "regular" sulfur grades, the previous No. 1-D and No. 2-D. S500 grades correspond to the previous "Low Sulfur"

TABLE 3—Detailed Requirements for Gas Turbine Fuel Oils at Time and Place of Custody Transfer to User[A,B]

Property	ASTM Test Method[C]	Grade[D]				
		No. 0-GT	No. 1-GT[E]	No. 2-GT[E]	No. 3-GT	No. 4-GT
Flash point °C (°F) min	D93	[F]	38 (100)	38 (100)	55 (130)	66 (150)
Water and sediment percent volume max	D2709	0.05	0.05	0.05	...	...
	D1796	...	...	...	1.0	1.0
Distillation						
Temperature	D86					
°C (°F)						
90 % volume recovered						
Min		...	...	282	...	...
Max		...	288	338	...	...
Kinematic viscosity						
2 mm/s[G]	D445					
AT 40°C (104°F) min		[F]	1.3	1.9	5.5	5.5
max			2.4	4.1	...	...
AT 100°C (212°F) max			...	...	50.0	50.0
Ramsbottom carbon residue on 10 % distillation Residue percent mass, max	D524	0.15	0.15	0.35	...	...
Ash percent mass, max	D482	0.01	0.01	0.01	0.03	...
Density at	D1298					
15°C kg/m³						
max		...	850	876	...	...
Pour point[G] °C (°F) max	D97	...	−18	−6	...	...

[A] To meet special operating conditions, modifications of individual limiting requirements may be agreed on between purchaser, seller, and manufacturer.
[B] Gas turbines with waste heat recovery equipment may require fuel sulfur limits to prevent cold end corrosion. Environmental limits may also apply to fuel sulfur in selected areas in the United States and in other countries.
[C] The test methods indicated are the approved referee methods. Other acceptable methods are indicated in 6.1.
[D] No. 0-GT includes naphtha, Jet B fuel, and other volatile hydrocarbon liquids. No. 1-GT corresponds in general to specification D 396 Grade No. 1 fuel and D 975 Grade 1-D diesel fuel in physical properties. No. 2-GT corresponds in general to Specification D 396 No. 2 fuel and D 975 Grade 2-D diesel fuel in physical properties. No. 3-GT and No. 4-GT viscosity range brackets specification D 396 Grades No. 4, No. 5 (light), No. 5 (heavy), and No. 6, and D 975 Grade No. 4-D diesel fuel in physical properties.
[E] Under United States regulations, Grades No. 1-GT and No. 2-GT are required by 40 CFR Part 80 to contain a sufficient amount of dye Solvent Red 164 so its presence is visually apparent. At or beyond terminal storage tanks, they are required by 26 CFR Part 48 to contain the dye Solvent Red 164 at a concentration spectrally equivalent to 3.9 lb per thousand barrels of the solid dye standard Solvent Red 26.
[F] When the flash point is below 38°C (100°F) or when kinematic viscosity is below 1.3 mm²/s at 40°C (104°F) or when both conditions exist, the turbine manufacturer should be consulted with respect to safe handling and fuel system design.
[G] For cold weather operation, the pour point should be specified 6°C below the ambient temperature at which the turbine is to be operated except where fuel heating facilities are provided. When a pour point less than −18°C is specified for Grade No. 2-GT, the minimum viscosity shall be 1.7 mm²/s and the minimum 90 % recovered temperature shall be waived.

grades. S15 grades are commonly referred to as "Ultra-Low Sulfur" grades or ULSD.

While the foregoing classifications deal chiefly with distillate fuels, residual fuels are used extensively in large, slow-speed, marine main propulsion engines and stationary land-based engines principally because of lower fuel cost. More recently, the lower cost aspect has promoted the increasing popularity of residual fuels in some smaller, medium-speed auxiliary engines. Mixtures of residual and distillate fuels are used in installations where a compromise is sought between the more desirable properties of the latter and the lower cost of the former.

The ISO Marine Fuels Specification, ISO 8217, has four categories of distillate fuels and 15 categories of fuels containing residual components. A typical heavy residual fuel is illustrated in the specifications established by ASTM D396 to define No. 6 fuel oil, which is often referred to as "Bunker C." These specifications, relatively few in number and broad in their limits, are shown in Table 5. Some diesel engines are

TABLE 4—Descriptions of the Fuels Covered by Specification D975

Grade	D975-48 T	D975-08a
No. 1-D S5000	A volatile distillate fuel oil for engines in service requiring frequent speed and load changes.	Grade No. 1-D S5000 comprises the class of volatile fuel oils from kerosine to the intermediate middle distillates. Fuels within this grade are applicable for use in high-speed diesel engines applications necessitating frequent and relatively wide variations in loads and speeds, and also for use in cases where abnormally low operating temperatures are encountered.
No. 1-D S500	Not Applicable	Grade No. 1-D S500 comprises the class of low-sulfur, volatile fuel oils from kerosine to the intermediate middle distillates. Fuels within this grade are applicable for use in *(1)* high-speed diesel engines that require low sulfur fuels, *(2)* in applications necessitating frequent and relatively wide variations in loads and speeds, and *(3)* in applications where abnormally low operating temperatures are encountered.
No. 1-D S15	Not Applicable	Grade No. 1-D S15 comprises the class of very low sulfur, volatile fuel oils from kerosine to the intermediate middle distillates. Fuels within this grade are applicable for use in *(1)* high-speed diesel engines and diesel engine applications that require ultra-low sulfur fuels, *(2)* applications necessitating frequent and relatively wide variations in loads and speeds, and *(3)* applications where abnormally low operating temperatures are encountered.
No. 2-D S5000	A distillate fuel oil of low volatility for engines in industrial and heavy mobile service.	Grade No. 2-D S5000 includes the class of middle distillate gas oils of lower volatility than Grade No. 1-D S5000. These fuels are applicable for use in *(1)* high-speed diesel engines in applications necessitating relatively high loads and uniform speeds, or *(2)* in diesel engines not requiring fuels having higher volatility or other properties specified for Grade No. 1-D S5000.
No. 2-D S500	Not Applicable	Grade No. 2-D S500 includes the class of low-sulfur, middle distillate gas oils of lower volatility than Grade No. 1-D S500. These fuels are applicable for use in *(1)* high-speed diesel engine applications that require low sulfur fuels, *(2)* applications necessitating relatively high loads and uniform speeds, or *(3)* diesel engines not requiring fuels having higher volatility or other properties specified for Grade No. 1-D S500.
No. 2-D S15	Not Applicable	Grade No. 2-D S15 includes the class of very low sulfur, middle distillate gas oils of lower volatility than Grade No. 1-D S15. These fuels are applicable for use in *(1)* high speed diesel engines and diesel engine applications that require ultra-low sulfur fuels, *(2)* applications necessitating relatively high loads and uniform speeds, or *(3)* diesel engines not requiring fuels having higher volatility or other properties specified in Grade No. 1-D S15.
No. 4-D	A fuel oil for low and medium speed engines.	Grade No. 4-D comprises the class of more viscous middle distillates and blends of these middle distillates with residual fuel oils. Fuels within this grade are applicable for use in low- and medium-speed diesel engines in applications necessitating sustained loads at substantially constant speed.

operated on fuels of even heavier than No. 6 fuel oil. However, it should be recognized that heavier fuels, while having the advantage of lower fuel cost, usually require special equipment and special lubricating oils to achieve satisfactory performance.

Additives may be used to improve diesel fuel performance. Cetane number improvers such as alkyl nitrates and nitrites can improve ignition quality. Pour-point depressants can improve low-temperature performance. Antismoke additives may reduce exhaust smoke, which is a growing concern as more and more attention is paid to atmospheric pollution. Antioxidant and sludge dispersants can minimize or prevent the formation of insoluble compounds. Fuels formulated with cracked stock components in particular may form such compounds, which could cause fuel line and filter plugging.

Nonaviation Gas Turbine Fuel

In ASTM D2880, Grade No. 0-GT includes naphthas, Jet B aviation fuel, and other light hydrocarbon liquids that characteristically have low flash points and low viscosities as compared with kerosene and fuel oils.

In ASTM D2880, Grade No. 1-GT is a light distillate oil suitable for use in nearly all gas turbines and corresponds in physical properties to No. 1 fuel oil and 1-D diesel fuel oil. Grade No. 2-GT is a heavier distillate grade than No. 1-GT and is for use in turbines not requiring the clean-burning characteristics of No. 1-GT. It is similar in properties to No. 2 fuel oil and 2-D diesel fuel oil.

Grade No. 3-GT is a low-ash fuel having the same viscosity range that brackets No. 4, 5, and 6 fuel oils and 4-D diesel fuel oil. Grade No. 3-GT may be a heavier distillate than that encompassing Grade No. 2-GT, a residual fuel that meets the low-ash requirements, or a blend of a distillate with a residual fuel oil. For gas turbines operating at turbine inlet temperatures below 649°C (1,200°F), the nonmandatory vanadium, sodium-plus-potassium, and calcium limits are not critical provided a silicon-based additive or equivalent is used to prevent excessive ash deposition in the turbine. Fuel-heating equipment will be required by the gas turbine in almost every installation using No. 3-GT fuel.

Grade No. 4-GT covers the same viscosity range as Grade No. 3-GT, but it has no restriction on the quantity of ash. Grade No. 4-GT has a suggested magnesium-to-vanadium weight ratio limit of 3.5 maximum. Nearly all residual fuels will satisfy this requirement, but an additive may be needed to inhibit the corrosive action of the vanadium. In general, Grade No. 4-GT will form ash deposits in the turbine so that, with continuous operation, there will be a progressive reduction in power output and thermal efficiency, necessitating periodic shutdown for cleaning.

FUEL PROPERTIES AND TESTS

Density

Density is a fundamental physical property that can be used in conjunction with other properties to characterize both the light and heavy fractions of petroleum and petroleum products. Accurate determination of the density of petroleum products is necessary for the conversion of measured volumes to volumes at the standard temperature of 15°C (60°F). Density is a factor governing the quality of crude petroleum;

TABLE 5—Detailed Requirements for Fuel Oils[A]

Property	ASTM Test Method[B]	No. 1 S500[B]	No. 1 S5000[B]	No. 2 S500[B]	No. 2 S5000[B]	Grade No. 4 (Light)[B]	No. 4	No. 5 (Light)	No. 5 (Heavy)	No. 6
Flash Point, °C, min	D93 – Proc. A	38	38	38	38	38	...	...	...	...
	D93 – Proc. B	...	...	...	...	...	55	55	55	60
Water and sediment, percent volume, max	D2709	0.05	0.05	0.05	0.05	...	...	...	...	...
	D95 + D473	...	...	...	...	(0.50)[C]	(0.50)[C]	(1.00)[C]	(1.00)[C]	(2.00)[C]
Distillation Temperature, °C	D86									
10 percent volume recovered, max		215	215	...	...					
90 percent volume recovered, min		...	...	282	282					
90 percent volume recovered, max		288	288	338	338					
Kinematic viscosity at 40°C, mm^2/s	D445									
min		1.3	1.3	1.9	1.9	1.9	>5.5	...	...	...
max		2.4	2.4	4.1	4.1	5.5	24.0[D]			
Kinematic viscosity at 100°C, mm^2/s	D445									
min		...	...	...	...	...	...	5.0	9.0	15.0
max		...	...	...	...	...	...	8.9[D]	14.9[D]	50.0[D]
Ramsbottom carbon residue on 10% distillation residue percent mass, max	D524	0.15	0.15	0.35	0.35	...	...	...	...	...
Ash, percent mass, max	D482	...	...	...	...	0.05	0.10	0.15	0.15	...
Sulfur, percent mass max[E]	D129	...	0.50	...	0.50	...	...	...	...	...
	D2622	0.05		0.05						

TABLE 5—Detailed Requirements for Fuel Oils[A] (Continued)

Property	ASTM Test Method[B]	No. 1 S500[B]	No. 1 S5000[B]	No. 2 S500[B]	No. 2 S5000[B]	Grade No. 4 (Light)[B]	No. 4	No. 5 (Light)	No. 5 (Heavy)	No. 6
Copper strip corrosion rating, max, 3 h at a minimum control temperature of 50°C	D130	No. 3	No. 3	No. 3	No. 3	...	...	...	...	...
Density at 15°C, kg/m³	D1298									
min		...	...	...	...	>876[F]	...	...	...	...
max		850	850	876	876	...	...	...	...	...
Pour Point °C, max[G]	D97	−18	−18	−6	−6	−6	−6	...	...	[H]

[A] It is the intent of these classifications that failure to meet any requirement of a given grade does not automatically place an oil in the next lower grade unless in fact it meets all requirements of the lower grade. However, to meet special operating conditions, modifications of individual limiting requirements may be agreed on among the purchaser, seller, and manufacturer.
[B] Under United States regulations, Grades No. 1 S5000, No. 1 S500, No. 2 S5000, No. 2 S500, and No. 4 (Light) are required by 40 CFR Part 80 to contain a sufficient amount of the dye Solvent Red 164 so its presence is visually apparent. At or beyond terminal storage tanks, they are required by 26 CFR Part 48 to contain the dye Solvent Red 164 at a concentration spectrally equivalent to 3.9 lb per thousand barrels of the solid dye standard Solvent Red 26.
[C] The amount of water by distillation by Test Method D 95 plus the sediment by extraction by Test Method D 473 shall not exceed the value shown in the table. For Grade No. 6 fuel oil, the amount of sediment by extraction shall not exceed 0.50 mass percent, and a deduction in quantity shall be made for all water and sediment in excess of 1.0 mass percent.
[D] Where low sulfur fuel oil is required, fuel oil falling in the viscosity range of a lower numbered grade down to and including No. 4 can be supplied by agreement between the purchaser and supplier. The viscosity range of the initial shipment shall be identified and advance notice shall be required when changing from one viscosity range to another. This notice shall be in sufficient time to permit the user to make the necessary adjustments.
[E] Other sulfur limits may apply in selected areas in the United States and in other countries.
[F] This limit ensures a minimum heating value and also prevents misrepresentation and misapplication of this product as Grade No. 2.
[G] Lower or higher pour points can be specified whenever required by conditions of storage or use. When a pour point less than −18°C is specified, the minimum viscosity at 40°C for grade No. 2 shall be 1.7 mm²/s and the minimum 90% recovered temperature shall be waived.
[H] Where low sulfur fuel oil is required, Grade No. 6 fuel oil will be classified as Low Pour (+15°C max) or High Pour (no max). Low Pour fuel oil should be used unless tanks and lines are heated.

however, it is an uncertain indication of petroleum product quality unless correlated with other properties.

API gravity (D1298) is another measure of the density of fuel that has been used for years. It is calculated with the following formula:

$$\text{API gravity, deg} = (141.5/\text{specific gravity } 60/60^\circ\text{F}) - 131.5 \quad (1)$$

As use of the metric (SI) system of units and measure becomes more common, use of API gravity decreases. However, it is still commonly used to describe the density of crude oil and in fuel terminals, pipeline operations, and similar custody transfer situations. The primary benefit of using API gravity, aside from habit, is that it permits calculations, descriptions, and transactions using whole numbers instead of just decimals.

Ignition and Combustion Characteristics (Cetane Number)

Diesel engine performance is a function of compression ratio, injection timing, the manner in which fuel and air are mixed, and the resulting ignition delay or time from the start of injection to the beginning of combustion. The nature of the fuel is an important factor in reducing ignition delay. Physical characteristics such as viscosity, gravity, and mid-boiling point are influential. Hydrocarbon composition is also important as it affects both the physical and combustion characteristics of the fuel. Straight-chain paraffins ignite readily under compression, but branched-chain paraffins and aromatics react more slowly. The first widely used measure of ignition quality was the diesel index. The diesel index was calculated as:

$$\text{Diesel Index} = [(\text{API Gravity})(\text{Aniline Point})]/100 \quad (2)$$

By the mid-1930s, it was determined that a better measurement of ignition quality was needed. The result was an engine test, ASTM D613, Standard Test Method for Ignition Quality of Diesel Fuels by the Cetane Method. This test involves operating a standard, single-cylinder, variable compression ratio engine using a specified fuel flow rate and time of injection (injection advance) for the fuel sample and each of two bracketing reference fuels of known cetane number. The engine compression ratio is adjusted for each fuel to produce a specified ignition delay, and the cetane number is calculated to the nearest tenth by interpolation of the compression ratio values.

The cetane number scale uses two primary reference fuels. One, *n*-hexadecane (normal cetane), has excellent ignition qualities and, consequently, a very short ignition delay. This fuel was arbitrarily assigned a cetane number of 100. The second fuel, α-methylnaphthalene, has poor ignition qualities and was assigned a cetane number of zero. The α-methylnaphthalene was later replaced with heptamethylnonane, which was calibrated against the original fuels and assigned a cetane number of 15. The cetane number scale is now defined by the following equation for volumetric blends of the two primary reference materials:

$$\text{Cetane Number} = \%n\text{-cetane} + 0.15(\text{percent heptamethylnonane}) \quad (3)$$

In practice, the primary reference fuels are only used to calibrate two secondary reference fuels. These are selected diesel fuels of mixed hydrocarbon composition, which are designated as "T" and "U." "T" fuel typically has a cetane number of approximately 75, while "U" fuel is usually in the low 20 cetane number range. Each set of "T" and "U" fuels are paired, and test engine calibrations define the cetane numbers for volumetric blends of these two secondary reference fuels.

Higher cetane number fuels tend to lessen combustion noise, increase engine efficiency, increase power output, start easier (especially at low temperatures), reduce exhaust smoke, and reduce exhaust odor. In order to ensure acceptable cold weather performance, most modern diesel engines require a minimum cetane number of 40 [2] and this is the requirement in D975.

Cetane Index

Methods for calculating approximate cetane numbers were also developed for times when performing the engine test was not feasible. The two ASTM methods are:

- ASTM D976: Standard Test Method for Calculated Cetane Index of Distillate Fuels
- ASTM D4737: Standard Test Method for Calculated Cetane Index by Four Variable Equation

Both standards use fuel density and distillation values in their calculations. Standard D4737 is the more widely used method because it is newer and better represents diesel fuels currently in the market place. The following are among the limitations of these calculated cetane index methods.

1. They are not applicable to fuels containing additives for raising cetane number.
2. They are not applicable to pure hydrocarbons, synthetic fuels, alkylates, or coal tar products.
3. They are not applicable to biodiesel or blends of biodiesel with petroleum diesel.

Volatility/Distillation

The distillation characteristics of a diesel fuel exert a great influence on its performance. Two methods are commonly used to measure distillation characteristics:

- ASTM D86: Standard Test Method for Distillation of Petroleum Products
- ASTM D2887: Standard Test Method for Boiling Range Distribution of Petroleum Fractions by Gas Chromatography

Method D86 is the specified method, in D975, for measuring distillation characteristics. Figure 3 is a plot of distillation data for a single, typical diesel fuel using both methods. It is obvious from this plot that the two methods can give quite different results. For most fuels, results like these are typical, with more deviation at the beginning and end of the distillation and less deviation at the center. For all fuel grades in D975, Test Method D2887 can be used as an alternate. Results from Test Method D2887 shall be reported as "Predicted D86" results by application of the correlation in Appendix X5 of Test Method D2887 to convert the values.

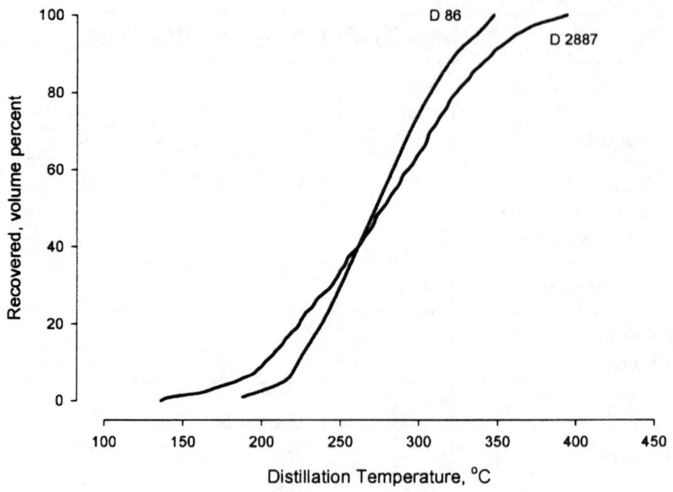

Fig. 3—Distillation curves for a typical No. 2 diesel fuel.

In case of dispute, Test Method D86 is the referee method. Grade No. 4-D does not have distillation requirements. The average volatility requirements of diesel fuels vary with engine speed, size, and design. However, fuels having too-low volatility tend to reduce power output and fuel economy through poor atomization, while those having too-high volatility may reduce power output and fuel economy through vapor lock in the fuel system or inadequate droplet penetration from the nozzle. In general, the distillation range should be as low as possible without adversely affecting the flash point, burning quality, heat content, or viscosity of the fuel. If the 10 % point is too high, poor starting may result. An excessive boiling range from 10 to 50 % evaporated may increase warm-up time. A low 50 % point is desirable to minimize smoke and odor. Low 90 % and end points tend to ensure low carbon residues and minimum crankcase dilution.

The temperature for 50 % evaporation, known as the mid-boiling point, is usually taken as an overall indication of the fuel distillation characteristics when a single numerical value is used alone. For example, in high-speed engines a 50 % point above 302°C might cause smoke formation, give rise to objectionable odor, cause lubricating oil contamination, and promote engine deposits. At the other extreme, a fuel with excessively low 50 % point would have too low a viscosity and heat content per unit volume. Therefore, a 50 % point in the range of 232 to 280°C is desirable for the majority of higher speed type diesel engines. This temperature range usually is broadened for larger, slower speed engines.

For the above reasons, some points on the distillation curve are considered more important and are included in fuel specifications more often. ASTM D975 contains only a limit on the 90 % point. Other specifications include requirements for initial boiling point (more so for gasoline), 10 %, 50 %, and, to a lesser degree, 95 % and final boiling point.

Viscosity

The method for measuring viscosity of diesel fuel is ASTM D445: Standard Test Method for Kinematic Viscosity of Transparent and Opaque Liquids (and the Calculation of Dynamic Viscosity). The unit of measurement for this method is squared millimeters per second (mm^2/s), also known as centistokes (cSt). This is the most widely used unit of measurement currently in use in the United States. Other units used in the past, and occasionally still used in the present, include Saybolt Universal Seconds (SUS) and centipoise.

For some engines it is advantageous to specify a minimum viscosity because of power loss due to injection pump and injector leakage. Maximum viscosity, on the other hand, is limited by considerations involved in engine design and size, and the characteristics of the injection system. Fuel viscosity exerts a strong influence on the shape of fuel spray. High viscosities can cause poor atomization, large droplets, and high-spray jet penetration. With high viscosities, the jet tends to be a solid stream instead of a spray of small droplets. As a result, the fuel is not distributed in, or mixed with, the air required for burning. This results in poor combustion, accompanied by loss of power and economy. In small engines, the fuel spray may impinge on the cylinder walls, washing away the lubricating oil film and causing dilution of the crankcase oil. Such a condition contributes to excessive wear.

Low fuel viscosities result in a spray that is too soft and does not penetrate far enough in the combustion chamber for good mixing. Combustion is impaired and power output and economy are decreased. Low viscosity can lead to excessive leakage past the injection pump plunger. Fuel metering becomes inaccurate and engine efficiency is reduced.

Fuel viscosity requirements for high-speed engines range from 1.8 to 5.8 cSt at 40°C. Usually the lower viscosity limit is established to prevent leakage in worn fuel injection equipment as well as to supply lubrication for injection system components in certain types of engines. During operation at low-atmospheric temperatures, the viscosity limit sometimes is reduced to 1.4 cSt at 40°C to obtain increased volatility and sufficiently low-pour point. Fuels having viscosities greater than 5.8 cSt at 40°C usually are limited in application to the slower-speed engines. The very viscous fuels commonly used in large stationary and marine engines normally require preheating for proper pumping, injection, and atomization.

Cloud Point

All diesel fuels contain dissolved paraffin wax. As the temperature of the fuel decreases, so does the solubility of the wax in the fuel. At some point wax crystals will begin to precipitate. If enough wax precipitates, the crystals can block fuel flow through screens, filters, and other restricted passages in the fuel system. The temperature at which the wax precipitation occurs depends on the origin, type, refining, and boiling range of the fuel. This temperature is known as the cloud point of the fuel. As the cloud point goes up, the suitability of the fuel for low-temperature operation decreases. The cloud point of the fuel can be measured by the following methods:

- ASTM D2500: Standard Test Method for Cloud Point of Petroleum Oils

The following methods are variations of D2500, including both automatic and automated methods.

- ASTM D3117: Standard Test Method for Wax Appearance Point of Distillate Fuels
- ASTM D5771: Standard Test Method for Cloud Point of Petroleum Products (Optical Detection Stepped Cooling Method)
- ASTM D5772: Standard Test Method for Cloud Point of Petroleum Products (Linear Cooling Rate Method)
- ASTM D5773: Standard Test Method for Cloud Point of Petroleum Products (Constant Cooling Rate Method)

For all grades of fuel listed in D975, any of these test methods may be used to measure cloud point. Method D2500 is the specified method and the others are considered alternates. In case of dispute, Test Method D2500 is the referee method.

Pour Point

Before a fuel can be burned in an engine, it must first be pumped from the fuel tank. The lowest temperature at which a fuel can be pumped is known as the pour point of the fuel. Although pour point is not specified in D975, test methods for measuring pour point include:

- ASTM D97: Standard Test Method for Pour Point of Petroleum Oils

The following methods are variations of D97, including both automatic and automated methods.

- ASTM D5949: Standard Test Method for Pour Point of Petroleum Products (Automatic Pressure Pulsing Method)
- ASTM D5950: Standard Test Method for Pour Point of Petroleum Products (Automatic Tilt Method)
- ASTM D5985: Standard Test Method for Pour Point of Petroleum Products (Rotational Method)

The pour point should be considered only a guide to the lowest temperature at which a fuel can be used. Some fuels, especially waxy fuels, will still flow at temperatures below their tested cloud point. In general, pour points are from 3 to 6°C below the cloud point for a given fuel; however, it is not uncommon for the difference to be as much as 11°C.

For any given fuel, there will be no wax precipitation problems at temperatures above the cloud point. At temperatures below the pour point, it is highly unlikely that the fuel will give satisfactory performance. It is not unusual to obtain satisfactory engine performance with a fuel at ambient temperatures between the cloud point and pour point. The degree of performance and the temperature depend on the engine, the vehicle design, and the fuel system configuration. Vehicles and fuel systems with small diameter lines, constrictions, small porosity strainers and filters, and fuel lines exposed to ambient temperatures or wind will tend toward poorer performance. Systems with insulation, supplemental heaters, or large sections of the fuel system in close proximity to engine heat can probably expect better performance at lower temperatures [1].

Low-Temperature Flow Test

As discussed earlier, the mere presence of wax crystals in a fuel does not guarantee the fuel will plug filters or other fuel system components. The tendency of a fuel to plug screens and filters at low temperatures is a dynamic property dependent on the size and shape of the wax crystals. [Vehicle fuel system design is also a factor.] For this reason, numerous dynamic tests for low-temperature operability have been developed. ASTM standardized one such test in 1985, ASTM D4539: Standard Test Method for Filterability of Diesel Fuels by Low-Temperature Flow Test (LTFT).

The LTFT was designed to yield results indicative of the low-temperature flow performance of the test fuel in some diesel vehicles.

Cold Filter Plugging Point

The Cold Filter Plugging Point (CFPP) was developed for use in Europe. The method is published by ISO as EN 116–IP 309. It is similar to the LTFT with two exceptions: (1) The fuel is cooled by immersion in a constant temperature bath, making the cooling rate nonlinear but comparatively much more rapid (about 40°C per hour); (2) The CFPP is the temperature of the sample when 20 mL of the fuel first fails to pass through a wire mesh in less than 60 seconds.

While the CFPP is the preferred method in Europe and is used in several European specifications, it appears to overestimate the benefit of using some additives, most especially for vehicles manufactured in North America [4,5].

Cleanliness

Diesel fuel cleanliness can mean many things to many people. It is safe to say that most users would consider any fuel that is visually free of water, sediment, and suspended matter to be a clean fuel. Indeed, this is the cleanliness (workmanship) requirement stated in D975. However, it is known that microscopic particulates in the fuel can lead to problems just as serious as the visible contaminants. The three most common methods of measuring cleanliness of diesel fuel are as follows:

- ASTM D2709: Standard Test Method for Water and Sediment in Middle Distillate Fuels by Centrifuge
- ASTM D4860: Standard Test Method for Free Water and Particulate Contamination in Mid-Distillate Fuels (Clear and Bright Numerical Rating)
- ASTM D6217: Standard Test Method for Particulate Contamination in Middle Distillate Fuels by Laboratory Filtration

D2709 is the method currently specified in D975. It is used to measure the amount of visible water, sediment, and suspended matter. This method gives no effective measurement of the presence or amount of microscopic particulates. Many would consider the level of contamination sufficient to produce readable results in this test to be gross contamination. However, this is an extremely sensitive test for contamination. The human eye is capable of seeing very small macroscopic particles, and the presence of one or two such particles could be considered a failure. In practice, the person conducting the test must exercise judgment based on experience and the requirements of the end use for the fuel.

D4860 was developed to give a means to quantify the level of cleanliness of a fuel. As with D2709, D4860 only provides a valid measurement of visible contamination. The benefit of this method is that the measurement is now an objective measure of cleanliness. While this test does provide a quantitative result, it is far less sensitive than D2709.

D6217 could be considered a combination of the best features of the two previous methods. It is sensitive to small amounts of particles and is even capable of detecting microscopic particulates. It is also a quantitative measure of the cleanliness of the fuel. The quantitative measure is in the form of milligrams of particulate (contamination) per liter of fuel. Currently there is no consensus standard with a specification limit for D6217. Many user specifications, including most federal and military diesel fuel specifications, include of limit of 10 mg/L for the results of this analysis. Most users have found fuel that meets this limit to give satisfactory performance in the vehicle.

The two methods for filterability of diesel fuels are as follows:

- ASTM D2068: Standard Test Method for Filter Plugging Tendency of Distillate Fuel Oils
- ASTM D6426: Standard Test Method for Determining Filterability of Distillate Fuel Oils

The primary weakness of visual and gravimetric methods is that there is no generally accepted correlation between the results of the test and the performance of the fuel in a vehicle fuel system. That is, how long could the vehicle operate on that fuel before the fuel filter plugs or the water separator fails? The British Royal Navy first developed D2068 as a dynamic test of the cleanliness of fuel for shipboard gas turbine engines [6]. The test was designed around the specific requirements of gas turbine–powered ships in the British Royal Navy. The most important requirement being that the fuel filters had a nominal porosity of 1 µm. As such, a glass fiber laboratory filter with pore size of 1 µm is used in the test. Over the years, the test apparatus was upgraded, making it more automated.

D6426 is a modification of Method D2068. The first difference between the two methods is the pump used in each. D2068 uses a piston pump, whereas D6426 uses a peristaltic pump. The second difference is the filter. D2068 uses a 13-mm-diameter, 1-µm pore size filter. D6426 uses a specially constructed specimen. The specimen is called an F-cell Filter Unit. It is a disposable, precalibrated assembly consisting of a shell and plug containing a 25-mm-diameter nylon membrane filter of nominal 5.0-µm pore size, nominal 60 % porosity, with a 17.7-mm^2 effective filtering area.

Despite the differences in equipment, both D2068 and D6426 have the same pass/fail criteria. A fuel fails the test if the pressure drop across the filter reaches 105 kPa (15 psi) before 300 mL of test fuel passes through the filter.

ASTM D4176, Standard Test Method for Free Water and Particulate Contamination in Distillate Fuels (Visual Inspection Procedures), is widely used in field situations for examining fuel for gas turbines, both marine and industrial gas turbines.

Stability

For the purposes of this discussion, *fuel stability* is defined as the resistance of the fuel to physical and chemical changes brought about by the interaction of the fuel with its environment.

There are three types of stability usually of concern for diesel fuel. They are thermal, oxidative, and storage. Each of these will be discussed separately.

THERMAL STABILITY

Thermal stability is the resistance of the fuel to change caused by thermal stress (elevated temperature). The ASTM method is D6468, Standard Test Method for High Temperature Stability of Distillate Fuels. Arguably, this test is the most often used method to monitor/predict fuel stability. The primary reason for its popularity is the short test time and simple equipment requirements. This has made the 150°C test, in one form or another, especially popular with many pipeline companies and others with the need to monitor the quality of fuel but to do it rapidly. No quantitative relationship exists between pad ratings and the gravimetric mass of filterable insolubles formed during the test. Additional information on the interpretation of results is found in Appendix XI of the test method.

OXIDATIVE STABILITY

Oxidative stability is the resistance of the fuel to change under severely oxidizing conditions. In addition to exposing the fuel to excess amounts of oxygen, oxidative stability tests usually incorporate elevated test temperatures to accelerate reaction rates. The method commonly used to measure oxidative stability of diesel fuel is ASTM D2274, Standard Test Method for Oxidation Stability of Distillate Fuel Oil (Accelerated Method). Because for many years this was the only standardized method to measure stability of middle distillate fuels, it became a part of numerous fuels specifications (mostly government or military) and was widely used as a predictor of fuel storage stability. Despite the wide acceptance as a storage stability test, most researchers agreed that, in fact, the results of D2274 had very poor correlation to actual, ambient storage of the fuel. In addition, the precision of the test was extremely poor. Because this poor correlation is now more widely known, the use of the method to assess the stability of diesel fuels is very low.

STORAGE STABILITY

Storage stability is generally defined as the fuel's resistance to change during storage at ambient temperature and conditions. Ambient storage is storage of the fuel in drums, storage tanks, vehicle fuel tanks, or similar containers outdoors. The two tests for storage stability are:

- ASTM D4625: Standard Test Method for Distillate Fuel Storage Stability at 43°C (110°F)
- ASTM D5304: Standard Test Method for Assessing Distillate Fuel Storage Stability by Oxygen Overpressure

All petroleum fuels undergo chemical reactions during storage. The results of these reactions are usually gums, acids, or particulates or a combination. The number and rate of these deleterious reactions depend on the concentrations of reaction precursors, the concentration of oxygen available for oxidation reactions, the presence of catalytic species such as metals, the amount of light, and the storage temperature. As mentioned earlier, the resistance of the fuel to these reactions is defined as its *storage stability*. Most researchers also go a step further and define *inherent storage stability* as the resistance to change in the absence of environmental factors such as metals and contaminants (i.e., in the absence of catalytic effects).

Ironically, that which makes the 43°C test such a useful research tool also makes it useless for quality assurance and quality control applications–that is, the long test times of the 43°C test make it useless as a quality control test. For this reason, much effort has been expended over the past 25 to 30 years to develop a rapid test for fuel storage stability. In this context, *rapid* is generally defined as test times of 16 hours or less. Very little success came of these efforts, with one exception. The U.S. Navy developed a test method that was eventually standardized as ASTM D5304 [7]. The applicability of the test method was demonstrated in a project conducted during 1992–1993 [8].

Note. A more detailed discussion of cleanliness and stability methods can be found in Chapter 11 of this manual.

Flash Point

Flash point is specified in D975 primarily for safety during transport, storage, and handling. A low flash point fuel can be a fire hazard, subject to flashing and possible continued ignition and explosion. Low flash point can also indicate contamination with low flash fuels such as gasoline. The flash point of a fuel has no significant relation to the performance of the fuel in the engine. Auto-ignition temperature is not influenced by variations in flash point.

There are three test methods approved for measuring flash point of diesel fuels:

- ASTM D56: Standard Test Method for Flash Point by Tag Closed Tester
- ASTM D93: Standard Test Methods for Flash Point by Pensky-Martens Closed Cup Tester
- ASTM D3828: Standard Test Methods for Flash Point by Small Scale Closed Tester

Each of these three test methods contains an introduction that is similar to the following:

> This flash point test method is a dynamic test method and depends on definite rates of temperature increases to control the precision of the test method. The rate of heating may not in all cases give the precision quoted in the test method because of the low thermal conductivity of certain materials. To improve the prediction of flammability, Test Method D3941, an equilibrium method was developed in which the heating rate is slower. This allows the vapor above the test specimen and the test specimen to be at about the same temperature.

ASTM D3941, Standard Test Method for Flash Point by the Equilibrium Method with a Closed-Cup Apparatus, covers the determination of the flash point of liquids in which the specimen and the air/vapor mixture above it are approximately in temperature equilibrium. The test method is limited to a temperature range from 32 to 230°F (0 to 110°C). Method D3941 should be used to measure and describe the properties of material, products, or assemblies in response to heat and flame under controlled laboratory conditions. It should not be used to describe or appraise the fire hazard or fire risk of materials, products, or assemblies under actual fire conditions. However, results of this test may be used as elements of a fire risk assessment that takes into account all of the factors that are pertinent to an assessment of the fire hazard of a particular end use.

It is important to note that, while D3941 may be an acceptable alternative method for some applications, the method(s) listed in a given specification, such as D975, should not be replaced without prior knowledge and consent of all interested parties. This is because the measured flash point of a fuel is dependent not only on the composition of the fuel but also on the apparatus used to make the measurement. *ASTM Manual 9* describes many flash point test methods, both United States and international [9]. It also lists pertinent specifications and standards worldwide.

Lubricity

Diesel fuel functions as a lubricant in certain items of fuel injection equipment such as rotary or distributor fuel pumps and injectors. In limited cases, fuel with very specific properties can have insufficient lubricating properties that can lead to a reduction in the normal service life of fuel pumps or injectors. Two fuel characteristics that affect equipment wear are low viscosity and lack of sufficient quantities of trace components, which have an affinity for metal surfaces. If fuel viscosity meets the requirements of a particular engine, a fuel film is maintained between the moving surfaces of the fuel system components. This prevents excessive metal-to-metal contact and avoids premature failure due to wear. Similarly, certain surface-active molecules in the fuel adhere to, or combine with, metallic surfaces to produce a protective film that also can protect surfaces against excessive wear. The concern about fuel lubricity is limited to situations in which fuels with lower viscosities than those specified for a particular engine are used or with fuels that have been processed in a manner that results in the elimination of the trace levels of the surface active species that act as lubricating agents. Presently the only fuels of the latter type shown to have lubricity problems resulted from sufficiently severe processing to reduce aromatics or sulfur levels [10].

The lubricity requirement in D975 is a maximum average wear scar diameter of 520 μm as determined by test method D6079, Standard Test Method for Evaluating Lubricity of Diesel Fuels by the High-Frequency Reciprocating Rig (HFRR). This is a less-restrictive limit compared to the ISO requirement of 460 μm, maximum, using the same test method. Despite being less restrictive, the limit in D975 has been generally found to meet the needs of engine and fuel injection equipment manufacturers.

Another test method for diesel fuel lubricity is D6078, Standard Test Method for Evaluating Lubricity of Diesel Fuels by the Scuffing Load Ball-on-Cylinder Lubricity Evaluator (SLBOCLE). Some users still find this test method applicable in selected situations; but, in general, this test method is rarely used because the specification test is D6079. Both D6078 and D6079 are bench test methods that provide a measure of the lubricity of a diesel fuel. A more accurate measure of the lubricity of a fuel in actual engine hardware can be obtained using D6898, Standard Test Method for Evaluating Diesel Fuel Lubricity by an Injection Pump Rig. This test method covers evaluating the lubricity of diesel fuels using a pump rig test and Stanadyne Model DB4427-4782 pumps. A poor result in such a test indicates that the fuel has low lubricity and may not be able to provide sufficient protection.

Aromatics

Aromatics warrant discussion for these reasons: (1) they have an effect on the combustion quality of the fuel, (2) typically, they are the only hydrocarbon type listed in diesel fuel specifications (including D975), and (3) increased amounts of aromatics can have a negative impact on vehicle emissions. It is well known that an increase in the total aromatics content of a diesel fuel can (and usually does) have an adverse effect on the ignition quality (i.e., cetane number of the fuel).

Several methods are available for the measurement of aromatic content:

- ASTM D1319: Standard Test Method for Hydrocarbon Types in Liquid Petroleum Products by Fluorescent Indicator Adsorption
- ASTM D2425: Standard Test Method for Hydrocarbon Types in Middle Distillates by Mass Spectrometry
- ASTM D5186: Standard Test Method for Determination of Aromatic Content and Polynuclear Aromatic Content of Diesel Fuels and Aviation Turbine Fuels by Supercritical Fluid Chromatography
- ASTM D5292: Standard Test Method for Aromatic Carbon Contents of Hydrocarbon Oils by High Resolution Nuclear Magnetic Resonance Spectroscopy

ASTM D1319 was applied to and specified for diesel fuel usually because no other suitable method was available. As suitable methods became standardized they grew in use but have not replaced D1319 in D975. This is because the requirement for aromaticity currently included in D975 comes from the requirement in 40 CFR Part 80. Because federal law requires D1319, it is the method listed in D975. Method D5186 is more appropriate for diesel fuel and is often used in place of D1319. However, in case of dispute, D1319, by virtue of its status as the legislated method, is considered the referee method. Method D2425 offers a more detailed analysis but requires considerable investment in instrumentation and sample preparation time. For these reasons, it is not typically used for routine analysis of diesel fuel. Method D5292 also offers more information than D1319 or D5186. However, the results are reported in mole percent rather than mass or volume percent, which are normally required in specifications. This test method covers the determination of the aromatic hydrogen content (Procedures A and B) and aromatic carbon content (Procedure C) of hydrocarbon oils using high-resolution nuclear magnetic resonance (NMR) spectrometers.

It should be noted that there are several standard methods for the analysis of aromatics. Each method yields a slightly different result and each is considered appropriate in different situations. One reason for this apparent inconsistency is that, because a single molecule can contain several chemical functional groups, it is possible to include it in several hydrocarbon classes. For example, a molecule could contain an aromatic ring, a paraffinic side chain, and a naphthenic ring. How should this molecule be classified? A hierarchy was established to address this situation. Under this hierarchy, aromatics are on top, then olefins, followed by naphthenes, and finally paraffins. Using this hierarchy, the example compound would be considered an aromatic compound.

The level of aromatics in the fuel is also important as it relates to the potential for elastomer and seal swell problems. This is especially true for older vehicles/fuel systems. Depending on the type of elastomer, prolonged exposure to relatively high levels of aromatics, followed by a sudden decrease in the amount of aromatics, can cause elastomeric seals to shrink and thus leak. If the elastomers are too old and have taken a set, they can also crack or break. This phenomenon was widely seen in late 1993 and early 1994 when mandated reductions in fuel sulfur and aromatics content went into effect. In most instances, the problems were solved by installing new seals made of less sensitive elastomer.

Heat Content

The heat content or heat of combustion of a fuel is the amount of heat produced when the fuel is burned completely. Gross and net heats of combustion are the two values measured for the heat of combustion.

The gross heat of combustion is the quantity of energy released when a unit mass of fuel is burned in a constant volume enclosure, with the products being gaseous, other than water that is condensed to the liquid state. The fuel can be either liquid or solid and contain only the elements carbon, hydrogen, nitrogen, and sulfur. The products of combustion, in oxygen, are gaseous carbon dioxide, nitrogen oxides, sulfur dioxide, and liquid water.

The net heat of combustion is the quantity of energy released when a unit mass of fuel is burned at constant pressure, with all of the products, including water, being gaseous. The fuel can be either liquid or solid, and contain only

TABLE 6—Conversion Factors for Heat of Combustion Values

1 cal (International Table calorie) = 4.1868 J
1 Btu (British thermal unit) = 1055.06 J
1 cal (I.T.)/g = 0.0041868 MJ/kg
1 Btu/lb = 0.002326 MJ/kg

the elements carbon, hydrogen, oxygen, nitrogen, and sulfur. The products of combustion, in oxygen, are carbon dioxide, nitrogen oxides, sulfur dioxide, and water, all in the gaseous state.

Heat of combustion is usually reported in units of mega-joules per kilogram (MJ/kg). Conversion factors to other units are given in Table 6. Heat of combustion can be estimated by calculation from selected properties or measured using bomb calorimetry. The methods typically used for diesel fuel are discussed below. ASTM D4868, Standard Test Method for Estimation of Net and Gross Heat of Combustion of Burner and Diesel Fuels, covers the estimation of the gross and net heat of combustion of petroleum fuel. The calculations use the fuel density, sulfur, water, and ash content. The equations for estimating net and gross heat of combustion used in this method were originally published by the NIST in Publication No. 97. The equations are as follow:

Calculate the gross heat of combustion of the fuel corrected for the sulfur, water, and ash content in accordance with the following equation:

$$Qv\ (\text{gross}) = \left(51.916 - 8.79d^2 \times 10^{-6}\right) \times [1 - (x + y + s)] + 9.420s \quad (4)$$

where:
Qv = gross heat of combustion at constant volume, MJ/kg,
d = density at 15°C, kg/m^3,
x = mass fraction of water (% divided by 100),
y = mass fraction of ash (% divided by 100), and
s = mass fraction of sulfur (% divided by 100).

Calculate the net heat of combustion of the fuel corrected for the sulfur, water and ash content in accordance with the following equation:

$$Q_P\ (\text{net}) = \left(46.423 - 8.792d^2 \times 10^{-6} + 3.170d \times 10^{-3}\right) \times [1 - (x + y + s)] + 0.420s - 2.449x \quad (5)$$

where:
Q_P = net heat of combustion at constant pressure, MJ/kg,
d = density at 15°C, kg/m^3,
x = mass fraction of water,
y = mass fraction of ash (% divided by 100), and
s = mass fraction of sulfur.

This test method is useful for estimating, using a minimum number of tests, the heat of combustion of burner and diesel fuels for which it is not usually critical to obtain very precise heat determinations. This test method is purely empirical. It is applicable only to liquid hydrocarbon fuels derived by normal refining processes from conventional crude oil. This test method is valid for those fuels in the density range from 750 to 1,000 kg/m^3 and those that do not contain an unusually high aromatic content. High aromatic content fuels will not normally meet fuel specification criteria for this method. This test method is not applicable to pure hydrocarbons. It is not intended as a substitute for experimental measurements of heat of combustion.

ASTM D240, Standard Test Method for Heat of Combustion of Liquid Hydrocarbon Fuels by Bomb Calorimeter, covers the determination of the heat of combustion of liquid hydrocarbon fuels ranging in volatility from that of light distillates to that of residual fuels. Under normal conditions, this test method is directly applicable to such fuels as gasoline, kerosene, Nos. 1 and 2 fuel oil, Nos. 1-D and 2-D diesel fuel, and Nos. 0-GT, 1-GT, and 2-GT gas turbine fuels. It is also worthy of mention that, despite the use of SI units throughout ASTM standards, the usage of the inch-pound units of Btu per pound/gallon/barrel remains common throughout much of the United States.

Total Sulfur

DIESEL ENGINES

Sulfur can cause wear in diesel engines as a result of the corrosive nature of its combustion by-products and increase the amount of deposits in the combustion chamber and on the pistons. The sulfur content of a fuel depends on the origin of the crude oil from which the fuel is made and on the refining methods used. Sulfur can be present in a number of forms—as mercaptans, sulfides, disulfides, or heterocyclic compounds such as thiophenes—all of which affect wear and deposits.

Fuel sulfur tolerance by a diesel engine depends largely on whether the engine is of the low- or high-speed type and the prevalent operating conditions. Low-speed engines can tolerate more sulfur than their high-speed counterparts because they operate under relatively constant speed and load conditions. Under these conditions lubricating oils, cooling water, and combustion zone temperatures show little fluctuation. These steady temperatures make low-speed engines more tolerant to sulfur.

High-sulfur fuels for diesel engines are undesirable from a purely technical standpoint regardless of engine type. However, less harm will occur from fuel sulfur when engines are operated at high-power outputs and operating temperatures than at lower temperatures. Under the lower-temperature conditions that result from stopping and starting or decrease of load or speed or both, moisture condensation is apt to occur within the engine. The sulfur in the fuel then combines with the water to form acid solutions that corrode metal components and increase wear of moving parts. Active sulfur in fuel tends to attack and corrode injection system components. Sulfur compounds also contribute to combustion chamber and injection system deposits.

The test methods for measuring total sulfur in diesel fuel, as prescribed in D975 are:

- ASTM D129: Standard Test Method for Sulfur in Petroleum Products (General Bomb Method)
- ASTM D1266: Standard Test Method for Sulfur in Petroleum Products (Lamp Method)
- ASTM D1552: Standard Test Method for Sulfur in Petroleum Products (High-Temperature Method)
- ASTM D2622: Standard Test Method for Sulfur in Petroleum Products by Wavelength Dispersive X-ray Fluorescence Spectrometry
- ASTM D3120: Test Method for Trace Quantities of Sulfur in Light Liquid Petroleum Hydrocarbons by Oxidative Microcoulometry
- ASTM D4294: Standard Test Method for Sulfur in Petroleum Products by Energy-Dispersive X-Ray Fluorescence Spectroscopy
- ASTM D5453: Standard Test Method for Determination of Total Sulfur in Light Hydrocarbons, Motor Fuels and Oils by Ultraviolet Fluorescence

TABLE 7—Sulfur Test Methods for the Various Grades and Sulfur Levels of Diesel Fuels Listed in D975

Sulfur Test Method	Range	Units	Grades
D129(referee)	>0.1 mass percent	D129 mass percent	No. 1-D S5000, No. 2-D S5000, No. 4-D
D1266	0.0005 to 0.4 mass percent 5 to 4000 mg/kg (wt ppm)	D1266 mass percent	No. 1-D S500, No. 2-D S500
D1552	>0.06 mass percent	D1552 mass percent	No. 1- D S5000, No. 2-D S5000, No. 4-D
D2622 (referee for S500 Grades)	0.0003 to 5.3 mass percent 3 to 53 000 mg/kg (wt ppm)	D2622 mass percent	All Grades
D3120	3.0 to 100 mg/kg (wt ppm)	D3120 ppm (μg/g)	No. 1-D S15, No. 2-D S15 No. 1-D S500, No. 2-D S500 (S500 grades must be diluted before testing)
D4294	0.0150 to 5.00 mass percent 150 to 50 000 mg/kg (wt ppm)	D4294 mass percent	No. 1- D S5000, No. 2-D S5000, No. 4-D

Table 7 shows the referee test methods and alternate test methods for sulfur, the range over which each test method applies and the corresponding fuel grades.

The sulfur content of diesel fuel is known to affect particulate matter (PM) exhaust emissions because some of the sulfur is converted to sulfate particles in the exhaust. The amount that is converted varies by engine, but reducing total sulfur produces a linear decrease in PM in nearly all engines. Fuel sulfur can also adversely affect cylinder wear (through the formation of acids) and deposit formation (many sulfur compounds are known deposit precursors).

The measurement of potential corrosivity is determined by means of a corrosion test such as the copper strip procedure described in ASTM Detection of Copper Corrosion from Petroleum Products by the Copper Strip Tarnish Test (D130/ IP 154). The quantitative determination is an indication of the corrosive tendencies of the fuel combustion products, while the potential corrosivity indicates the extent of corrosion to be anticipated from the unburned fuel, particularly in the fuel injection system.

GAS TURBINES

Sulfur by itself has little corrosive effect on vanes and blades in the turbine section. However, in the presence of alkali metals, sulfur reacts to form alkali sulfates, which do promote corrosion at high temperatures. At metal surface temperatures above about 760°C, very little sulfur is needed to form the sulfates.

At temperatures between 593°C (1,100°F) and 760°C (1,400°F), the concentration of sulfur trioxide in the gaseous combustion products has a far stronger effect because, in order to have corrosion in this temperature range, it is necessary to form sulfates of nickel or cobalt. These compounds, together with the alkali sulfates, form eutectic mixtures having low melting point temperatures, thus promoting corrosion at the lower temperatures.

In the exhaust section and particularly in installations where waste heat boilers are used, sulfur trioxide together with water vapor can condense as sulfuric acid if the system temperature drops below the acid dew point temperature.

In general, however, the sulfur concentration in a fuel is limited by emission requirements rather than by such technical considerations as those identified above. Methods for determining sulfur are indicated in ASTM Specification for Gas Turbine Fuel Oils (D2880).

Red-Dyed Diesel Fuel

Under United States regulations, if Grades No. 1–D S500 or No. 2–D S500 are sold for tax exempt purposes then, at or beyond terminal storage tanks, they are required by 26 CFR Part 48 to contain the dye Solvent Red 164 at a concentration spectrally equivalent to 3.9 lb per thousand barrels of the solid dye standard Solvent Red 26, or the tax must be collected. Under United States regulations, Grades No. 1–D S5000, No. 2–D S5000, and No. 4–D are required by 40 CFR Part 80 to contain a sufficient amount of the dye Solvent Red 164 so its presence is visually apparent. At or beyond terminal storage tanks, they are required by 26 CFR Part 48 to contain the dye Solvent Red 164 at a concentration spectrally equivalent to 3.9 lb per thousand barrels of the solid dye standard Solvent Red 26. The Clean Air Act Amendments of 1990 established standards for highway diesel fuel that, in part, made it illegal as of October 1, 1993, to manufacture, sell, supply, or offer for sale diesel fuel for highway use that has a sulfur content greater than 0.05 % by weight (this amount is also commonly expressed as 500 ppm). Similarly, it is illegal for any person to use fuel that has sulfur content greater than 0.05 % by weight in any on-highway vehicle.

Carbon Residue

DIESEL ENGINES

Carbon residue is the residue formed by evaporation and thermal degradation of a carbon-containing material. The residue is not composed entirely of carbon but is a coke that can be further changed by carbon pyrolysis. The term *carbon residue* is retained in deference to its wide common usage. The test method for carbon residue, as listed in the diesel fuel specification, is D524, Standard Test Method for Ramsbottom Carbon Residue of Petroleum Products.

The ASTM Test for Conradson Carbon Residue of Petroleum Products (D189/IP 13) and the Micro-Carbon Residue Test are also used widely. The carbon residue is a measure of the carbonaceous material left in a fuel after all the volatile components are vaporized in the absence of air. At one time, there was believed to be a definite correlation between Conradson carbon results in diesel fuels and deposit formation on injector nozzles, but this view now is thought to be an oversimplification. The type of carbon formed is as important as the amount. Small quantities of hard, abrasive deposits can do more harm than larger amounts of soft, fluffy deposits. The latter can be eliminated largely through the exhaust system.

Carbon residue tests are used primarily on residual fuels because distillate fuels that are satisfactory in other respects do not have high Conradson carbon residue. Because of the considerable difference in Conradson carbon residue results between distillate and residual fuels, the test can be used as an indication of contamination of distillate fuel by residual fuel.

The significance of the Conradson carbon test results also depends on the type of engine in which the fuel is being used. Fuels with up to 12 % weight Conradson carbon residue have been used successfully in slow-speed engines.

GAS TURBINES

In gas-turbine fuels, carbon residue is a rough approximation of the tendency of a fuel to form carbon deposits in the combustor. Combustion systems designed for use on Grade Nos. 3-GT and 4-GT are insensitive to this problem, but other gas turbines may require a limit on the carbon residue.

Carbon deposits in gas turbines are undesirable because they form heat-insulated spots in the combustor that become very hot. Adjacent metal is kept at a relatively low temperature by the cooling air. The "hot spot" formed by carbon deposit creates a large temperature gradient with resultant high stress, distortion, and perhaps eventual cracking of the combustor shell.

Carbon deposits may also contribute to nonuniformity of operation and flow pulsation. If pieces of the carbon deposit are broken off and carried through the turbine, blade erosion (efficiency loss) or partial blocking of the nozzles may occur. A smoky discharge and a low value for the heat release factor usually accompany carbon deposits in a combustor. Carbon deposits also occur on the fuel injectors. This disrupts the mixture formation and, consequently, combustion.

Ash

Small amounts of nonburnable material are found in fuels in two forms: (1) solid particles, and (2) oil or water-soluble metallic compounds. The solid particles are for the most part the same material that is designated as sediment in the water and sediment test. These two types of nonburnable material may be oxidized or otherwise modified during the combustion of the fuel.

The test method for ash content is D482 Standard Test Method for Ash from Petroleum Products. In this test, a small sample of fuel is burned in a weighed container until all of the combustible matter has been consumed. The amount of unburnable residue is the ash content, and it is reported as a percentage by weight of the fuel.

DIESEL ENGINES

Because diesel fuel injection components are made to extremely close tolerances, they are very sensitive to any abrasive material in the fuel. Depending on their size, solid particles can contribute to wear in the fuel system and plugging of the fuel filter and fuel nozzle. In addition, abrasive ash materials can cause wear within the engine by increasing the overall deposit level and adversely affecting the nature of the deposits.

GAS TURBINES

The soluble metallic compounds have little or no effect on wear or plugging, but they can contain elements that produce turbine corrosion and deposits.

The ash in distillate fuels is typically so low that it does not adversely affect gas turbine performance, unless such corrosive species as sodium, potassium, lead, or vanadium are present. Grade No. 4-GT fuels, however, may have considerable quantities of ash-forming constituents, and these may be augmented by the presence of corrosion-inhibiting additives, for example, magnesium, which are used to inhibit corrosion caused by vanadium compounds. In such cases, ash can accumulate on stationary and rotating airfoils, thus restricting gas flows and raising the compressor discharge pressures above the design limits. In addition, the accumulation of ash deposits on rotating airfoils compromises the ability of the turbine to extract work from the expanding combustion gases, and thermal efficiency is consequently reduced.

Acid Number

ASTM Test for Acid and Base Number by Color-Indicator Titration (D974/IP 139) is a measure of the inorganic and total acidity of the fuel and indicates its tendency to corrode metals that it may contact.

Vanadium, Sodium, Potassium, Calcium, Lead

GAS TURBINES

Vanadium can form low-melting compounds, such as vanadium pentoxide, which melts at 691°C and causes severe corrosive attack on all of the high-temperature alloys used for gas-turbine blades and diesel engine valves. For example, to reduce the corrosion rate at 871°C on AISI Type 310 steel to a level comparable with the normal oxidation rate, it is necessary to limit vanadium in the fuel to less than 2 ppm. At 10 ppm, the corrosion rate is three times the normal oxidation rate, and, at 30 ppm, it is 13 times the normal oxidation rate. However, if there is sufficient magnesium in the fuel, it will combine with the vanadium to form compounds with higher melting points and thus reduce the corrosion rate to an acceptable level. The resulting ash will form deposits in the turbine, but the deposits are self-spalling when the turbine is shut down. For gas turbines operating below 649°C, the corrosion of the high-temperature alloys is of minor importance, and the use of a silicon-base additive will further reduce the corrosion rate by absorption and dilution of the vanadium compounds.

Sodium and potassium can combine with vanadium to form eutectics, which melt at temperatures as low as 565°C, and with sulfur in the fuel to yield sulfates with melting points in the operating range of the gas turbine. These compounds produce severe corrosion, and, for turbines operating at gas inlet temperatures above 649°C, no additive has been

TABLE 8—Trace Metal Limits of Fuel Entering Turbine Combustor(s)[A,B]

Designation	Trace Metal Limits, mg/kg			
	Vanadium (V)	Sodium plus Potassium (Na + K)	Calcium (Ca)	Lead (Pb)
No. 0-GT	0.5	0.5	0.5	0.5
No. 1-GT	0.5	0.5	0.5	0.5
No. 2-GT	0.5	0.5	0.5	0.5
No. 3-GT	0.5	0.5	0.5	0.5
No. 4-GT	(Consult turbine manufacturers.)			

[A] Test Method D3605 may be used for determination of vanadium, sodium, calcium, and lead.
[B] Test Method D6728 may be used for determination of vanadium, sodium, potassium, calcium, and lead.

found that successfully controls such corrosion without forming tenacious deposits at the same time. Accordingly, the sodium-plus-potassium level must be limited, but each element is measured separately. Some gas turbine installations incorporate systems for washing oil with water to reduce the sodium-plus-potassium level. In installations where the fuel is moved by sea transport, the sodium-plus-potassium level should be checked prior to use to ensure that the oil has not become contaminated with sea salt. For gas turbines operating below 649°C, the corrosion due to sodium compounds is of minor importance and can be further reduced by silicon-base additives. A high-sodium content is beneficial even in these turbines, because it increases the water-solubility of the deposits and thereby increases the ease with which gas turbines can be water washed to obtain recovery of the operating performance.

Calcium is not harmful from a corrosion standpoint; in fact, it serves to inhibit the corrosive action of vanadium. However, calcium can lead to hard-bonded deposits that are not self-spalling when the gas turbine is shut down and not readily removed by water washing of the turbine. The fuel washing systems used at some gas-turbine installations to reduce the sodium and potassium level also will lower significantly the calcium content of fuel oil.

Lead can cause corrosion, and, in addition, it can spoil the beneficial inhibiting effect of magnesium additives on vanadium corrosion. Because lead is found only rarely in significant quantities in crude oils, its presence in the fuel oil is primarily the result of contamination during processing or transportation.

As a result of these concerns, limits are suggested in the Appendix of ASTM D2880 Specification for Gas Turbine Fuel Oils (Table 8).

The heavier fuels in Grade No. 4-GT, that is, whose viscosities approach the maximum 638 cSt at 50°C (50 cSt at 100°C) permitted by the specification, are usually washed, inhibited, and analyzed prior to combustion. However, such practices are seldom practiced with the lighter distillate fuels, such as Grade No. 2-GT fuels. Rather, the distillate fuels are kept usable by practices given in ASTM Practice for the Receipt, Storage, and Handling of Fuels for Gas Turbines (D4418). The practices recommended therein attempt to prevent the introduction of contaminant during transportation and storage. In addition, the proper maintenance of fuel storage tanks and the drainage of accumulated water from such storage tanks can be very effective in maintaining the cleanliness of distillate fuels.

Vanadium levels in blended marine fuels are limited in specification ISO 8217 depending on the category of fuel.

Alternative Fuels for Diesel Engines

BIODIESEL FUEL

Biodiesel is a blendstock for use with middle-distillate petroleum fuels. Biodiesel is covered under ASTM D6751, Specification for Biodiesel Fuel Blend Stock (B100) for Middle Distillate Fuels. (Chapter 5 of this manual contains a more detailed discussion of biodiesel.) Biodiesel is comprised of mono-alkyl esters of long chain fatty acids derived from vegetable oils or animal fats, designated B100. It is registered with the U.S. Environmental Protection Agency as a fuel and a fuel additive under Section 211(b) of the Clean Air Act. A biodiesel blend is a mixture of biodiesel fuel with petroleum-based diesel fuel; and is designated BXX, where XX is the volume percent of biodiesel. D975 allows up to 5 volume percent biodiesel as a blend with petroleum diesel (B5) with the proviso that diesel fuel oil containing up to 5 volume percent biodiesel shall meet the requirements for the appropriate grade No. 1-D or No. 2-D fuel, as listed in Table 1 in D975.

B100 has good lubricity properties and contains essentially no sulfur or aromatics. However, it has a relatively high pour point, which could limit its use in cold weather. Biodiesel is biodegradable, but this property may lead to increased contamination from biological growth during long-term storage. Biodiesel is also typically more susceptible to oxidative degradation than petroleum diesel, but it usually responds well to antioxidant additives.

Other alternative fuels that have been or are currently being considered or discussed, or both, in ASTM include:

- e-diesel (blend of ethanol and conventional petroleum diesel)
- gas-to-liquids fuels (Fischer-Tropsch fuels)
- emulsified fuels (emulsified blends of water and petroleum diesel)
- ethers
- alcohols
- naphtha
- and various gaseous fuels

Each of these has some advantage (such as reduced engine emissions or reduced fuel consumption or reduced engine deposits) associated with its use.

ASTM Standards

Number	Title
D56	Test Method for Flash Point by Tag Closed Tester
D86	Test Method for Distillation of Petroleum Products
D93	Test Methods for Flash Point by Pensky-Martens Closed Cup Tester
D97	Test Method for Pour Point of Petroleum Products
D129	Test Method for Sulfur in Petroleum Products (General Bomb Method)
D130	Test Method for Detection of Copper Corrosion from Petroleum Products by the Copper Strip Tarnish Test
D189	Test Method for Conradson Carbon Residue of Petroleum Products
D240	Test Method for Heat of Combustion of Liquid Hydrocarbon Fuels by Bomb Calorimeter
D396	Specification for Fuel Oils
D445	Test Method for Kinematic Viscosity of Transparent and Opaque Liquids (and the Calculation of Dynamic Viscosity)
D482	Test Method for Ash from Petroleum Products
D613	Test Method for Ignition Quality of Diesel Fuels by the Cetane Method
D974	Test Method for Acid and Base Number by Color-Indicator Titration
D975	Specification for Diesel Fuel Oils
D976	Test Method for Calculated Cetane Index of Distillate Fuels
D1266	Test Method for Sulfur in Petroleum Products (Lamp Method)
D1298	Test Method for Density, Relative Density (Specific Gravity), or API Gravity of Crude Petroleum and Liquid Petroleum Products by Hydrometer Method
D1319	Test Method for Hydrocarbon Types in Liquid Petroleum Products by Fluorescent Indicator Adsorption
D1552	Test Method for Sulfur in Petroleum Products (High-Temperature Method)
D2068	Test Method for Filter Plugging Tendency of Distillate Fuel Oils
D2274	Test Method for Oxidation Stability of Distillate Fuel Oil (Accelerated Method)
D2425	Test Method for Hydrocarbon Types in Middle Distillates by Mass Spectrometry
D2500	Test Method for Cloud Point of Petroleum Products
D2622	Test Method for Sulfur in Petroleum Products by Wavelength Dispersive X-ray Fluorescence Spectrometry
D2709	Test Method for Water and Sediment in Middle Distillate Fuels by Centrifuge
D2880	Specification for Gas Turbine Fuel Oils
D2887	Test Method for Boiling Range Distribution of Petroleum Fractions by Gas Chromatography
D3117	Test Method for Wax Appearance Point of Distillate Fuels
D3828	Test Methods for Flash Point by Small Scale Closed Tester
D3941	Test Method for Flash Point by the Equilibrium Method with a Closed-Cup Apparatus
D4052	Test Method for Density and Relative Density of Liquids by Digital Density Meter
D4294	Test Method for Sulfur in Petroleum Products by Energy-Dispersive X-Ray Fluorescence Spectroscopy
D4418	Practice for Receipt, Storage, and Handling of Fuels for Gas Turbines
D4539	Test Method for Filterability of Diesel Fuels by Low-Temperature Flow Test (LTFT)
D4625	Test Method for Distillate Fuel Storage Stability at 43°C (110°F)
D4737	Test Method for Calculated Cetane Index by Four Variable Equation
D4860	Test Method for Free Water and Particulate Contamination in Mid-Distillate Fuels (Clear and Bright Numerical Rating)
D4868	Test Method for Estimation of Net and Gross Heat of Combustion of Burner and Diesel Fuels
D5186	Test Method for Determination of Aromatic Content and Polynuclear Aromatic Content of Diesel Fuels and Aviation Turbine Fuels by Supercritical Fluid Chromatography
D5292	Test Method tor Aromatic Carbon Contents of Hydrocarbon Oils by High Resolution Nuclear Magnetic Resonance Spectroscopy
D5304	Test Method for Assessing Distillate Fuel Storage Stability by Oxygen Overpressure
D5453	Test Method for Determination of Total Sulfur in Light Hydrocarbons, Motor Fuels and Oils by Ultraviolet Fluorescence
D5771	Test Method for Cloud Point of Petroleum Products (Optical Detection Stepped Cooling Method)
D5772	Test Method for Cloud Point of Petroleum Products (Linear Cooling Rate Method)
D5773	Test Method for Cloud Point of Petroleum Products (Constant Cooling Rate Method)
D5949	Test Method for Pour Point of Petroleum Products (Automatic Pressure Pulsing Method)
D5950	Test Method for Pour Point of Petroleum Products (Automatic Tilt Method)
D5985	Test Method for Pour Point of Petroleum Products (Rotational Method)
D6078	Test Method for Evaluating Lubricity of Diesel Fuels by the Scuffing Load Ball-on-Cylinder Lubricity Evaluator (SLBOCLE)

Number	Title
D6079	Test Method for Evaluating Lubricity of Diesel Fuels by the High-Frequency Reciprocating Rig (HFRR)
D6217	Test Method for Particulate Contamination in Middle Distillate Fuels by Laboratory Filtration
D6426	Test Method for Determining Filterability of Distillate Fuel Oils
D6428	Test Method for Total Sulfur in Liquid Aromatic Hydrocarbons and Their Derivatives by Oxidative Combustion and Electrochemical Detection

References

[1] Ryan, T. W. III, "Diesel Fuel Combustion Characteristics," Chap. 27, *Fuels and Lubricants Handbook: Technology, Properties, and Testing, Manual 37,* G.E. Totten, ed. ASTM International, West Conshohocken, PA, 2003.

[2] Clerc, J. C., "Cetane Number Requirements of Light Duty Diesel Engines at Low Temperatures," SAE No. 861525, Society of Automotive Engineers, Warrendale, PA, 1986.

[3] Westbrook, S. R., et al., "Fuel System Design Considerations for Diesel and Gas Turbine Engine Powered Military Vehicles," *Proceedings of the Second International Conference on Long-Term Storage Stabilities of Liquid Fuels,* Southwest Research Institute, San Antonio, TX, 1986, pp. 416–425.

[4] "The Cold Filter Plugging Point of Distillate Fuels," a European Test Method, CEC Report No. P-01-74, 1974.

[5] Coley, T., et al., "New Laboratory Test for Predicting Low-Temperature Operability of Diesel Fuels," *J. Inst. Petroleum,* Vol. 52, No. 510, June 1966.

[6] Hiley, R. W., "Filterability of Degraded Fuels," *Proceedings of the 2nd International Conference on Long-Term Storage Stabilities of Liquid Fuels,* San Antonio, TX, July 29–Aug. 1, 1986.

[7] Hardy, D. R., Hazlett, R. N., Beal, E. J., and Burnett, J. C., *Energy Fuels,* Vol. 3, No. 20, 1989.

[8] Turner, L. M., et al., "Use of ASTM D5304 in Assessing Unstable Diesel Fuel," *Proceedings of the Fifth International Conference on Stability and Handling of Liquid Fuels,* Rotterdam, the Netherlands, Oct., 3–7, 1994.

[9] Wray, H. A., *Manual on Flash Point Standards and Their Use,* Manual 9, ASTM, Philadelphia, PA, 1992.

[10] Excerpted from Appendix X3 of ASTM Standard Specification D975-98b, *Annual Book of ASTM Standards,* Vol. 05.01.

5

Biodiesel

Steve Howell[1]

FOR PURPOSES OF THIS CHAPTER, THE TERM "biodiesel" is a fuel comprised of mono-alkyl esters of long chain fatty acids derived from vegetable oils or animal fats meeting ASTM D6751 specifications and is designated as B100. Biodiesel is an ultra-low sulfur, renewable fuel that can be used as a neat (pure) fuel but is most often blended into petroleum diesel (petrodiesel). To distinguish the amount in the blend, the convention prescribed in D6751 is to state the amount of biodiesel in the blend by the volume percentage preceded by the uppercase letter "B." Therefore, a 5 % blend of biodiesel in petrodiesel is B5 and a 20 % blend of biodiesel in petrodiesel is B20. Biodiesel blends can be used in many applications where petroleum middle distillate (i.e., diesel fuel) products are used, such as on-road and off-road diesel, home heating oil and boiler fuel, marine diesel fuel, and nonaviation gas turbine fuel. There may be some limitations on the blend percentages used in unmodified diesel engines, and the original equipment manufacturer (OEM) should be consulted, especially if using higher blend ratios. Most biodiesel made today is not suitable for use in kerosene lamps due to its higher viscosity and surface tension, resulting in less than optimum wicking properties, although it can be used in heating applications where wicking is not important. The biodiesel industry focus in the United States over the past 10 years has been the on-road and off-road diesel market and the home heating oil market. Biodiesel is most commonly used as a blend of B5 with conventional petrodiesel in the existing equipment that more traditionally has operated solely on petrodiesel. Blends in the range of B6 to B20 are also in regular usage with heavier duty diesel trucks and buses. Other sections of this manual cover in more detail both general and special engine and burner equipment considerations when using petrodiesel. These same considerations are also valid for the use of biodiesel blends and will not be covered further in this chapter.

Biodiesel's long straight chain hydrocarbon structure with a limited number of double bonds (directly related to the structure of the natural oils and fats from which it is produced) coupled with an ester linkage with a short chain alcohol (usually methanol or ethanol) provides for naturally high cetane, good lubricity, biodegradability, and lower emissions [1]. Combustion of biodiesel in a diesel engine results in lower emissions of unburned hydrocarbons, carbon monoxide, and particulate matter (i.e., black smoke) when used in unregulated engines and compared to conventional petrodiesel [1]. For some applications biodiesel also provides reductions in nitric oxide (NOx) due to its oxygen content (i.e., 20 % NOx reduction with B20 for home heating oil), while in other applications the use of biodiesel can be NOx neutral or have a slightly higher NOx (i.e., 10 % NOx increase for B100 in on-road diesel engines) depending on the amount of biodiesel, the duty cycle and the engine technology [1]. Beginning in 2010, new diesel engines will be fitted with catalytic technology to reduce NOx from both biodiesel and diesel fueled engines by over 90 %, so any detriment of biodiesel on NOx will not be an issue with post-2010 diesel engines.

The 2008 volatility in the cost of a barrel of crude oil, with a high of over \$140/bbl, and increased concern over global warming has caused the interest in biomass-based fuels, and biodiesel in particular, to skyrocket. Federal legislation in the United States provides for a \$1.00 per gallon tax incentive for B100, and there are some state incentives and some state usage requirements. The state of Minnesota requires 5 % biodiesel in most of the diesel fuel used in the state, and the states of Washington and Oregon also require biodiesel to be blended into diesel. Additionally, the U.S. Congress, as part of the Energy Independence and Security Act of 2007, passed the second version of a national Renewable Fuel Standard (RFS2), which requires over 1 billion gallons of biomass-based diesel be used in the U.S. diesel fuel pool by 2012. These state and anticipated federal mandate volumes, specific to biodiesel, have caused a dramatic increase in the building of production capacity and overall product volumes for biodiesel (see Figs. 1 and 2).

ASTM's support and achievements in integrating biodiesel as a blend stock into diesel have been an integral part of the development of the biodiesel industry. A key part of the recent legislation supporting biodiesel was that an ASTM standard be issued and in use in commerce. This requirement for an ASTM Standard Specification for biodiesel is embedded in the tax incentive and RFS2 language as a condition of the legislation and participation in the tax incentive programs.

Biodiesel so far introduced to the market has, for the most part, been made with the oils and fats that were readily available in the commercial marketplace in the United States, none of which were originally tailored or designed for fuel use. In the United States, the primary oils and fats for biodiesel are soybean oil, animal fats or tallow (beef, pork, poultry), and used restaurant frying oils (referred to as yellow grease, which comes in a variety of quality levels). All of these oils and fats are produced as minor by-products of growing food or animal feed or, in the case of used cooking oil, as a second-use product. Soybeans are grown primarily as a high protein animal feed because they are 80 % high protein meal and only 20 % oil. Beef, hogs, and chickens are grown for meat for human consumption and yield only

[1] *MARC-IV Consulting, Inc., Kearney, MO.*

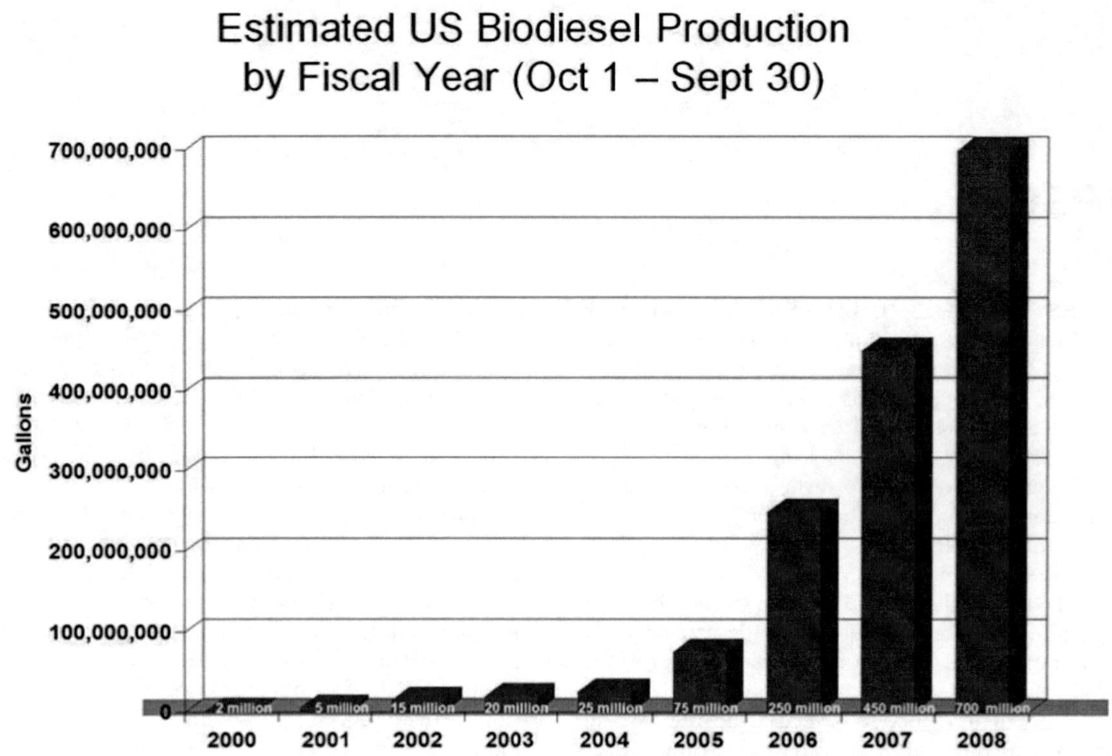

Fig. 1—Estimated US biodiesel production by fiscal year (Oct. 1–Sept. 30). (Published with permission from www.biodiesel.org.)

around 10 % fat. These existing oils and fats can be used to produce a product meeting the ASTM specifications for biodiesel. The chemical structure of these existing oils and fats, however, can result in fuel with some deficiencies compared to conventional petrodiesel. Examples of this are long-term stability and cold flow properties. These deficiencies can be overcome by choice of feedstock and process technology or blending with petrodiesel. In some cases, these deficiencies can also be ameliorated or corrected by the selective and judicious use of additives (e.g., using an antioxidant additive for storage improvement).

The increase in interest in biodiesel, and long-term mandates for a minimum of 1 billion gallons per year of biodiesel use, is providing the economic incentives for further development and improvement of oilseed crops for biodiesel purposes (i.e., breeding and genetics to provide oils with lower polyunsaturates and lower saturates for improved stability, shorter chain lengths for improved cold flow properties) as well as a plethora of new production technologies (solid catalysts, use of resins for filtering and drying, further processing and separation to produce a jet fuel–like product) and relatively new or novel oil sources such as jatropha, algae, enzymatic production of triglycerides from cellulose, use of microorganisms for direct production of methyl esters from sugar, and even production of biodiesel from the municipal sludge. Some of these so-called second- or third-generation biodiesel routes can provide a biodiesel that has superior cold flow properties and stability even compared to petrodiesel, while keeping the already beneficial biodegradability, high cetane, improved lubricity and emissions reductions associated with current first-generation biodiesel produced from the traditional oils and fats by transesterification to the methyl and ethyl esters. Over time, it is expected that biodiesel processing and the finished product will improve and its usage rate will ultimately be driven by economics and market forces but also requirements to meet mandates on carbon dioxide (CO_2) to ameliorate climate change. The need to meet sustainability goals will also likely play a role in the future of biodiesel.

ASTM BIODIESEL TASK FORCE HISTORY

In 1993, a Task Force was formed within ASTM Committee D02 on Petroleum Products and Lubricants to begin development of an industry consensus standard for biodiesel. The first step undertaken by the Task Force was the determination of the philosophy to be used for the standard. Various options were considered, from adding a section to the existing ASTM petrodiesel standard (D975), to development of a standard for a blend with petrodiesel, to a stand-alone standard. The following was agreed on by the Biodiesel Task Force and subsequently by the membership of ASTM.

1. Develop a stand-alone specification for pure biodiesel. It was assumed the biodiesel would most likely be produced by a different commercial entity than the petrodiesel refinery and then blended at terminals with petrodiesel. For such third-party blending, a trading standard for the pure biodiesel would be needed.
2. If biodiesel meets its B100 specification, it can be blended with petrodiesel in any percentage. This is similar to status of No. 1 (kerosene type) diesel product being blended into No. 2 grade fuels as long as they meet the respective specifications within ASTM D975.
3. Base the development of the standard for the end product on performance tests needed for a "fit for purpose fuel" in existing diesel engines, not on the source or processing used to make the biodiesel. This is similar to how the petrodiesel specifications were developed. Use physical and chemical tests as deemed essential in defining the product, or where it makes sense to use such tests in place of a performance test.

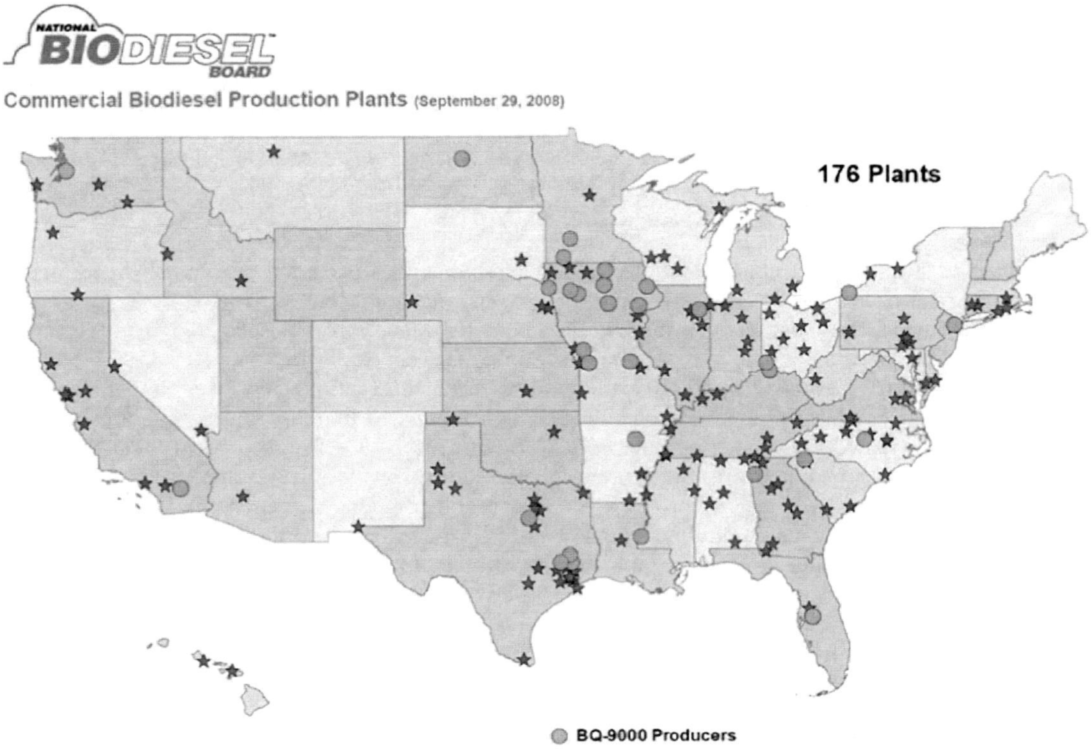

Fig. 2—Commercial biodiesel production plants in the United States as of September 29, 2008. (Published with permission from www.biodiesel.org.)

4. Begin with the existing D975 petrodiesel specification as the baseline for a "fit for purpose fuel" for use in conventional diesel engines.
5. For the biodiesel blend stock specification, eliminate items not applicable to biodiesel such as the distillation curve, cetane index, and aromatics content.
6. Extend the biodiesel specification to address biodiesel-specific quality properties not in D975 and required for acceptable performance, such as acid value, total and free glycerin, and phosphorous content.
7. Extend the biodiesel specification to new characteristics being considered for D975, and as D975 is updated add characteristics such as lubricity and conductivity.

This philosophy formed the basis used as the biodiesel standards progressed through the ASTM development and balloting process and led to the formal issuance of ASTM D6751, Standard Specification for Biodiesel Fuel Blend Stock (B100) for Middle Distillate Fuels, and incorporation of up to 5 volume percent biodiesel into ASTM D975, Standard Specification for Diesel Fuel Oils, ASTM D396, Standard Specification for Fuel Oils, and ASTM D7467, Standard Specification for Diesel Fuel Oil, Biodiesel Blend (B6 to B20).

Many references to "biofuels" and to "biodiesel" fuels can be found in the technical literature, but the definitions are not always clear. In literature outside ASTM, the term "biodiesel" has been associated with fuel-like materials such as pure vegetable oils, mixtures of vegetable oils and petrodiesel, partially esterified natural oils, and mixtures of esters with petrodiesel. In the extreme, a case has been made for coal slurry being a "biodiesel" on the grounds that it is derived from long-decayed biomass. There has been significant negative experience with the use of unprocessed or raw vegetable oils in existing diesel engines in the past [2]. After discussions with the engine and vehicle manufacturers regarding their negative experience with a variety of these materials in the past, and the generally positive experience with the methyl esters of vegetable oils in the United States and Europe, it was apparent that once the philosophy for the development of the standard was agreed on, a written description narrowing the scope of what could be called biodiesel was the next essential step.

The ASTM Biodiesel Task Force therefore adopted the following description of biodiesel:

biodiesel, n—fuel comprised of mono-alkyl esters of long chain fatty acids derived from vegetable oils or animal fats, designated B100.

Discussion-biodiesel, as defined above, is registered with the U.S. Environmental Protection Agency (EPA) as a fuel and a fuel additive under Section 211(b) of the Clean Air Act. There is, however, other usage of the term "biodiesel" in the marketplace. Due to its EPA registration and the widespread commercial use of the term "biodiesel" in the U.S. marketplace, the term "biodiesel" will be maintained for this specification.

Discussion—Biodiesel is typically produced by a reaction of a vegetable oil or animal fat with an alcohol such as methanol or ethanol in the presence of a catalyst to yield mono-alkyl esters and glycerin, which is removed. The finished biodiesel derives approximately 10 % of its mass from the reacted alcohol. The alcohol used in the reaction may or may not come from renewable resources.

The first key point in this definition is that biodiesel is a mono-alkyl ester. Conventionally, biodiesel is produced through a transesterification reaction of a natural oil triglyceride

TABLE 1—Typical Biodiesel Reaction

Catalyst						
100 pounds	+	10 pounds	=	10 pounds	+	100 pounds
Triglyceride		Alcohol		Glycerin		Mono-alkyl esters
(Soybean oil)		(Methanol)				(Biodiesel)

(animal fat or vegetable oil) with a short chain alcohol (typically methanol) in the presence of a catalyst (usually sodium or potassium hydroxide). The reaction occurs stepwise, with one fatty acid chain being removed from the glycerin backbone first (forming one mono-alkyl ester and a diglyceride), the second fatty acid removed next (forming two molecules of mono-alkyl esters and a monoglyceride), and last, reaction of the third fatty acid. The resulting products are three mono-alkyl esters (biodiesel) and glycerin. Glycerin is removed as a co-product and can be upgraded to a valuable pharmaceutical grade. The reaction is depicted in Table 1.

The mono-alkyl ester definition, therefore, eliminates pure vegetable oils as well as monoglycerides and diglycerides from consideration as biodiesel. During the 1970s and 1980s, research was conducted with pure vegetable oils and partially esterified oils in their neat (or pure) form as well as with blends of petrodiesel. The use of these pure or partially esterified oils caused a variety of engine and injector problems and should not be confused with biodiesel meeting today's ASTM specification for B100 blend stock, ASTM D6751.

The second key point is that biodiesel is produced from a vegetable oil or animal fat. This eliminates some of the confusion over other materials referenced as "biodiesel" in the past. Another key point is biodiesel's intended use in compression ignition (diesel) engines. Biodiesel is not suitable for use in gasoline engines–it is strictly intended for use in diesel engines.

The ASTM Biodiesel Task Force determined that the critical items in the determination of biodiesel quality are as follows:

1. Complete reaction to the mono-alkyl esters
2. The removal of free glycerin
3. The removal of residual processing catalyst
4. The removal of reactant alcohol
5. The absence of free fatty acids

The B100 standard was developed to address each of these quality assurance needs.

While the initial proposal for the biodiesel specification at ASTM was focused on B100 as a stand-alone fuel, experience of the fuel in use with blends above B20 was insufficient to provide the technical data needed to secure approval from the ASTM members for B100 as a stand-alone fuel. Based on this, efforts after 1994 were focused on defining the properties for pure biodiesel needed to provide a "fit for purpose" fuel for use in existing diesel engines at a B20 or lower level blend.

The biodiesel industry was extremely small in the United States during the 1993–1998 period, and efforts focused on securing the technical data on the use of biodiesel, primarily in bus fleets, as well as emissions and health effects information needed for EPA registration as a legal fuel or fuel additive in the United States. Over $50 million in research and development investments occurred to secure this information, with the majority of that funding coming from American soybean farmers through the soybean check-off program. One-half of 1 % of the purchase price of a bushel of soybeans is provided by each soybean farmer into the soybean check-off fund each year.

Representatives are elected from the farmers inputting into the check-off program, and they invest the funds in research, marketing, new uses, and promotion of soybeans as a means to increase the profitability of soybean farming. Soybeans are 80 % high protein meal, primarily used as feed for hogs (pigs) and poultry, and 20 % oil. Because the demand for meal was increasing faster than the demand for oil, large excess supplies of soybean oil were a consistent problem for the soybean industry over the years. Farmers viewed this excess supply of oil as an increasing threat to profitability in the future, because demand for meal would likely increase even more as countries like China and India gradually include more meat, especially hogs and poultry, in their diet, which today is mostly grain based.

As interest in cleaner burning fuels increased in the late 1990s, biodiesel volumes began to grow and there was increasing interest in finalizing ASTM specifications for biodiesel. A provisional specification, ASTM PS121, for B100 as a blend stock was approved by ASTM in 1999. The primary reason for the provisional specification was to secure a formal ASTM test method for the GC method for measurement of the total and free glycerin–one of the most critical specifications for biodiesel–because it was necessary to include the test method in a mandatory appendix for PS121. After securing an ASTM test method for total and free glycerin, ASTM D6584, the first full specification was developed and approved in 2001 and released for use in 2002 as ASTM D6751, Standard Specification for Biodiesel Fuel Blend Stock (B100) for Middle Distillate Fuels.

The philosophy used to approve D6751 as a blend stock was the same as that used for the blending of kerosene-type fuel, No. 1 grade, into a No. 2 grade of fuel within the conventional specification, ASTM D975. If the parent fuels meet their respective specifications, then the two can be blended in any percentage and used in conventional diesel engines. These same conditions hold true for biodiesel; if biodiesel meets ASTM D6751 and conventional diesel meets ASTM D975, the two can be blended and used in conventional engines. Users should note there may be OEM warranty and usage recommendations, which usually include an upper limit of biodiesel content in the finished fuel (i.e., B5 or B20) with either no restrictions or some minor service modifications such as limits on oil drain intervals.

The most important aspect of successful use of B20 and lower blends over the past 5 years has been to ensure B100 meets D6751 prior to blending. While this philosophy of meeting a base stock specification for blending has served the U.S. market well, there has been substantial effort since 2003 to develop and formally approve specifications for the finished blends of biodiesel and conventional diesel fuel. In addition, several improvements and changes to D6751 were also undertaken, some as a result of changes needed to secure approval of the finished blended biodiesel specifications.

BIODIESEL STANDARDS

ASTM D6751-09 is the most recent version of the ASTM specifications for pure biodiesel, B100, as a blend component with middle distillate fuels. The specification contains two grades of biodiesel, S15 and S500, although almost all of the biodiesel in use in the United States is of the S15 variety. Natural vegetable oil feedstocks have virtually no sulfur (usually less than 1 or 2 ppm), but some animal fats or yellow grease-based biodiesel may have sulfur slightly higher than 15 ppm due to the presence of hair or hide materials from the animal fat rendering process or the frying of foods high in sulfur like onion rings.

D6751 has been officially approved by ASTM only for use in blends up to B20, as noted in the requirements section of the standard states:

NOTE 2–A considerable amount of experience exists in the U.S. with a 20 % blend of biodiesel, primarily produced from soybean oil, with 80 % diesel fuel (B20). Experience with biodiesel produced from animal fat and other oils is similar. Experience with B20 and lower blends in other applications is not as prevalent. Although biodiesel (B100) can be used, blends of over 20 % biodiesel with diesel fuel (B20) should be evaluated on a case by case basis until further experience is available.

NOTE 3–The user should consult the equipment manufacturer or owner's manual regarding the suitability of using biodiesel or biodiesel blends in a particular engine or application.

Table 2 is taken from ASTM D6751-09.

TABLE 2—Detailed Requirements for Biodiesel (B100) (All Sulfur Levels)

Property	Test Method[A]	Grade S15 Limits	Grade S500 Limits	Units
Calcium and Magnesium, combined	EN 14538	5 max	5 max	ppm (μg/g)
Flash point (closed cup)	D93	93 min	93 min	°C
Alcohol control				
One of the following must be met:				
1. Methanol content	EN 14110	0.2 max	0.2 max	mass percent
2. Flash point	D93	130 min	130 min	°C
Water and sediment	D2709	0.050 max	0.050 max	volume percent
Kinematic viscosity, 40°C	D445	1.9–6.0[B]	1.9–6.0[B]	mm^2/s
Sulfated ash	D874	0.020 max	0.020 max	mass percent
Sulfur[C]	D5453	0.0015 max (15)	0.05 max (500)	mass percent (ppm)
Copper strip corrosion	D130	No. 3 max	No. 3 max	
Cetane number	D613	47 min	47 min	
Cloud point	D2500	Report[D]	Report[D]	°C
Carbon residue[E]	D4530	0.050 max	0.050 max	mass percent
Acid number	D664	0.50 max	0.50 max	mg KOH/g
Cold soak filterability	Annex A1	360 max[F]	360 max[F]	seconds
Free glycerin	D6584	0.020 max	0.020 max	mass percent
Total glycerin	D6584	0.240 max	0.240 max	mass percent
Phosphorus content	D4951	0.001 max	0.001 max	mass percent
Distillation temperature, atmospheric equivalent temperature, 90 % recovered	D1160	360 max	360 max	°C
Sodium and potassium, combined	EN 14538	5 max	5 max	ppm (μg/g)
Oxidation stability	EN 14112	3 min	3 min	hours

[A] The test methods indicated are the approved referee methods. Other acceptable methods are indicated in 5.1.
[B] See X1.3.1. The 6.0 mm^2/s upper viscosity limit is higher than petroleum-based diesel fuel and should be taken into consideration when blending.
[C] Other sulfur limits can apply in selected areas in the United States and in other countries.
[D] The cloud point of biodiesel is generally higher than petroleum-based diesel fuel and should be taken into consideration when blending.
[E] Carbon residue shall be run on the 100 % sample (see 5.1.11).
[F] B100 intended for blending into diesel fuel that is expected to give satisfactory vehicle performance at fuel temperatures at or below −12°C shall comply with a cold soak filterability limit of 200 s maximum.

The key issues with regard to biodiesel quality identified by the ASTM Biodiesel Task Force are controlled by parameters in the specification in the following way:

1. Conversion of the fat or oil to mono-alkyl esters is ensured through measurement of the total glycerin (D6584), which includes all bound glycerin (i.e., mono-, di-, and triglycerides) as well as the unbound free glycerin.
2. Removal of the unbound glycerin is ensured through measurement of the free glycerin (D6584).
3. Removal of the catalyst used in the production of biodiesel is ensured through the measurement of the sulfated ash (D874), and is further controlled by limiting the levels of combined sodium and potassium (EN 14538).
4. Removal of the alcohol (usually methanol) is ensured through direct measurement of methanol by either GC (EN14110) or by use of a high flash point value (D93), or by use of a high flash point value if alcohols other than methanol are used.
5. The absence of fatty acids is ensured through measurement of the acid number (D664).

There are several key distinctions between petrodiesel and biodiesel that are evidenced in the biodiesel standard and the testing methods employed. Perhaps the most important is that the cetane number, ASTM D613, commonly referred to as the "cetane engine test," must be used. You cannot use the Calculated Cetane Index method for biodiesel. The Calculated Cetane Index (CCI), ASTM D4737, is based on historical data for the distillation curve of petroleum diesel and is not applicable to biodiesel. If a Calculated Cetane Index is performed for either biodiesel or a biodiesel blend, the result will be artificially lower than the cetane number derived from the cetane engine test. If one only desires a go/no-go answer for a minimum cetane value, it is possible to use the Calculated Cetane Index on biodiesel or a biodiesel blend for that purpose because the real cetane value will be higher.

The lack of applicability of the cetane index is primarily due to the lack of a "distillation curve" for biodiesel. Where petrodiesel is comprised of hundreds of compounds boiling at differing temperatures (determined by the petroleum refining process), biodiesel contains only a few compounds–primarily C16 to 18 carbon chain length alkyl esters (determined by the feedstock) along with minor variations in carbon double bonds, but these compounds all boil at approximately the same temperature. Most biodiesel processes today produce what petroleum refiners would call a "whole cut" product, where virtually all of the feedstock is made into one biodiesel product without further separation or distillation of the individual fatty acid esters that make up biodiesel. Biodiesel, therefore, exhibits more of a tight series of boiling points over a short temperature range than a "distillation curve." The fats and oils commonly used today have a very similar fatty acid profile, as seen in Table 3, giving biodiesel a very tight boiling range regardless of the feedstock.

The molecular weight and composition of biodiesel also account for its high flash point, which gives an additional safety margin in enclosed areas such as underground mines.

There are some additional important differences in the testing methods necessary for measurement of biodiesel:

- The carbon residue must be run on the 100 % sample, not the 10 % residue after distillation as done with petrodiesel. It is difficult to leave only 10 % of the sample upon distillation because biodiesel all boils at about the same temperature.

TABLE 3—Oil Types and Length of Fatty Acid Chains in U.S. Fats and Oils[A]

No. of Carbons in Fatty Acid Chain	<14	16	18	>20	
Oil type					
Corn		12	88		
Tallow		5	32	64	
Peanut			12	81	7
Used frying oil[B]	4	19	76	1	
Rapeseed		5	94	1	
Pork	2	27	70	1	
Soybean		13	87		

[A] Source: Procter and Gamble.
[B] Source: Fats and Proteins Research Foundation.

- The ash content is changed from oxidative ash (ASTM D482) to sulfated ash (ASTM D874) to ensure more accurate measurement of sodium or potassium, which could be present as residual catalyst from the transesterification reaction in the manufacture of the biodiesel B100.
- The flash point is substantially higher (93°C minimum) than that for ASTM D975 (52°C minimum for No. 2, and 38°C minimum for No. 1). All properly processed biodiesel from available fats and oils already meets the higher flash point value, and 93°C (200°F) minimum was chosen to coincide with the value that places B100 in a nonhazardous shipping classification in the United States.
- A vacuum distillation (ASTM D1160) with a T-90 maximum of 360°C is used to ensure the absence of contaminants that would otherwise go undetected, such as used motor oil.
- The cetane number is set substantially higher (47 minimum) than that in D975 (40 minimum). Properly processed biodiesel in the United States will naturally meet this higher cetane and the 47 value was selected based on that being the minimum cetane number for "premium diesel fuel" as determined by the National Conference on Weights and Measures (NCWM) in the United States.
- Combined sodium and potassium, as well as combined calcium and magnesium, are both controlled to 5 ppm where there are no such controls in ASTM D975.
- Phosphorous is controlled to 10 ppm, whereas there is no such control for ASTM D975. Phosphorous is a natural element found in all plant matter and is a concern in fuels where it can react adversely with exhaust catalysts causing some deactivation of the catalyst, referred to as poisoning of the catalyst.
- Stability is controlled to 3 hours minimum, where there are no such controls in ASTM D975. The 3-hour minimum was selected to coincide with acceptable ASTM D4625 long-term storage results for B20 blends representing approximately 6 months or longer under normal storage conditions [3]. Certain applications may require a higher level of stability, which can be attained through selection of various processing or feedstock options or by using an antioxidant additive.

- A cold soak filterability parameter was recently included to control minor compounds at the B100 level, which are not detected or controlled by other means within the spec and which may cause some biodiesel blends to exhibit filter clogging at temperatures above the cloud point of the blend. There are no such controls in ASTM D975.

ASTM D6751 has undergone several improvements since its passage in 2001 and it will continue to undergo improvements over time, especially improvements in analytical methods and techniques.

BIODIESEL BLEND SPECIFICATIONS

Immediately after the passage of ASTM D6751 in 2001 as a blend stock, the ASTM Biodiesel Task Force began efforts to secure standards for finished blends of biodiesel with diesel fuel. Passage of D6751 and the common operating practice of ensuring the B100 met D6751 prior to blending and having biodiesel or petrodiesel blends treated similar to the way No. 1 and No. 2 diesel fuels are treated (if the parent fuels meet their specifications, they can be blended and used in diesel engines with no need to test or measure the blend) allowed the industry to grow while deliberations on finished specifications were conducted. However, large biodiesel blend purchasers and users such as the Defense Energy Support Center and the military, and fuel enforcement entities like NCWM representatives, and engine companies desired a set of properties that could be measured for the finished blends.

Having one specification for the finished blends would facilitate the purchasing of biodiesel blends (only need to specify or monitor one specification versus two specifications for the components of a blend), it would facilitate engine testing (because the engine sees the finished blend not the components individually), and it would facilitate fuel quality enforcement because regulators may not have access to samples of the blend components.

As Subcommittee E began the deliberations on the blended fuel specifications in the 2001–2003 time frame, there were several new fuels under consideration that claimed to be usable in existing diesel engines with little or no modification. Biodiesel blend, ethanol-diesel blends, and water-emulsified diesel blends all had active task forces or working groups within Subcommittee E. The idea of a "Fill and Go" concept as a means to incorporate all these nonconventional fuels into ASTM D975 came into being during this time period as a mechanism by which all these "new" fuels could potentially be incorporated into ASTM D975. There were several key aspects that gained momentum at the June 2003 ASTM meeting:

- A new category that all new Fill and Go fuels will fall under "should be developed."
- While used in the same engine and vehicle as D975 fuel, Fill and Go fuels may need new methods, different limits, or different parameters than currently exist in D975.
- B20 and lower should be part of D975. D975 is embedded into a variety of state and federal laws, regulations, safety codes, tax codes, etc. and it would be very beneficial to have B20 covered by D975 rather than adding a new ASTM specification everywhere D975 is already codified.
- Specs should cover all blends B20 and lower (i.e., B2, B5, B11, B20).
- Finished properties should be set based on satisfactory engine performance.
- Existing D975 properties covered B5 and lower, if B100 meets D6751.
- Additional properties are needed for B6 to B20.

It was originally proposed to incorporate B20 and lower blends into the specification for conventional petroleum-based diesel fuel, ASTM D975. After some deliberation, in June 2005, Subcommittee E settled on the following general guidelines for incorporation of biodiesel, and other new blending components, in order to be balloted into D975:

- Identify an ASTM specification the blend component must meet.
- Specify upper limit of component allowed.
- Set the test method for determining the level of the component in the finished D975 blend.
- Use existing test methods and limits for D975.
- Expanding limits or test method is NOT acceptable, unless it is also acceptable to change the limit for the D975 fuel(s).

With this much agreed, it was decided the blended fuel specification should be set in such a fashion so that if the biodiesel and diesel fuel both met their parent specifications the blended fuel would also always meet its specification [3]. This will prevent the need for reanalysis after blending, except to confirm proper blending and to monitor for contamination or changes in the fuel over time. Based on this, ballots moved forward to incorporate B5 and lower blends of biodiesel into the conventional on/off road diesel spec, ASTM D975, and the conventional heating oil specification, ASTM D396.

Blends between 6 and 20 % biodiesel would be balloted under a new specification, as in the end it was determined that two additional parameters were needed for a B6 to B20 blend specification that are not needed for petrodiesel. These were acid number and stability. It was also agreed to increase the T90 distillation temperature of the B6 to B20 blend by 5°C in order to keep the principle of the blend always meeting specification if two in-spec parent fuels were used, because B100 has a slightly higher T90 than petrodiesel. A higher T90 for petrodiesel is viewed by some as problematic, as it can increase particulate emissions and therefore affect long-term durability of engine components and exhaust emissions, but because biodiesel reduces particulates the higher T90 for blends was acceptable to the engine community–provided the diesel fuel was qualified to its T90 prior to blending. Wording explaining this is embedded it the B6 to B20 specification D7467.

In June 2008, ASTM's Committee D02 on Petroleum Products and Lubricants passed the incorporation of B5 into both the D396 Fuel Oil specification and the D975 Diesel Fuel specification. No changes in Table 1 for either the test methods or the specifications were involved–the finished blend of biodiesel blend stock and petrodiesel, and biodiesel blend stock and heating oil, just needs to meet the same respective properties in Table 1 that conventional diesel fuel and heating oil meets. The wording below is contained in the Section 7 Requirements of ASTM D975-09 and similar wording is contained in ASTM D396-08:

> 7.3 *Fuels Blended with Biodiesel*–The detailed requirements for fuels blended with biodiesel shall be as follows:
>
> 7.3.1 *Biodiesel for Blending*–If biodiesel is a component of any diesel fuel, the biodiesel shall meet the requirements of Specification D6751.

7.3.2 Diesel fuel oil containing up to 5 vol% biodiesel shall meet the requirements for the appropriate grade No. 1-D or No. 2-D fuel, as listed in Table 1.
7.3.3 Test Method EN 14078 shall be used for determination of the vol% biodiesel in a biodiesel blend.
7.3.4 Diesel fuels containing more than 5 vol% biodiesel component are not included in this specification.
7.3.5 Biodiesel blends with No. 4-D fuel are not covered by this specification.

One of the key aspects of the blended fuel specifications is the necessity that the biodiesel meet its specification, ASTM D6751, prior to blending. Several requirements were added to ASTM D6751 as part of the ballot negotiating process in order to help ensure that if biodiesel meets the ASTM D6751 specification, then no other requirements for the blend were needed beyond those already in ASTM D975. The acid number was lowered, new specifications for combined calcium and magnesium and combined sodium and potassium were added (for adequate exhaust diesel particulate filter life on new engines on sale after model year 2007), a stability specification was added to the B100, and most recently a new cold soak filtration test was added.

The last change, the addition of the cold soak filtration, was somewhat unexpected as very few problems in the field had been seen with biodiesel meeting all the other parameters already in ASTM D6751. However, a small amount of B100 was found to contain trace amounts of saturated monoglycerides, soaps, and sterol glucosides that are not specifically controlled by the other parameters in the specification. The presence of these minor compounds, and potentially other as yet not identified compounds, in certain levels may cause the cloud point of the blended fuel to no longer be predictive of filter clogging and operability in cold winter conditions. The presence of allowable amounts of water can exacerbate the phenomenon. These minor compounds (saturated monoglycerides, soaps, and sterol glucosides) are found in natural oils and fats and can remain in biodiesel B100 product as trace contaminants. But they are not found in crude oils and therefore are not picked up by any of the tests in D975. Before ASTM members would allow up to 5 % biodiesel in the fungible diesel fuel pool, a performance test was required to be developed (the new cold soak filterability test) to address this issue at the B100 level so that no additional testing would be needed for the finished blend other than those tests already in D975.

The B6 to B20 specification presented new challenges, particularly how to handle the differences between the No. 1 and No. 2 grades of fuels. Because both No. 1 and No. 2 are used interchangeably in today's diesel engines, in the end it was decided the new B6 to B20 specification (ASTM D7467-08) would be stripped of the No. 1/No. 2 designation and create grades only based on sulfur level. ASTM D7467 was largely based on the D975 performance requirements, with the widest value of the parameters between the No. 1 and No. 2 grades of D975 used for each D975 parameter with the exception of the T90, which is allowed to be 5°C higher for B6 to B20. An acid number of 0.3 mg KOH/g maximum was added as an additional and convenient control for stability, and a stability parameter of 6 hours minimum was added. Table 1 from D7467 and the footnotes are given in Table 4.

IMPORTANCE OF GENERAL FUEL CHARACTERISTICS

Biodiesel or biodiesel blends described in this chapter (up to B5 in ASTM D975 and B6 to B20 covered by ASTM D7467) are used in the same equipment as conventional diesel fuels, often without modification or restrictions. Therefore, the same general fuel characteristics and considerations for engines and equipment with conventional diesel fuels also applies to the use of biodiesel and biodiesel blends. The reader is referenced to those sections in this manual for more information and this will not be repeated here. We will instead focus the next information on the considerations that vary from those of conventional petrodiesel.

Sulfated Ash

Ash-forming materials may be present in biodiesel in three forms: (1) abrasive solids, (2) soluble metallic soaps, and (3) unremoved catalysts. Abrasive solids and unremoved catalysts can contribute to injector, fuel pump, piston and ring wear, and engine deposits. Soluble metallic soaps have little effect on wear but may contribute to filter plugging and engine deposits. The ash-forming materials may also contribute to exhaust catalyst plugging and additional deposition in diesel particulate filters (sometimes referred to as particulate traps).

Sulfur

B100 is essentially sulfur free, although some animal fat–based biodiesel has been found with up to 100 ppm sulfur (a result of hides and hair from the animal fat–rendering process) and some yellow grease–based biodiesel has been found with similar levels (a result of frying foods high in sulfur like onion rings). Test Method ASTM D5453 should be used with biodiesel. Use of other test methods may provide falsely high results when analyzing B100 with extremely low sulfur levels (less than 5 ppm). Biodiesel sulfur analysis from RR: D02-1480, Biodiesel Fuel Cetane Number Testing Program, January–April, 1999, using Test Method D2622 yielded falsely high results due to the presence of the oxygen in the biodiesel. Sulfur results using Test Method D2622 were more accurate with B20 than with B100 due to the lower oxygen content of B20. Potential future improvements to Test Method D2622 may provide more accurate values.

Cetane Number

Cetane number is a measure of the ignition quality of the fuel and influences startability, white smoke, and combustion roughness. The cetane number requirements depend on engine design, size, nature of speed and load variations, and starting and atmospheric conditions. The calculated cetane index, Test Method D976 or D4737, may not be used to approximate the cetane number with biodiesel or its blends as it will result in falsely low values. There is, as yet, no substantiating data to support the calculation of cetane index with biodiesel or biodiesel blends.

Carbon Residue

Carbon residue gives a measure of the carbon depositing tendencies of a fuel. While not directly correlating with engine deposits, this property is considered an approximation. Although biodiesel is in the petroleum diesel boiling range, most biodiesels boil at approximately the same temperature and it is difficult to leave a 10 % residual upon distillation. Thus, a 100 % sample is used to replace the 10 % residual sample.

TABLE 4—Detailed Requirements for B6 to B20 Biodiesel Blends

Property	Test Method	Grade		
		B6 to B20 S15	B6 to B20 S500[A]	B6 to B20 S5000[B]
Acid umber, mg KOH/g, max	D664	0.3	0.3	0.3
Viscosity, mm^2/s at 40°C	D445	1.9–4.1[C]	1.9–4.1[C]	1.9–4.1[C]
Flash point, °C, min	D93	52[D]	52[D]	52[D]
Cloud point, °C, max or LTFT/CFPP, °C, max	D2500, D4539, D6371	[E]	[E]	[E]
Sulfur content (μg/g)	D5453	15	. . .	. . .
mass percent, max	D2622	. . .	0.05	. . .
mass percent, max	D129	. . .	. . .	0.50
Distillation temperature, °C, 90 % vol recovered, max	D86	343	343	343
Ramsbottom carbon residue on 10 % bottoms, mass percent, max	D524	0.35	0.35	0.35
Cetane number, min	D613[F]	40[G]	40[G]	40[G]
One of the following must be met: (1) Cetane index, min.	D976-80[H]	40	40	40
(2) Aromaticity, % vol, max	D1319-03[H]	35	35	. . .
Ash content, mass percent, max	D482	0.01	0.01	0.01
Water and sediment, volume percent, max	D2709	0.05	0.05	0.05
Copper corrosion, 3 h at 50°C, max	D130	No. 3	No. 3	No. 3
Biodiesel content, % (V/V)	D7371	6. - 20.	6. - 20.	6. - 20.
Oxidation stability, hours, min	EN 14112	6	6	6
Lubricity, HFRR at 60°C, micron (μm), max	D6079	520[I]	520[I]	520[I]

[A] Under United States of America regulations, if Grades B6-20 S500 are sold for tax exempt purposes then, at, or beyond terminal storage tanks, they are required by 26 CFR Part 48 to contain the dye Solvent Red 164 at a concentration spectrally equivalent to 3.9 lb per thousand barrels of the solid dye standard Solvent Red 164, or the tax must be collected.

[B] Under United States of America regulations, Grades B6-20 S5000 are required by 40 CFR part 80 to contain a sufficient amount of the dye Solvent Red 164 so its presence is visually apparent. At or beyond terminal storage tanks, they are required by 26 CFR Part 48 to contain the dye Solvent Red 164 at a concentration spectrally equivalent to 3.9 lb per thousand barrels of the solid dye standard Solvent Red 26.

[C] If Grade No. 1-D or blends of Grade No. 1-D and Grade No. 2-D diesel fuel are used, the minimum viscosity shall be 1.3 mm^2/s.

[D] If Grade No. 1-D or blends of Grade No. 1-D and Grade No. 2-D diesel fuel are used, or a cloud point of less than –12°C is specified, the minimum flash point shall be 38°C.

[E] It is unrealistic to specify low temperature properties that will ensure satisfactory operation at all ambient conditions. In general, cloud point (or wax appearance point) Low Temperature Flow Test, and Cold Filter Plugging Point Test may be useful to estimate vehicle low temperature operability limits but their use with B6 to B20 has not been validated. However, satisfactory operation below the cloud point (or wax appearance point) may be achieved depending on equipment design, operating conditions, and the use of flow-improver additives as described in X3.1.2. Appropriate low temperature operability properties should be agreed upon between the fuel supplier and purchaser for the intended use and expected ambient temperatures. Test Methods D4539 and D6371 may be especially useful to estimate vehicle low temperature operability limits when flow improvers are used but their use with B6 to B20 from a full range of biodiesel feedstock sources has not been validated. Due to fuel delivery system, engine design, and test method differences, low temperature operability tests may not provide the same degree of protection in various vehicle operating classes. Tenth percentile minimum air temperatures for U.S. locations are provided in Appendix X3 as a means of estimating expected regional temperatures. The tenth percentile minimum air temperatures may be used to estimate expected regional target temperatures for use with Test Methods D2500, D4539, and D6371. Refer to X3.1.3 for further general guidance on test application.

[F] Calculated cetane index approximation, Test Method D4737, is not applicable to biodiesel blends.

[G] Low ambient temperatures, as well as engine operation at high altitudes, may require the use of fuels with higher cetane ratings. If the diesel fuel is qualified under Table 1 of Specification D975 for cetane, it is not necessary to measure the cetane number of the blend. This is because the cetane number of the individual blend components will be at least 40, so the resulting blend will also be at least 40 cetane number.

[H] These test methods are specified in 40 CFR Part 80.

[I] If the diesel fuel is qualified under Table 1 of Specification D975 for lubricity, it is not necessary to measure the lubricity of the blend. This is because the lubricity of the individual blend components will be less than 520 micron (μm) so the resulting blend will also be less than 520 (μm).

Acid Number

The acid number is used to determine the level of free fatty acids or processing acids that may be present in biodiesel. Biodiesel with a high acid number has been shown to increase fueling system deposits and may increase the likelihood for corrosion.

The acid number measures a different phenomenon for biodiesel than petroleum based diesel fuel. The acid number for biodiesel measures free fatty acids or degradation byproducts not found in petroleum-based diesel fuel. Increased fuel temperatures in some new fuel designs due to fuel recycling from common rail injector systems, may accelerate fuel degradation, which could result in high acid values and increased filter plugging potential.

Free Glycerin

The free glycerin method is used to determine the level of glycerin in the fuel. High levels of free glycerin can cause injector deposits, as well as clogged fueling systems, and result in a build up of free glycerin in the bottom of storage tanks and fueling systems.

Total Glycerin

The total glycerin method is used to determine the level of glycerin in the fuel and includes the free glycerin and the glycerin portion of any unreacted or partially reacted oil or fat. Low levels of total glycerin ensure that high conversion of the oil or fat into its mono-alkyl esters has taken place. High levels of monoglycerides, diglycerides, and triglycerides can cause injector deposits and may adversely affect cold weather operation and filter plugging.

Phosphorus Content

Phosphorus, a natural element in all plants that is also found in vegetable oils, can affect the conversion rates in diesel exhaust catalytic converters used to control emissions. Accordingly, the phosphorous level should be kept low. Catalytic converters are increasingly being used globally on diesel-powered equipment as emissions standards are tightened. Biodiesel produced from U.S. sources has been shown to have low phosphorus content (below 1 ppm) and the specification value of 10 ppm maximum is not problematic. Biodiesel from other sources may or may not contain higher levels of phosphorus, and this specification was added to ensure that all biodiesel, regardless of the source, has low phosphorus content.

Reduced Pressure Distillation

Biodiesel exhibits a series of close boiling points rather than a distillation curve. The fatty acids chains in the raw oils and fats from which biodiesel is produced are mainly comprised of straight chain hydrocarbons with 16 to 18 carbons that have similar boiling temperatures. The atmospheric boiling point of biodiesel generally ranges from 330 to 357°C, thus the specification value of 360°C is not problematic. This specification was incorporated as an added precaution to ensure the fuel has not been adulterated with high boiling contaminants.

Density

The density of biodiesel meeting the specifications in Table 1 (Table 1 in D396 and D975) falls between 0.86 and 0.90, with typical values falling between 0.88 and 0.89. Because biodiesel density falls between 0.86 and 0.90, a separate specification is not needed. The density of raw oils and fats can be similar to biodiesel and the use of density as an expedient check of fuel quality may not be as useful for biodiesel as it is for petroleum-based diesel fuel.

Lubricity

In certain fuel injection equipment in compression ignition engines, such as rotary/distributor fuel pumps and injectors, the fuel functions as a lubricant. Blending biodiesel fuel with petroleum-based compression-ignition fuel typically improves fuel lubricity. No specification is needed for biodiesel lubricity as values are lower than a 300 μ Wear Scar Diameter (WSD) using HFRR with B100.

Alcohol Control

Alcohol control is to limit the level of unreacted alcohol remaining in the finished fuel. This can be measured directly by the volume percent alcohol or indirectly through a high flash point value.

The flash point specification, when used for alcohol control for biodiesel, is intended to be 100°C minimum, which has been correlated to 0.2 volume percent alcohol. Typical values are over 160°C. Due to high variability with Test Method D93 as the flash point approaches 100°C, the flash point specification has been set at 130°C minimum to ensure an actual value of 100°C minimum. Improvements and alternatives to Test Method D93 are being investigated. Once complete, the specification of 100°C minimum may be reevaluated for alcohol control.

Calcium and Magnesium

Calcium and magnesium may be present in biodiesel as abrasive solids or soluble metallic soaps. Abrasive solids can contribute to injector, fuel pump, piston, and ring wear, as well as to engine deposits. Soluble metallic soaps have little effect on wear, but they may contribute to filter plugging and engine deposits. High levels of calcium and magnesium compounds may also collect in the exhaust catalyst and in the diesel particulate filter (DPF). These compounds are not typically removed from the diesel particulate filter during passive or active regeneration, and may result in ash accumulation on the catalyst or in the DPF, producing increased back pressure and the potential for reduced time between service intervals.

Sodium and Potassium

Sodium and potassium may be present in biodiesel as abrasive solids or soluble metallic soaps. Abrasive solids can contribute to injector, fuel pump, piston and ring wear, and also to engine deposits. Soluble metallic soaps have little effect on wear, but they may contribute to filter plugging and engine deposits. High levels of sodium or potassium compounds may also collect in the exhaust catalysts and in the DPF. These compounds are not typically removed from the diesel particulate filter during passive or active regeneration and may result in ash accumulation on the catalyst or in the DPF producing increased back pressure and the potential for reduced time between service intervals.

Oxidation Stability

Products of oxidation in biodiesel can take the form of various acids or polymers, which, if in high enough concentration, can cause fuel system deposits and lead to filter clogging and fuel system malfunctions. Additives designed to terminate reactions leading to the formation of peroxides that precede the formation of polymers and gums can significantly improve the oxidation stability performance of biodiesel.

ASTM Test Methods

Number	Title
D93	Test Methods for Flash Point by Pensky-Martens Closed Cup Tester
D130	Test Method for Corrosiveness to Copper from Petroleum Products by Copper Strip Test
D189	Test Method for Conradson Carbon Residue of Petroleum Products
D445	Test Method for Kinematic Viscosity of Transparent and Opaque Liquids (and Calculation of Dynamic Viscosity)
D524	Test Method for Ramsbottom Carbon Residue of Petroleum Products
D613	Test Method for Cetane Number of Diesel Fuel Oil
D664	Test Method for Acid Number of Petroleum Products by Potentiometric Titration
D874	Test Method for Sulfated Ash from Lubricating Oils and Additives
D974	Test Method for Acid and Base Number by Color-Indicator Titration
D975	Specification for Diesel Fuel Oils
D976	Test Method for Calculated Cetane Index of Distillate Fuels
D1160	Test Method for Distillation of Petroleum Products at Reduced Pressure
D1266	Test Method for Sulfur in Petroleum Products (Lamp Method)
D1796	Test Method for Water and Sediment in Fuel Oils by the Centrifuge Method (Laboratory Procedure)
D2274	Test Method for Oxidation Stability of Distillate Fuel Oil (Accelerated Method)
D2500	Test Method for Cloud Point of Petroleum Products
D2622	Test Method for Sulfur in Petroleum Products by Wavelength Dispersive X-Ray Fluorescence Spectrometry
D2709	Test Method for Water and Sediment in Middle Distillate Fuels by Centrifuge
D2880	Specification for Gas Turbine Fuel Oils
D3117	Test Method for Wax Appearance Point of Distillate Fuels
D3120	Test Method for Trace Quantities of Sulfur in Light Liquid Petroleum Hydrocarbons by Oxidative Microcoulometry
D3242	Test Method for Acidity in Aviation Turbine Fuel
D3828	Test Methods for Flash Point by Small Scale Closed Cup Tester

Number	Title
D4057	Practice for Manual Sampling of Petroleum and Petroleum Products
D4177	Practice for Automatic Sampling of Petroleum and Petroleum Products
D4294	Test Method for Sulfur in Petroleum and Petroleum Products by Energy Dispersive X-Ray Fluorescence Spectrometry
D4530	Test Method for Determination of Carbon Residue (Micro Method)
D4737	Test Method for Calculated Cetane Index by Four Variable Equation
D4865	Guide for Generation and Dissipation of Static Electricity in Petroleum Fuel Systems
D4951	Test Method for Determination of Additive Elements in Lubricating Oils by Inductively Coupled Plasma Atomic Emission Spectrometry
D5452	Test Method for Particulate Contamination in Aviation Fuels by Laboratory Filtration
D5453	Test Method for Determination of Total Sulfur in Light Hydrocarbons, Spark Ignition Engine Fuel, Diesel Engine Fuel, and Engine Oil by Ultraviolet Fluorescence
D5773	Test Method for Cloud Point of Petroleum Products (Constant Cooling Rate Method)
D6217	Test Method for Particulate Contamination in Middle Distillate Fuels by Laboratory Filtration
D6300	Practice for Determination of Precision and Bias Data for Use in Test Methods for Petroleum Products and Lubricants
D6450	Test Method for Flash Point by Continuously Closed Cup (CCCFP) Tester
D6469	Guide for Microbial Contamination in Fuels and Fuel Systems
D6584	Test Method for Determination of Free and Total Glycerin in B-100 Biodiesel Methyl Esters by Gas Chromatography
D6890	Test Method for Determination of Ignition Delay and Derived Cetane Number (DCN) of Diesel Fuel Oils by Combustion in a Constant Volume Chamber
D7039	Test Method for Sulfur in Gasoline and Diesel Fuel by Monochromatic Wavelength Dispersive X-Ray Fluorescence Spectrometry
2.2	*Government Standard:*
40 CFR Part 79	Registration of Fuels and Fuel Additives Section 211(b) Clean Air Act[A]
2.3	*Other Documents:*
UOP 389	Trace Metals in Oils by Wet Ashing and ICP-OES

Number	Title
UOP 391–91	Trace Metals in Petroleum Products or Organics by AAS
EN 14112	Fat and Oil Derivatives—Fatty Acid Methyl Esters (FAME)—Determination of Oxidation Stability (Accelerated Oxidation Test)
EN 14110	Fat and Oil Derivatives—Fatty Acid Methyl Esters (FAME)—Determination of Methanol Content
EN 14538	Fat and Oil Derivatives—Fatty Acid Methyl Esters (FAME)—Determination of Ca, K, Mg and Na Content by Optical Emission Spectral Analysis with Inductively Coupled Plasma (ICP OES)

[A] Available from U.S. Government Printing Office Superintendent of Documents, 732 N. Capitol St., NW, Mail Stop: SDE, Washington, DC 20401.

References

[1] Hoekman, S. K., Gertler, A., Broch, A., and Robbins, C., "Investigation of Biodistillates as Potential Blendstocks for Transportation Fuels," CRC Project No. AVFL-17 Final Report, June 2009.

[2] Engine Manufacturers Association, Alliance of Automobile Manufacturers, "Use of Raw Vegetable Oil or Animal Fats in Diesel Engines," March 2006.

[3] McCormick, R. L., Alleman, T. L., Waynick, J. A., Westbrook, S. R., and Porter, S., "Stability of Biodiesel and Biodiesel Blends: Interim Report," prepared under Task No. FC06.9400, NREL/TP-540-39721, April 2006.

6

Burner, Heating, and Lighting Fuels[1]

C. J. Martin[2] and Lindsey Hicks[3]

ALTHOUGH MOST PETROLEUM PRODUCTS CAN BE used as fuels, the term "fuel oil," if used without qualification, may be interpreted differently in various countries. For example, in Europe, fuel oil generally is associated with the black, viscous, residual material that remains as the result of refinery distillation of crude oil, either alone or in a blend with lighter components, and it is used for steam generation for large slow-speed diesel engine operation and industrial heating and processing. In the United States, the term "fuel oil" is applied to both residual and middle distillate type products, such as domestic heating oil, kerosine, and burner fuel oils.

Because fuel oils are complex mixtures of compounds of carbon and hydrogen, they cannot be classified rigidly or defined exactly by chemical formulas or definite physical properties. For purposes of this chapter, the term "fuel oil" will include all petroleum oils heavier than gasoline that is used in burners. Because of the wide variety of petroleum fuel oils, the arbitrary divisions or classifications, which have become widely accepted in industry, are based more on their application than on their chemical or physical properties. Thus, it is not uncommon to find large variations in properties among petroleum products sold on the market for the same purpose. However, two broad classifications are generally recognized: (1) "distillate" fuel oils and (2) "residual" fuel oils. The latter are often referred to as heavy fuel oils and may contain cutter stock or distillates.

Middle distillate fuel oils are petroleum fractions that have been vaporized and condensed. They are produced in the refinery by a distillation process in which petroleum is separated into fractions, according to their boiling range. These middle distillate fuel oils may be produced not only directly from crude oil, that is, "straight-run," but also from subsequent refinery conversion processes, such as thermal or catalytic cracking. Domestic heating oils and kerosine are examples of middle distillate fuel oils. Common terms for the lighter (distillate) products include range oil, stove oil, and furnace oil, with range and stove oil being the lighter of the products.

On the other hand, residual or heavy fuel oils are composed wholly or in part of undistillable petroleum fractions from crude oil distillation (atmospheric or vacuum tower bottoms), vis-breaking, or other refinery operations. The various grades of heavy fuel oil are generally produced to meet definite specifications to ensure suitability for their intended purpose. Residual oils are classified usually by viscosity in contrast with distillates, which normally are defined by boiling range.

BIODIESEL ALLOWANCE IN HEATING FUELS

The increased emphasis on use of alternative fuels and renewable energy has caused the ASTM International to revise ASTM D396, Standard Specification for Fuel Oils, to include an allowance for up to 5 volume percent biodiesel for the appropriate Grade No. 1 or No. 2 as listed in the Table 1 of ASTM D396. Fuel oil containing more than 5 volume percent biodiesel component are not covered by the specification. Biodiesel blends with Grades 4, 5, or 6 are not covered by ASTM D396.

ASTM International also developed ASTM D6751-08, Specification for Biodiesel Fuel Blend Stock (B100) for Middle Distillate Fuels, to define pure biodiesel (B100) quality prior to blending with conventional diesel and burner heating type fuels.

ENVIRONMENTAL AND OTHER ISSUES

In recent years, the nature of processing, transporting, and storing fuel oil has changed significantly. Factors driving these changes include changes in refineries, evolving environmental regulations, and emerging energy conservation programs. Refineries have been modified to expand processing units, add or change catalytic processes to accommodate different crude oils, and address shifting product demands. In the United States and many other industrialized countries, environmental regulations have been established to address concerns about air pollution emissions. Regulations affecting fuel oils include control of sulfur content. In the United States, individual state governments have promulgated regulations affecting fuel oil sulfur content applicable to specific geographic locations. ASTM D396, Standard Specification for Fuel Oils, defines various grades of fuel oil and addresses their intended uses, sulfur content, and the need for consulting appropriate regulatory agencies.

As a result of the Clean Air Act Amendments of 1990, the U.S. Environmental Protection Agency (EPA) and the Internal Revenue Service (IRS) require red dye to be added to certain classes of distillate fuel. The EPA requires high sulfur distillate fuels above a certain sulfur content (current limit is above 0.05 mass percent or 500 ppm) and used for off-highway purposes to contain red dye. This red dye requirement applies to diesel fuel, the four grades of middle distillate fuel oils (No. 1 S5000, No. 2 S5000, No. 1 S500, and No. 2 S500 for use in

[1] In preparation of this chapter, the contents of the seventh edition were drawn on. The authors acknowledge one of the previous authors of this chapter in the seventh edition: Regina Gray, Defense Energy Support Center. The authors would also like to acknowledge the valuable assistance of Victor Turk (R. W. Becket Corporation) for his review and contribution. The current addition will review and update the topics as addressed by the previous authors and introduce the new allowance of biodiesel. Tables 1 and 2 were used as references for preparation of this chapter.

[2] *Martin and Associates, Fort Washington, MD.*
[3] *Defense Energy Support Center, Fort Belvoir, VA.*

domestic and small industrial burners), and kerosine used for off-highway purposes. On-highway middle distillates must contain lower sulfur content (0.05 mass percent max) and are not required to be dyed. Other environmental regulatory changes impacting heating fuels are the Tier II Highway Diesel Fuel Sulfur Control Requirements–Final Rule, which took effect in 2006. This rule limits the sulfur content in highway diesel fuel to 15 ppm. The reason this regulation is important to heating fuels is because it also establishes a new definition for diesel, which now reads, "kerosine and any other distillate product,

TABLE 1—Detailed Fuel Oils Requirements[A]

Property	ASTM Test Method[B]	No. 1 S500[B]	No. 1 S5000[B]	No. 2 S500[B]	No. 2 S5000[B]	Grade No. 4 (Light)[B]	No. 4	No. 5 (Light)	No. 5 (Heavy)	No. 6
Flash point, °C, min	D93 – Proc. A	38	38	38	38	38	...	...	...	...
	D93 – Proc. B	...	...	...	...	...	55	55	55	60
Water and sediment, percent volume, max	D2709	0.05	0.05	0.05	0.05	...	...	...	...	...
	D95 + D473	...	...	...	...	(0.05)[C]	(0.05)[C]	(1.00)[C]	(1.00)[C]	(2.00)[C]
Distillation temperature, °C	D96									
10 percent volume recovered, max		215	215	...	...					
90 percent volume recovered, min		...	...	282	282					
90 percent volume recovered, max		288	288	338	338					
Kinematic viscosity at 40°C, mm^2/s	D445									
min		1.3	1.3	1.9	1.9	1.9	>5.5	...	...	...
max		2.4	2.4	4.1	4.1	5.5	24.0[D]			
Kinematic viscosity at 100°C, mm^2/s	D445									
min		...	...	...	...	...	...	5.0	9.0	15.0
max		...	...	...	...	...	...	8.9[D]	14.9[D]	50.0[D]
Ramsbottom carbon residue on 10 % distillation residue percent mass, max	D524	0.15	0.15	0.35	0.35	...	...	...	...	...
Ash, percent mass, max	D482	...	...	...	...	0.05	0.10	0.15	0.15	...
Sulfur, percent mass max[E]	D129	...	0.50	...	0.50	...	...	...	...	...
	D2622	0.05		0.05						
Copper strip corrosion rating, max, 3 h at a minimum control temperature of 50°C	D130	No. 3	No. 3	No. 3	No. 3	...	...	...	...	...

(Continued)

TABLE 1—Detailed Fuel Oils Requirements[A] (Continued)

Property	ASTM Test Method[B]	No. 1 S500[B]	No. 1 S5000[B]	No. 2 S500[B]	No. 2 S5000[B]	Grade No. 4 (Light)[B]	No. 4	No. 5 (Light)	No. 5 (Heavy)	No. 6
Density at 15°C, kg/m³	D1298									
min		...	...	...	...	>876[F]	...	...	...	...
max		850	850	876	876	...	...	...	...	...
Pour point °C, max[G]	D97	−18	−18	−6	−6	−6	−6	...	...	[H]

[A] It is the intent of these classifications that failure to meet any requirement of a given grade does not automatically place an oil in the next lower grade unless in fact it meets all requirements of the lower grade. However, to meet special operating conditions, modifications of individual limiting requirements may be agreed upon among the purchaser, seller, and manufacturer.
[B] Under United States regulations, Grades No. 1 S5000, No. 1 S500, No. 2 S5000, No. 2 S500, and No. 4 (Light) are required by 40 CFR Part 80 to contain a sufficient amount of the dye Solvent Red 164 so its presence is visually apparent. At or beyond terminal storage tanks, they are required by 26 CFR Part 48 to contain the dye Solvent Red 164 at a concentration spectrally equivalent to 3.9 lb per thousand barrels of the solid dye standard Solvent Red 26.
[C] The amount of water by distillation by Test Method D 95 plus the sediment by extraction by Test Method D473 shall not exceed the value shown in the table. For Grade No. 6 fuel oil, the amount of sediment by extraction shall not exceed 0.50 mass percent, and a deduction in quantity shall be made for all water and sediment in excess of 1.0 mass percent.
[D] Where low sulfer fuel oil is required, fuel oil falling in the viscosity range of a lower numbered grade down to and including No. 4 can be suplied by agreement between the purchaser and supplier. The viscosity range of the initial shipment shall be identified and advance notice shall be required when changing from one viscosity range to another. This notice shall be in sufficient time to permit the user to make the necessary adjustments.
[E] Other sulfur limits may apply in selected areas in the United States and in other countries.
[F] This limit ensures a minimum heating value and also prevents misrepresentation and misapplication of this product as Grade No. 2.
[G] Lower or higher pour points can be specified whenever required by conditions of storage or use. When a pour point less than −18°C is specified, the minimum viscosity at 40°C for grade No. 2 shall be 1.7 mm²/s and the minimum 90 % recovered temperature shall be waived.
[H] Where low sulfur fuel oil is required, Grade No. 6 fuel oil will be classified as Low Pour (+15°C max) or High Pour (no max). Low Pour fuel oil should be used unless tanks and lines are heated.

that meets the definition of motor vehicle diesel fuel, is subject to the standards and requirements under 40 CFR 80.520." In other words, fuels like kerosine and Burner Fuel Oil Grades 1 and 2, including Nos. 1 and 2 S500 grades, must meet the 2006 maximum sulfur limit of 15 ppm, if these fuels are used as highway diesel fuel or blending components in diesel.

KEROSINE

Kerosine is defined in ASTM D3699, Standard Specification for Kerosine, as a refined petroleum distillate consisting of a homogeneous mixture of hydrocarbons. Kerosine was a major petroleum product, which was used initially for illumination and subsequently for heating. While modern technology has diminished the importance of kerosine, it is still used as a primary source of light in some lesser-developed countries and for standby or emergency lighting in the more advanced areas of the world. Because kerosine is reasonably affordable, portable, and flexible, it is also used for domestic space and water heating, refrigeration, heating of garages and greenhouses, incubation and chicken brooders, and cooking. The discussion of kerosine in this chapter will focus on its two primary uses–illumination and heating. ASTM D3699 covers two grades of kerosine suitable for use in critical kerosine burner applications. Number 1-K is special low-sulfur kerosine suitable for use in non–flue-connected kerosine burner appliances and for use in wick-fed illuminating lamps. Number 2-K is a regular grade kerosine suitable for use in flue-connected burner appliances and for use in wick-fed illuminating lamps.

For heating and illumination purposes, kerosine is obtained by fractional distillation of crude oil between approximately 150°C (300°F) and 300°C (575°F) and constitutes a potential 5–20 percent volume yield, depending on the crude source. (In the United States less than 2 % of the crude oil is used for kerosine.) To decrease smoking, paraffinic stocks are normally used in the manufacture of kerosine for lighting and heating. For the same reason, aromatic stocks and cracked components are avoided.

When low-sulfur paraffinic crudes are fractionated to yield the proper boiling range cut for kerosine, only a drying operation may be required to make the product suitable for sale. Usually, however, some objectionable odors (mercaptans) are present, and these are removed by caustic washing or converted to odorless compounds by refining sweetening processes.

Kerosine from naphthenic or high-sulfur crudes requires hydro-treating, acid treatment, and water washing, or extraction with a solvent and caustic wash and clay brightening to remove undesirable aromatics or sulfur compounds. Following treatment, the kerosine streams are blended to meet specifications, and the finished product is ready for marketing.

DOMESTIC HEATING OILS

Domestic heating oils are also middle distillate products. Their approximate boiling range generally is from 150°C (300°F) to 400°C (755°F). They can include components from three sources: (1) virgin stocks distilled directly from crude, (2) oils manufactured by catalytic cracking of heavier stocks, and (3) thermally cracked streams. The percentages of these components blended into the heating oil pool will vary considerably among individual refineries and between various countries. For example, domestic heating oil in many areas outside

TABLE 2—Detailed Requirements for Kerosine

Property	ASTM Test Method	Limit[A]
Flash point °C, min	D56	38
Distillation temperature, °C	D86	
10 percent volume recovered, max		205
Final boiling point, max		300
Kinematic viscosity at 40°C, mm^2/s	D445	
min		1.0
max		1.9
Sulfur, percent mass	D1266	
No. 1-K, max		0.04
No. 2-K, max		0.30
Mercaptan sulfur, percent mass, max[B]	D3227	0.003
Copper strip corrosion rating max, 3 h at 100°C	D130	No. 3
Freezing point, °C, max	D2386	−30
Burn quality		
Time of burning	D187	Minimum 16 h continuous after first weighing
Rate of burning	IP 10	18 to 26 g/h after first weighing
Chimney appearance	D187	Maximum light white deposit (at end of test)
Flame characteristics	D187	Maximum variance of flame width – 6 mm
(comparison of properties from beginning and end of test)		Maximum variance of flame height lowered – 5 mm
Saybolt color, min	D156	+16[C]

[A] To meet special operating conditions, modifications of individual limiting requirements, except sulfur, can be agreed upon among purchaser, seller, and manufacturer.
[B] The mercaptan sulfur determination can be waived if the fuel is considered sweet by Test Method D4952.
[C] Appendixes X1.1 and X1.12 contain additional information on color, red dye, and potential application problems.

North America normally consists of straight-run gas oil from the distillation of the crude; in the United States, the straight-run gas oil fraction is usually blended with the appropriate boiling-range material from catalytic cracking processes.

With few exceptions, the components distilled to the proper boiling range for domestic heating oil require some chemical treatment. For example, additional treatment may be required to remove or convert mercaptans to nonodorous disulfides. Following the chemical treatment, the oils are usually washed with water to remove all traces of the treating chemicals. As a finishing step, the oil is dried by clay filtration or coalescing.

Domestic heating oils derived from heavier catalytic and thermally cracked stocks may require more severe treatment to saturate olefins and remove or convert other unstable compounds. This is accomplished commonly by catalytic hydrogen treating (hydro-treating).

Wherever applicable, the various heating oil streams finally are blended to meet manufacturing specifications. At this point, a stabilizing inhibitor (antioxidant) is often added. To provide the flow properties needed in cold climates, pour-point depressant additives or wax crystal modifiers may be used, or low pour-point kerosine may be blended with the fuel oil.

TRANSPORTATION OF DISTILLATE HEATING FUELS

The primary means of transportation of heating fuels to retail market is by refined product pipeline, although ship, barge, or rail may be used. Final delivery is usually by tank truck. Generally, in the United States, common carrier pipelines publish specifications for heating oil grades. Different grade specifications may be published to accommodate varying sulfur content, cloud point, pour point, or dye requirements. Some heating oil shipped by pipeline may be required to meet the U.S. IRS limit of 11.1 mg/L (ppm) of red dye, while other heating oil grades may only require visual presence of red dye according to the U.S. EPA regulations. Pipelines that ship only visibly dyed heating oil may inject dye at the destination to ensure compliance with the higher IRS dye requirement. Pipelines may also require that heating oil meet a higher flash point limit than the ASTM minimum specification, because flash point may degrade in transit through the pipeline to the destination. To protect the pipeline system, some pipelines may inject corrosion inhibitor into the heating oil, while others may require the producer of the heating oil to inject the corrosion inhibitor. Pipelines may also inject pipeline drag reducing agent

(PDRA) to enhance flow rates in the pipeline system. PDRA is a high-molecular-weight polymer that, when injected in small amounts, reduces energy loss in a flowing stream.

RESIDUAL OR HEAVY FUEL OILS

Residual fuel oils have a minimum flash point of 60°C (140°F) and viscosities that generally vary between 60 and 650 mm^2/s (centistokes, cSt) at 50°C. The oil viscosity required to obtain satisfactory atomization varies according to the type of burner used and is approximately 80 cSt for steam atomizing burners and 40–45 cSt minimum for mechanical atomizing burners. Therefore, it is necessary to heat the oil in order to achieve the required viscosity reduction. This is done in a heat exchanger, consisting of a tube assembly through which the oil flows, placed inside a shell containing the heating medium. Usually, the medium is steam, at a pressure of 138 kPa (20 psig) or less. Generally, the heating is done in two steps. The first step heats the oil as it is withdrawn from the storage tank, or in some cases, the whole tank of oil is heated to about 38°C (100°F) so that its viscosity is within the pumping range of 850–1,100 cSt. In the second step, transfer pumps deliver the oil from storage to the burner heating and pumping equipment, at which point it is raised to the required temperature and pressure for atomization.

USED LUBRICATING OIL AS BURNER FUEL

Another increasing trend in the fuel oil arena is the blending of used lubricating oils in burner oils to make industrial fuel oils. ASTM D6448, Standard Specification for Industrial Burner Fuels from Used Lubricating Oils, defines four grades of fuel oils made in whole or in part with hydrocarbon-based used or reprocessed lubricating oil or functional fluids, such as preservative and hydraulic fluids. The grades of fuel are intended for use in various types of fuel oil burning industrial equipment under various climatic and operating conditions. These fuels are not intended for use in residential heaters, small commercial boilers, or combustion engines. Grades RFO4, RFO5L, RFO5H, and RFO6 are used lubricating oil blends, with or without distillate or residual fuel oil, or both, of increasing viscosity and are intended for use in industrial boilers equipped to handle these types of recycled fuels. The designation RFO identifies them as Reprocessed Fuel Oils. "L" and "H" stand for "Light" and "Heavy," respectively.

BURNING EQUIPMENT

To understand fuel oil tests and their significance, it is necessary to have a fundamental knowledge of the various types of burning equipment associated with kerosine, domestic heating oils, and residual oils. The following paragraphs provide a brief review of the subject.

Kerosine Burners

Although the appliances in which kerosine is used vary widely in design and efficiency, there are three main burner types: wick-fed yellow flame, wick-fed blue flame, and the pressure burner.

The wick-fed yellow-flame type burner consists essentially of an oil reservoir into which one end of a wick is immersed. The other end passes through a wick guide and projects upward from it. A draft deflector and a chimney with air inlet are provided. Oil flows by capillary action to the top of the wick, where it burns and produces a luminous yellow flame. The flame size may be adjusted by turning the wick up or down, thus exposing more or less area. Either flat or tubular wicks are used. Apart from its domestic use in kerosine lamps and small portable stoves for heating and cooking, the wick-fed yellow-flame type burner in simple form is still used in brooders and incubators in the poultry industry.

In the wick-fed blue-flame type burner, a long drum or chimney is mounted over the burner to induce the air for combustion. The burner itself includes a circular wick and a flame spreader. A well-known application of the wick-fed blue-flame method is the perforated sleeve vaporizing burner. This design is known commonly as the "range burner," because of its early widespread use in kitchen ranges. (Kerosine is sometimes referred to as "range oil.")

A typical range burner consists of a flat, cast iron, or pressed steel base with concentric interconnected grooves and concentric perforated metal sleeves where combustion takes place. Kerosine is maintained at a depth of 3–6 mm ($^1/_8$–$^1/_4$ in.) in the grooves. Ceramic fiber wicks, lighted manually, are used to provide heat for oil vaporization and ultimate ignition of the oil vapors. As the base heats up, the entering oil vaporizes from the surface, and the flame lights from the wicks. Combustion air is induced by natural draft. The flame is blue, and operation is essentially silent, odorless, and smokeless. A flue usually is provided to remove combustion products. Constant level valves, or other devices such as an inverted bottle based on the "chicken feeder" principle, are required for the kerosine feed.

Appliances of the wick-fed, blue-flame type are used primarily for heating and cooking purposes. They also are used with incandescent mantles for lighting. The design is such that intimate mixing of air and oil vapor takes place. This results in more complete combustion, whereby the oil burns with an almost nonluminous blue flame.

Blue-flame type kerosine water heaters and vaporizing pot burners will be discussed in the section on domestic heating oil burners, because they can also burn domestic heating oils.

With the pressure burner design, the oil reservoir is fitted with a pump, which enables pressure to be maintained. This pressure forces the oil up into a central tube, through a previously heated vaporizing coil, and out through a jet. The issuing oil vapor then mixes with air drawn in from the outside, and the mixture passes to the burner where combustion takes place.

The principle of the pressure-type burner is employed in the Primus stove, various kinds of blow-lamps, certain pressure burners fitted with mantles for illumination, and for a variety of minor industrial applications.

Domestic Heating Oil Burners

For domestic heating oil to burn readily and satisfactorily, fuel must be vaporized and mixed with the amount of air required to provide a proper combustible mixture. This vaporization can be accomplished either through the application of heat or atomization into very fine particles. Domestic heating oil burners are classified according to the manner in which the vaporization and subsequent burning occurs. One of the simplest domestic burners is the vaporizing pot burner. Although no longer manufactured, these burners are still in use in room space heaters, water heaters, and for central heating in warm air furnaces. Both natural and mechanical drafts are employed.

A typical pot burner consists of a metal pot perforated with holes for the entrance of combustion air. Oil flows into the bottom of the pot by gravity and is vaporized from the hot surface. After mixing with primary air that enters the lower ports and additional secondary air in the upper section of the pot, the vapors finally burn at the top of the vessel. Between periods of demand or "high-fire" operation, most pot burners idle at a "low-fire" condition. Using only primary air, the flames burn in the bottom section of the pot at a fraction of the high-feed rate and provide the ignition source required for automatic operation.

A typical water heater fired with a natural draft-pot burner requires a relatively high draft [15 Pa (0.06 in. of water) air pressure or more] to provide adequate combustion air. The thermostat controls the fuel flow to give high-fire operation for heating and low-fire operation when the water is at design temperature.

By far the most common of the domestic heating oil burners in the United States and Europe is the high-pressure atomizing gun burner. It is the most robust and among the cheapest of the domestic heating oil burners. In this type of burner, oil is supplied to a high-pressure simplex nozzle, at 552–1380 kPa (80–200 psi). The high-pressure oil is accelerated in tangential slots to a high velocity in the nozzle's swirl chamber. Upon leaving the nozzle's orifice, the oil breaks into a fine, cone-shaped spray of droplets. Besides the atomizing function, the nozzle also meters the oil. These burners are manufactured in either a yellow-flame or a blue-flame configuration.

In a yellow-flame high-pressure gun burner, pump pressure is controlled by a pressure-regulating valve, which bypasses excess oil back to the pump inlet or tank. A fan mounted on the same shaft as the pump supplies combustion air. The air flows through the blast tube in which the nozzle is centered. At the exit end of the blast tube, a "spinner" or retention head is mounted to aerodynamically stabilize the resulting flame. The ignition electrodes are located slightly behind the spray cone. When power is supplied, the fan and pump begin to rotate and a high-voltage spark begins between the electrodes. The regulating valve opens after the atomizing pressure is reached. By this time, the air stream is blowing the center of the spark into the oil spray, and the oil is ignited.

Blue-flame high-pressure gun burners are typically fitted with a downstream flame tube with perforations around the base of the flame. The purpose is to induce recirculation of the combustion gases that allow the fuel to vaporize and burn completely with a blue flame. Air and fuel pressures tend to be higher than those required for a yellow-flame burner.

The low-pressure atomizing burner is a far less common and more expensive device but offers some advantages over the high-pressure gun. Atomization is accomplished in a siphon or air-atomizing nozzle. Oil is supplied at relatively low pressure [7–103 kPa (1–15 psi)]. The high velocity of the primary air stream is used to shear off the incoming oil at right angles. A finer spray results than is obtainable from high-pressure nozzles, and this creates a lower smoking tendency. Between 2 and 15 % of the combustion air is supplied as primary air. Secondary air usually is provided around the nozzle, as in the high-pressure gun burner. The oil is metered by means of a metering pump or an orifice in the oil line.

Because the oil passages in a low-pressure siphon nozzle are not as small as those required for the high-pressure simplex nozzle (atomizer), the low-pressure type of burner is less prone to plug with particulate contamination (usually unfiltered fuels). This enables the low-pressure burner to function at 1–2 L/h (0.25–0.50 gal), while high-pressure burners, because of a tendency for particulate contamination, are generally not operated below 2 L/h (0.50 gal).

Residual Oil Burning Equipment

In general, residual oil burning equipment is categorized according to the method by which the fuel is atomized. Described in the following paragraphs are air, steam, mechanical atomizing burners, and rotary-type burners.

Air atomizing burners are used largely in industrial furnaces. Because atomizing air mixes intimately with the oil and comprises part of the required combustion air, burning begins more quickly and is completed sooner with this type than with any other type of burner. This results in a shorter flame and permits a smaller combustion space.

Inside-mix air atomizing burners are designed so that all atomizing air enters at the air inlet and is controlled simultaneously with the oil by a single control lever. Primary air enters a swirl chamber tangentially, which imparts a rotating motion to the air. As the air approaches the oil nozzle, it attains maximum velocity because of the venturi shape of the throat. The impact of the air atomizes the oil as it emerges from the nozzle. The resultant air/oil mixture leaves the mixing nozzle in the form of a divergent cone. It comes into contact with the secondary air from an outer nozzle and is further atomized, thus assisting the mixing process.

Steam atomizing burners are two basic types: inside mix and outside mix. As the names suggest, the primary difference lies in where the mixing occurs. In the inside-mix type, the steam and oil mix within the burner nozzle prior to entering the furnace. Additionally, the flame from inside-mix burners may be either flat or conical, depending on disposition of the burner openings, while the outside-mix burners, which were the first type of heavy oil burners, have a flat flame.

Both the inside and outside-mix burners have relatively fixed-ratio nozzles, which operate best at relatively constant fuel and air supply rates. Because of this limitation, simple inside and outside-mix burners are being used less frequently. More sophisticated atomizing burners that yield efficient combustion over a range of firing rates and loads, even with very low fuel grades, have become available.

In the mechanical atomizing burner, atomization of the fuel oil is achieved by forcing high-pressure, high-velocity oil into the furnace through a small bore orifice or sprayer plate. Required pressure is approximately 2,070 kPa (300 psi) or more.

The important part of any mechanical atomizing burner is the sprayer plate. The oil passes through slots in the plate at high velocity. The tangential arrangement of the slots imparts a rotating motion to the oil. The resulting centrifugal force causes the oil to break up into a hollow atomized cone as it enters the furnace through the central orifice.

The firing range of boilers equipped with mechanical atomizing burners can be changed by (1) changing the number of burners in service, (2) replacing the sprayer plates with plates having another size central orifice, (3) altering the burner design so that more or less oil is allowed to flow through the same size sprayer plate, and (4) changing the oil pressure and method of control in conjunction with a burner design alteration similar to that described in the third point.

In the rotary-type burners, atomization is achieved through centrifugal force imparted to the oil by a cup rotating at high speed. The atomized oil issues from the rotating cup in the form of a hollow cone. The primary air supply enters the furnace concentrically with this cone and mixes with the oil to form a conical spray. Rapid mixing of the atomized oil and primary air is obtained. The air supplied by the primary air fan is a minor part of the air required for combustion. The secondary air is supplied by either the draft induced by the stack or an additional force-draft fan and enters the furnace through an air register in the furnace wall.

The rotary burner may consist of a drive motor connected to a shaft that drives the rotating cup, the fuel oil pump, and the primary air fan, or the primary air and fuel oil may be supplied by an independently driven unit. The oil viscosity required for the rotating cup burner varies between 32–76 cSt at 50°C.

FUEL OIL CLASSIFICATION AND SPECIFICATIONS

Because the quality and general performance requirements for various applications of fuel oils differ widely, many countries have adopted general quality limitations for various fuel grades. These serve as guides in the manufacture, sale, and purchase of the fuel oils. While these quality definitions are sometimes called "specifications," they are more properly "classifications" because of their broad, general nature. In addition to these classifications or general specifications, there may be more precise specifications of quality requirements for any given application. These may be dictated by competitive considerations, customer needs, or government agencies.

In general, specifications or classifications are usually based on minimum equipment manufacturers' performance requirements and serve primarily as a common basis for agreement between producer, distributor, and consumer. These classifications divide the basic types of oils into broadly defined grades. Broad tolerances permit a customer to select a grade suitable for the intended purpose, but they do not define the ideal fuel for a particular application. At the same time, the consumer is assured of, at least, minimum performance quality, and there is no undue restriction on the supply of acceptable products.

As an example, the detailed requirements for ASTM Standard Specification for Fuel Oils (D396) are shown in Table 1. Because ASTM D396 is revised periodically, the current ASTM Standard should be consulted if an up-to-date specification is needed.

FUEL OIL LABORATORY TESTS AND THEIR SIGNIFICANCE

Any property prescribed in a product's specification should be related to product performance or of value in the refining or handling of the product. Specific tests have been designed to determine to what degree a given product meets stated specifications. The test procedures commonly applied to heating and power generation fuels and their significance are summarized in the following paragraphs.

API and Specific Gravity or Density

The gravity of a fuel oil is an index of the weight of a measured volume of the product. There are two scales in use in the petroleum industry: specific gravity or density: ASTM Standard Test Method for Density, Relative Density (Specific Gravity), or API Gravity of Crude Petroleum and Liquid Petroleum Products by Hydrometer Method (D1298/IP 160) and American Petroleum Institute (API) Gravity: ASTM Standard Test Method for API Gravity of Crude Petroleum and Petroleum Products (Hydrometer Method) (D287/IP 192).

Specific gravity of a fuel oil is the ratio of the weight of a given volume of the material at a temperature of 15.6°C (60°F) to the weight of an equal volume of distilled water at the same temperature, both weights being corrected for the buoyancy of air. Specific gravity is seldom used in the United States, but it is in general use in some foreign countries. API gravity of a fuel oil is based on an arbitrary hydrometer scale, which is related to specific gravity in accordance with the formula

$$\text{API gravity, deg} = \frac{141.5}{@15^\circ \text{ density k/m}^3} - 131.5 \qquad (1)$$

This scale generally is used for most transactions in the United States as well as in refinery practice.

ASTM Standard Guide for Petroleum Measurement Table (D1250/IP 200) contains tables showing equivalent specific gravity, lb/U.S. gal, and U.S. gal/lb at 15.6°C (60°F) for each tenth of a degree API from 0 to 100° API.

Density is the mass (weight in vacuo) of a unit volume of fuel oil at any given temperature [15.6°C (60°F) as determined in ASTM D1298/IP 160]. ASTM D4052, Standard Test Method for Density and Relative Density of Liquids by Digital Density Meter, can be used as an alternative.

Gravity by itself is of limited significance as an indication of fuel oil quality. It is used by the refiner in the control of refinery operations and has significance to a customer who needs information on net or gross energy release to use in combustion system calculations.

On a weight basis, the heating value of petroleum fuels decreases with increasing specific gravity or density (decreasing API gravity), because the weight ratio of carbon (low-heating value) to hydrogen (high-heating value) increases as the specific gravity increases. On a volume basis, the increasing specific gravity more than compensates for the decreasing heating value per unit weight, with the net result that fuels having high specific gravity yield more heat energy per unit volume than those of low specific gravity. Because many fuels are purchased on a volume basis, the relationship of higher density fuels containing more energy is very significant to some purchasers.

Flash and Fire Point

The flash point of a fuel is a measure of the temperature to which the fuel must be heated to produce an ignitable vapor-air mixture above the liquid fuel when exposed to an open flame. Flash point is normally included in industry specifications for fuel oil. The minimum flash point for many fuel oils is 38°C or 40°C, for safety reasons. The "fire point" of a fuel is defined as the temperature when an oil in an open container gives off vapor at a sufficient rate to continue to burn after a flame is applied.

Flash point is used primarily as an index of fire hazards. As an example, the U.S. Department of Transportation shipping regulations use flash point as the criterion to establish labeling requirements. Consequently, most industry specifications or classifications place limits on the flash point to ensure compliance with fire regulations, insurance, and legal requirements. All fuel oils described in this section should have a minimum flash point of 38°C or higher. The purpose of this minimum flash point restriction is to reduce fire or explosion hazards during handling, transportation, and in the event of a leak or spill in the vicinity of furnaces or flames.

Fuels with low flash points (below 38°C), such as gasoline, must be used in burners designed for normal fuel oils because of the risk of explosion, especially during initial ignition of the burner. Similarly, fuel oils contaminated with a small amount of gasoline, resulting in a low flash point, may present a serious safety hazard in handling and use. Therefore, determination of flash point can be useful in detecting such contamination and, thereby, avoiding serious safety hazards.

A furnace fuel oil with a very high flash point, such as 90 to 120°C, may be difficult to ignite in some burner applications such as home furnaces. Thus, some fuel oil specifications may include a maximum flash point restriction to ensure adequate lightability.

Depending on the apparatus and method of test used, the determined flash point of a fuel may vary. In all cases, however, it is an empirical value because the conditions of the test do not match all possible conditions of commercial handling of the product. Most flash point tests are corrected to standard conditions of temperature and pressure, and assume normal atmospheric oxygen concentration. Thus, many flash point tests may have useful correlation to actual fire hazards in normal situations. However, there may be little or no correlation of flash test results to fire hazards where oxygen concentration is appreciably different from normal, such as at high elevations (lower oxygen concentration) or at subsurface situations, such as deep mines, or applications under increased pressure. Higher oxygen pressure may significantly reduce actual flash point temperatures and, thus, increase fire or explosion hazards.

The ASTM Standard Test Method for Flash Point by Pensky-Martens Closed Tester (D93/IP 34) and the ASTM Standard Test Method for Flash Point by Tag Closed Tester (D56) are normally used for determining the flash point of fuel oils, because these procedures more nearly approach the conditions of storage of the fuel in tanks. ASTM D56 may be used as an alternative for Grades 1 and 2, provided the flash point is below 93°C and the viscosity is below 5.5 mm^2/s at 40°C. However, D56 will give slightly lower values. ASTM D3828, Standard Test Method for Flash Point by Small Scale Closed Tester, is another test that can be used for all fuel oil grades. Flash point by ASTM D93 is cited by some fire safety regulations.

Viscosity

The viscosity of a fluid is a measure of its resistance to flow and is generally expressed as kinematic viscosity in mm^2/s (1 mm^2/s = 1 cSt). Alternate units include Saybolt Universal seconds and Saybolt Furol seconds.

The currently accepted procedure for determining viscosity is kinematic, ASTM Standard Test Method for Kinematic Viscosity of Transparent and Opaque Liquids (and the Calculation of Dynamic Viscosity) D445/ IP 71, with a test temperature of 40°C (KMT) or 50°C (122T). ASTM D445/IP 71 is shown as the preferred viscosity procedure in ASTM D396. D396 specifies 40°C for Grades 1, 2, and 4 and 100°C for Grades 5 and 6.

The determination of residual fuel oil viscosity is complicated by the fact that some fuel oils containing significant quantities of wax do not behave as simple Newtonian liquids in which the rate of shear is directly proportional to the shearing stress applied. At temperatures in the region of 38°C (100T), residual fuels tend to deposit wax from solution. This wax deposition exerts an adverse effect on the accuracy of the viscosity result, unless the test temperature is raised sufficiently high for all wax to remain in solution. By using the kinematic system, the petroleum industry today has moved toward 100°C (212T) as the standard test temperature for the majority of residual fuels for viscosity determination.

Viscosity is one of the more important heating oil characteristics. It is indicative of the rate at which the oil will flow in fuel systems and the ease with which it can be atomized in a given type of burner. Because the viscosities of heavier residual fuel oils are high, this property tends to be particularly relevant to their handling and use. The viscosity of a heavy fuel decreases rapidly with increasing temperature. For this reason, heavy fuels can be handled easily and atomized properly by preheating before use. If no preheating facilities are available, lighter or less viscous oils must be used, and, if the preheating equipment is inadequate, it may be necessary to burn a lighter oil during cold weather.

Overly viscous oil can produce problems throughout the system. Besides being difficult to pump, the burner may be hard to start, and flashback or erratic operation may be encountered. Viscosity also affects the output or delivery of a spray nozzle and the angle of spray. With improper viscosity at the burner tip, poor atomization can result in carbonization of the tip, carbon deposition on the walls of the fire box, or other conditions leading to poor combustion. High viscosity could result in poor atomization (larger droplets). Such coarser droplets could impinge on the wall of the fire box, resulting in a localized "hot spot," leading to failure of the fire box wall.

Typically, the lower limit of viscosity for easy pumpability is reached at about 1,100 cSt. Difficulties in moving the oil in fuel oil systems will be encountered below the temperature at which this viscosity is reached.

For good atomization in pressure and steam atomizing burners, a viscosity range of 21–43 cSt at the burner is considered desirable. Frequently, a viscosity at the burner of 16–18 cSt is recommended with low-pressure air atomizing burners. Rotary cup burners are less critical to atomizing viscosity and can operate satisfactorily with viscosities at the burner cup as high as 75–87 cSt.

Pour Point

Pour point, ASTM Standard Test Method for Pour Point of Petroleum Oils (D97/IP 15), is defined as the lowest temperature at which the oil will just flow under standard test conditions. Automatic test methods D5949, Standard Test Method for Pour Point of Petroleum Products–Automatic Pressure Pulsing Method, D5950, Standard Test Method for Pour Point of Petroleum Products–Automatic Tilt Method, and D5985, Standard Test Method for Pour Point of Petroleum Products–Rotational Method, can be used for testing pour point for all fuel oil grades. Anticipated storage conditions and fuel application are usually the primary considerations in the establishment of pour point limits. Storage of higher-viscosity fuel oils in heated tanks will permit higher pour points than would otherwise be possible.

The pour point is important for deliverability of lighter fuel oils that are not normally heated, such as home heating oil. While it does take time for a large volume of oil to cool in very cold weather, trying to deliver furnace fuel oil at temperatures below the pour point of the fuel could result in the fuel "waxing" or "gelling" (that is, turning solid) in the tank truck or manifold, so that it cannot be pumped off a delivery truck into a fuel storage tank.

While the failure to flow at the pour point normally is attributed to the separation of wax from the fuel, it also can be due to the effect of high viscosity in the case of very viscous oils. In addition, pour points, particularly in the case of residual fuels, may be influenced by the previous thermal history of the oils. As an example, any loosely knit wax structure built up on cooling of the oil can normally be broken by the application of relatively little pressure. Thus, the usefulness of the pour point test in relation to residual fuel oil is open to question, and the tendency to regard pour point as the limiting temperature at which a fuel will flow can be misleading, unless correlation is made with low temperature viscosity.

Although the pour point test is still included in many specifications, it is not designated for the heavier fuels in ASTM D396 Grades 5 (Light and Heavy) and Grade 6. Higher pour point fuels may be allowed where heated storage and adequate piping facilities are provided. The technical limitations of pour point have motivated efforts to devise a satisfactory alternative for the pour point test in assessing the low temperature pumpability characteristics of heavy fuel oils. ASTM D3245, Standard Test Method for Pumpability of Industrial Fuel Oils, is such an alternative, which is designed to give an indication of minimum handling temperatures for Grades 4 (Light), 4, 5 (Light), 5 (Heavy), and 6, or similar fuels.

Cloud Point

Middle distillate fuels, especially, begin to form wax crystals and become cloudy in appearance as they are cooled toward the pour point. The temperature at which this begins to occur is called the cloud point, ASTM Standard Test Method for Cloud Point of Petroleum Oils (D2500/IP 219). Cloud points often occur at 4–5°C (7–9°F) above the pour point, and temperature differentials of 8°C (15°F) or more are not uncommon. Basically, the temperature differential between cloud and pour point depends on the nature of the fuel components, but the use of wax crystal modifiers or pour depressants tends to accentuate these differences.

As the temperature continues to decrease below the cloud point, the formation of wax crystals is accelerated. These crystals may clog fuel filters and lines and, thus, reduce the supply of fuel to the burner. Because the cloud point is at a higher temperature than the pour point, it can be considered even more important than the pour point in establishing distillate fuel specifications for cold weather usage, especially with newer, high-pressure burners equipped with fine filtration (e.g., 10-μm filters).

Sediment and Water Content

The tests for water and insoluble solid content of middle distillate fuel oils are important to the consumer and among those most commonly applied. Contamination by water and sediment can lead to filter and burner problems and the production of emulsions, which are removable only with difficulty. The corrosion of storage tanks also may be associated with water bottoms that accumulate from atmospheric condensation and water contamination. In addition, the presence of water promotes microbial contamination (refer to ASTM D6469, Standard Guide to Microbial Contamination in Fuels and Fuel Systems). Water in fuel oil can lead to delivery or flow problems when the temperature of fuel drops below 0°C. Free water will freeze, and may freeze a valve shut so it cannot be opened, or may form ice crystals that will plug a screen in the off-loading line. Similarly, water that was dissolved in fuel at a warmer temperature, such as +5 to +15°C, may separate out as ice crystals as the fuel is cooled to low temperatures, such as −20 or −30°C. Note that the solubility of water in fuel oils is related to temperature: hot oil will dissolve much more water (perhaps 100–300 ppm), whereas cold fuels may only be able to hold 10–20 ppm dissolved water.

Water content can be determined by a distillation procedure, ASTM Standard Test Method for Water in Petroleum Products and Bituminous Materials by Distillation (D95/IP 74), and sediment can be estimated by a method, ASTM Standard Test Method for Sediment in Crude and Fuel Oils by Extraction (D473/IP 53), which involves toluene extraction of the oil through a refractory thimble where the insoluble sediment is retained in the thimble. A density of 1.0 kg/L shall be used for ASTM D95. Total water and sediment can be determined together by a centrifuge procedure for Grades 1 and 2, as well as Grades 1 and 2 Low Sulfur, using ASTM Test Method for Water and Sediment in Middle Distillate Fuel Oils by Centrifuge (D2709), but separate determinations of water and sediment generally are more accurate. ASTM D2709 determines the gross volume of free water and sediment in middle distillate fuels. Using a specifically designed centrifuge tube, a 100-mL sample of the undiluted fuel is centrifuged at a relative centrifugal force of 800 for 10 min at 21–32°C (70–90°F) in a centrifuge tube readable to 0.005 (mL) and measurable to 0.01 mL. After centrifugation, the volume of water and sediment that has settled into the tip of the centrifuge tube is read to the nearest 0.005 mL and reported as the volumetric percent age of water and sediment by centrifuge.

Ash Content

Ash content may be defined as that inorganic residue that remains after combustion of the oil in air at a specified, high temperature. To measure ash content, a joint ASTM/IP test method, ASTM Standard Test Method for Ash from Petroleum Products (D482/IP 4), is used.

Ash-forming materials found in residual fuels are derived normally from the metallic salts and organometallic compounds found in crude oils. Because ash-forming constituents of crude oil ultimately concentrate in the distillation residue, distillate fuels tend to contain only negligible amounts of ash. However, both distillate and residual fuels may pick up ash contributors during transportation from the refinery. Water transportation, in particular, presents many opportunities for fuel oils to be contaminated with ash producers, such as seawater, dirt, scale, rust, etc.

The total ash content in different residual fuels is normally less than 0.2 % by weight. In composition, the ash will vary among residual fuels, largely as a function of their crude oil antecedents, but metallic compounds such as sodium, vanadium, nickel, iron, and silica generally are present.

Depending on the use of the fuel, ash composition has a considerable bearing on whether detrimental effects will occur. Ash in heavy fuel oils can cause slagging or deposits and high-temperature corrosion in boilers; it may attack refractories in high-temperature furnaces, kilns, etc., and it may affect the finished product in certain industrial processes, such as ceramic and glass manufacture.

Because most of the ash in heavy fuels occurs naturally, it usually is difficult for the refiner to remove it economically. Purchasing different crude oils with lower ash content can be very expensive. Therefore, methods have been developed for counteracting the effects of ash. These include the use of additives, modifications in equipment design, and the application of fuel processing methods such as water washing.

Carbon Residue

The ASTM Standard Test Method for Ramsbottom Carbon Residue of Petroleum Products (D524/IP 14), sometimes called "carbon residue," measures the relative coke-forming tendency or the carbonaceous residue (or both) and mineral matter remaining after destructive distillation of a fuel oil under specified conditions. The carbon residue of burner fuel serves as a rough approximation of the tendency of the fuel to form deposits in vaporizing pot-type and sieve-type burners, where the fuel is vaporized in an air-deficient atmosphere.

Little is known concerning the relationship of carbon residue to fuel performance when applied to residual fuel oils. Pressure jet and steam atomizing type burners are not very sensitive to the carbon residue of the fuel used. In well-designed installations incorporating such burners, and where combustion efficiency is maintained at a high level, it is unlikely that difficulties would normally arise in burning residual fuel oils. Under such circumstances, therefore, it is debatable whether the carbon residue test has any real significance relative to the combustion characteristics of residual fuel.

With middle distillate fuel oils, high-carbon residue appears to cause rapid carbon buildup and nozzle fouling in certain types of smaller automatic heating units. In the vaporizing pot-type burner described previously, the oil is brought into contact with a hot surface, and the oil vapor subsequently is mixed with combustion air. Any carbonaceous residue formed by the decomposition of the oil or by any incomplete vaporization is deposited in or near the vaporizing surface, with resultant loss in burner efficiency. Such a burner, therefore, can be operated satisfactorily only on distillate fuel oils having low carbon-forming tendencies.

To improve the accuracy of carbon residue determination for light distillate fuel oils, which form only small amounts of carbonaceous deposits, the carbon residue value is measured on the 10 % residue obtained by an adaptation of a standard distillation procedure, ASTM Standard Test Method for Distillation of Petroleum Products (D86/IP 123). This procedure is applicable to light distillate oils, such as ASTM Nos. 1 and 2 fuel oils, where the light fuel oil is concentrated by distilling off 90 % of the sample and performing the carbon residue test on the remaining "10 % bottoms" portion from the distillation.

The carbon residue test is, therefore, a useful means for approximating the deposit-forming tendencies of distillate type fuel oils used in home heating installations. For this purpose, maximum values for Ramsbottom carbon residue (10 % residue) are given as 0.15 weight percent for ASTM Grades No. 1 Low Sulfur and No. 1, and 0.35 weight percent for ASTM Grades No. 2 Low Sulfur and No. 2, as shown in Table 1.

Sulfur Content

The sulfur content of fuel oils can be determined by a variety of methods. The ASTM Standard Test Method for Sulfur in Petroleum Products (General Bomb Method) (D129/IP 61) and ASTM Standard Test Method for Sulfur in Petroleum Products (High Temperature Method) (D1552) or Quartz-Tube Method (IP 63) have long been established. Other more rapid techniques are also available. These include ASTM D1552, ASTM D2622, Standard Test Method for Sulfur in Petroleum Products (X-Ray Spectrographic Method), and ASTM D4294, Standard Test Method for Sulfur in Petroleum Products by Energy Dispersive X-Ray Fluorescence Spectroscopy. In addition, ASTM D1266 can be used for Grades 1 and 1 Low Sulfur, but only with samples having sulfur contents of 0.4 mass percent and less (down to 0.01 mass percent).

Fuel oils contain varying amounts of sulfur (as organic compounds of sulfur), depending on the crude source, refining processes, and fuel grade. The high boiling range fractions and the residual fuels usually contain higher amounts of sulfur, which is generally regarded as an undesirable constituent because of its potential to create corrosion and pollution problems.

In boiler systems, the conversion of even a small fraction of the sulfur to sulfur trioxide during combustion of the fuel can cause low temperature corrosion problems if this gas is allowed to condense and form corrosive sulfuric acid on cool metal surfaces of the equipment (exhaust piping). In combination with sodium and vanadium complexes, the sulfur from the fuel contributes to the formation of deposits on external surfaces of superheater tubes, economizers, and air heaters. These deposits cause corrosion of equipment and loss of thermal efficiency (heat transfer).

While desulfurization is being used to reduce the sulfur limits of residual fuel oils, efforts have also been directed toward other means of reducing the effects of acid condensation. One effective procedure is to operate with a minimum of excess air, provided that the air/fuel mixture is not too close to the smoke point. Another means of reducing low temperature corrosion is to maintain the cooler metal surfaces of the equipment above the acid dew point; however, this reduces thermal efficiency. Injection of ammonia into the flue gases and the addition of neutralizers are also methods used to minimize corrosion, but these are costly when used in the required concentrations.

Combustion of sulfur-containing fuel oils produces sulfur oxides, which have been identified as atmospheric pollutants. To meet clean air standards in densely populated industrial areas, stack emission control devices and sulfur scrubbing procedures may be required.

ASTM D396 sets maximum sulfur limits for the four types of Grades 1 and 2 fuels, but does not include any such limits for the heavier grades. Limited sulfur content of fuel oil can be required to reduce heat exchanger and flue corrosion and deposition, and for special uses in connection with heat treatment, nonferrous metal, glass, and ceramic furnaces, or to meet federal, state, or local legislation or regulations. Where sulfur content is considered critical, agreement on limits generally is made between the supplier and consumer to meet locally established standards.

When refinery desulfurization processes are used to lower the sulfur content of residual fuels, a lower viscosity fuel (Grades 4 or 5) is usually produced. Also, to meet sulfur limits, fuels with higher pour points, such as 15°C (60°F) or higher, are being marketed. When the latter fuels are used, adequate preheating facilities are required to ensure fluidity and pumpability.

Distillation

The distillation procedure (ASTM D86/IP 123) measures the amount of liquid vaporized and subsequently condensed as the temperature of the fuel in the distillation flask is raised at a prescribed rate. A record is made of the volume of distillate collected at specified temperatures or, conversely, the temperature at each increment of volume distilled (usually 10 % increments). The temperature at which the first drop of condensate is collected is called the "initial boiling point." The "end point" usually is the highest temperature recorded as the bottom of the flask becomes dry. If oils are heated above 370°C (700°F), they tend to crack and give unreliable results. The test usually is stopped when this point is reached. Some distillations may be run under a high (10 mm Hg) vacuum in order to avoid cracking, such as using ASTM Standard Test Method for Distillation of Petroleum Products at Reduced Pressure (D1160). Under these conditions, heavy fuel oils may be distilled up to temperatures equivalent to 510°C (950°F) at atmospheric pressure. However, distillation tests generally are not applied to the heavy fuel oils.

The distillation test is significant for the middle distillate fuels. When heating installations that use vaporizing burners or atomizing burners are operated with middle distillate fuel oils, it is essential that the fuels contain sufficient volatile components to ensure that ignition and flame stability can be accomplished easily. In addition, the volatility of the fuel must be uniform from batch to batch if frequent resetting of burner controls is to be avoided and maximum performance and efficiency are to be maintained.

Table 1 outlines distillation limits for all U.S. domestic fuel oils–Grades 1 and 2–which includes Nos. 1 and 2 Low Sulfur Grades. No distillation limits are established for the other grades of fuels.

Corrosion

Tests for corrosion are of a qualitative type and are made to ascertain whether fuel oils are free of a tendency to corrode copper fuel lines and brass or bronze parts used in the burner assemblies. The method specified for Grade 1 (distillate) burner fuel is ASTM Standard Test Method for Copper Corrosion from Petroleum Products by the Copper Strip Tarnish Test (D130/IP 154).

The copper strip corrosion test is conducted by immersing a polished copper strip in a sample of fuel contained in a chemically clean test tube. The tube is then placed in a bath maintained at a temperature of 50°C (122°F) for 3 h. After washing, the strip is then examined for evidence of corrosion and judged by comparison with the corrosion scale in the standard.

Additional Tests

HEAT CONTENT

The heat content or thermal value of the fuel is the amount of heat given off as a result of its complete combustion. The results are usually expressed as kilogram-calorie (kg-cal), British thermal unit (Btu)/lb, or mega joules/liter (MJ/L). One kg-cal is equivalent to 1.8 Btu/lb. Heat content determination is made in a bomb calorimeter under specified conditions, ASTM Standard Test Method for Heat of Combustion of Liquid Hydrocarbon Fuels by Bomb Calorimeter (D240/IP 12).

The energy content of petroleum fuels is measured by the heat of combustion. For most applications, such as fuels used in engines or furnaces, the *net* heat of combustion is the appropriate measure of energy. This value is sometimes called the lower heating value (LHV), net heating value, net calorific value, or specific energy. It assumes that water produced from combustion goes "up the stack" as water vapor. The *Gross* Heat of Combustion, also called the higher heating value (HHV), requires that water vapor be condensed back to liquid water to recover the "heat of vaporization" of water. For most fuels, such as diesel fuel, furnace oil, stove oil, gasoline, and aviation fuels, the net heat of combustion can be estimated directly from the density of the fuel. The density or API gravity should be available on the bill of lading or from the local supplier. While "Btu per Imperial gallon" used to be the measure of heat content, MJ/L is the metric unit.

Experimental heat content or thermal value tests involve careful standardization of equipment and rigid adherence to prescribed details of operation. When properly run, such tests can yield highly accurate results. However, it is time consuming and costly to perform on a routine basis.

ASTM D4868 (Standard Test Method for Estimation of Net and Gross Heat of Combustion of Burner and Diesel Fuels) is useful for estimating the heat of combustion of burner fuels, where it is not critical to obtain very precise heat determinations. This is because this test method uses a minimum number of tests. For normal purposes, the energy content of a particular fuel can be calculated with sufficient accuracy from other routinely measured data. For example, the heat of combustion of middle distillate fuels has been related to the density (specific gravity), sulfur content, and aniline point. The energy content of residual fuels is also related to the gravity corrected for water, ash, and sulfur content. D4868 is also applicable to petroleum fuels that are low in sulfur (below 0.5 mass percent), with negligible water or ash. The heat of combustion on a volumetric basis is used rather than net heat of combustion on an weight (or mass) basis. In North America, most fuel is bought and sold on a volumetric basis, rather than on a weight basis. Also, many fuel injectors inject fuel on a volumetric basis rather than a weight basis. Using ASTM D4868, the energy content of two fuels can be compared readily. For example, two diesel fuels having densities of 0.865 kg/L and 0.835 kg/L will have corresponding energy content values of 36.83 MJ/L and 35.85 MJ/L, respectively. This quick comparison shows a density difference of 0.030 kg/L (or about 3.5 %) and energy content difference of 0.98 MJ/L (or about 2.5 %) between the two fuels.

STABILITY

In essence, stability can be defined as the capability of a fuel to resist change in composition. Instability is manifested by a change in color, the formation of gummy materials or insoluble solids, waxy sludge, or asphaltic deposition on the bottom of storage tanks, etc.

The storage stability of fuel oils may be influenced by many factors. Among these are crude oil origin, hydrocarbon composition, refinery treatment, water, and other contaminants. Fuels containing unsaturated hydrocarbons (olefins) and catalytically cracked components are inherently less stable chemically and have a greater tendency to form sediment on aging rather than straight-run fuels. The presence of reactive compounds of sulfur, nitrogen, and oxygen may also contribute to a fuel oil's instability.

Much speculation exists regarding the mechanism of sludge deposition. It may be a consequence of such factors as oxidation, polymerization, and the method of production of the fuel, which can result in the formation of insoluble compounds that eventually settle to the tank bottoms and form sludge. Situations of such deposition in light residual fuel oils, used in large heating installations, may be determined by clogging of external cold filters, blockage or restriction of pipelines, deposit formation on burner nozzles, and combustion difficulties.

Asphaltic deposition may result from the mixing of fuels of different origin and treatment, each of which may be perfectly satisfactory when used alone. Such fuels are said to be incompatible. Straight-run fuels from the same crude oil normally are stable and mutually compatible. Fuels produced from thermal cracking and visbreaking operations may be stable by themselves, but can be unstable or incompatible if blended with straight-run fuels and vice versa.

Instability or incompatibility can be a serious problem that ranges in severity from simple customer dissatisfaction with an off-color fuel to a cause of serious equipment malfunctions. Therefore, test procedures are necessary to predict fuel stability and ensure a satisfactory level of performance by the fuel oil.

A common test is ASTM D2274, Standard Test Method for Oxidation Stability of Distillate Fuel Oil (Accelerated Method).

This test method measures the inherent stability of middle distillate petroleum fuels under specified oxidizing conditions. It should be recognized that any correlation between this test and field storage may vary significantly under different field conditions or with middle distillates from different sources. This test is not applicable to fuels containing residual oil or significant amounts of components derived from nonpetroleum sources. Another test method is D5304 (Standard Test Method for Assessing Distillate Fuel Storage Stability by Oxygen Overpressure), but it too has limitations because results are not intended to correlate to actual fuel storage stability.

Although no universally accepted procedures are available for assessing the tendency of fuel oils to deposit sludge while in storage, various empirical methods are available within individual petroleum companies.

SPOT TEST FOR COMPATIBILITY

A method for measuring compatibility of residual fuels is ASTM D4740, Standard Test Method for Cleanliness and Compatibility of Residual Fuel by Spot Test.

NITROGEN

To comply with environmentally permitting requirements for nitrogen oxide emissions, more state and local governments are requiring certain combustion facilities to meet specified fuel nitrogen content. For purposes of these nitrogen oxide emission regulations, distillate fuels, low nitrogen residual fuels, and high nitrogen residual fuels have been defined by their nitrogen content. Installations must meet different emission standards based on the type of fuel being used.

MICROBIAL CONTAMINATION

Standard Guide for Microbial Contamination in Fuels and Fuel Systems (D6469) is a useful guide, which provides fuel-handling personnel with a better understanding of the causes, occurrence, and the consequences of chronic microbial contamination. Such contamination can occur in all fuels, including fuel oils and kerosine. The right conditions are needed for microbes to form, namely temperature changes, macro- and micronutrients, and water, the essential ingredient. Even negligible traces of water are sufficient to support microbial populations. Microbes can be introduced in refinery tanks, as fuel is transported in tankers and exposed in contaminated vessel compartments, and anywhere along the fuel distribution system where fuel comes in contact with air and water. Strategies for controlling microbial growth include keeping fuel systems as dry as possible by having an effective tank design, good housekeeping practices, appropriate sampling and analysis programs, and remediation procedures, which may include use of biocides. Because biocides are very toxic, users are advised to review and comply with the applicable safety, handling, and disposal requirements provided in each product's material safety data sheet (MSDS) and technical literature.

DOMESTIC HEATING OIL PERFORMANCE EVALUATION

To evaluate domestic heating oil performance, two ASTM methods are available. Each method provides a means for comparison of burning equipment and for assessment of burning characteristics of the oils. Both can be used either as laboratory or as field procedures.

ASTM D2156, Standard Test Method for Smoke Density in the Flue Gases from Distillate Fuels, is used to evaluate smoke density in the flue gases from burning middle distillate fuels. Because excessive smoke density adversely affects the efficient operation of domestic heating installations, the test is necessary if the smoke production is to be held at an acceptable level.

The second test method, ASTM D2157, Standard Test Method for Effect of Air Supply on Smoke Density in Flue Gases from Burning Distillate Fuels, is applicable only to equipment incorporating pressure atomizing and rotary wall-flame burners. In this test method, efficiency of operation is related to clean burning of the fuel. The extent to which combustion air can be reduced without producing an unacceptable level of smoke is indicative of the maximum efficiency for a particular installation at any acceptable smoke level.

KEROSINE PERFORMANCE EVALUATION

Selection of Burner

Although the performance of a particular fuel oil depends to some extent on its inherent properties, with normal kerosine, performance is more dependent on burner design than on hydrocarbon-type composition. Consequently, selection of a burner in which to assess the burning quality of an oil assumes great importance. ASTM D3699 specification for kerosine is shown in Table 2.

For kerosine performance evaluation, simple wick-fed yellow-flame burners are used because kerosine type and quality are more critical in these units. Kerosines that are mainly paraffinic burn well in wick-fed yellow-flame lamps with a poor draft, while kerosine containing high proportions of aromatics and naphthenes burn with a reddish or even smoky flame under the same conditions. Predominantly aromatic kerosines can burn brilliantly in a wick-fed blue-flame lamp with good draft characteristics, while the paraffinic type may burn with a flame of comparatively low-illuminating value.

Because combustion is more complete in the wick-fed blue-flame burner than in the yellow-flame type, the former shows less differentiation between kerosine types. Even less differences between oil types are exhibited with pressure burners, which can operate satisfactorily with a wide range of kerosines.

Burning Characteristics

INITIAL FLAME HEIGHT SMOKE POINT

As the wick of a yellow-flame type lamp is turned up, a point is reached where smoking occurs. Therefore, the degree of illumination possible depends mainly on the height of the nonsmoking flame obtainable. This height varies according to the hydrocarbon-type composition of the kerosine. The maximum height of flame obtainable without smoking, termed the "smoke point," is greatest with paraffins, considerably lower with naphthenes, and much lower still with aromatics.

Even if full advantage is not taken to use maximum nonsmoking flame heights, the characteristics of high-smoke point ensure that there will be less tendency for smoking to occur in the event that a sudden draft causes extension in flame height.

Although a low-smoke point is undesirable, in that it may not yield a satisfactory range of smokeless performance, a high-smoke point alone is no guarantee that a kerosine has generally satisfactory burning characteristics.

ASTM D1322, Standard Test Method for Smoke Point of Aviation Turbine Fuels, is the only test for the determination of smoke point used in the United States. This method consists of burning a sample in an enclosed lamp with a scale. The maximum flame height that can be achieved without smoking is estimated to the nearest 0.5 mm.

CONSTANCY OF FEED TO WICK

The maintenance of the initial degree of illumination in a lamp depends on the constancy of flow of the kerosine to the wick, as well as on the conditions of the wick. The quantity of oil flowing up a wick is related to the height of the top of the wick above the level of oil in the container, and the viscosity and surface tension of the oil. Viscosity is more significant in this respect than surface tension, because viscosity varies more in magnitude with different kerosines and changes in temperature.

When a kerosine warms up during the initial burning period, the flame size tends to increase slightly as a result of increase in evaporation rate and decrease in viscosity. In the case of lamps not provided with constant level feed, the viscosity of the fuel becomes significant, because the height of un-immersed wick increases with the consumption of the kerosine. If the viscosity is too high, the feed of kerosine can be restricted, leading to the flame height and stability being significantly reduced. The presence of moisture in a wick also hinders the upward flow of kerosine and causes a drop in flame height and a decrease in kerosine consumption.

FORMATION OF A CHAR ON WICK

After a kerosine lamp has been burning for some time, the condition of the exposed wick of the lamp begins to be affected by the formation of a carbonaceous incrustation or char. This is not significant unless it affects the flame or the mechanism for adjusting the wick.

The char may be either hard and brittle or soft, and the amount and appearance will vary according to the nature and properties of the kerosine burned. It may be of irregular formation, which causes flame distortion by producing localized deposits known as "mushrooms" on the surface of the wick, or it may be formed in such an amount and manner as to restrict the size of the flame and, in serious cases, extinguish it.

The formation of char depends mainly on the chemical composition and purity of the kerosine and can be affected by the nature of the wick, and the design and operating conditions of the lamp. Possible causes of high-char formation may be:

- Insufficient refining, resulting in the presence of deleterious impurities in the oil.
- The presence of high-boiling residues that do not vaporize easily on the top of the wick and causes decomposition and carbonization.
- Contamination with even minute amounts of heavier products, such as lubricants or heavy fuel oil.

FORMATION OF LAMP-GLASS DEPOSITS

There should not be any appreciable formation of deposits or "bloom" on the lamp-glass chimney during burning under normal operating conditions. Such bloom, when it does appear, may be either white, gray, brown, or blue and should not be confused with the brownish-black or black carbon deposits caused by a smoky flame.

Certain factors such as the design, composition, temperature of the glass chimney, and the purity of the atmosphere can have a bearing on bloom formation. However, this formation is due primarily to the deposition of sulfur compounds derived from the sulfur content of the oil.

Standard Burning Tests

Burning quality is an indication of the kerosine performance in critical applications. The inherent burning quality potential of the bulk fuel, as determined by conventional parameters such as smoke point, luminometer number, or hydrogen content, cannot always be fully realized because of adverse effects of trace quantities of certain sulfur, oxygen, or nitrogen compounds that can be present in some kerosines. Therefore, the burning quality of kerosine must be evaluated by designation of a suitable bench-type burning test.

ASTM Standard Test Method for Burning Quality of Kerosine (D187/IP 10) provides a good assessment of the burning characteristics discussed previously, with the exception of maximum flame height. The tests are applied generally to kerosine used as an illuminant or as a fuel for space heaters, cookers, incubators, etc. The most important features of this ASTM D187 are the shape and size of the flame. Changes in flame size or shape are generally caused by changes in the portion of the wick adjacent to the flame. Burning tests are essentially performance tests and are a direct method for determining the quality of the kerosine, based on its intended use. However, it is not possible to have tests for all types of commercial equipment or under all conditions, such as location, time, temperature, humidity, air currents, and cleanliness. These difficulties are partially overcome by selecting equipment for the burning test that is known to be severe using ASTM D187, and by extending the test beyond the typical time interval between cleanings of the lamp in its usual service.

In ASTM D187, the oil is burned for 24 h in a standard lamp with a flame initially adjusted to specified dimensions. The details of operation are carefully stated, as well as cover test room conditions, volume of sample, wick nature, pretreatment of the wick and glass chimney, method of wick trimming, and the removal of the char.

At the conclusion of the test, the kerosine consumption is calculated, and a qualitative assessment of the appearance of the glass chimney is made. In the IP method, the amount of char formed on the wick is determined, and the char value is calculated as mg/kg of kerosine consumed. The ASTM method uses only qualitative assessment of the wick, and no quantitative determination of char value is made.

Aside from the intrinsic significance of char value with respect to oil quality, real differences in such values in a series of kerosines enable a relative comparison of burning quality to be made. Even traces of high-boiling contaminants greatly increase the char-forming tendencies of kerosines.

For an assessment of the maximum flame height at which a kerosine will burn, a Smoke Point Test (IP 57) may be used. In this test, a sample is burned in a special lamp in which flame height is varied against the background of a graduated scale and adjusted until the smoky tail of the flame just disappears.

Burning Test for Long-Time Burning Oils

Burning tests for long-time burning oils are used to evaluate the burning properties of oils for use in railway signal lamps and similar kinds of lamps for which long periods of uninterrupted and unattended burning may be necessary.

In Burning Test-7 Day (IP 11), the oil is burned for 7 days in one of two specified signal lamps under strictly controlled conditions. The flame height and condition of the lamp are recorded every 24 h, during the test. At the end of the test, the total consumption and, if required, the char value are determined.

OTHER KEROSINE LABORATORY TESTS AND THEIR SIGNIFICANCE

Color

ASTM D156, Test for Saybolt Color of Petroleum Products–Saybolt Chromometer Method, is an indication of the overall purity of the product and is a very useful parameter in determining if the product has been contaminated with heavy products, which can result in kerosine being unsuitable for designated critical applications.

Mercaptan Sulfur and Doctor Test

Mercaptans are toxic sulfur compounds with an objectionable odor, occurring in crude oils, and are removed from most petroleum products by refining. Because of the adverse effect on fuel system components and/or kerosine burner equipment, mercaptans (ASTM D3227, Standard Test Method for Mercaptan Sulfur in Gasoline, Kerosine, Aviation Turbine, and Distillate Fuels–Potentiometric Method) are limited to preclude undesirable side-reactions and to minimize the unpleasant odor. The Doctor Test (ASTM D4952, Standard Test Method for Qualitative Analysis for Active Sulfur Species in Fuels and Solvents–Doctor Test) is an indirect indication of mercaptan levels.

Sulfur

Limited sulfur content of kerosine may be required for special uses or to meet federal, state, or local regulatory requirements for sulfur dioxide and other air pollutant emissions. ASTM D1266 is the referee method in the United States, and other acceptable test methods include D2622 and D4294.

Distillation

Distillation (ASTM D86) is an indication of the volatility of a fuel. The maximum 10 % distillation temperature ensures adequate volatility for ignition. The final boiling point limits heavy components to minimize soot and char. While not limited by most specifications, a fuel oil should normally have a smooth and continuous distillation curve. There should not be a large gap in the distillation, as may occur with the "dumb-bell blending," which is the blending of a very light stream or component with a very heavy component, but no material boiling in the middle range. Such blends may give rise to operational problems under certain conditions.

Note that the standard distillation test, ASTM D86, Standard Test Method for Distillation of Petroleum Products at Atmospheric Pressure, is a very simple distillation ("single plate distillation"). If a particular fuel is giving operational problems and contamination by heavy components is suspected, such heavy contaminants may not show up in a D86 test. It may be necessary to perform a gas chromatographic test, such as ASTM D2887 (Standard Test Method for Boiling Range Distribution of Petroleum Fractions by Gas Chromatography), which gives much greater resolution and separation of components in a fuel sample (equivalent to a 200 + plate distillation). This test may detect low levels (less than half a percent) of very high boiling contaminants, which could be causing the operational problems.

Flash Point

The flash point of kerosine is used primarily as an index of fire hazards. The minimum permissible flash point (usually 38°C) is generally regulated by federal, state, or municipal laws and is based on accepted practice in handling and use. ASTM D56 is the method typically used, and ASTM D3828 may also be used.

Freezing Point

Freezing point is the temperature at which crystals of hydrocarbons formed on cooling disappear when the temperature of the fuel is allowed to rise. This is an important parameter for critical kerosine burner applications, because waxy crystals may clog the wick in wick-fed systems and can block filter passages in fuel-handling systems. ASTM D2386, Standard Test Method for Freezing Point of Aviation Fuels, has historically been used; however, automatic tests D5901, Standard Test Method for Freezing Point of Aviation Fuels–Automatic Optical Method, and D5972, Standard Test Method for Freezing Point of Aviation Fuels–Automatic Phase Transition Method, are becoming more and more prevalent today, because of the ease and faster response time when using automatic test equipment.

Corrosion

Corrosion (ASTM D130) is an indication of the tendency to corrode copper and copper-alloy components that may be present in the kerosine handling and burner systems. Fuels with a copper corrosion rating of 1A or IB should be suitable for extended use without corroding copper or copper alloy parts. ASTM D130 primarily detects copper corrosion caused by aggressive sulfur compounds (H_2S or elemental sulfur are the most common culprits). While other contaminants in fuel may attack copper, causing the copper to turn black or another color, they will not cause the copper strip to progress through the range of colors in the D130 rating standard.

Applicable ASTM Specifications

Number	Title
D396	Specification for Fuel Oils
D3699	Specification for Kerosine
D6448	Specification for Industrial Burner Fuels from Used Lubricating Oils

Applicable ASTM/IP Standards

ASTM	IP	Title
	11	Burning Test-7 Day
D56		Flash Point by Tag Closed Tester
	57	Smoke Point
D86	123	Distillation of Petroleum Products at Atmospheric Pressure
D93	34	Flash Point by Pensky-Martens Closed Tester
D95	74	Water in Petroleum Products and Bituminous Materials by Distillation
D97	15	Pour Point of Petroleum Products
D129	61	Sulfur in Petroleum Products (General Bomb Method)
D130	54	Detection of Copper Corrosion from Petroleum Products by the Copper Strip Tarnish Test

ASTM	IP	Title
D156		Saybolt Color of Petroleum Products (Saybolt Chromometer Method)
D187	10	Burning Quality of Kerosine
D240	12	Heat of Combustion of Liquid Hydrocarbon Fuels by Bomb Calorimeter
D287	192	API Gravity of Crude Petroleum and Petroleum Products (Hydrometer Method)
D445	1	Kinematic Viscosity of Transparent and Opaque Liquids (the Calculation of Dynamic Viscosity)
D473	53	Sediment in Crude Oils and Fuel Oils by the Extraction Method
D482	4	Ash from Petroleum Products
D524	14	Ramsbottom Carbon Residue of Petroleum Products
D1160		Distillation of Petroleum Products at Reduced Pressure
D1266		Sulfur in Petroleum Products (Lamp Method)
D1250	200	Guide for Petroleum Measurement Tables
D1298	160	Practice for Density, Relative Density (Specific Gravity), or API Gravity of Crude Petroleum and Liquid Petroleum Products by Hydrometer Method
D1322		Smoke Point of Kerosine and Aviation Turbine Fuel
D1405		Estimation of Net Heat of Combustion of Aviation Fuels
D1552		Sulfur in Petroleum Products (High-Temperature Method)
	63	Sulfur in Petroleum Oils (Quartz-Tube Method)
D2156		Smoke Density in the Flue Gases from Burning Distillate Fuels
D2157		Effect of Air Supply on Smoke Density in Flue Gases from Burning Distillate Fuels
D2161		Practice for Conversion of Kinematic Viscosity to Saybolt Universal Viscosity or to Saybolt Furol Viscosity
D2274		Oxidation Stability of Distillate Fuel Oil (Accelerated Method)
D2386		Freezing Point of Aviation Fuels
D2500	219	Cloud Point of Petroleum Oils
D2622		Sulfur in Petroleum Products (X-Ray Spectrographic Method)
D2709		Water and Sediment in Middle Distillate Fuels by Centrifuge
D2887		Boiling Range Distribution of Petroleum Fractions by Gas Chromatography
D3227		(Thiol Mercaptan) Sulfur in Gasoline, Kerosine, Aviation Turbine, and Distillate Fuels (Potentiometric Method)
D3828		Flash Point by Small Scale Closed Tester
D4052		Density and Relative Density of Liquids by Digital Density Meter
D4294		Sulfur in Petroleum and Petroleum Products by Energy-Dispersive X-Ray Fluorescence Spectrometry
D4740		Cleanliness and Compatibility of Residual Fuels by Spot Test
D4952		Qualitative Analysis for Active Sulfur Species in Fuels and Solvents (Doctor Test)
D5304		Assessing Distillate Fuel Storage Stability by Oxygen Overpressure
D5453		Test Method for Determination of Total Sulfur in Light Hydrocarbons, Spark Ignition Engine Fuel, Diesel Engine Fuel, and Engine Oil by Ultraviolet Fluorescence
D5901		Freezing Point of Aviation Fuels (Automatic Optical Method)
D5949		Pour Point of Petroleum Products (Automatic Pressure Pulsing Method)
D5950		Pour Point of Petroleum Products (Automatic Tilt Method)
D5972		Freezing Point of Aviation Fuels (Automatic Phase Transition Method)
D5985		Pour Point of Petroleum Products (Rotational Method)
D6469		Guide to Microbial Contamination in Fuels and Fuel Systems
D6751		Specification for Biodiesel Fuel Blend Stock (B100) for Middle Distillate Fuels

Readings

Francis, W., *Fuels and Fuel Technology, 2* Volumes, Bergman Press, Oxford, UK, 1965.

Kewley, S. and Gilbert, C. L., *Kerosine,* Vol. IV, Science of Petroleum, 1938.

Modern Petroleum Technology, Institute of Petroleum, London, UK, 3rd ed., 1962.

Petroleum Products Handbook, V. B. Guthrie, Ed., New York, 1960.

7

Aviation Fuels[1]

John Rhode[2]

TO DISCUSS AVIATION FUELS PROPERLY, IT IS best to review briefly the development of the different types of fuel and describe the quality requirements posed by the various engines and aircraft. The resulting specifications define the required fuel qualities and specify the standard methods to be used. The international acceptance and enforcement of these specifications ensure the availability of fuels for all types of aircraft on a worldwide basis.

It is neither feasible nor desirable to cover in detail all international specifications in this chapter. Instead, the chapter is based on the fact that all major specifications measure and control similar properties. Typical examples of the physical and chemical requirements in current specifications are included for each of the major aviation gasoline (AVGAS used in piston-engine aircraft) and jet fuel grades.

HISTORICAL DEVELOPMENT OF AVIATION FUELS

Automobiles and aircraft used the same straight-run gasoline fuel until World War I (WWI). The U.S. Navy developed the first motor gasoline specification in 1907 designated as U.S. Navy Specification 24G, 4 years after the success of the Wright Flyer. The connection between fuel quality and engine performance did not evolve until WWI, when reliability and high power output became critical.

Aviation gasolines for spark ignition engines reached their development peak in the 1939–1945 war years due to military interest in aviation in both the United States and Europe. In 1944, the U.S. military issued a specification for Grade 115/145, which had the highest antiknock rating to obtain maximum performance from high performance engines. After that time, there was little additional piston engine development because the development efforts switched to gas turbine engines. Although aviation gasoline demand is expected to continue for years, quality requirements are unlikely to change significantly, except for the increasing pressure to remove lead from this last lead-containing fuel in the petroleum fuel inventory.

The first gas turbine engines were regarded to have no critical fuel requirements. Because ordinary illuminating kerosene was the original development fuel, the first turbine fuel requirements were written around the properties and test methods of this well-established product. Those properties of aviation gasoline deemed important for all aviation fuels were also included. With the escalating complexity and increasingly demanding operating conditions of both engines and aircraft, fuel specifications inevitably became more complicated and rigorous. Current demands for improved performance, economy, and overhaul life will continue to influence the trend toward additional requirements; nevertheless, the optimum compromise between fuel quality and availability has been largely achieved by current fuel specifications.

AVIATION GASOLINE

Composition and Manufacture

Aviation gasoline is one of the most restrictive fuels produced in a refinery. Quality control parameters have been gradually added and refined from the early specifications of the past century to give a highly developed performance fuel where production must meet stringent aviation requirements to ensure a high level of quality control, cleanliness, and traceability from refinery to aircraft. Strict process control is required to ensure that the stringent (and sometimes conflicting) requirements are met for antiknock ratings, volatility, and calorific values. Careful handling is essential during storage and distribution to guard against various forms of contamination.

Aviation gasoline consists substantially of hydrocarbons. Sulfur- and oxygen-containing impurities are strictly limited by the specifications, and only certain additives are permitted. (Refer to the section on "Aviation Fuel Additives.") The main component of high-octane aviation gasoline is isooctane produced in the alkylation process by reacting refinery butanes with isobutene over acid catalysts. To meet minimum volatility requirements of the final blend, a small proportion of isopentane (obtained by the superfractionation of light straight-run gasoline) is added. The aromatic component required to improve the rich rating is usually a catalytic reformate consisting primarily of toluene. The amount of aromatic component is limited by the high gravimetric calorific value (specific energy) requirement and the distillation end point. The low freezing point is the specification parameter that excludes benzene. All blending components must have high-octane values.

Only the low-octane grade can include a proportion of straight-run gasoline, because such gasolines that contain various amounts of paraffins, naphthenes, and aromatics lack the necessary branched paraffins (isoparaffins) required to produce a high-octane fuel.

Specifications

CONTENT

Aviation gasoline specifications generally cover composition and chemical and physical tests. The composition section

[1] In preparation of this chapter, the contents of the sixth and seventh editions were drawn upon. The author acknowledges the authors of the sixth edition, Geoffrey J. Bishop of Shell International Petroleum Company, London, UK and Cyrus P. Henry, Jr., of DuPont Company, Deepwater, NJ and author of the seventh edition Kurt H. Strauss retired. The current edition will review and update the topics as addressed by the previous authors, introduce new technology that has been developed, and include up-to-date references.

[2] *Air BP, Warrenville, IL*

TABLE 1—Aviation Gasolines, Main International Specification Grades, and Current Specifications

Identifying Color	Nominal Antiknock Characteristics Lean/Rich	NATO Code Number	DefStan 91/90 British Ministry of Defence	ASTM D910	Use
Colorless	73	F13[A]	...	...	Obsolete
Colorless	80	...	...	...	Obsolete
Red	80/87	F-12[A]	80	80	Minor civil
Purple	82	...	...	82UL[B]	New engine fuel
Blue	91/96	F-15[A]	...	...	Obsolete
Blue	100/130	F-18[A]	100LL	100LL	Major civil
Green	100/130	...	100	100	Minor civil/military
Brown	91/98	...	91	91	Minor civil
Brown	108/135	...	...	...	Obsolete
Purple	115/145	F-22[A]	...	...	Military—obsolete

[A] Obsolete designation.
[B] ASTM Specification D6227.

stipulates that the fuel must consist entirely of hydrocarbons, except trace amounts of specified additives including tetraethyl lead antiknock additive, oxidation inhibitors, and conductivity improvers. Nonhydrocarbon blending components, such as oxygenates, are not permitted. The chemical and physical test section is the one most familiar to users, because it carefully defines the allowable limits for the properties as well as the test methods to measure and control these properties.

FUEL GRADES

As many as six grades were in use up to the end of World War II. In the following decade, the turbine engine became the primary military engine with commercial aviation following shortly afterward. This increase in turbine engines caused decreased AVGAS demand, which led to the elimination of both the high- and low-octane grades. This was also facilitated by the fact that only the octane requirement and the permitted tetraethyl ethyl-lead (TEL) content differed between the various grades. Fewer grades allowed the reduction of manufacturing, storage, and handling costs with subsequent benefits to the consumer. (At many commercial airports, the AVGAS tanks were converted to jet storage.) Although three grades–80, 91, 100, and 100LL–are listed in the ASTM Specification for Aviation Gasoline (D910), only the 100LL grade is predominant in the United States and much of the rest of the world.

Various bodies have drawn up specifications covering the various grades. The most commonly quoted specifications are issued by ASTM (D910) and the British Ministry of Defence (DefStan 91/90). Table 1 lists grades in former and in current use and indicates their identifying colors and present status.

Due to the international nature of aviation activities, the technical requirements of Western specifications are virtually identical, and only differences of a minor nature exist between the specifications issued in the major countries. Russian GOST (gosudarstvennyy standard or state standard later changed to Gosstandart) specifications differ in the grades covered and in respect to some of the limits applied, but in general the same properties are used and most test methods are basically similar to their Western equivalents [ASTM and Institute of Petroleum (IP) standards]. Russian aviation gasoline grades are summarized in Table 2.

Table 3 provides the detailed requirements for aviation gasoline contained in the ASTM Specification for Aviation Gasoline (D910). In general, the main technical requirements of all other Western specifications are virtually identical to those in Table 3, although differences can occur in the number of permitted grades and the amount of maximum permitted TEL content. Within the specification, the various grades differ only in certain vital respects such as color, antiknock rating, and TEL content. The two remaining grades in the GOST specification are subdivided into a regular and a premium grade with differing limits for aromatics, olefins, sulfur, and acidity.

The limits for Western aviation gasoline were, in most cases, originally dictated by military aircraft engine requirements. Since then, the performance requirements for civil and military engines have changed very little. However, improved manufacturing techniques and the reduced demand for certain grades have allowed fuel suppliers to produce modified fuel grades more suitable to the market. The primary result of this trend has been the 100LL grade, which is certified for both low- and high-output piston engines.

Characteristics and Requirements

ANTIKNOCK PROPERTIES

The various grades are classified by their "antiknock" characteristics measured in special laboratory engines.

TABLE 2—Russian Aviation Gasoline Grades

Specification	Grade	Color	Use
Tu 38.10913-82	B70	Colorless	Obsolete
GOST-1012	B91/115[A]	Green	Current
GOST-1012	95/130	Yellow	Current

[A] Of regular quality.

TABLE 3—Detailed Requirements for Aviation Gasolines ASTM Specification D910[A]

Requirement		Grade 80	Grade 91	Grade 100LL	Grade 100
Knock value, lean mixture					
Motor octane number	min	80.7	90.8	99.6	99.6
Aviation lean rating	min	80.0	91.0	100.0	100.0
Knock value, rich mixture					
Octane number	min	87	98	...	...
Performance number	min	...	...	130.0	130.0
Tetraethyl-lead, mL					
TEL/L	max	0.13	0.53	0.53	1.06
gPb/L	max	0.14	0.56	0.56	1.12
Color dye content					
Blue dye, mg/L	max	0.2	3.1	2.7	2.7
Yellow dye, mg/L	max	None	None	None	2.8
Red dye, mg/L	max	2.3	2.7	None	None
Orange dye, mg/L	max	None	6.0	None	None

Requirements for All Grades		
Density at 15°C, kg/m^3		Report
Distillation		
Initial boiling point, °C		Report
Fuel evaporated		
10 volume percent at °C	max	75
40 volume percent at °C	min	75
50 volume percent at °C	max	105
90 volume percent at °C	max	135
Final boiling point, °C	max	170
Sum of 10 % + 50 % evaporated temperatures, °C	min	135
Recovery volume percent	min	97
Residue volume percent	max	1.5
Loss volume percent	max	1.5
Vapor pressure, 38°C, kPa	min	38.0
	max	49.0
Freezing point, °C,	max	–58
Sulfur, mass percent	max	0.05
Net heat of combustion, MJ/kg	min	43.5
Corrosion, copper strip, 2 h @ 100°C	max	No. 1
Oxidation stability 5 h aging (16 h aging)		
Potential gum, mg/100 mL	max	6 (max 10)
Lead precipitate, mg/100 mL	max	3 (max 4)

(*Continued*)

TABLE 3—Detailed Requirements for Aviation Gasolines ASTM Specification D910[A] (Continued)

Requirements for All Grades		
Water reaction		
Volume change, mL	max	±2
Electrical conductivity, pS/m	max	450

[A] For additional requirements contained in specification footnotes, refer to Table I in ASTM D910.

Knock, or detonation, is a form of abnormal combustion where the air/fuel charge in the cylinder ignites spontaneously in a localized area instead of being consumed by the spark-initiated flame front. Knocking combustion can damage the engine and cause serious power loss if allowed to persist. For a gasoline engine to work effectively, the fuel must ignite at the correct moment and burn smoothly delivering power to the piston. The various grades were designed to guarantee knock-free operation for engines ranging from those used in light aircraft to those in high-powered transports and military aircraft. The fact that higher-octane fuels than those required for an engine can be used without problems has been a major factor in the historical elimination of several grades.

The first aviation gasoline specification written by the U.S. Navy on September 1, 1917, was to standardized the fuel. The fuel was highly volatile with a low distillation end point and lacked performance parameters. By 1922, it was recognized that performance parameters were necessary to regulate vapor lock and carburetor icing, and methods were needed to quantify fuel combustion quality to control knock. Various methods were developed to quantify fuel combustion quality such as the "toluene scale," but industry harmonized on the "octane scale" to prevent disputes over quality, and by 1929 the Cooperative Fuels Research (CFR) engine was developed. Antiknock ratings of aviation gasolines are determined in single-cylinder ASTM laboratory engines by matching a fuel's knock resistance against reference blends of pure isooctane (2,2,4-trimethylpentane), assigned an octane rating of 100, and *n*-heptane with a rating of 0. A fuel's rating is given as an octane number (ON), which is the percentage of isooctane in the matching reference blend. Fuels of higher antiknock performance than pure isooctane are rated against isooctane containing various percentages of TEL additive. The ratings of such fuels are expressed as performance numbers (PN), which are defined as the percentage of maximum knock-free power output obtained from the fuel compared to the power obtained from unleaded isooctane.

Two different engine methods are used to rate a fuel. Early on, knock was detected under cruise conditions where the fuel portion of the mixture was decreased as much as possible to improve efficiency. This condition, known as the lean or weak mixture method, is measured by the ASTM Test for Knock Characteristics of Motor and Aviation Fuels by the Motor Method (D2700/IP 236). Knocking conditions are obtained by increasing engine compression ratio under constant conditions in the engine described by this method. At the beginning of World War II, newly designed, high-power-output, supercharged engines also were found to knock under engine takeoff conditions. Here, mixture strength is increased (richened) with the additional fuel acting as a coolant. This suppresses knocking combustion and results in higher power output, until ultimately knock occurs under these conditions also. To duplicate these conditions, a different single-cylinder engine was developed from a Pegasus radial engine with supercharging and variable fuel/air ratio. ASTM Test for Knock Ratings of Aviation Fuels by the Supercharge Method (D909/IP 119) produces the resulting "rich or supercharged" rating.

Until 1975, ASTM Specification D910 designated aviation gasoline grades with two ratings, such as 100/130, in which the first number was the lean and the second number the rich rating. Although the specification now uses only one number (the lean rating) to designate a grade, some other specifications use both. However, both ratings are required to meet the specification.

It is important to note that the operating conditions of both laboratory engines were developed to match the knock performance of full-scale engines in service during the World War II period. Since then, considerable engine development has taken place in the smaller in-line engines, so that the relationship between current full-scale and laboratory engines may be different from that which paced the original laboratory engine development. As a result, the Federal Aviation Administration (FAA) is conducting an extensive program of rating the knock resistance of current production engines to reestablish the relationship with the laboratory engines. Other work has also indicated that modern, in-line piston engines are not knock-limited under takeoff conditions, compared to the older, larger radial engines.

VOLATILITY

All internal combustion engine fuels must be convertible from the liquid phase in storage to the vapor phase in the engine to allow the formation of the combustible air/fuel vapor mixture, because liquid fuels must evaporate to burn. If gasoline volatility is too low, liquid fuel enters the cylinders and washes the lubricating oil off the walls. This increases engine wear and causes dilution of the crankcase oil. Low volatility can also give rise to critical maldistribution of mixture strength between cylinders. Too high a volatility causes fuel to vaporize too early in the fuel compartments and distribution lines, giving undue venting losses and possible fuel starvation through "vapor lock" in the fuel lines. The cooling effect due to rapid evaporation of highly volatile material can also cause carburetor icing, which is due to moisture in the air freezing on the carburetor under certain conditions of humidity and temperature. Many modern engines, therefore, have anti-icing devices on the engines, including carburetor heating.

Volatility is measured and controlled by the gasoline distillation and vapor pressure. Distillation characteristics are determined with a procedure (ASTM D86/IP 123) in which a fuel sample is distilled and the vapor temperature is recorded for the percentage of evaporated or distilled fuel throughout the boiling range. The following distillation points are selected to control volatility for the reasons indicated.

1. The percentage evaporated at 75°C (167°F) controls the most volatile components in the gasoline. Not less than

10 % but no more than 40 % must evaporate at that temperature. The minimum value ensures that volatility is adequate for normal cold starting. The maximum value is intended to prevent vapor lock, fuel system vent losses, and carburetor icing.

2. The requirement that at least 50 % of the fuel be evaporated at 105°C (221°F) ensures that the fuel has even distillation properties and does not consist of only low boiling and high boiling components ("dumb-bell" fuel). This provides control over the rate of engine warm-up and stabilization at slow running conditions.
3. The requirement that the sum of the 10 % plus the 50 % evaporated temperatures exceed 135°C (307°F) also controls the overall volatility and indirectly places a lower limit on the 50 % point. This clause is another safeguard against excessive fuel volatility.
4. The requirement that a minimum of 90 % of the fuel be evaporated at 135°C (275°F) controls the portion of less volatile fuel components and, therefore, the amount of unvaporized fuel passing through the engine manifold into the cylinders. The limit is a compromise between ideal fuel distribution characteristics and commercial considerations of fuel availability, which could be adversely affected by further restrictions on this limit.
5. The final distillation limit of 170°C (338°F) maximum limits undesirable heavy materials, which could cause maldistribution, crankcase oil dilution, and in some cases combustion chamber deposits.

All spark ignition fuels have a significant vapor pressure, which is another measure of the evaporation tendency of the more volatile fuel components. Additionally, when an aircraft climbs rapidly to high altitudes, the atmospheric pressure above the fuel is reduced and may become lower than the vapor pressure of the fuel at that temperature. In such cases, the fuel will boil and considerably more quantities of fuel will escape through the tank vents.

Vapor pressure for aviation gasoline is controlled and determined by any of three methods, consisting of ASTM D323/IP 69, Test for Vapor Pressure of Petroleum Products (Reid Method), ASTM D5190, Test for Vapor Pressure of Petroleum Products (Automatic Method), and D5191/IP 394, Test for Vapor Pressure of Petroleum Products (Mini Method). In case of disputes, D5191 is designated the referee method. Allowable limits are between 38 and 49 kPa (5.5–7.1 psi). The lower limit is an additional check on adequate volatility for engine starting, while the upper limit controls excessive vapor formation during high-altitude flight and "weathering" losses in storage.

A review of the aviation gasoline specification reveals that volatility, unlike that for motor gasoline, contains no adjustments for differing climatic condition but is uniform and unchanging wherever the product is used.

DENSITY AND SPECIFIC ENERGY

No great variation in either density or specific energy occurs in modern aviation gasolines because these properties depend on hydrocarbon composition, which is already controlled by other specification properties. However, the specific energy requirement limits the aromatic content of the gasoline. Both properties have greater importance for jet fuels, as discussed later.

FREEZING POINT

Maximum freezing point values are set for all aviation fuels as a guide to the lowest temperature at which the fuel can be used without risking the separation of solidified hydrocarbons. Such separation could lead to fuel starvation through clogged fuel lines or filters, or loss in available fuel load due to retention of solidified fuel in aircraft tanks. The low freezing point requirement also virtually precludes the presence of benzene, which, while a high-octane material, has a very high freezing point.

The standard freezing point test involves cooling the fuel until crystals form throughout the fuel and then rewarming the fuel and calling the temperature at which all crystals disappear the freezing point. The freezing point, therefore, is the lowest temperature at which the fuel exists as a single phase. Freezing points are determined by ASTM Test for Freezing Point of Aviation Fuels (D2386/IP 16).

STORAGE STABILITY

Aviation fuel must retain its required properties for long periods of storage in all kinds of climates. Unstable fuels oxidize and form polymeric oxidation products that remain as a resinous material or "gum" on induction manifolds, carburetors, valves, etc., when the fuel is evaporated. Formation of this undesirable gum must be strictly limited and is assessed by the existent and accelerated (or potential) gum tests.

The existent gum value is the amount of gum actually present in fuel at the time of the test. It is determined by the ASTM Test for Existent Gum in Fuels by Jet Evaporation (D381/IP 131). The potential gum test, ASTM Test for Oxidation Stability of Aviation Fuels (Potential Residue Method) (D873/IP 138), predicts the possibility of gum formation during protracted storage.

To ensure that the strict limits of the stability specification are met, aviation gasoline components are given special refinery treatments to remove the trace impurities responsible for instability. In addition, controlled amounts of oxidation inhibitors are normally added. Currently, little trouble is experienced with gum formation or degradation of the antiknock additive.

SULFUR CONTENT

Total sulfur content of aviation gasoline is limited to 0.05 % mass maximum, because most sulfur compounds have a deleterious effect on the antiknock effect of alkyl lead compounds. If sulfur content were not limited, specified antiknock values would not be reached for highly leaded grades of aviation gasoline. The sulfur content is measured by ASTM Test for Sulfur in Petroleum Products (Lamp Method) (D1266/IP 107) or by ASTM Test for Sulfur in Petroleum Products by X-Ray Spectrometry (D2622/IP 447).

Some sulfur compounds can have a corroding action on the various metals in the engine system. Effects vary according to the chemical type of sulfur compound present. Elemental sulfur and hydrogen sulfide are particularly implicated. Because copper is considered the most sensitive metal, fuel corrosivity toward copper is measured in ASTM Test for Detection of Copper Corrosion from Petroleum Products by the Copper Strip Test (D130/IP 154).

WATER REACTION

The original intent of the water reaction test was to prevent the addition of high-octane, water-soluble compounds, such as alcohol, to aviation gasoline. The test method involves shaking 80 mL of fuel with 20 mL of buffered water under standard conditions and observing phase volume changes.

Some specifications for aviation gasoline now have interface conditions and phase separation requirements, in addition to volume changes. The Test for Water Reaction of Aviation Fuels (D1094/IP 289) rates all three of these criteria.

Unleaded Aviation Gasolines

Up to this point, the discussion has dealt with aviation gasolines containing TEL per ASTM D910. Leaded aviation gasolines have outlived other lead-containing fuels. At the time of this writing, they are the only lead-containing fuel in the fuels inventory of the United States and many other countries. Although aviation gasolines are currently exempted from regulations prohibiting leaded fuels, such an exemption is based on the realization that no suitable unleaded high-octane fuel is available for much of the general aviation fleet. Two approaches are intended to alleviate this condition. An FAA/industry research project is engaged in identifying possible high-octane candidate fuels for high-output, in-line engines. A parallel effort is to establish the octane appetite of these engines to obtain ultimately a match between practical fuel candidates and existing engines.

Current activities include a formal collaborative industry research program in addition to FAA-sponsored research. The industry collaborative research program is currently led by the Coordinating Research Council (CRC) Unleaded AVGAS Development Group. Working in parallel with this group is the CRC Aviation Engine Octane Rating Group. The Unleaded Aviation Gasoline Development Group was formed with the objective of conducting research and testing that will facilitate development of the next-generation aviation gasoline–a high-octane unleaded aviation gasoline as an environmentally compatible, cost-effective replacement for the current ASTM D910 100LL fuel. The CRC Aviation Engine Octane Rating Group has the objective of developing a method to consistently rate aircraft engine octane requirements under harsh repeatable conditions and to determine the general aviation fleet octane requirements. The FAA William J. Hughes Technical Center has been instrumental in providing test facilities and funding and the U.S. Environmental Protection Agency (EPA) has continued to monitor the industry program. It should be noted that an industry standard for octane rating aircraft engines did not previously exist; to alleviate this dilemma two ASTM standard practices, ASTM D6424 and ASTM D6812, were developed and implemented. ASTM D6424 was implemented in 1999 and applies to the octane rating of normally aspirated aircraft engines. ASTM D6812 was released in 2002 and applies to the octane rating of turbocharged engines. Engines representative of the general aviation fleet were octane rated using these ASTM procedures to determine the unleaded octane requirement. Test results indicated a minimum unleaded octane requirement greater than 100 Motor Octane Number (MON) for naturally aspirated engines and higher for turbocharged engines depending on engine power output and configuration. Such findings are consistent with the observed difference between leaded and unleaded fuels to achieve octane satisfaction in full-size engines as observed in earlier historical work. CRC research into unleaded aviation gasoline alternatives has focused on meeting engine octane requirements, and blend properties were not controlled for agreement with the ASTM D910 specification as the primary focus was engine octane satisfaction. Research results to date reflecting the unleaded fuel blends tested have not identified a transparent replacement for the 100LL AVGAS product. Although full-scale engine tests indicated some blends were capable of providing knock-free operation in the test engine, these blends represented the use of specialty chemicals that require further evaluation with respect to environmental impact. Depending on engine power output and configuration, high-performance aviation engines can require unleaded fuels in excess of 100 MON to achieve octane satisfaction. Leaded AVGAS 100LL or 91/98 offers greater octane satisfaction in full-size engines compared to unleaded products of similar laboratory MON. CRC test results are indicative of the significant challenge regarding a high-octane unleaded AVGAS formulation and further serve as a reminder that aviation fuels represent specialized products optimized over many years to maximize performance and flight safety. Through the CRC, a broad range of industry expertise and facilities has been made available to investigate this issue. Such groups, with input from all parties and working in collaboration with industry, offer a viable means of conducting meaningful research. The goal remains a viable solution that ensures performance and flight safety for both the existing and future general aviation fleets.

For new engines with lower octane appetites, ASTM published D6227 in 1998, Standard Specification for Grade 82 UL Aviation Gasoline. Aviation Gasoline 82 UL is an aviation-grade product that can be blended from automotive gasoline streams and is based on the requirements of piston engines designed and approved for such a fuel. D6227 differs from D910 Aviation Gasoline in distillation characteristics, oxygenate and TEL content, and vapor pressure limits. Grade 82 unleaded aviation gasoline (82UL) is cited as a partial replacement for aviation gasoline Grade 100LL. 82UL was developed by the aviation industry as a possible alternative to MOGAS (automotive gasoline used in piston-engine aircraft) for use in supplemental type certificate (STC) aircraft or for those engines seeking an aviation-grade MOGAS product. The D6227 specification also states that the fuel is not considered suitable for engines certified on gasoline meeting D910 and, thus, is intended for engines with lower-power output currently under development. The specification is summarized in Table 4. A number of requirements are similar to D910, but the volatility requirements differ from those for aviation gasoline and those for motor gasoline. Thus, the distillation and allowable vapor pressure of 82 UL describe a more volatile product than D910 but less volatile than permitted for motor gasoline. The specification specifically prohibits the use of oxygenates or any additives not approved for aviation use. The absence of a rich rating in D6227 is based on the finding that such a requirement is not needed for low power in-line engines. Use of the fuel in radial engines is not anticipated because these engines have high supercharge octane requirements not required by this specification. The lower specific energy requirement, compared to D910, permits fuels with higher aromatic content. The aviation market has been cautious in adopting 82UL due to limited applicability, limited manufacturing volumes, two grades of AVGAS storage problems, risk of misfuelings, performance versus Grade 100LL, and engine/aircraft certification. To distinguish it from other unleaded as well as leaded fuels, 82 UL is dyed purple. D6227 82UL represents an aviation-grade alternative to MOGAS developed by the industry for use in certain aircraft types, not a replacement for standard AVGAS grades. Lack of market acceptance has highlighted key issues that must be addressed for the success of any future product.

TABLE 4—Requirements for Unleaded Aviation Gasoline (82 UL)[A] from ASTM Specification D6227

Property		Requirement
Knock value, lean mixture		
Motor method octane number	min	82.0
Color		Purple
Dye content		
Blue dye, mg/L	max	7.5
Red dye, mg/L	max	1.9
Distillation temperature, °C (°F) at % evaporated		
10 volume percent	max	70 (158)
50 volume percent		66 (150)–121 (250)
90 volume percent	max	190 (374)
End point	max	225 (437)
Residue, volume percent	max	2
Distillation recovery, volume percent	min	95
Distillation loss, volume percent	max	3.0
Net heat of combustion, MJ/kg (BTU/lb)	min	40.8 (17,540)
Freezing point, °C (°F)	max	–58 (–72)
Vapor pressure, kPa (psi)	max	62 (9.0)
KPa (psi)	min	38 (5.5)
Lead content, g/L (g/U.S. gal)	max	0.013 (0.05)
Corrosion, copper strip, 3 h @ 50°C (122°F)	max	No. 1
Sulfur, mass percent	max	0.07
Potential gum (5 h aging), mg/100 mL	max	6
Alcohol and ether content		
Total combined methanol and ethanol, mass percent	max	0.3
Combined aliphatic ethers, methanol, and ethanol, as mass percent oxygen	max	2.7

[A] For additional requirements contained in specification footnotes, refer to Table 1 in ASTM D6227.

Automotive (Motor) Gasoline—Use in Aircraft

In general, at the time of this printing, reciprocating aviation engines and their fuel systems are certified to operate on one of the grades in D910 or the 82 UL grade in D6227. Most major piston engine manufacturers specifically exclude motor gasoline from their list of approved fuels. Because of that position, many fuel manufacturers also disapprove of the use of motor gasoline in any aircraft. Some reasons for this position follow.

Motor gasoline can vary in composition and quality from supplier to supplier, from country to country, and, in temperate climates, from season to season; in comparison to aviation gasoline, motor gasoline is not a closely or uniformly specified product. A particularly troublesome variable in recent years is the increasing inclusion of strong detergent additives and of alcohols or other oxygenates in motor gasoline. Differences in the handling and quality control of motor gasoline may involve risks that a potential user should assess.

Availability and cost considerations have encouraged many owners of light aircraft to seek acceptance of motor gasoline as an alternative to aviation gasoline. In recognition of this trend, and to maintain regulation and control over the use of motor gasoline, various civil aviation regulatory agencies around the world have extended supplemental or special certification provisions to permit the use of motor gasoline in a limited number of specified aircraft types, whose design features are considered to be less sensitive to fuel characteristics. In the United States, the gasoline types permitted by the STCs depend on the specific engine/aircraft combination. They may be permitted to use leaded motor gasoline or unleaded gasoline meeting the requirements of ASTM D4814, Specification for Spark Ignition Engine Fuel, or the 82 UL grade cited earlier. Restrictions also exist on the minimum permitted octane. Alcohol, which is included in ASTM D4814, is not permitted for aviation use.

The compositional and property differences between motor gasoline and aviation gasoline are detailed next, listing their potential adverse effects on engine/aircraft operation and flight safety:

1. The normally reported motor gasoline octanes (R + M)/2 are not comparable to aviation gasoline ratings. Thus, preignition or detonation conditions could develop with motor gasoline if its use is based on improper octane number comparisons. In addition, motor gasolines have a wider distillation range than aviation fuels. This could promote poor distribution of the high antiknock components of the fuel in some carbureted engines.
2. Higher volatilities and vapor pressures of motor gasolines could overtax the vapor handling capability of certain engine-airframe fuel systems and could lead to vapor lock or carburetor icing. Fire hazards could also be increased.
3. Motor gasoline has a shorter storage stability lifetime because of seasonal changeovers. As a result, it could form gum deposits in aviation systems, causing poor mixture distribution and other mechanical side effects, such as intake valve sticking.
4. Due to higher aromatic content and the possible presence of oxygenates, motor gasoline could have solvent characteristics unsuitable for some engine/airframe combinations. Seals, gaskets, flexible fuel lines, and some fuel tank materials could be affected.
5. Motor gasoline may contain additives, which can prove incompatible with certain in-service engine or airframe components. Detergents, required to meet the requirements of advanced automotive fuel injection systems, can cause operating difficulties by preventing normal water separation in storage systems. Alcohols or other oxygenates could increase the tendency to hold water, either in solution or in suspension. In the presence of sufficient water, it will combine with alcohol and remove this octane enhancer from the gasoline. Other additives, not detailed here, could also lead to problems not specifically addressed in this document.
6. The testing and quality protection measures for automotive gasoline are much less stringent than those for aviation fuels. There is a greater possibility of contamination occurring and less probability of it being discovered. Because motor gasolines meet less-stringent requirements, compositional extremes still meeting ASTM D4814 might cause undefined difficulties in certain aircraft. Furthermore, ASTM D4814 is continually revised.
7. The antiknock compounds in leaded motor gasolines contain an excess of chlorine or bromine-containing lead scavengers, while aviation gasolines contain lesser concentrations of bromine compounds only. Chlorine compounds result in more corrosive combustion products. Lead phase-down in some countries can result in motor gasoline containing insufficient lead to prevent valve seat wear in certain engines.

These factors illustrate that the use of motor gasoline in aircraft may involve certain risks that the potential user should assess before using the product. The adoption of automotive gasoline, where development and specifications are focused on fuel supply for surface vehicles, is inappropriate for aircraft operations.

AVIATION TURBINE FUELS (JET FUELS)

Fuel and Specification Development

Military jet fuel development has been somewhat dissimilar in Europe and America. The U.S. specification was most likely derived from the aviation gasoline specification, while the British specification reflected the properties of illuminating kerosene.

Because of differences in early development philosophies, a brief historical review is a valuable preamble to the discussion of the test requirements and their significance. This review also reflects the chronological order of development, with the military demands preceding civil ones by more than two decades.

BRITISH MILITARY FUELS

The British jet fuel specification DERD 2482, issued shortly after World War II, was based on operating experience with illuminating kerosene. The first British jet engine fuel specification, RDE/F/KER (Provisional), was introduced by the end of World War II and covered what was virtually an illuminating kerosene. It was rather restrictive on aromatics (12 % max.), sulfur content (0.1 % max.), and calorific value (18,500 BTU/lb min.) but contained no burning quality requirements. Although further experience permitted relaxation of some early requirements, it became necessary to introduce new limitations and to amend some existing specification limits as new service problems were encountered. For example, the development of more powerful turbine-powered aircraft with greater range and higher altitude capability made the −40°C freezing point inadequate during extensive cold soaking at altitude. After a few amendments, RDE/F/KER was superseded in 1947 by D.Eng.RD. (DERD) 2482 and this was in turn reissued from time to time with increasingly stringent requirements. This specification became obsolete in 1965 when it was replaced by D.Eng.RD. 2494. DERD 2494, the replacement specification, issued in 1957, incorporated a freezing point of −50°C (−58°F). This fuel quality remained the optimum compromise between engine requirements, fuel cost, and strategic availability until recently. A minimum flash point of 38°C (100°F) was specified in both specifications, more for fiscal than technical reasons. One oil company introduced a −50°C (−58°F) freezing point kerosene in 1956 in anticipation of long-range, high-altitude operations. Some years later, this grade was covered by the British specification DERD 2494, which has evolved into the current Defence Standard 91-91 specification for Jet A-1.

It is interesting to note here that the British Ministry of Defence is responsible for the entire aviation specification system for both military and commercial fuels. In the United States, these requirements are handled by completely different entities, with the Department of Defense for military and ASTM International for civil or commercial fuels.

While DERD 2494 (now termed Def Stan 91-91) is the standard British civil jet fuel, a new DERD 2453 (now Def Stan 91-87) was issued in 1967 for military use, incorporating fuel system icing inhibitor and corrosion inhibitor additives in line with the latest military and NATO requirements. During 1980, a freezing point relaxation to −47°C was permitted in both specifications to increase availability.

High flash point kerosene was introduced as early as 1948 to reduce the fire risk aboard aircraft carriers. The first specification for this grade was RDE/F/KER 203 and called for a flash point similar to light diesel fuel. The 65°C flash point was later amended to 60°C in DERD 2488 because it was too restrictive. Defence Standard 91-86 (DERD 2452) is the current British military specification for high flash kerosene. In answer to a need for improved low-temperature performance, a later specification, DERD 2498, dropped the maximum freezing point to −48°C (−55°F). In 1966, the freezing point was changed to −46°C (−51°F) max. Ultimately in 1976, DERD 2452 (now Def Stan 91-86) was issued to bring the British high flash naval fuel in line with U.S. military and NATO standards.

Because crude oils with high gasoline yields are not in abundant supply, wide boiling range jet fuel was never used in the United Kingdom to the extent it was in the U.S. military. However, in the interests of commonality, DERD 2486 was issued to correspond to the U.S. Grade JP-4 (MIL-T-5624). Ultimately, this grade was brought completely in line with JP-4 with DERD 2454 (now Def Stan 91-88) by incorporating fuel system icing inhibitor and corrosion inhibitor. Table 5 lists current British and corresponding U.S. military specifications.

AMERICAN MILITARY JET FUELS

In the United States, jet fuel progress followed a different pattern. In 1944, the United States published specification AN-F-32 (which later changed to MIL-T-5616) for JP-1, a −60°C (−76°F) freezing point paraffinic kerosene. This very restrictive requirement drastically limited fuel availability, and the grade soon became obsolete (although the term JP-1 is still used incorrectly to describe any kerosene-type jet fuel). The MIL-T-5616 was superseded by various wide-cut fuels: JP-2 (1945), JP-3 (1947), and JP-4 (1951, AVTAG [Aviation Turbine Fuel Wide Cut], NATO F-40). These wide-cut fuels are mixtures of naphtha and kerosene that greatly increase availability because of the gasoline component in the product.

The first wide-cut grade (JP-2) had a vapor pressure of 14 kPa (2.0 psi) max., obtained by the addition of heavy gasoline fractions to kerosene. Experience soon indicated that an increase in vapor pressure would facilitate low temperature starting. The resulting fuel (JP-3) had a vapor pressure range of 35–49 kPa (5–7 psi), similar to aviation gasoline. However, excessive venting losses occurred in the high-powered F-100 fighter and other Century fighters, due to fuel boiling during rapid climb. Reducing the vapor pressures to 14–21 kPa (2.0–3.0 psi) corrected this problem. With slight modifications and the inclusion of certain additives, this fuel called JP-4 (MIL-PRF-5624) was the mainstay of the U.S. Air Force and of the air forces of many countries until it was replaced by JP-8 (MIL-DTL-83133).

The first U.S. Navy aircraft used aviation gasoline, but the lead in the fuel attacked the hot section components in the engine. One proposed approach was to blend aviation gasoline with kerosene to form Jet Mix, a product similar to JP-4. JP-5 (AVCAT, NATO F-44), a high flash point kerosene developed by the Navy for use in Jet Mix, that was first covered by the specification MIL-F-7914 in 1952. Subsequently, JP-5 was included in MIL-F-5624B in 1953. Although considerable work was done on Jet Mix, this product was never used operationally and JP-5 remains the primary jet fuel for most navies around the world. Grade JP-5, a low-volatility fuel in carrier use by naval aircraft with its high minimum flash point of 60°C (140°F) is dictated by shipboard combat conditions, while its low freezing point of −46°C (−51°F) is based on aircraft demands. There are several other special military grades of aviation kerosene that exist today or have been made redundant for specific reasons. JP-6 was a kerosene fuel developed in 1956 for the supersonic bomber XB-70 aircraft. JP-6 was similar to JP-5 but with a lower freezing point and improved thermal oxidative stability. The cancellation of the XB-70 program resulted in the cancellation of the JP-6 specification, MIL-J-25656 in 1956. JPTS is a special purpose jet fuel developed in 1956 to power the high-flying U-2 reconnaissance aircraft. JPTS is an extremely thermally stable jet fuel with a low freezing point to support this type of mission. JPTS, produced to specification MIL-DTL-25524, is still used today in the U-2 and the newer TR-1 aircraft flown by the National Aeronautics and Space Administration (NASA). The development of the SR-71 Blackbird in the late 1960s required a new fuel having low vapor pressure and excellent thermal oxidative stability to meet the requirements of high altitude and Mach 3+ cruising. JP-7 is not a distillate fuel like most other jet fuels but is composed of special blending stocks to produce a very clean hydrocarbon mixture low in aromatics (typically <3 %) and nearly void of the sulfur, nitrogen, and oxygen impurities found in other fuels. The combustion characteristics are also tightly specified to ensure adequate combustor life, initially specified in terms of luminometer number, but later changed to hydrogen content. A high net heat of combustion was also specified. The JP-7 specification, MIL-DTL-38219, was first published in 1970. The SR-71 was retired twice and is not in the U.S. Air Force (USAF) inventory.

JP-8 (AVTUR, NATO F-34), a kerosene fuel very similar to commercial Jet A-1, was developed by the USAF to reduce

TABLE 5—U.S. and British Military Fuel and Related Specifications

			British		
Designation	U.S. Specification	NATO No.	Designation	DefStan Specification	Description
JP4	MIL-PRF-5624	F-40	AVTAG/FSII	91/88	Wide-cut fuel
JP5	MIL-PRF-5624	F-44	AVCAT/FSII	91/86	High flash kerosene
JP8	MIL-PRF-83133	F-34	AVTUR/FSII	91/87	Standard military kerosene
...	...	F-35	AVTUR	91/91	Standard civil kerosene
FSII	MIL-DTL-27686	F-1745	FSII	68/252	DiEGME
Corr./lubricity improver	MIL-PRF-25017	S-1747	...	68/251	Corrosion inhibitor/lubricity improver

the fire hazards associated with wide-cut fuels, which became apparent during the Southeast Asian conflict. JP-8 replaced JP-4 as the primary military jet fuel for USAF operations in Great Britain in 1979 and is currently the primary jet fuel for NATO. The USAF completed its conversion to JP-8 in 1995. JP-8 is covered by the specification MIL-DTL-83133 and British Defence Standard 91-87. Although JP-8 has replaced JP-4 in most every case, the potential need for JP-4 under emergency situations necessitates maintaining this grade in specifications MIL-DTL-5624 and Defence Standard 91-88.

After extensive service trials, the USAF started a changeover to JP-8 MIL-DTL-83133 and British Defence Standard 91-87, a kerosene type product, beginning in the late 1970s. The changeover to JP-8 from JP-4 was essentially completed by 1995 (UK in 1979, NATO in 1988, Pacific and Continental U.S. in 1995). JP-8 (AVTUR, NATO F-34), a kerosene fuel very similar to commercial Jet A-1, was developed by the USAF to reduce the fire hazards associated with wide-cut fuels. Small arms fire accounted for 63 % of USAF aircraft losses while the U.S. Navy under similar flight conditions did not experience similar losses. The USAF issued a Required Operational Capability in 1967 for all USAF aircraft to have the capability to fly on JP-8. JP-8's primary difference with JP-4 is its decreased volatility and considerably higher freezing point. The volatility change improved ground-handling and combat safety, but significant changes were needed to obtain adequate low temperature starting with the lower-volatility, higher-viscosity fuel in some older aircraft. The adoption of JP-8 in aircraft became an important logistic improvement, because it allowed JP-8 to become the single battlefield fuel in the air and on the ground where diesels and gas turbines took the place of gasoline-powered vehicles. Having the same base fuel as commercial airlines has allowed the military to use the commercial fuel transportation system by incorporating the military additives at the point of entry into the military system. As mentioned, Table 5 lists U.S. military specifications for jet fuels and some related products.

AMERICAN CIVIL JET FUELS

In the United States, ASTM created the first commercial jet fuel specification drawing on both the British Ministry of Defence and USAF specifications for experience in relating fuel properties to performance. The effect of kinetic heating due to increases in aircraft speed reduced the need for a −50°C (−58°F) freezing point and allowed this parameter to be relaxed slightly to relieve problems with availability. The current ASTM D1655 specification covers two commercial fuel grades: Jet A and Jet A-1. Jet A and Jet A-1 are kerosene fuels that are essentially identical except for their freezing points; Jet A has a maximum freezing point of −40°C, whereas Jet A-1's freezing point is −47°C (−53°F). Jet A is used almost exclusively by commercial airlines operating within the continental United States, while Jet A-1 is used in most other countries. Jet B is a wide-cut distillate fuel that is the commercial equivalent of JP-4 but without the mandatory additives. Jet B is not widely used as its volatility makes it less safe than Jet A or Jet A-1 and certain aircraft engines are not certified to operate with this fuel. Prior to 2000, Jet B was included in D1655 but since then was transferred to a separate specification (ASTM D6615, Specification for Jet B Wide-Cut Aviation Turbine Fuel). Details of the two kerosene grades in D1655, as well as the characteristics of Jet B in D6615, are contained in Table 6. Jet A with its −40°C freezing point is the general domestic jet fuel in the United States and accounts for about half the civil jet fuel used throughout the world. It satisfies the requirements of both domestic flights and most of the international flights originating in the United States. The Jet A-1 freezing point of −50°C was originally intended to satisfy the unusual demands of long-range, high-altitude flights, but

TABLE 6—Detailed Requirements of Aviation Turbine Fuels[A]

Property		Jet A or Jet A-1 D1655		Jet B D6615
COMPOSITION				
Acidity, total mg KOH/g	max	0.10		...
Aromatics, volume percent D1319	max	25		25
D6379	max	26.5		26.5
Sulfur, mercaptan,[C] mass percent	max	0.003		0.003
Sulfur, total mass percent	max	0.30		0.30
VOLATILITY				
Distillation temperature, °C:		D86	D2887	D86
10 % recovered, temperature	max	205	185	...
20 % recovered, temperature	min	...	...	90
	max	...	...	145
50 % recovered, temperature		report	report	
	min	...	...	110
	max	...	...	190

(Continued)

TABLE 6—Detailed Requirements of Aviation Turbine Fuels[A] (Continued)

Property		Jet A or Jet A-1 D1655		Jet B D6615
90 % recovered, temperature		report	report	
	max	. . .	. . .	245
Final boiling point, temperature	max	300	340	1.5
Distillation residue, vol percent	max	1.5	. . .	1.5
Distillation loss, vol percent	max	1.5	. . .	1.5
Flash point, °C	min	38		. . .
Density at 15°C, kg/m³		775 to 840		751 to 802
Vapor pressure, 38°C, kPa		. . .		14 to 21
FLUIDITY				
Freezing point, °C	max	−40 Jet A		−50
	max	−47 Jet A-1		
Viscosity—20°C, mm²/s	max	8.0		. . .
COMBUSTION				
Net heat of combustion, MJ/kg	min	42.8		42.8
One of the following requirements shall be met:				
(1) Smoke point, mm, or	min	25		25
(2) Smoke point, mm, and	min	18		18
Naphthalenes, vol percent	max	3.0		3.0
CORROSION				
Copper strip, 2 h at 100°C	max	No. 1		No. 1
THERMAL STABILITY				
D3241 (2.5 h at control temperature of 260°C min)				
Filter pressure drop, mm Hg	max	25		25
Tube deposits less than		3		3
		No *Peacock* or *Abnormal* Color Deposits		
CONTAMINANTS				
Existent gum, mg/100 mL	max	7		7
Microseparaometer, rating				
Without electrical conductivity	min	85		85
With electrical conductivity	min	70		70
ADDITIVES		See specification		See specification
Electrical conductivity, pS/m				
if used	max	600		450
if specified at point of delivery		50 to 600		50 to 450

[A] For additional requirements contained in specification footnotes, refer to Table 1 in ASTM D1655 or Table 1 in ASTM D6615.

in 1980 the freezing point was raised to −47°C to respond to availability concern and to take advantage of better definitions of long-range flight requirements. For international and domestic flights outside the United States, Jet A-1 is the standard fuel. Although Jet A would meet many local requirements, the design of most airport fuel systems limits them to a single grade. To differentiate commercial from military grades (which often contain additives not found in civil fuel), the terms "Jet A-1" and "Jet B" are used worldwide to describe civil fuels, although Jet B usage is extremely limited.

TABLE 7—Russian Jet Fuel Specifications

Specification	Grade	Type	Use
GOST 10227	TS-1 (premium)	Kerosene (SR)[A]	Most common civil
GOST 10227	TS-1 (regular)	Kerosene (SR)[A]	Most common civil
GOST 10227	T-1 (regular)	Kerosene	Common civil
GOST 10227	T-1S (special)	Kerosene	Special application
GOST 10227	T-2	Wide cut	Standby (reserve) fuel
GOST 10227	RT	Kerosene (HT)[B]	Military/occasionally civil
GOST R 52050-2006	Jet A-1	Kerosene	Commercial

[A] SR = straight-run.
[B] HT = hydrotreated.

Major U.S. aircraft engine manufacturers and certain airlines also issue jet fuel specifications. These are either similar to the ASTM specification or possibly less restrictive than one or more of the ASTM grades. Should a manufacturer's specification be more restrictive than ASTM, it would create major problems because the manufacturer's specification is normally used for certification and would, therefore, have to be followed by the users. In turn, the ASTM specification would become an unused piece of paper in such cases.

RUSSIAN JET FUELS

Several jet fuels covered by various GOST specifications are manufactured for both civil and military use. The main grades are also covered by specifications issued by a number of east European countries, although a number of these countries are changing to Western specifications as they purchase and operate Western aircraft. While Russian fuel characteristics in some cases differ considerably from those of fuels made elsewhere, the main properties are controlled by test methods similar to their ASTM/IP equivalents. A few additional test methods, such as iodine number (related to olefin content), hydrogen sulfide content, ash content, and naphthenic soaps, are sometimes included. Thermal stability is usually specified but by completely different test procedures. However, a recent research program sponsored by the International Air Transport Association (IATA) is intended to establish the relationship between Russian and Western test methods.

Brief details are given in Table 7. TS-1 and sometimes RT are the only grades normally offered to international airlines at civil airports. Both the RT grade and the more common TS-1 Premium normally satisfy current Jet A-1 specification requirements, with the exception of a flash point minimum of 28°C (82°F). However, Western engine manufacturers are also concerned about the thermal stability of Russian jet fuels because of basic differences in test methods. Efforts are currently under way to reconcile specification limits set by ASTM D3241, the thermal oxidation stability test method procedure, and GOST 11802-88, the Russian test method.

Other National Specification

Several other counties also issue jet fuel specifications, and the most important are listed in Table 8. In most cases, these specifications are identical with their U.S. or British counterparts, particularly for countries committed to multinational military standardization agreements, such as NATO. In many of these countries, little or no use is made of the national specification, and most fuels are manufactured as Jet A-l to the commercially accepted Joint Check List (see later). However, a few countries, including Brazil, Canada, France, and Sweden, make considerable use of their national standards.

TABLE 8—Other National Aviation Fuel Specifications

Country/Issuing Agency	Kerosene	Wide-Cut	High-Flash Kero
Australia/A. Dept. Defence	DEF (AUST) 5240 QAV-1	DEF (AUST) 5280 QAV-4 ...	DEF (AUST) 5207
Canada/CAN/CGSB	3.23	3.22	3.24
Peoples Republic of China	GB 1778 (No. 2 Jet Fuel)	SH 0348 (No. 4 Jet Fuel)	GBJ 560A (No. 5 Jet Fuel)
	GB 6537 (No. 3 Jet Fuel)	GJB 2376 (No. 4 Jet Fuel)	
France/Service des Essences des Armees	DCSEA 134 Jet A-1	AIR 3407	AIR 3404
Germany	Joint Check List Jet A-1	DefStan 91/88	...
Japan/Japan Defense Agency	JIS K 2209 Jet A-1	DSP K 2206 (JP-4)	DSP K 2206 (JP-5)
Sweden/SDMA	FSD8607	FSD8608	...

INTERNATIONAL STANDARD SPECIFICATIONS

Modern civil aviation recognizes few frontiers, and a need, therefore, exists to have aviation fuels of similar characteristics available in all parts of the world. This is especially important for jet fuels used by the international airlines. An early attempt to simplify the specification picture was the establishment of a checklist to be used by eleven major fuel suppliers where more than two suppliers furnished fuel to commingled terminals or airports. This checklist is formally termed "The Aviation Fuel Quality Requirements for Jointly Operated Systems (AFQRJS)" and applies outside the United States. This checklist included the most severe requirements of ASTM Jet A-1, Def Stan 91-91, and the IATA guidance material for Jet A-1 grade. A major shortcoming of this approach has been that over time more and more suppliers, such as government-owned oil companies, manufactured jet fuel but were not part of the group issuing the checklist. IATA has, therefore, issued "guidance material for aviation fuel" in the form of four specifications. Included are the domestic U.S. fuel (Grade Jet A) based on ASTM D1655, the internationally supplied Jet A-1 grade meeting Def Stan 91-91 and ASTM Jet A-1, and the Russian specification TS-1. Although the first three are all kerosene-type fuels, which are basically similar, the differences between specifications are sufficient to prevent combining them into a single grade. Thus, Jet A differs from the others in having a higher freezing point, while the Russian fuel has both a lower flash and freezing point. An international airline is, therefore, likely to obtain Jet A in the United States, TS-1 in Russia and some other Eastern countries, and Jet A-l in the rest of the world. Jet B is included because of its use in a few northern locations where an airline might have to take the fuel on an emergency basis. Table 9 summarizes some of the significant differences between the various major specifications. Over the past few years, a great deal of effort has gone into harmonizing the worlds jet fuel specifications to reduce the confusion caused by having different limits for the same quality parameters. This effort is particularly important today as air travel is on the rise and the barriers that prevented global travel have all but disappeared. The *International Air Transport Association (IATA) Guidance Material for Aviation Turbine Fuels Specifications (Edition 6)* has been substantially revised to include the requirements for Jet A, Jet A-1, and the Russian civil fuel TS-1. With this change, the guidance material now includes all grades of jet fuel used for civil aviation. Although every effort was made to harmonize the kerosene grades, differences still exist, which will most likely remain for the near future.

Illuminating kerosene was the first true aviation turbine fuel developed more 50 years ago. Today, after all the amazing technological advances, the fuels used by the world's jet aircraft are still kerosene based, whether they are Jet A in the United States, TS-1 in Russia, or Jet A-1 elsewhere in the world. Minor changes to specification properties will result over time as dictated by new equipment requirements and environmental pressures to reduce aircraft emissions. Due in part to the long service life of aircraft and the lack of commercially viable alternatives, jet fuel, much as we know it today, will remain the principal aviation fuel well into the 21st century.

Composition and Manufacture

Aviation turbine fuels are manufactured predominantly from straight-run (noncracked) kerosene obtained by the atmospheric distillation of crude oil. Straight-run kerosene from some sweet crudes meet all specification requirements without further processing, but for the majority of crudes certain trace constituents have to be removed before the product

TABLE 9—Comparison of Critical Properties Among Major Specifications

	ASTMD1655	DefStan 91-91	GOST 10227		ASTM D6615	MIL-PRF-5624
Property	Jet A	Jet A-l	TS-1		Jet B	JP-5
			Prem.	Reg.		
Flash point, °C, min	38	38.0	28		<18	60
Vapor pressure, kPA @38°C	Approx. 0.28–0.62	Approx. 0.28–0.62	Approx. 0.48–1.38		14–21	<1
Freeze point, °C, max	–40	–47	–60		-50	–46
Density @ 15°C, kg/m^3	775–840	775.0–840.0	780 min	775 min	751–802	788–845
Smoke point, mm	25 min or	25.0 min or	25 min		25 min or	19.0 min
Or smoke point, min + naphthalenes, vol %	18	19.0			18	
	3.0 max	3.00 max			3.0 max	13.4^A
Aromatics, vol percent, max D6379 max	25.0 or 26.5	25.0 or 26.5	22 m %		25.0 or 26.5	25.0
Distillation, °C 10 % recov., max flash boiling point max	205 300	205 300.0	165 250^B		90 min^C 145 max^C 245^D	206 300

^{A} Percent hydrogen.
^{B} 98 % recovered.
^{C} 20 % recovered.
^{D} 90 % recovered.

meets aviation fuel specifications. This is normally done by contacting the component with hydrogen in the presence of a catalyst (hydrotreating or hydrofining) or by a wet chemical process such as Merox treating. Further details on composition and constituent removals are covered in the following section on specification requirements.

Traditionally, jet fuels have been manufactured only from straight-run (noncracked) components, because the inclusion of raw thermally or catalytically cracked stocks would invariably produce an off-specification fuel. In recent years, however, hydrocracking processes have been introduced that provide high-quality kerosene fractions ideal for jet fuel blending.

Fuels from Nonconventional Sources

Jet fuel containing synthetic hydrocarbons have been previously allowed under ASTM D1655; however, the fraction of these hydrocarbons was not limited, and no requirements were placed on the final blend. It has been recognized that synthetic blends and bio-derived jet fuels may be divergent from the fuel properties that went into development of Table 1 in ASTM D1655 and a long-term strategy must be developed to handle these alternative fuels. The current ASTM plan is to progress generic approval of synthetic iso-paraffinic kerosene (SPK) in jet at up to 50 % based on approval of technical merit through research reports featuring five SPK fuels (CTL – Sasol (South African Coal, Oil and Gas Corp.), GTL – Syntroleum, Shell, Sasol 1, and Sasol 2). In the interim, a new specification will be introduced to capture fuels made from nonconventional sources. This specification will have annexes specific to different feedstocks and offer flexibility for inclusion of future biofuels. The new specification will capture any unusual testing requirements without undue limitations or complication of ASTM D1655. Proposals to add additional flash point, volatility, lubricity, and aromatic constraints to D1655 product manufactured from crude oil and related sources have been dropped. The additional features were seen as specific to SPKs and an unnecessary limitation on conventional fuel. The product manufactured to the new specification can be recertified to D1655 jet and enter the conventional jet pool. Currently, SASOL semisynthetic fuel, a blend of conventional produced jet and a synthetic kerosene, is specified in Defence Standard 91-91/Issue 5, dated February 8, 2005, and is recognized as meeting the ASTM D1655 specification requirements.

Defence Standard 91-91 protocol for fully synthetic fuels was also considered and ASTM sought clarification from the original equipment manufacturers (OEMs) regarding the approved product.

Fuels for Nonconventional Aircraft

Some aviation authorities around the world have type certified aircraft with diesel engines to run on jet fuel. As more and more spark-ignition engines (AVGAS powered) are being changed out for diesel engines, the possibility for misfueling increases. The diesel and AVGAS aircraft look identical. Additionally, the engine certification process did not fully consider all relevant properties of jet fuel being used in diesel aircraft engines.

A limited review of jet cetane number (CN) has been undertaken to determine the possible minimum market quality that might be received by new aviation diesel engines. A 37 CN had been observed and some suggested 32 CN was possible–evidence of this was requested. A Cessna representative commented that a test with a U.S. West Coast jet/65°F ambient engine would not start. Data will be reviewed by the FAA and shared with European Aviation Safety Agency (EASA) and other regulators. Work is continuing with the FAA in regard to diesel powered aircraft using turbine jet fuel.

Specification Requirements

The requirements for jet fuels stress a different combination of properties and tests than those for aviation gasoline. Some tests are used for both fuels, but the majority of jet fuel requirements fit into different categories, as will be seen.

COMPOSITION

Jet fuels are required to consist entirely of hydrocarbons, except for trace quantities of sulfur compounds and approved additives. As mentioned earlier, these fuels are made mostly from straight-run kerosene and hydrocracked streams, and satisfactory operating experience has been based on this manufacturing pattern. This experience has resulted in specifications in which the test requirements can be divided into two arbitrary groups. The first group can be called bulk properties because a significant change in composition is required to change the property. Bulk properties have a major effect on availability, that is, the amount of jet fuel obtainable from a barrel of crude. Trace properties, on the other hand, are affected by small changes in composition, sometimes as little as 1 ppm. These properties do not affect availability but are in the specification to prevent or solve specific operating problems. The following sections will elaborate further on these themes. As will be seen, certain cleanliness factors are also in use but are not included in all specifications.

BULK PROPERTIES

The following groups of properties bear directly on availability:

- Volatility
- Low temperature properties
- Combustion
- Density
- Specific heat
- Aromatic content

Volatility

Volatility is the major difference between kerosene and wide-cut fuels and is described by three tests. Kerosene-type fuel volatility is controlled by flash point and distillation, and the more volatile wide-cut fuels by vapor pressure and distillation. Flash point is a guide to the fire hazard associated with the fuel and can be determined by several standard methods, which are not always directly comparable. In each method, the fuel is warmed in a closed container under controlled conditions, and the vapor space flammability is periodically tested with a flame or spark. The flash point is the temperature at which enough vapor is formed to be ignitable, but not enough to keep burning. Differences in apparatus, vapor-to-liquid ratio, heating rate, and other test variables are responsible for the disagreements between methods. Unfortunately, these methods are old and have become embedded in all types of handling regulations, making the adoption of a single international method unlikely. ASTM and U.S. JP-8 military specifications call for the use of the Tag Closed Cup Tester (D56) or the Seta Closed Cup

(D3828/IP 303). British specifications usually require the Abel Flash Tester (IP 170). High flash point JP-5 fuels call for the use of the Pensky-Martens Closed Tester (D93/IP 34), which is also the referee method for JP-8 along with D4052. As noted, the various flash point methods can yield different numerical results. In the case of the most commonly used methods (Abel and TAG), the former (IP 170) has been found to give results up to 1–2°C lower than the latter method (D56). Seta flash values tend to be very close to Abel results. Various studies have shown the flash point of kerosene-type fuels to be one of the critical limitations on the amount of aviation kerosene obtainable from crude oil.

Vapor pressure is the major volatility control for wide-cut fuels. Flash point methods are not directly applicable, because these fuels are ignitable at room temperature and, therefore, cannot be heated under controlled conditions before a flame is applied. (Vapor pressure is not a suitable control for kerosene fuels, because their vapor pressure at 38°C is too low to be measured accurately in the Reid vapor pressure method.) As with aviation gasoline, minimum vapor pressure affects low temperature and in-flight starting, while the maximum allowable vapor pressure limits tank venting losses, as well as possible vapor lock at altitude.

Distillation points of 10, 20, 50, and 90 % are specified in various ways to ensure that a properly balanced fuel is produced with no undue proportion of light or heavy fractions. The distillation end point limits heavier material that might give poor vaporization and ultimately affect engine combustion performance. In some specifications, the standard distillation (ASTM D86) can be replaced by a gas chromatographic method (ASTM D2887), but different distillation limits are then specified. Jet fuel distillation limits are not nearly as limiting to the refiner as the distillation limits for aviation gasoline. Instead, front-end volatility for kerosene is controlled by flash point, while wide-cut volatility is limited by vapor pressure.

Low Temperature Properties

Jet fuels must have acceptable freezing points and low temperature pumpability characteristics, so that adequate fuel flow to the engine is maintained during long cruise periods at high altitudes. Normal paraffin compounds in fuels have the poorest solubility in jet fuel and are the first to come out of the solution as wax crystals when temperatures are lowered. The ASTM Freezing Point of Aviation Fuels (D2386/IP 16) and its associated specification limits guard against the possibility of solidified hydrocarbons separating from chilled fuel and blocking fuel lines, filters, nozzles, etc. In addition to ASTM D2386, a manual method, which is the referee method, three automatic freezing point methods are permitted at this time (2009) by ASTM D1655, Def Stan 91-91, and MIL-DTL-81383 (JP-8). These are ASTM D5972/IP 435, Automatic Phase Transition Method, ASTM D7153/IP 529, Automatic Laser Method, and ASTM, D7154/IP 528 Automatic Fiber Optical Method. The text in D1655 (Section 10.1.4) allows for all four methods but recommends that D5972 or D7153, or both, be used to certify and recertify fuel on the basis of the reproducibility and cross-contamination detection reported in RR: D02-1572. The cause of freezing point results outside specification limits by automated methods should be investigated, but such results do not disqualify the fuel from aviation use if the results from the referee test (Test Method D2386) are within the specification limit. An additional test not included in the three aviation specifications is ASTM D4305/IP 422, Filter Flow of Aviation Fuels at Low Temperatures. This test gives results similar to D2386 but can only be run on fuels with viscosities below 5.0 mm^2/s at −20°C, because higher viscosities can show filter plugging without any wax precipitation.

Extensive studies have shown the −40°C freezing point of Jet A to be limiting aircraft performance on very long flights over the North Pole, particularly flying in the westerly direction. Considerable work on this problem continues, including measuring the freezing point of the fuel at the point of aircraft loading. Supply system constrictions normally prevent furnishing both Jet A and Jet A-l at the same airport. On the other hand, supplying Jet A-1 instead of Jet A throughout the entire U.S. system involves a very significant product loss, indicating the important role of freezing point in maintaining fuel availability.

Fuel viscosity at low temperature is limited to ensure that adequate fuel flow and atomization are maintained under all operating conditions and that fuel injection nozzles and system controls will operate to design conditions. The primary concern is over engine starting at very low temperatures, either on the ground or at altitude relight. Fuel viscosity can also significantly influence the lubricating property of the fuel that, in turn, can affect the fuel pump service life. Viscosity is measured by ASTM Determination of Kinematic Viscosity of Transparent and Opaque Liquids (and the Calculation of Dynamic Viscosity) (D445/IP 71).

Combustion Quality

Combustion quality is largely a function of fuel composition. Paraffins have excellent burning properties, in contrast to those of aromatics–particularly the heavy polynuclear types. Naphthenes have intermediate burning characteristics closer to those of paraffins. Because of compositional differences, jet fuels of the same category can vary widely in burning quality as measured by engine smoke formation, carbon deposition, and flame radiation.

One of the simplest and oldest laboratory burning tests is the smoke point, determined by the Smoke Point of Aviation Turbine Fuels (D1322/IP 57). This test uses a modified kerosene lamp and measures the maximum flame height obtainable without the appearance of smoke. However, the test is not universally accepted as the sole criterion for engine combustion performance. An early alternative was the Test for Luminometer Number of Aviation Turbine Fuels (D1740), but this test has been dropped from the jet fuel specification ATSM D1655. An acceptable alternative to the smoke point alone is a combination of smoke point and naphthalenes content, as measured by the Test for Naphthalene Content of Aviation Turbine Fuels by Ultraviolet Spectroscopy (D1840). Several chromatographic methods are currently under consideration for the measurement of aromatics and naphthalenes. Another alternative, used in some specifications, is hydrogen content (D3701/IP 338).

However, the relationship of all these tests to engine combustion performance parameters is completely empirical and does not apply equally to different engine designs, particularly where major differences in engine operating conditions exist.

Emissions

Exhaust gas composition is part of the combustion process, but fuel quality has varying effects. Carbon or soot formation tends to correlate inversely with the above combustion tests, but other carbon-containing emissions, such as carbon monoxide or carbon dioxide, are engine functions and are little affected by fuel quality. Sulfur oxides (SO_x) are directly proportional to fuel total sulfur content and can be decreased by reducing fuel sulfur content. Nitric oxides (NO_x), on the other hand, depend on combustion conditions and are not affected by jet fuel characteristics, with fuel nitrogen content being extremely low for other reasons.

Density and Specific Heat (formerly Heat of Combustion)

Fuel density is a measure of fuel mass/unit volume. It is important for fuel load calculations, because weight or volume limitations may exist according to the type of aircraft and flight pattern involved. Because it is normally not possible to supply a special fuel of closely controlled density for specific flights, flight plans must be adjusted to include the available fuel density.

Density and specific energy (calorific value) vary somewhat according to crude source. Paraffinic fuels have slightly lower density but higher gravimetric calorific value than those of naphthenic fuels (J/kg or Btu/lb). On the other hand, naphthenic fuels have superior calorific values on a volumetric basis (joules/liter or Btu/gallon).

Because density changes with temperature, it is specified at a standard temperature, the most common being 15°C (60°F). Density at 15°C in units of kg/m^3 is now becoming the most widely used standard for fuel density worldwide, although some specifications still use relative density (or specific gravity) at 15.6°C/15.6°C (60°F/60°F). Relative density is the ratio of a mass of a given volume of fuel to the same volume of water under standard conditions. The Test for Density, Relative Density (Specific Gravity) or API Gravity of Crude Petroleum and Liquid Petroleum Products by Hydrometer Method (D1298/IP 160) may be used to determine density and relative density. An alternate method, Test for Density and Relative Density by Digital Density Meter (D4052/IP 365), is also acceptable for aviation fuels.

Specific energy, formerly heat of combustion, is the quantity of heat liberated by the combustion of a unit quantity of fuel with oxygen. Heat of combustion directly affects the economics of engine performance. The specified minimum value is normally a compromise between the conflicting requirements of maximum fuel availability and good fuel consumption characteristics. The Test for Heat of Combustion by Bomb Calorimeter (Precision Method) (D4809) is a direct measure of specific energy. Test results are corrected for the heat generated by the combustion of any sulfur compounds. Because this method is cumbersome, two alternative methods are permitted for the calculation of specific energy using other fuel characteristics.

The "aniline-gravity" method is based on the arithmetic product of fuel density and aniline point, the aniline point being the lowest temperature at which the fuel is miscible with an equal volume of aniline. This temperature is inversely proportional to the aromatic content. ASTM D4529/IP 381, Test for Estimation of Heat of Combustion of Aviation Fuel, gives the relationship between the aniline-gravity product and the heat of combustion with corrections for sulfur content.

In another empirical method, the heat of combustion (ASTM D3338) is calculated from the fuel's density, the 10, 50, and 90 % distillation temperatures, and the aromatic content. This method avoids the use of aniline, a highly toxic reagent, and uses characteristics that are measured as part of specification compliance. Resolution of any disputes requires the use of the bomb calorimeter.

TRACE PROPERTIES

As pointed out earlier, these properties are primarily intended to solve operating problems of various types and have no direct effect on availability. The following properties are involved:

- High temperature stability
- Storage stability
- Corrosion
- Compatibility with system materials
- Electrical conductivity
- Lubricity
- Contaminants

High Temperature Stability

The ability of fuel not to "break down" under engine operating conditions is critical in today's engines. The engine designer uses fuel as a heat sink to carry away heat from various lubricating oil systems and aircraft operating systems. Additionally, the engine fuel pump rejects heat into the fuel as excess fuel is bypassed back from the fuel control and is recirculated through the pump. A final heat source is the hot compressor discharge air that surrounds the nozzle feed arms ahead of the combustion chamber. Fuel temperature is also influenced strongly by the mass of fuel passing through the system. Flow is maximum at aircraft takeoff and is minimum at the end of cruise and the beginning of descent when fuel flow is cut back to flight idle. Thus, the highest fuel temperatures occur at the end of cruise. In this challenging environment, fuel must not form lacquers or deposits that could adversely affect fuel/oil heat exchangers, metering devices, fuel filters, and injection nozzles. More efficient engines, the constant goal of engine design, use less fuel and, therefore, cause more heat rejection per mass of fuel, higher fuel temperatures, and greater heat stress on the fuel.

Research on the problem has shown it to be one of high temperature oxidation. In Western specifications, that property is measured by a dynamic test, the Test for Thermal Oxidation Stability of Aviation Fuels (D3241/IP 323). In this procedure, fuel is pumped over a heated aluminum tube and through a very fine, heated stainless steel screen. Fuel performance is based on the color of tube deposits and the final pressure drop across the screen. Russian specifications use a static heating test. Work under way at the time of this writing is intended to establish the relationship between the two tests.

Storage Stability

Unlike aviation gasoline, straight-run jet fuel has good storage stability, as it does not readily oxidize under normal storage conditions. However, high-pressure hydrotreating or hydrocracking destroys the sulfur- and nitrogen-containing heteroatoms, which act as natural oxidation inhibitors, so that such fuels can form peroxides as part of the oxidation process. These peroxides, in turn, attack nitrile rubber components in the fuel system. Military and some civil specifications prevent the problem by the mandatory addition of oxidation inhibitors at the refinery.

Corrosion

Direct corrosion of metals, particularly copper, has been attributed to the presence of hydrogen sulfide or elemental sulfur at levels of 1 ppm or less. Rather than analyze for these materials, the fuel is exposed to copper strips heated to 100°C for 2 h. Copper strip appearance is then compared with a color chart, ASTM D130/IP 154, which is the Test for the Detection of Copper Corrosion by Petroleum Fuels by the Copper Strip Tarnish Test, with the color chart an adjunct to the method. Corrosion by organic acids in the fuel is limited by measuring and controlling the acidity of fuels by ASTM D3242/IP 354, the Test for Acidity in Aviation Fuels.

Early jet engines experienced hot section corrosion through attack by sulfur compounds in the exhaust stream. Improved high-temperature engine materials have eliminated this problem. However, sulfur compounds are limited in jet fuel and are measured by ASTM 1266/IP 107, D1552, D2622, or D4294.

Compatibility with System Materials

Aside from the corrosion of metals, compatibility with other materials has involved primarily the interaction between fuel constituents and system elastomers. Elastomers are designed to swell a certain amount in the presence of fuel to seal systems. Fuel aromatics have played a key role in this regard, although the role of specific aromatics has not been well identified. Some concerns have arisen over possible seal shrinkage with fuels with zero aromatic content, but a minimum aromatic content requirement to prevent this possibility has not been enacted.

Specific sulfur compounds (i.e., mercaptans) are limited to 0.001–0.005 % by mass because of objectionable odor, adverse effects on certain elastomers, and corrosiveness of certain fuel system materials, particularly cadmium. Mercaptan sulfur content is determined by the Test for Mercaptan Sulfur in Gasoline, Kerosene, Aviation Turbine, and Distillate Fuels (Potentiometric Method) (D3227/IP 342) or by the qualitative Doctor Test (D4952/IP 30).

Electrical Conductivity

Hydrocarbons are poor conductors of electricity, with the result that charges of static electricity, generated by fuel, travel through the distribution system, may accumulate, and take significant time to leak off to the ground. In some cases, such charges have discharged as high energy sparks that have caused fires or explosions under certain air/fuel vapor conditions. This is particularly true for modern jet fuels because of their high purity, the high pumping velocities used, and the use of microfiltration capable of producing a high rate of charge separation and static buildup in the fuel. Measures must be taken to prevent such possibilities, one being the inclusion of a conductivity-improving additive. Many fuel specifications require the use of static dissipator additive (see later) to improve handling safety. In such cases, the specification defines both minimum and maximum electrical conductivity. The minimum level ensures adequate charge relaxation, while the maximum prevents too high a conductivity that can upset capacitance-type fuel gages in some aircraft. Other measures like increased relaxation time can be taken as well. All are described in greater detail in the Guide for Generation and Dissipation of Static Electricity in Petroleum Fuel Systems (D4865).

The standard field test for electrical conductivity has been the Test for Electrical Conductivity of Aviation and Distillate Fuels (D2624/IP 274). Although the method is intended for the measurement of conductivity with the fuel at rest in storage tanks, it can also be used in a laboratory. However, the method discourages the shipment of samples, because of container and storage effects. If needed, a more precise laboratory method for fuels of very low conductivities is the Test for Electrical Conductivity of Liquid Hydrocarbons by Precision Meter (D4308).

Lubricity

Under a combination of high loads and sliding action, such as between gear teeth, metal-to-metal separation must be maintained to prevent scuffing or seizing. Straight-run fuels appear to include enough heteroatoms containing sulfur or nitrogen compounds to act as a surface film that separates the metal surfaces. The property of maintaining this separation is known as lubricity, and the heteroatoms are considered natural lubricity agents. However, when fuel has been processed under conditions that destroy these agents, the resultant fuel has poor lubricity and is sometimes called a "hard" or "dry" fuel. Poor lubricity can be corrected by the addition of as little as 10 % straight-run fuel or by the addition of an approved lubricity additive. Most likely, the extensive mixing of jet fuel in the U.S. supply system has prevented lubricity problems here. However, where a refinery making hard fuel is the only supplier to an airport, and aircraft there operate mostly on such fuel, lubricity problems such as fuel pump or engine control failures have occurred, and fuel corrections must be made. Engine and accessory manufacturers are continuing to design their equipment to operate on hard fuels. Operating problems, therefore, have occurred mostly in older equipment.

Lubricity is measured with the Ball on Cylinder Evaluator (BOCLE) (D5001). A hardened cylinder is rotated at constant speed while it dips into a sample of test fuel. A ball bearing is pressed against the wetted cylinder under load for a specified period of time. During the entire test, the apparatus is kept under a temperature and humidity controlled atmosphere. The resultant wear scar on the ball is measured under a microscope and reported in millimeters. The test is complicated and difficult to run and would be burdensome if required on every refinery batch as part of acceptance testing. At this time, British specification writers have introduced a proposed limit of 0.85 mm maximum into Def Stan 91-91. The limit would apply when the fuel is more than 95 % hydroprocessed material, and at least 20 % is severely hydroprocessed. ASTM is closely following this work and expects to take action as appropriate.

Contaminants

Modern aircraft fuel systems demand a fuel be maintained in as clean a condition as possible to be relatively absent of water, dirt, and foreign contaminants. To deliver clean and dry fuel, multistage filtration systems are used at terminals and airports and on the delivery vehicles. Particularly in the United States, jet fuel is widely delivered from refineries to terminals through large, very long pipelines that also handle other products. As a consequence, contamination of jet fuel by water, solids, and additive traces is inevitable and must be removed by ground filtration systems. Additives can be surface-active and interfere with the proper operation of filtration systems by dispersing water and dirt. Surfactant removing filters (clay filters) are a common constituent in U.S. cleanup systems at terminals and sometimes at airports.

Airport control of cleanliness is determined by various quality control documents such as ATA 103 and IATA Guidance Material for Aviation Turbine Fuels Part III–Cleanliness and Handling that determine acceptance or use limits. Testing for contaminants of various types occurs at many points in the distribution system. During aircraft fueling, jet fuel appearance is tested for "clear and bright" by visually examining a sample using ASTM D4176. Delivered fuel must also contain less than 1 mg/L of particulates and less than 30 mg/L of free water per U.S. military specifications. For civil fuels, cleanliness requirements tend to be a matter of contractual agreement between supplier and user.

ASTM Test for Particulate Contaminant in Aviation Turbine Fuels (D2276/IP 216) provides a quantitative measure of dirt mass by filtration through a membrane. It can be supplemented by comparing the color of a membrane after test against the color standards in Appendix XI of ASTM D2276/IP 216. However, no direct relationship exists between particulate mass and membrane color, and field experience is required to assess the results by either method.

Free water dispersed in jet fuel can be detected with a variety of field kits developed over the years by major oil companies. These tests generally rely on color changes produced when chemicals on a filter go into aqueous solution. The Test for Undissolved Water in Aviation Turbine Fuels (D3240) has been standardized and uses a device called the Aquaglo II, which is capable of more precise quantitative results than the chemical tests, although test simplicity is sacrificed.

The total water content of aviation fuels (free plus dissolved water) can be measured with the ASTM Test Method for Determination of Water in Petroleum Products, Lubricating Oils, and Additives by Coulometric Karl Fisher Titration (D6304). However, this is a laboratory procedure requiring careful sample handling, and results are difficult to compare with the free water tests mentioned earlier.

WATER RETENTION AND SEPARATING PROPERTIES

Because of higher density and viscosity, jet fuels tend to suspend fine particulate matter and water droplets much longer than does aviation gasoline. Jet fuels also tend to vary considerably in their tendencies to pick up water droplets and to hold them in suspension, depending on the presence or absence of trace surface-active impurities (surfactants). Some of these materials–such as sulfonic or naphthenic acids and their sodium salts–may result from the crude source or certain refinery processes. Others may be picked up by contact with other products during transportation to the airport, particularly in multiproduct pipelines. These surfactants may be natural constituents of other, less refined products (for example, heating oils) or may consist of additives trailing back from other products.

Surfactants tend to impair the performance of water-separating equipment (filter-separators) intended to remove traces of free or undissolved water. While some additives specified for jet fuels, including corrosion inhibitors and static dissipator additives, also have surface-active properties, jet fuel filtration equipment is designed to operate with these approved additives. However, very small traces of free water can adversely affect jet engine and aircraft operation, particularly by ice formation. The water retention and separating characteristics have, thus, become a critical quality consideration. Tests to measure and control these properties have been mentioned earlier under "Trace Properties."

Microbial growth activity is another type of contamination, which can give rise to various service problems. Difficulties can usually be avoided by the adoption of good housekeeping techniques, but major incidents in recent years have led to the development of microbial biocides, as well as microbial monitoring tests for jet fuels. Fuel in tropical areas is particularly at risk, because elevated fuel temperatures favor microbial growth. An excellent discussion of the subject will be found in ASTM D6469, Guide for Microbial Contamination in Fuels and Fuel Systems.

MISCELLANEOUS PROPERTIES

Special tests may be in proprietary specifications, but are not necessarily in industry specifications. These include color limits by the Saybolt Color Method (D156) or Color by the Automatic Tristimulus Method (D6045). Although not normally a specification item, color deterioration can be a useful indication of interproduct contamination or instability (gum formation).

INSPECTION DATA ON AVIATION TURBINE FUELS

Many airlines, government agencies, and petroleum companies make detailed studies of inspection data provided on production aviation turbine fuels. Because a large number of inspections are generally involved, these studies are frequently made with the aid of computers. Without a standardized format for reporting data from different sources, transcribing the reported data for computer programming is laborious.

To facilitate the reporting of inspection data on aviation turbine fuels, ASTM has established a standardized report form. It appears as Appendix X3 to ASTM D1655 and is attached here as Fig. 1.

AVIATION FUEL SAMPLING

Sampling of aviation products is normally carried out by following ASTM D4057, Practice for Manual Sampling of Petroleum and Petroleum Products. However, certain properties of jet fuels are very sensitive to trace contamination that can originate from sample containers. These properties include thermal stability, water separation, electrical conductivity, and lubricity. For recommended sample containers, refer to ASTM D4306, Practice for Aviation Fuel Sample Containers for Test Affected by Trace Contamination.

AVIATION FUEL ADDITIVES

General

Only a limited number of additives are permitted in aviation fuels, and for each fuel grade, the type and concentration are closely controlled by the appropriate fuel specification. Additives may be included for a number of reasons, but, in every case, the specification defines the requirements as follows:

Mandatory–must be present between minimum and maximum limits.

Optional–may be added by fuel manufacturer's choice up to a maximum limit.

Permitted–may be added only by agreement of user/purchaser within specified limits.

Not allowed additives not listed in specifications cannot be added to aviation fuels.

INSPECTION DATA ON AVIATION TURBINE FUEL
(Items in **bold** type are referenced in the specification)

MANUFACTURER/SUPPLIER ________	DATE SAMPLED ________
PRODUCT CODE/GRADE ________	DATE RECEIVED AT LAB ________
SPECIFICATION ________	CONTRACT NO. ________
SAMPLE NUMBER ________	ORDER NO. ________
DATE SAMPLED ________	TANK NO. ________
SAMPLING LOCATION ________	DESTINATION ________
BATCH NO. ________	CRUDE SOURCE ________
QUANTITY LITRES @ 15°C ________	PROCESSING METHOD ________
QUANTITY U.S. GALLONS @ 60°F ________	REMARKS ________
LABORATORY ________	

Method			Result
		APPEARANCE	
010	D 156	Color (Saybolt)	◊◊◊
020	D 6045	Color (Saybolt)	◊◊◊
030	D 4176	Visual ("Pass" or "Fail")	◊◊◊◊
		COMPOSITION	
100C	**D 3242**	**Acidity, Total (mgKOH/g)**	◊•◊◊◊
110	**D 1319**	**Aromatics (vol %)**	◊◊•◊
112	**D 6379**	**Aromatics (vol %)**	◊◊•◊
115	D 1319	Olefins (vol %)	◊•◊
120	**D 1840**	**Naphthalene (vol %)**	◊•◊◊
130	**D 3227**	**Sulfur, Mercaptan (mass %)**	◊•◊◊◊◊
140	D 4952	Doctor Test (P = pos, N = neg)	◊
150A	D 129	Sulfur, Total (mass %)	◊•◊◊
150B	**D 1266**	**Sulfur, Total (mass %)**	◊•◊◊
150D	**D 2622**	**Sulfur, Total (mass %)**	◊•◊◊
150E	D 3120	Sulfur, Total (ppm)	◊◊◊◊
150F	**D 4294**	**Sulfur, Total (mass %)**	◊•◊◊
150G	**D 5453**	**Sulfur, Total (ppm)**	◊◊◊◊
160A	D 3343	Hydrogen Content (mass %)	◊◊•◊◊
160B	D 3701	Hydrogen Content (mass %)	◊◊•◊◊
		VOLATILITY	
200A	**D 86**	**Distillation by Auto/Manual (°C)**	◊
200B	D 2887	Distillation by GC (°C)	◊
201		Distillation by Initial BP (°C)	◊◊◊•◊
202		Distillation by 10 % Rec (°C)	◊◊◊•◊
203		Distillation by 20 % Rec (°C)	◊◊◊•◊
204		Distillation by 50 % Rec (°C)	◊◊◊•◊
205		Distillation by 90 % Rec (°C)	◊◊◊•◊
206		Distillation by 95 % Rec (°C)	◊◊◊•◊
211		Distillation by Final BP (°C)	◊◊◊•◊
213		Residue (vol %)	◊•◊
214		Loss (vol %)	◊•◊
220A	**D 56**	**Flash Point, Tag Closed (°C)**	◊◊•◊
220B	D 93	Flash Point, PM Closed (°C)	◊◊•◊
220C	**D 3828**	**Flash Point, Setaflash (°C), Meth A**	◊◊•◊
220D	D 3828	Flash Point, Setaflash (°C), Meth B	◊◊•◊
221	D 3828	Flash Point, Setaflash (Flash/No Flash)	◊
230A	**D 1298**	**Density @ 15°C (kg/m^3)**	◊◊◊•◊
230B	**D 4052**	**Density @ 15°C (kg/m^3)**	◊◊◊•◊
231A	D 1298	API Gravity @ 60°F	◊◊•◊
240A	D 323	Vapor Pressure, Reid (kPa)	◊◊•◊
240B	D 4953	Vapor Pressure, Dry Method (kPa)	◊◊•◊
240C	D 5190	Vapor Pressure, Automatic Method (kPa)	◊◊•◊
240D	D 5191	Vapor Pressure, Mini Method (kPa)	◊◊•◊
		FLUIDITY	
300A	**D 2386**	**Freezing Point (°C)**	-◊◊•◊◊
300B	**D 5901**	**Freezing Point (°C)**	-◊◊•◊◊
300C	**D 5972**	**Freezing Point (°C)**	-◊◊•◊◊
300D	**D 4305**	**Freezing Point (°C)**	-◊◊•◊◊
300F	**D 7154**	**Freezing Point (°C)**	-◊◊•◊◊
310	**D 445**	**Viscosity @ -20°C (mm^2/s)**	◊◊•◊◊◊
311	D 445	Viscosity at other temp (mm^2/s)	◊◊•◊◊◊
312	D 445	Temp (°C) of Item 311	◊◊◊◊

Method			Brand	Result
		COMBUSTION		
400A	D 240	Net Heat of Combustion (MJ/kg)		◊◊•◊◊◊
400B	D 1405	Net Heat of Combustion (MJ/kg)		◊◊•◊◊◊
400C	**D 3338**	**Net Heat of Combustion (MJ/kg)**		◊◊•◊◊◊
400D	**D 4529**	**Net Heat of Combustion (MJ/kg)**		◊◊•◊◊◊
400E	**D 4809**	**Net Heat of Combustion (MJ/kg)**		◊◊•◊◊◊
410	**D 1740**	**Luminometer No.**		◊◊
420	**D 1322**	**Smoke Point (mm)**		◊◊•◊
		CORROSION		
500	**D 130**	**Copper Strip**		◊◊
510	IP 227	Silver Strip		◊
		STABILITY		
601A	**D 3241**	**JFTOT ΔP (mm Hg) @ other temp**		◊◊•◊
602A	**D 3241**	**JFTOT Tube Deposit @ other temp**		◊◊◊◊
603A	**D 3241**	**JFTOT TDR Spun Rating @ other temp**		◊◊
604A		**Temperature (°C) of above JFTOT**		◊◊◊◊
601B	**D 3241**	**JFTOT ΔP (mm Hg) @ 260°C**		◊◊•◊
602B	**D 3241**	**JFTOT Tube Deposit Rating @ 260°C**		◊◊◊◊
603B	**D 3241**	**JFTOT TDR Spun Rating @ 260°C**		◊◊
		CONTAMINANTS		
700	IP 225	Copper Content (mg/kg)		◊•◊◊
710	**D 381**	**Existent Gum (mg/100 mL)**		◊◊◊
710A	**IP 540**	**Existent Gum (mg/100 mL)**		◊◊◊
720A	D 2276	Particulate (mg/L)		◊•◊◊
720B	D 5452	Particulate (mg/L)		◊•◊◊
730		Filtration time (minutes)		◊◊
740	**D 1094**	**Water Reaction Interferance Rating**		◊◊
750	D 3948	MSEP (With SDA)		◊◊◊
751	D 3948	MSEP (Without SDA)		◊◊◊
		ADDITIVES	**Brand**	
800		Antioxidant (mg/L)	[ ]	◊◊•◊
810		Metal Deactivator (mg/L)	[ ]	◊•◊
820		Static Dissipator Additive (mg/L)	[ ]	◊•◊
830A	(D5006)	FSII (vol%)	[ ]	◊•◊◊◊
830B	(FTM5327)	FSII (vol%)	[ ]	◊•◊◊◊
830C	(FTM5340)	FSII (vol%)	[ ]	◊•◊◊◊
840		Corrosion Inhibitor (mg/L)	[ ]	◊◊•◊
		OTHER TESTS		
900	**D 2624**	**Conductivity (pS/m)**		◊◊◊◊
901	**D 2624**	**Conductivity Test Temperature (°C)**		◊◊◊

Comments and/or Additional Tests:
____
____
____
____
____
____
____
____

CERTIFIED BY ________

Fig. 1—Standard form for reporting inspection data on aviation turbine fuels. (Source: ASTM D1655-02.)

As part of this process, the fuel manufacturer, blender, or handling agent is required to declare the type of additive and its concentration in the fuel. This documentation should accompany the fuel throughout its movement to the airport.

In the case of aviation gasolines, there is little variation in the types and concentrations of additives normally present in each standard grade, but considerable variations occur in the additive content of jet fuels, depending on the country of origin and whether they are for civil or military use. Table 10 summarizes the most usual additive content of aviation fuels on a worldwide basis (except for Russian grades). Many exceptions occur, and reference to the specification is recommended.

TABLE 10—Summary of Additive Requirements for U.S. and British Aviation Fuels

Additive	Aviation Gasoline	Civil Jet Fuels	Military Jet Fuels
Tetraethyl-lead	Optional[A]	Not allowed	Not allowed
Color dyes	Mandatory	Not allowed	Not allowed
Antioxidant	Optional	Optional[B]	Optional[B]
Metal deactivator	Not allowed[C]	Optional	Optional[D]
Corrosion inhibitor/lubricity improver	Optional	Optional	Mandatory
Fuel system icing inhibitor (FSII)	Optional[E]	Optional[D]	Mandatory
Conductivity improver	Optional[F]	Optional	Mandatory
Leak detector	Not allowed[D]	Optional[D]	Permitted[D]

Note. For detailed additive requirements and limitations, refer to individual specification.
[A] Mandatory for ASTM D910 and Def Stan 91-90, max unintentional limit of 0.013 g of lead/L allowed in ASTM D6227.
[B] Mandatory for hydroprocessed fuels in British, major U.S. military, and international civil Jet A-1 fuel.
[C] Optional in ASTM D6227.
[D] By customer agreement only.
[E] User option but, if required, normally added by aircraft operator.
[F] Mandatory in Canada.

Additive Types

Additives may be included in aviation fuels for various reasons. While their general purpose is to improve certain aspects of fuel performance, they usually achieve the desired effect by suppressing some undesirable fuel behavior, such as corrosion, icing, oxidation, detonation, etc. Additive effectiveness is due to their chemical nature and the resulting interaction with fuel constituents, usually on the trace level. During additive approval, it is important to establish not only that the additive achieves the desired results and is fully compatible with all materials likely to be contacted, but also to ensure that it does not react in other ways to produce adverse side effects (possibly by interfering with the actions of other additives). Individual aircraft and engine manufacturers, generally called OEMs, normally carry out the approval testing of aviation additives. Their results and conclusions appear in company documents and are then approved by appropriate governmental certifying agencies. Once this process is completed, international specification groups can review this approval for adoption into specifications. Although additives for civil fuels are listed in industry specifications following consensus decision, additive listing in ASTM specifications does not constitute ASTM approval, because only the equipment manufacturer has the legal authority for additive approval. However, it is up to ASTM to ensure that approvals have been obtained from all pertinent manufacturers before the additive is listed. For military fuels, additive approvals rest with the military authority and are often designed to satisfy specific military considerations. In some cases, military experience is cited as a reason for approving civil use of an additive. However, civil approval still has to go through the formal process outlined earlier.

To rationalize the expensive approval procedure for aviation fuel additives, ASTM Practice for Evaluating the Compatibility of Additives with Aviation Turbine Fuels and Aircraft System Materials (D4054) was created. It was used in conjunction with ASTM Guidelines for Additive Approval (Research Report D02-1125), and Compatibility Testing with Fuel System Materials (Research Report D02-1137), the procedure offered the possibility of testing by a single manufacturer with the results acceptable to others. At the time of this writing (2009), ASTM D4054 has been completely rewritten and was approved at the December 2009 ASTM meeting in Anaheim, California, as a replacement for the old specification. The new specification will retain the same ASTM D4054 designation with the following new title: Standard Practice for Qualification and Approval of New Aviation Turbine Fuels and Fuel Additives and, Research Report D02-1137 and Research Report D02-1137 will be withdrawn. The subject of this ballot is a complete rewrite of ASTM D4054. The standard practice provides a framework for the qualification and approval of new fuels and new fuel additives for use in commercial and military aviation gas turbine engines. The practice was developed as a guide by the aviation gas-turbine engine and airframe OEMs with FAA and ASTM International member support. The intent of this standard practice is to streamline the approval process. The objective is to permit a new fuel or additive to be evaluated and transitioned into field use in a cost-effective and timely manner. Its purpose is to guide the sponsor of a new fuel or new fuel additive through a clearly defined approval process that includes the prerequisite testing and required interactions with the engine and airframe manufacturers, the FAA, EASA, and listing bodies. This standard practice provides a basis for calculating the volume of additive or fuel required for assessment, insight into the cost associated with taking a new fuel or new fuel additive through the approval process, and a clear path forward for introducing a new technology for the benefit of the aviation community.

The following paragraphs describe the aviation fuel additives in current use. Table 10 lists the additive types and an indication whether the additives are optional, mandatory, or allowed with specific limitations. No attempt is made to list the various chemical and trade names of all approved materials, as these will be found in the individual specifications.

TETRAETHYL-LEAD (TEL)

TEL is used widely to improve the antiknock characteristics of aviation gasoline. An adverse side effect of this material is the deposition of solid lead compounds on engine parts, leading to spark plugs fouling and corrosion of cylinders, valves, etc. To alleviate this potential problem, a scavenging chemical–ethylene dibromide–is always mixed with the TEL. Ethylene dibromide largely converts the lead oxides into volatile lead bromides, which are expelled with the exhaust gases. As a compromise between economic considerations and the avoidance of side effects, the maximum level of TEL is carefully controlled in specifications by using tests for Lead in Gasoline (ASTM D5059 or D3341). TEL is not permitted in jet fuels, because lead compounds, even in trace amounts, could damage turbine blades and other hot engine parts.

COLOR DYES

Dyes are required in all leaded fuels as a toxicity warning. They are also used in aviation gasoline to identify the different grades. The required colors are achieved by the addition of up to three special anthraquinone-based and azo dyes (blue, yellow, and red). The amounts permitted are controlled between closely specified limits to obtain the desired colors. The Test Method for Color of Dyed Aviation Gasolines (D2392) is used to determine minimum required color levels, while maximum color is controlled by dye concentration.

In general, dyes are not permitted in jet fuels, except in special circumstances.

ANTIOXIDANTS (GUM INHIBITORS)

Antioxidant additive is normally added to aviation gasoline to prevent the formation of gum and precipitation of lead compounds. The additive type and concentration are controlled closely by specifications.

Jet fuels are inherently more stable than aviation gasoline. Antioxidants are optional, but not mandatory in all cases. To combat the problem of peroxide formation mentioned earlier, some specifications require the addition of oxidation inhibitors to all hydrogen-treated fuels. Antioxidant use in all hydrogen-treated fuels is probably unnecessary, but it is easier to add the antioxidant to all such fuels than to establish which fuels need the additive and which fuels do not. A maximum concentration of 24.0 mg/L applies for all jet fuels, with a minimum of 17.2 mg/L when the additive is mandatory.

Antioxidants are defined by composition. A wide range of antioxidants is approved with some variations of chemical types among specifications. Hindered phenols predominate among various specifications.

METAL DEACTIVATOR

One approved metal deactivator (*N,N′*-disalicylidene-1,2-propane diamine) is permitted in jet fuels but not in aviation gasoline. The purpose of the additive is to passivate certain dissolved metals, which degrade the storage stability or thermal stability of the fuel by catalytic action. Copper is the worst of these materials and is sometimes picked up during distribution from the refinery to the airport. Copper-containing heating coils in some marine tankers have been identified as one copper source. If thermal stability has been degraded by such copper pickup, it can sometimes be restored by doping the fuel with metal deactivator additive (MDA).

On initial manufacture of fuel at the refinery, MDA content is limited to 2.0 mg/L, not including the weight of solvent. Higher initial concentrations are permitted in circumstances when copper contamination is suspected to occur during distribution. Cumulative concentration of MDA after re-treating the fuel shall not exceed 5.7 mg/L.

CORROSION INHIBITORS/LUBRICITY IMPROVERS

The corrosion inhibitors/lubricity improvers (CI/LIs) are used, when specifically authorized, in jet engine fuels, for the prevention of corrosion in fuel handling, transportation, and storage equipment. Certain of the inhibitors are also used in automotive gasoline, diesel fuel, and related petroleum products. CI/LI additives are used to protect fuel handling infrastructure from corrosion, but that purpose is usually secondary in aviation fuels. Their primary role is to improve the lubricating properties of the fuel. Lubricity additives may be blended into Jet A-1 per Def Stan 91-91 without prior customer notification to correct a lubricity problem, but use of these additives in Jet A/Jet A-1 per ASTM D1655 is by agreement of the purchaser. Aircraft and engine fuel system components and fuel control units rely on the fuel to lubricate their moving parts. The effectiveness of a jet fuel as a boundary lubricant in such equipment is referred to as its lubricity. Differences in fuel system components design and materials result in varying degrees of equipment sensitivity to fuel lubricity. Similarly, jet fuels vary in their level of lubricity. In-service problems experienced have ranged in severity from reductions in flow to unexpected mechanical failure leading to in-flight engine shutdown. The chemical and physical properties of jet fuel cause it to be a relatively poor lubricating material under high temperature and high load conditions. Severe hydroprocessing removes trace components resulting in fuels tend, tend to have lower lubricity than other fuels, such as straight-run, wet-treated, or mildly hydrogen-treated fuels. Certain additives, for example, corrosion inhibitors, can improve the lubricity and are widely used in military fuels. They have been used occasionally in civil jet fuel to overcome aircraft problems but only as a temporary remedy while improvements to the fuel system components or changes to fuel were achieved. Because of their polar nature, these additives can have adverse effects on filtration systems qualified to API 1581 Third Edition or older. The API 1581 Fifth Edition has extensive qualification tests using a full package of additive including a CI/LI and has much improved fuel/water separation characteristics with surfactant-laden fuel. Some modern aircraft fuel systems components have been designed to operate on low lubricity fuel. Other aircraft may have fuel system components that are sensitive to fuel lubricity. In these cases, the manufacturer can advise precautionary measures, such as use of an approved lubricity additive to enhance the lubricity of a particular fuel. Problems are most likely to occur when aircraft operations are confined to a single refinery source where fuel is severely hydroprocessed and where there is no co-mingling with fuels from other sources during distribution between refinery and aircraft. ASTM D5001 (BOCLE) is a test for assessing fuel lubricity and is used for in-service troubleshooting, for lubricity additive evaluation, and in the monitoring of low lubricity test fluid during endurance testing of equipment. However, because the BOCLE may not accurately model all types of wear that cause in-service problems, other methods may be developed to better simulate the type of wear most commonly found in the field.

Both U.S. and British military specifications require the additives on a mandatory basis. U.S. and British military authorities publish specifications for corrosion inhibitors/lubricity agents, the U.S. specification being MIL-PRF-25017, while the British specification is Def Stan 68/251. Approved additives for each specification are in Qualified Products Lists (QPL), the U.S. list being QPL 25017 and the British list, QPL 68/251. Additives on these lists are approved as individual proprietary materials, and the QPLs show relative effective minimum and maximum allowable concentrations for each additive. As required by MIL-PRF-25017, the Relative Effective Concentration (REC) was determined by the Rusting Test Method; the Minimum Effective Concentration was determined by either the BOCLE or 1.5 × REC; and the Maximum Allowable Concentration was determined by the lowest of the following: 54 g of inhibitor/m^3 of fuel, 4 × REC, microseparometer rating, or the change in electrical conductivity with fuels containing static dissipater additive. There is currently an international effort to create a single list of approved additives, and ASTM is expected to adopt this coordinated listing into civil jet fuel specifications.

FUEL SYSTEM ICING INHIBITORS (ANTI-ICING ADDITIVE)

A fuel system icing inhibitor (FSII) was developed originally to overcome fuel system icing problems in USAF aircraft. Most commercial aircraft and many British military aircraft heat the fuel ahead of the main engine filter to prevent the formation of ice by water precipitated from fuel in flight. To maximize aircraft performance, many U.S. military aircraft do not have such heaters, and FSII is required to prevent icing problems. FSII is designed to lower the freezing point of water to such a level that no ice formation occurs.

FSII is now a mandatory requirement in most military fuels, especially those covered by NATO standards. The original FSII was ethylene glycol monomethyl ether (EGME), known also as methyl cellosolve, methyl oxitol, and 2-methoxyethanol by various manufacturers. When this additive was added to jet fuel for naval aircraft (JP-5/Avcat), it was sometimes difficult to meet the minimum 60°C flash point, due to the low flash point of EGME (about 40°C). Consequently, a new type of FSII was introduced into military fuels consisting of diethylene glycol monomethyl ether (diEGME) with a higher flash point (about 65°C) and lower health and safety risks. However, both glycols suffer from poor solubility in jet fuel that has to be overcome by thorough mixing, and they also have a high partition coefficient that causes ready additive extraction by free water. Additive concentration is required to be between 0.10 and 0.15 % by volume. Following the introduction of diEGME, approval of EGME as an icing inhibitor was rescinded due to environmental concerns.

Shortly after introducing FSII to combat icing problems, the USAF experienced a great reduction in the number of microbiological contamination problems in both aircraft tanks and ground storage systems. Studies confirmed that this improvement was due to the biocidal nature of the additive. It is now generally accepted that EGME and diEGME are effective biostats if used continually in fuel.

With minor exceptions, commercial aircraft heat the fuel ahead of the engine filter and have no requirement for FSII. A few turbine-powered helicopters and corporate aircraft do not have fuel heaters, and most operators make their own arrangements for additive injection into their fuel. In tropical areas, some civil aircraft operators require fuel with FSII for its biocidal properties. In these cases, local arrangements tend to be made to inject the additive at the airport.

Although primarily a jet fuel additive, EGME or diEGME is sometimes used as an anti-icer in aviation gasoline for fuel-injected engines. However, for such aircraft, it is more common to use isopropyl alcohol (IPA). The Specification for Fuel System Icing Inhibitors (D4171) defines the properties of all these materials. Concentration limits for the additives are given in the pertinent fuel specification. In addition, the Test Method for Measurement of Fuel System Icing Inhibitors (Ether type) in Aviation Fuels (D5006) provides a field method for measuring the concentration of FSII.

It has been observed that when isopropyl alcohol is added to Grade 100 AVGAS, the antiknock rating of the fuel may be significantly reduced. Typical performance number reductions with the addition of 1 volume percent of IPA have been about 0.5 PN for the lean rating and 3.0 to 3.5 PN on the rich rating. Nonetheless, there have been no field reports of engine distress resulting from these effects. Specification for Aviation Gasoline (D910) contains cautionary statements and gives further details on the phenomenon in Appendix XI.

STATIC DISSIPATOR ADDITIVE (CONDUCTIVITY IMPROVER ADDITIVE)

Static charges can build up during movement of fuel and can lead to high-energy spark discharges. Static dissipator additives (SDAs) are designed to prevent this hazard by increasing the electrical conductivity of the fuel, which, in turn, promotes a rapid relaxation of any static charge. Almost all jet fuel specifications permit the optional use of SDA, but many make it mandatory. SDA is now mandatory in U.S. military grades of JP-8 and JP-4, as well as in Def Stan 91-91 and 91-97. International Jet A-l specifications also contain the requirement. Only U.S. domestic jet fuel leaves the additive as optional, and most such fuel does not contain the additive.

In Canada and the United States, SDA is optional in aviation gasoline because the hazards of static discharges are particularly severe under very low ambient conditions.

The only static dissipator additive currently available for use in aviation fuels is Innospec's Stadis® 450 additive. Its composition is proprietary. The additive is used at very low dosage levels, being limited to 3 mg/L at the time of fuel manufacture and a cumulative total of 5 mg/L after re-treatment. Additive concentration is not measured in the field; instead, additive presence is checked by conductivity measurements by D2624 to ensure that fuel electrical conductivity is within specification limits.

LEAK DETECTOR

Leaks in underground portions of fuel systems have long presented detection problems, particularly where such leaks were small but allowed fuel to accumulate underground. Recent regulations have made periodic system leak checks mandatory. One way of conducting such checks is by the use of a leak detection additive. The only such additive approved for aviation fuel depends on a unique composition (sulfur hexafluoride) and its identification in ground samples to establish the existence of a leak. The additive was developed by the Tracer Research Company and is available as Tracer A®. Its presence is limited to 1 mg/kg of fuel. The method to detect the additive in ground samples is proprietary.

THERMAL STABILITY IMPROVER ADDITIVE (JP-8 PLUS 100 ADDITIVE)

Standard engine design parameters limit maximum fuel temperatures to 163°C (325°F). High-temperature deposits in some current military engines and anticipated higher fuel temperatures in future aircraft have caused the USAF to develop a thermal stability improver that increases the allowable fuel temperature limit by 100°F (60°C). To find a solution to the thermal breakdown of fuel, the Wright Laboratory established as a goal the development of an additive or additive package to improve the thermal stability of JP-8 by 100°F; hence, the project was named JP-8 + 100. Basically, the additive package consists of an approved antioxidant and metal deactivator, as well as a proprietary dispersant and detergent combination. The approved additives are surface active agents (surfactants) that disable the API 1581 Third Edition filter water separator elements used to filter particulate matter and separate water from fuel. Consequently, the thermal stability additives were injected into a dedicated fleet of refuelers to fuel USAF jet fighters. The additive package had little effect on API 1583 absorbent monitor filtration and USAF refuelers handling JP-8 + 100 were converted from filter separators to monitor filtration. The original problem of disarming filter water separators was resolved in December 1999 when Facet successfully qualified to API 1581 Fourth Edition to M100 testing and the other major filter companies followed with their own qualifications. The original intent, when the program started in May 1996, was to only target USAF jet fighter aircraft. To accomplish that goal, special handling procedures were implemented to avoid giving the product to aircraft outside the JP-8 + 100 program. From 1996 to 2000, more than 100 USAF and Air National Guard bases were converted to JP-8 + 100 and eventually the program was expanded to the U.S. Army. Approved additives are listed in MIL-DTL-83133 (JP-8). As of April 1999, one proprietary additive package, available from two suppliers, has been approved. Other additives have been produced but approval testing has not been completed by the USAF.

Currently, the additive is mainly in U.S. military use, but other NATO countries have had aircraft operating on the additive package. The major engine builders have approved the additive for civil engines, but so far, there has been only limited interest for such use.

NONSPECIFICATION ADDITIVES

No additives except those mentioned earlier are listed in current fuel specifications, but there are others that are sometimes used for special purposes. However, before they can be used, all such additives require approval by the OEMs and the agreement of the user. Only two of these additives (Biobor JF and Kathon FP 1.5) have had significant use in commercial aircraft, but several others merit attention.

BIOCIDES

Use of biocides is for the eradication of microbial growth in contaminated fuel systems and to provide long-term protection to prevent the growth of microorganisms. Biocides normally have two dose rates: a higher curative dose to kill the microbes and a lower maintenance dose for prevention. The two most widely used are Biobor JF and Kathon FP 1.5. Biobor JF is a fuel-soluble mixture of dioxaborinanes that prevent microbial growth in hydrocarbon fuels. Approval by most engine and aircraft manufacturers is limited to intermittent or noncontinuous use in concentrations not to exceed 270 mg/L (20 ppm elemental boron). Biobor JF is normally used to "disinfect" aircraft during a period of at least 24 h when the aircraft can be left standing filled or partially filled with doped fuel. Depending on additive concentration, the fuel may have to be drained and replaced with uninhibited fuel, or it can be burned in the engines. To prevent the possible deposition of boron compounds in the engine, the treatment is only permitted at infrequent intervals. Biobor JF is no longer authorized for use in some European countries.

Kathon FP 1.5 consists of two quaternary ammonium compounds in a glycol solvent. The maximum permitted dose is 100 ppm, including the solvent. It is intended to be used in intermittent fashion similar to Biobor JF. Kathon FP 1.5 acts in equilibrium in both the fuel and water phases but will partition to the water phase and is biodegradable at very low levels. Both additives are approved for jet fuel only.

This additive type is not listed in specifications at the time of this writing for several reasons. One concern is the lack of a mechanism of ensuring that the total additive concentration remains at or below the maximum permitted level if fuel in storage tanks and aircraft is treated simultaneously. A second reason is possible overuse of an additive that has restricted approval. As a result, several major airlines consider it their function to maintain control by having to agree to the use of any biocide ahead of or at the airport.

PIPELINE DRAG REDUCER ADDITIVE

Pipeline drag reducer (PDR) is not permitted in jet fuel. After an exhaustive study of the effects of PDR on jet fuel, CRC Report No. 642 (CRC Project No. CA-68-97) "Investigation of Pipeline Drag Reducers in Aviation Turbine Fuels" (Stanford P. Seto, May 2005 review) concluded that the PDR in jet fuel, in concentrations of 8.8 ppm up to 32.0 ppm (active ingredient), is not acceptable for use. Engine, auxiliary power unit (APU) and airframe OEMs, and industry representatives reviewed the results from the GE Transportation (GE Aircraft Engines) PDR in aviation turbine fuels (jet or jet fuel) testing for cold weather and altitude combustor ignition. The review combustor sector-test data indicate a significant loss in both cold ignition on the ground from sea level pressure to 15,000 ft (4,572 m) and in altitude ignition capability from 5000 ft (1,524 m) to above 25,000 ft (7,620 m). Loss in capability could be from 15 % up to 55 %. The presence of fully sheared PDR in the jet fuel did have the effect of diminishing the fuel-spray angle and atomization capability of several engine-type fuel nozzles and injectors at cold conditions but did not seriously compromise thermal stability, filter ability, or other tested qualities. The presence of unsheared PDR did impair filter ability.

Testing of the jet fuel did not progress beyond the sector ignition work. Consequently, there is no finding with regard to the effects of PDR on engine hot-section carboning or exhaust emissions. The following is the historical background that prompted the CRC Report No. 642

investigation into the feasibility of using PDR in jet fuel. Several large U.S. pipelines had reached or were approaching their maximum flow capacity. To increase product throughput in such lines required additional pumping stations, or even adding more lines in parallel. However, another solution was the addition of PDR additive, which decreases pipeline drag or flow resistance some 30–40 % and can, therefore, increase line capacity proportionately. PDR additives are being added to crude oils and distillate products, such as gasoline and middle distillates. The problem of limited capacity is critical in several pipelines supplying jet fuel to airports, and the use of PDR additives in jet fuel appears to be the most practical solution.

In 1997, Buckeye pipeline approached CRC to launch an investigation, which resulted in Report No. 642. The aim of the investigation was to obtain aircraft equipment manufacturers' approval of these additives. Two forms of drag reducers were evaluated: a hydrocarbon-based gel and an aqueous slurry. In both cases, the active ingredients were high-molecular-weight olefins to be added to jet fuel at a maximum total concentration of 8.8 ppm. The cooperative industry effort included equipment manufacturers, pipelines, and jet fuel shippers.

IGNITION CONTROL ADDITIVE

To minimize the adverse effect of spark plug deposits in gasoline engines, several phosphorus-containing additives have been developed. Typical of these is tricresyl phosphate (TCP), which modifies lead compounds so that they do not cause pre-ignition. Spark plug fouling was pronounced in certain older types of aircraft piston engines, and TCP was used to overcome the problem. As these engines were withdrawn from service and as TEL content of aviation gasoline was reduced over time, the problem diminished. Now it is doubtful whether the additive has any significant use.

ADDITIVE TESTS

Although the type and amount of each permitted additive is strictly limited, test methods for checking additive concentrations are not always specified. Where tests are not called for, a written statement of the additives' addition is accepted as evidence of its presence. The following paragraphs recap the tests for additives discussed previously.

Tetraethyl-Lead

In aviation gasolines, the TEL content has such a critical influence on the antiknock properties and deposit-forming tendencies of the fuel that a test for TEL content is included in all routine laboratory tests. There are two alternative test methods for lead in gasoline–D5059/IP 228 and D3341/IP 270.

Color of Aviation Gasoline

After the specified dye has been added, the minimum color is checked by ASTM D2392. Maximum color is controlled by dye concentration. Lovibond color (IP 17) is required in some specifications.

Antioxidant, Metal Deactivator, and Corrosion Inhibitor/Lubricity Additive

After the required amounts of antioxidants and metal deactivators have been added to fuels, checks on the concentration are not required; therefore, no test methods are included in the specification. However, the refiner or blender is required to list the amount of each additive on the Certificate of Quality. When corrosion inhibitors are required, the same procedure holds. However, when these materials are added as lubricity agents, fuel lubricity can be checked by the BOCLE test, ASTM D5001. Should actual additive concentration be required, several analytical methods have been published but not standardized.

Fuel System Icing Inhibitor

FSII in jet fuel can be lost through evaporation but more likely through extraction by water that contacted the fuel during transportation. ASTM D5006 is designed for the quantitative determination of FSII in fuel. In the method, the additive is extracted and its concentration is measured by refractometer. The instrument can be calibrated for either EGME or diEGME but is not designed to measure mixtures of these two materials.

Static Dissipator Additive

This additive is added in such low concentrations that it is extremely difficult to detect by any standard analytical procedure. Therefore, it is controlled by measuring the resultant electrical conductivity of the fuel. Meters described in ASTM D2624/IP 264 are commonly used for the purpose. A new test method has been developed to detect SDA by HPLC and has been published as IP 568. Currently, this test method is in the ballot process for ASTM adoption of a new test method.

Leak Detector

Tracer A® is also added in very low concentrations. Control is maintained through a document trail that gives the user the assurance that the maximum permitted dosage has not been exceeded.

Pipeline Drag Reducer

No methods for PDR content have been standardized. Progress on obtaining an ASTM method subsided when CRC Report 642 basically outlawed the use of PDR in jet fuel.

Thermal Stability Additive

At this time, no analytical procedure for the quantitative determination of this additive has been published. Proper additive concentration depends on monitoring of injection equipment, but one or more analytical procedures can be expected.

Biocides

Biobor JF can be detected by the measurement of boron by a number of analytical procedures. There is no published or standardized method for the determination of Kathon FP 1.5 in jet fuel.

Applicable ASTM Specifications

Number	Title
D910	Specification for Aviation Gasolines
D1655	Specification for Aviation Turbine Fuels
D4171	Specification for Fuel System Icing Inhibitors
D6227	Specification for Grade 82 Unleaded Aviation Gasoline
D6615	Specification for Wide Boiling Aviation Turbine Fuel

Applicable ASTM/IP Test Methods, Practices, and Guides

ASTM	IP	Standard
D56		Test Method for Flash Point by Tag Closed Tester
D86	123	Test Method for Distillation of Petroleum Products at Atmospheric Pressure
D93	34	Test Method for Flash Point by Pensky-Martens Closed |Cup Tester
	170	Test Method for Flash Point by Abel Apparatus
D130	154	Test Method for Corrosiveness to Copper from Petroleum Products by the Copper Strip Test
D156		Test Method for Saybolt Color of Petroleum Products
	17	Test Method for Color by Lovibond Tintometer
D240	12	Test Method for Heat of Combustion of Liquid Hydrocarbon Fuels by Bomb Calorimeter
D323	69	Test Method for Vapor Pressure of Petroleum Products (Reid Method)
D381	131	Test Method for Gum Content in Fuels by Jet Evaporation (Steam)
	540	Determination of the Existent Gum Content of Aviation Turbine Fuel—Jet Evaporation Method (Air and Steam)
D445	71	Test Method for Kinematic Viscosity of Transparent and Opaque Liquids (and Calculation of Dynamic Viscosity)
D611	2	Test Method for Aniline Point and Mixed Aniline Point of Petroleum Products and Hydrocarbon Solvents
D873	138	Test Method for Oxidation Stability of Aviation Fuels (Potential Residue Method)
D909	119	Test Method for Supercharge Rating of Spark Ignition Aviation Gasoline
D974	139	Test Method for Acid and Base Number by Color-Indicator Titration
	225	Copper in Aviation Turbine Fuels and Light Petroleum Distillates
D1094	289	Test Method for Water Reaction of Aviation Fuels
D1266	107	Test Method for Sulfur in Petroleum Products (Lamp Method)
D1298	160	Test Method for Density, Relative Density (Specific Gravity) or API Gravity of Crude Petroleum and Liquid Petroleum Products by Hydrometer Method
D1319	156	Test Method for Hydrocarbon Types in Liquid Hydrocarbon Products by Fluorescent Indicator Adsorption
D1322	57	Test Method for Smoke Point of Kerosene and Aviation Turbine Fuels
D1405		Test Method for Estimation of Net Heat of Combustion of Aviation Fuels
D1552		Test Method for Sulfur in Petroleum Products (High Temperature Method)
D1740		Test Method for Luminometer Numbers of Aviation Turbine Fuels
D1840		Test Method for Naphthalene Hydrocarbons in Aviation Turbine Fuels by Ultraviolet Spectrophotometry
D2276	216	Test Method for Particulate Contaminant in Aviation Fuel by Line Sampling
D2386	16	Test Method for Freezing Point of Aviation Fuels
D2392		Test Method for Color of Dyed Aviation Gasoline
D2622	447	Test Method for Sulfur in Petroleum Products by Wavelength Dispersive X-Ray Fluorescence Spectrometry
D2624	274	Test Method for Electrical Conductivity of Aviation and Distillate Fuels
D2700	236	Test Method for Motor Octane Number of Spark Ignition Engine Fuel
D2887		Test Method for Boiling Range Distribution of Petroleum Fractions by Gas Chromatography
D3227	342	Test Method for Thiol (Mercaptan) Sulfur in Gasoline, Kerosene, Aviation Turbine and Distillate Fuels (Potentiometric Method)
D3240		Test Method for Undissolved Water in Aviation Turbine Fuels
D3241	323	Test Method for Thermal Oxidation Stability of Aviation Turbine Fuels
D3341	270	Test Method for Lead in Gasoline-Iodine Monochloride Method
	224	Test Method for Trace Amounts of Lead in Aviation Turbine Fuels and Light Petroleum Distillates
D3343		Test Method for Estimation of Hydrogen Content of Aviation Fuels
D3701	338	Test Method for Hydrogen Content of Aviation Turbine Fuels by Low Resolution Nuclear Magnetic Resonance Spectrometry
D3703		Test Method for Peroxide Number of Aviation Turbine Fuels
D3828	303	Test Method for Flash Point by Small Scale Closed Cup Tester
D3948		Test Method for Determining Water Separation Characteristics of Aviation Turbine Fuels by Portable Separometer

ASTM	IP	Standard
D4052	365	Test Method for Density and Relative Density of Liquids by Digital Density Meter
D4054		Practice for Qualification and Approval of New Aviation Turbine Fuel and Fuel Additives
D4057		Practice for Manual Sampling of Petroleum and Petroleum Products
D4176		Test Method for Free Water and Particulate Contamination in Distillate Fuels (Visual Inspection Procedures)
D4294	336	Test Method for Sulfur in Petroleum Products by Energy Dispersive X-Ray Fluorescence Spectroscopy
D4305	422	Test Method for Filter Flow of Aviation Fuels at Low Temperatures
D4306		Practice for Aviation Fuel Sample Containers for Tests Affected By Trace Contamination
D4308		Test Method for Electrical Conductivity of Liquid Hydrocarbons by Precision Meter
D4529	381	Test Method for Estimation of Net Heat of Combustion of Aviation Fuels
D4809		Test Method for Heat of Combustion of Liquid Hydrocarbon Fuels by Bomb Calorimeter (Precision Method)
D4865		Guide for Generation and Dissipation of Static Electricity in Petroleum Fuel Systems
D4952	30	Test Method for Qualitative Analysis for Active Sulfur Species Fuels and Solvents (Doctor Test)
D4953		Test Method for Vapor Pressure of Gasoline and Gasoline-Oxygenate Blends (Dry Method)
D5001		Test Method for Measurement of Lubricity of Aviation Turbine Fuels by Ball-On-Cylinder Lubricity Evaluator (BOCLE)
D5006		Test Method for Measurement of Fuel System Icing Inhibitor (Ether Type) In Aviation Fuels
	424	Determination of Fuel System Icing Inhibitor Content of Aviation Turbine Kerosines by High Performance Liquid Chromatography
D5059	228	Test Method for Lead in Gasoline by X-Ray Spectroscopy
D5190		Test Method for Vapor Pressure of Petroleum Products (Automatic Method)
D5191	394	Test Method for Vapor Pressure of Petroleum Products (Mini Method)
D5452	423	Test Method for Particulate Contamination in Aviation Fuels by Laboratory Filtration
D5453		Test Method for Determination of Total Sulfur in Light Hydrocarbons, Spark Ignition Engine Fuel, Diesel Engine Fuel, and Engine Oil by Ultraviolet Fluorescence
D5901	434	Test Method for Freezing Point of Aviation Fuels (Automatic Optical Method)
D5972	435	Test Method for Freezing Point of Aviation Fuels (Automatic Phase Transition Method)
D6304	438	Determination of Water in Petroleum Products, Lubricating Oils and Additives by Coulometric Karl Fischer Titration
D6379		Test Method for Determination of Aromatic Hydrocarbon Types in Aviation Fuels and Petroleum Distillates (High Performance Liquid Chromatography Method with Refractive Index Refraction)
D6446		Test Method for Estimation of Net Heat of Combustion (Specific Energy) of Aviation Fuels
D6469		Guide to Microbial Contamination in Fuels and Fuel Systems

Bibiliography

Ebbinghaus, A., Bauldreay, J. M., and Grandvallet, T, "Lubricity Survey of Hydrotreated and Hydrocracked Jet Fuel Components," *Proceedings of the Seventh International Conference on Stability and Handling of Liquid Fuels*, Graz, Austria, September, 2000.

Executive Summary, CRC Research Results, Unleaded High Octane Aviation Gasoline, A Report to the CRC Unleaded AVGAS Development Panel, Project No. AV-7-07, April 2008.

Handbook of Aviation Fuel Properties, CRC Report 530, Coordinating Research Council, Atlanta, GA, 1983.

Heneghan, S. P. and Harrison, W. E., III, "JP-8 + 100: The Development of High Thermal Stability Jet Fuel," *Proceedings of the Sixth International Conference on Stability and Handling of Liquid Fuels*, Vancouver, Canada, October, 1997.

Ogsten, R., "A Short History of Aviation Gasoline Development, 1903–1980," SAE Paper 810848, Society of Automotive Engineers, Warrendale, PA, July 1981.

Smith, M., *Aviation Fuels*, G. T. Foulis and Co., Ltd., UK, 1970.

Stanford P. Seto, "Investigation of Pipeline Drag Reducers in Aviation Turbine Fuels," CRC Report No. 642 (CRC Poject No. CA-68-97), May 2005.

Strauss, K. H., "Low Temperature Properties for Jet Fuels, Problems and Solutions," *Proceedings of the Seventh International Conference on Stability and Handling of Liquid Fuels*, Graz, Austria, September, 2000.

8

Crude Oils

Harry N. Giles[1]

CRUDE OILS ARE A HIGHLY COMPLEX COMBINA-tion of hydrocarbons; heterocyclic compounds of nitrogen, oxygen, and sulfur; organometallic compounds; inorganic sediment; and water. Approximately 600 different hydrocarbons have been positively identified in crude oil, and it is likely that thousands of compounds occur, many of which probably will never be identified. In a study sponsored by the American Petroleum Institute (API), nearly 300 individual hydrocarbons were identified in Ponca City, Oklahoma, crude oil [1,2]. Some 200 individual sulfur compounds were identified in a 20-year systematic study of four crude oils [3]. Not only is the composition of crude oil highly complex, it is also highly variable from field to field, and even within a given field, it is likely to exhibit inhomogeneity from reservoir to reservoir. Physical and chemical characterization of this complex mixture is further complicated for the analyst by the fact that crude oils are not pure solutions but commonly contain colloidally suspended components, dispersed solids, and emulsified water.

Compared to refined products such as gasoline and aviation turbine fuel, there is relatively little in the literature on the analysis and characterization of crude oils. Indeed, for many years, there were relatively few ASTM methods specific to crude oils, although a number of ASTM methods had been adapted for their analysis. This situation may have resulted, at least in part, from the historical tendency of refinery chemists to independently develop or modify analytical methods specific to their needs and, subsequently, for the methods to become company proprietary. In recent years, the unique problems associated with sampling and analysis of crude oils have received more attention, and more methods for determining selected constituents and characteristics of crude oils have been standardized.

A series of articles [4–9] illustrate the diversity of crude oil assay practices employed by major refiners in the United States and Austria. The dissimilarity of published results [10] and as provided by a number of companies on their Web sites [11] is a reflection of this independent development of analytical schemes, even though standardized approaches to crude oil analysis have been published [12–15]. Despite the complexity of crude oil composition and the diversity of analytical methodology, probably more crude oil analyses are routinely performed on a daily basis using inherently similar methods than are analyses on any single refined petroleum product except, possibly, gasoline.

The overriding issue when performing comprehensive crude oil assays is economics. Crude oils are assayed to determine (1) the slate of products that can be produced with a given refinery's process technology; (2) the processing difficulties that may arise as a result of inherent impurities and contaminants; and (3) the downstream processing and upgrading that may be necessary to optimize yields of high-value, specification products. Today, analytical data are typically stored in an electronic database that can be accessed by computer models that generate refinery-specific economic valuations of each crude oil or crude slate, that is, a mixture of crude oils processed together. Linear programming (LP) models are available from several commercial vendors, but a number of companies have developed their own models to meet the needs of their specific refinery configurations.

Analyses are also performed to determine whether each batch of crude oil received at the refinery gate meets expectations. Does the crude receipt match the database assay so that the projected economic valuations and operational strategies are valid? Has any unintentional contamination or purposeful adulteration occurred during gathering, storage, or transport of the crude oil that may increase the processing cost or decrease the value of the refined products? The information needed to answer these questions is often refinery specific—a function of the refinery's operating constraints and product slate.

To obtain the desired information, two different analytical schemes are commonly used, namely, an inspection assay and a comprehensive assay. Inspection assays usually involve determination of a few key whole crude oil properties such as API gravity, sulfur content, and pour point—principally as a means of determining if major changes in a crude oil stream's characteristics have occurred since the last comprehensive assay was performed. Additional analyses may be performed to help ensure that quality of the cargo or shipment received is that which is expected; to ascertain the quantity of impurities such as salt, sediment, and water; and to provide other critical refinery-specific information. Inspection assays are routinely performed on all shipments received at a terminal or refinery. The comprehensive assay, on the other hand, is complex, costly, and time consuming and is normally performed only when a new field comes on stream, or when the inspection assay indicates that significant changes in the stream's composition have occurred. Except for these circumstances, a comprehensive assay of a particular crude oil stream may not be updated for several years.

Moreover, many major pipeline companies require a comprehensive assay when accepting a new crude oil stream for transportation in their system on a common stream basis. Thereafter, an inspection assay is used for checking the quality of shipments.

[1] *PetroStorTech LLC, Arlington, VA.*

INSPECTION ASSAYS

Inspection assays comprise a limited number of tests generally restricted to the whole crude oil. Based on published data, there is little agreement as to what constitutes an inspection assay. As the data are primarily for intra-company use, there is little driving force for a standard scheme. At a bare minimum, API gravity, sulfur content, and sediment and water are usually determined, although it is useful to also know the pour point, which provides some basic perception of the crude oil's fluidity and aromaticity. A more detailed inspection assay might consist of the following tests: API gravity (or density or relative density), total sulfur content, pour point, viscosity, salt content, total acid number (neutralization number), and water and sediment content. Individual refiners may substitute or add tests, for example, trace metals or organic halides, that may be critical to their operations. Coupling the results from these few tests of a current crude oil batch with the archived data from a comprehensive assay, the process engineer will be able to estimate generally the product slate that the crude will yield and any extraordinary processing problems that may be encountered.

In the early 1990s, the API formed the Ad Hoc Crude Oil Quality Task Force. The report of this task group recommends a set of crude oil quality testing procedures that, if adopted by a shipper or refiner, would help ensure the quality of crude oil from the wellhead to the refinery [16]. These procedures include tests for API gravity, sediment and water, organohalide compounds, salt, sulfur, and neutralization number, among others. While not a standard, the report is an important aid to members of the petroleum industry in protecting the quality of common stream crude petroleum from contamination by foreign substances or crude petroleum of unspecified makeup. It is also a useful guide for an inspection program using mostly standardized procedures widely accepted in the industry for monitoring the quality of mercantile commodity.

It is important to note that, in the following discussion of test methods, "crude oil" may not be included in the title or even in the scope. Many test methods have, however, been adapted to and are widely used and accepted for crude oil analysis.

API Gravity

Accurate determination of the gravity of crude oil is necessary for the conversion of measured volumes to volumes at the standard temperature of 15.56°C (60°F) using ASTM D1250, Petroleum Measurement Tables. API gravity is a special function of relative density (specific gravity) represented by the following:

$$\text{API gravity, degrees} = \left(\frac{141.5}{\text{specific gravity } 60/60°\text{F}}\right) - 131.5 \qquad (1)$$

No statement of reference temperature is required, as 60°F is included in the definition.

Gravity is also a factor reflecting the quality of crude oils. Generally, the heavier (lower the API gravity) the crude oil, the greater is the quantity of heavier components that may be more refractory and require greater upgrading or more severe cracking to produce salable products. Conversely, the lighter the crude oil the greater the quantity of distillable products.

The relative density (specific gravity) or density of a crude oil may also be reported in analyses. Relative density is the ratio of the mass of a given volume of liquid at a specific temperature to the mass of an equal volume of pure water at the same or a different temperature. Both reference temperatures must be explicitly stated. Density is simply mass per unit volume at a specified temperature.

API gravity, or density or relative density, can be determined easily using one of two hydrometer methods [ASTM D287, Test Method for API Gravity of Crude Petroleum and Petroleum Products (Hydrometer Method), or ASTM D1298, Test Method for Density, Relative Density (Specific Gravity), or API Gravity of Crude Petroleum and Liquid Petroleum Products by Hydrometer Method]. Many laboratories are now using an instrumental method (ASTM D5002, Test Method for Density and Relative Density of Crude Oils by Digital Density Analyzer) rather than the hydrometer methods. A third hydrometer method (ASTM D6822, Test Method for Density, Relative Density, and API Gravity of Crude Petroleum and Liquid Petroleum Products by Thermohydrometer Method) is applicable to field applications where limited laboratory facilities are available.

Sulfur Content

The sulfur content of a crude oil, which may vary from less than 0.1 to over 5 mass percent, is one of its most important quality attributes. Sulfur compounds contribute to corrosion of refinery equipment and poisoning of catalysts, cause corrosiveness in refined products, and contribute to environmental pollution as a result of emission of sulfur oxides from combustion of fuel products. Sulfur compounds may be present throughout the boiling range of crude oils, although, as a rule, they are more abundant in the heavier fractions. In some crude oils, thermally reactive sulfur compounds can decompose on heating to produce hydrogen sulfide, which is highly toxic and very corrosive. Consequently, in reporting the hydrogen sulfide content of a crude oil, it is important to distinguish between that which is dissolved and that which is evolved on heating or distillation. The mercaptans usually present in a crude oil can impart a foul odor, depending on the species. Butyl mercaptan, a compound naturally present in many crude oils, is the odorant commonly used in natural gas. The fetid smell in the secretion ejected by skunks is also due to this compound.

Until relatively recently, one of the most widely used methods for determination of total sulfur content has been combustion of a sample in oxygen to convert the sulfur to sulfur dioxide, which is collected and subsequently titrated iodometrically or detected by nondispersive infrared. This is commonly referred to as the "Leco" technique, but in its standard form is ASTM D1552, Test Method for Sulfur in Petroleum Products (High-Temperature Method). An even older method involving combustion in a bomb with subsequent gravimetric determination of sulfur as barium sulfate [ASTM D129, Test Method for Sulfur in Petroleum Products (General Bomb Method)] is not as accurate as the high-temperature method, possibly because of interference from the sediment inherently present in crude oil.

These older techniques are rapidly being replaced by instrumental methods. Among these are ASTM D4294, Test Method for Sulfur in Petroleum Products by Energy-Dispersive X-Ray Fluorescence Spectroscopy, and ASTM D2622, Test Method for Sulfur in Petroleum Products by X-Ray Spectrometry. ASTM D4294 has slightly better repeatability and reproducibility than the high-temperature method and is adaptable to field applications; however, this method can be affected by some commonly present interferences such as halides. Of the two methods, ASTM D2622 has better precision and the capability of correcting for interferences but is currently limited to laboratory use, and the equipment is more expensive. Sediment and water commonly present in crude oil samples will interfere in sulfur determination by both of the X-ray methods. These should be removed from the sample by centrifugation or settling prior to analysis, but care must be taken that sample integrity is not compromised.

Hydrogen sulfide and mercaptans are commonly determined by non-aqueous potentiometric titration with silver nitrate (UOP 163, Hydrogen Sulfide and Mercaptan Sulfur in Liquid Hydrocarbons by Potentiometric Titration). Hydrogen sulfide is highly reactive, however, and unless precautions are taken in the collection and preservation of samples, results will not be representative. A test kit has been developed that is very useful for rapidly determining hydrogen sulfide concentration in liquid samples in the field [17]. This kit has an accuracy of about ±20 % for hydrogen sulfide.

Salt Content

The salt content of crude oil is highly variable and results principally from production practices used in the field and, to a lesser extent, from its handling aboard tankers bringing it to terminals. The bulk of the salt present will be dissolved in coexisting free water and can be removed in desalters, but small amounts of salt may be dissolved in the crude oil itself. Salt may be derived from reservoir or formation waters or from other waters used in secondary recovery operations. Aboard tankers, ballast water of varying salinity may also be a source of salt contamination.

Salt in crude oil may be deleterious in several ways. Even in small concentrations, salts will accumulate in stills, heaters, and exchangers, leading to fouling that requires expensive cleanup. More important, during flash vaporization of crude oil, certain metallic salts can be hydrolyzed to hydrochloric acid according to the following reactions:

$$2\ NaCl + H_2O \rightarrow HCl + Na_2O \quad (2)$$

$$MgCl_2 + H_2O \rightarrow 2\ HCl + MgO \quad (3)$$

The hydrochloric acid evolved is extremely corrosive, necessitating the injection of a basic compound, such as ammonia, into the overhead lines to minimize corrosion damage. Salts and evolved acids can also contaminate both overhead and residual products, and certain metallic salts can deactivate catalysts. A thorough discussion of the effects of salt on crude processing is included in a manual on impurities in petroleum [18].

For many years, the salt content has been routinely determined by comparing the conductivity of a solution of crude oil in a polar solvent to that of a series of standard salt solutions in the same solvent [ASTM D3230, Test Method for Salts in Crude Oil (Electrometric Method)]. This test method provides an approximate measure of the chloride content of the crude oil being tested based on measurement of its conductivity. The chloride content is obtained by reference to a calibration curve prepared using a given mixture of salts. As conductivity varies with varying salt composition, unless the composition of salts in the sample being tested is the same as the calibration mixture, results will be affected. Also, other conductive materials present in the crude oil sample will affect results. These factors contribute to the relatively poor precision of the method. ASTM D6470, Test Method for Salt in Crude Oils (Potentiometric Method), is less affected by salt composition and has considerably better precision than the older method. Regardless of the method used, it is necessary to use other methods, such as atomic absorption, inductively coupled argon plasma spectrophotometry, or ion-chromatography to determine the composition of the salts present.

Water and Sediment

The water and sediment content of crude oil, like salt, results principally from production and transportation practices. Water, with its dissolved salts, may occur as easily removable suspended droplets or as an emulsion. The sediment dispersed in crude oil may be comprised of inorganic minerals from the production horizon or from drilling fluids, as well as from scale and rust from pipelines and tanks used for oil transportation and storage. Usually water is present in far greater amounts than sediment, but, collectively, it is unusual for them to exceed 1 % (v/v) of the crude oil on a delivered basis. Like salt, water and sediment can foul heaters, stills, and exchangers and can contribute to corrosion and to deleterious product quality. Also, water and sediment are principal components of the sludge that accumulates in storage tanks and must be disposed of periodically in an environmentally acceptable manner.

Further, water bottoms in storage tanks can promote microbiological activity, and, if the system is anaerobic, production of corrosive acids and hydrogen sulfide can result. This is not usually a problem with crude oils, as stocks are normally rotated on a regular basis. Nevertheless, anaerobic degradation of crude oil stocks and production of hydrogen sulfide has been known to happen, and the operator must be aware of the potential for this to occur and the analyst must take this into consideration in evaluating results.

Knowledge of the water and sediment content is also important in accurately determining net volumes of crude oil in sales, taxation, exchanges, and custody transfers. When a significant amount of free water is present in marine cargo, identification of its probable source should be a major consideration. Guidelines that include basic sampling, testing, and analytical procedures and interpretation and presentation of results for this process have been published [19].

A number of test methods exist for the determination of water and sediment in crude oil. Centrifugal separation of the water and sediment [ASTM D4007, Test Method for Water and Sediment in Crude Oil by the Centrifuge Method (Laboratory Procedure)] is rapid and relatively inexpensive, but, almost invariably, the amount of water detected is lower than the actual water content. A more accurate method for sediment entails extraction with hot toluene in a refractory thimble (ASTM D473, Test Method for Sediment in Crude Oils and Fuels Oils by the Extraction Method). A somewhat less time-consuming method of determining sediment involves dissolving a sample in hot toluene and filtering the solution under gravity through a membrane filter (ASTM D4807, Test Method for Sediment in Crude Oil by Membrane Filtration). The use of toluene in laboratories is coming under increasing scrutiny by safety and health groups, however, and a ban on its use is not inconceivable. Improved techniques for measuring water content include heating under reflux conditions with a water immiscible solvent that distills as an azeotrope with the water (ASTM

D4006, Test Method for Water in Crude Oil by Distillation), potentiometric titration (ASTM D4377, Test Method for Water in Crude Oils by Potentiometric Karl Fischer Titration), or the more generally preferred coulometric titration (ASTM D4928, Test Method for Water in Crude Oils by Coulometric Karl Fischer Titration). The latter two Karl Fischer methods include a homogenization step designed to redisperse any water that has separated from the crude oil while the sample has been stored.

Fluidity—Pour Point and Viscosity

Pour point and viscosity determinations of crude oils are performed principally to ascertain their handling characteristics at low temperatures. There are, however, some general relationships about crude oil composition that can be derived from pour point and viscosity data. Commonly, the lower the pour point of a crude oil, the more naphthenic or aromatic it is, and the higher the pour point, the more paraffinic it is. There are numerous exceptions to this rule-of-thumb, and other data must be used to verify a crude oil's character. Viscosity is also a function of the aromaticity or paraffinicity of the sample. Those crude oils with a greater concentration of paraffins generally have a higher viscosity than crude oils having a relatively large proportion of aromatic and naphthenic compounds.

Pour point is determined by cooling a preheated sample at a specified rate and examining its flow characteristics at intervals of 3°C. ASTM D97, Test Method for Pour Point of Petroleum Products, is the most widely used procedure for this measurement, even though crude oils are not mentioned in the method's scope. An alternative procedure specifically for testing the pour point of crude oils is described in ASTM D5853, Test Method for Pour Point of Crude Oils.

Viscosity is determined by measuring the time for a volume of liquid to flow under gravity through a calibrated glass capillary viscometer [ASTM D445, Test Method for Kinematic Viscosity of Transparent and Opaque Liquids (and the Calculation of Dynamic Viscosity)]. While the preferred unit of kinematic viscosity is millimeter squared per second (mm^2/s), many older analyses report it in centistokes (cSt). These units are equivalent, with 1 mm^2/s equaling 1 cSt. At one time the petroleum industry measured viscosity by means of the Saybolt viscometer, and expressed values in units of Saybolt Universal Seconds (SUS) or Saybolt Furol Seconds (SFS). This practice is now largely obsolete in the industry. ASTM D2161, Practice for Conversion of Kinematic Viscosity to Saybolt Universal Viscosity or to Saybolt Furol Viscosity, establishes equations that may be used for calculating kinematic viscosities from SUS and SFS data that appear in older literature. By determining viscosity at two temperatures such as 25 and 40°C, viscosity at any other temperature *over a limited range* may be interpolated or extrapolated using viscosity-temperature charts (ASTM D341, Viscosity-Temperature Charts for Liquid Petroleum Products). It must be kept in mind that these charts are not linear. Also, the lowest temperature at which viscosity is determined must be several degrees higher than the pour point. Otherwise, the crude oil will not exhibit Newtonian behavior.

Characterization Factor

Probably the most widely used index of composition is the Characterization or Watson K-Factor [20], which was originally defined as the cube root of the average molal boiling point in °F absolute (Rankine) temperature divided by the specific gravity, at 60/60°F. It has conveniently been related to viscosity and API gravity (UOP Method 375, Calculation of UOP Characterization Factor and Estimation of Molecular Weight of Petroleum Oils). Typically, paraffin base crude oils have $K \geq 12.0$, with lower values indicating crude oils of a more aromatic or naphthenic character [21]. These values provide a general rule-of-thumb on product yields; the paraffin base crude oils will give the highest gasoline yields, while the aromatic base feedstocks will be the most refractory and require a greater degree of upgrading.

Trace Elements

A number of trace elements have been detected in crude oil, with nickel and vanadium commonly being the most abundant. Until recently, however, relatively little systematic analytical work has been carried out on other trace elements. With heightened environmental awareness and susceptibility of many catalysts to poisoning or deactivation by trace metals, more work is being done on determining their presence in crude oils. Published reports indicate that over 30 trace metals definitively occur in crude oils [22,23]. An extensive review of the literature published through 1973 provides information on the occurrence and concentration of 45 trace elements [24]. Using highly sophisticated techniques such as neutron activation analysis, and with the greatly improved sensitivity of modern detectors, it is likely that even more elements will be found, but probably in subparts-per-billion concentrations. Knowledge of the trace element constituents in crude oil is important because they can have an adverse effect on petroleum refining, product quality, and the environment. Among the problems associated with trace elements are catalyst poisoning in the refinery and excessive atmospheric emissions in combustion of fuels. Elements such as iron, arsenic, and lead are catalyst poisons. Vanadium compounds can cause refractory damage in furnaces, and sodium compounds have been found to cause superficial fusion on fire brick [25]. Some organometallic compounds are volatile, which can lead to contamination of distillate fractions [26] and a reduction in their stability or malfunctions of equipment when they are combusted. Concentration of the nonvolatile organometallics in heavy products such as premium coke can have a significant impact on price, marketability, and use. Knowledge of trace element concentrations is also useful in exploration in correlating production from different wells and horizons in a field [27].

A number of trace metals are of considerable interest because of their potential impact on the environment, resulting from atmospheric emissions when fuels are burned or from discharge of process streams or disposal of wastes. In support of the North Sea Action Plan to reduce emissions, a detailed study of crude oils imported into the Netherlands was conducted [28]. It was found that cadmium, zinc, and copper were not indigenous to the crude oils but were the result of contamination with associated water or particles, or both, from the producing wells. Chromium was found to be indigenous for the most part and associated with the hydrocarbon matrix. Some inorganic chromium was thought to be present as a contaminant. The study was unable to determine the origin of arsenic found in the crude oils, but it was considered to probably be a contaminant. The intention to study mercury was abandoned because a reliable analysis technique was not found.

Two metals of considerable environmental concern are mercury and selenium, both of which occur naturally in crude oil at varying concentrations. Mercury is of concern as both an air and water pollutant, and selenium is of concern as a water pollutant.

There is substantial evidence indicating that mercury can occur in crude oil as volatile, dissolved, and particulate (suspended) species, all of which differ considerably in their behavior. Supporting the presence of volatile species, elemental mercury has been found condensed in cooler regions in refinery distillation towers and in cryogenic heat exchangers that liquefy petroleum gases. Further, replicate laboratory analyses on the same sample have found decreases in concentration over time [29]. Mercury has also been found in sludge that accumulates in strategic stockpiles of crude oil, clearly indicating the occurrence of particulate or suspended species [30]. Finally, mercury can be present in various petroleum distillation fractions across a broad boiling range.

Selenium has become a priority pollutant because of its high toxicity to aquatic wildlife. In refineries, it partitions into wastewater streams and can be discharged from treatment plants into the environment where it rapidly bioaccumulates. As with mercury, selenium can be present in different species that behave differently and complicate identification and remediation.

A number of trace metals are now customarily included in crude oil analyses. Among these are calcium, copper, iron, mercury, nickel, selenium, sodium, and vanadium. The suite of elements determined will be dictated by refinery processes, product slate, regulation, and environmental considerations. Several analytical methods are available for the routine determination of many trace elements in crude oil. Some of these allow direct aspiration of the samples (diluted in a solvent) instead of the time-consuming sample preparation procedures such as wet ashing (acid decomposition), or flame or dry ashing (removal of volatile/combustible constituents). Among the techniques used for trace element determinations are flameless and flame atomic absorption (AA) spectrophotometry (ASTM D5863, Test Method for Determination of Nickel, Vanadium, Iron, and Sodium in Crude Oils and Residual Fuels by Flame Atomic Absorption Spectrometry), and inductively coupled argon plasma spectrophotometry [ASTM D5708, Test Method for Determination of Nickel, Vanadium, and Iron in Crude Oils and Residual Fuels by Inductively-Coupled Plasma (ICP) Atomic Emission Spectrometry]. A modified version of ASTM D5185, Test Method for Determination of Additive Elements, Wear Metals, and Contaminants in Used Lubricating Oils and Determination of Selected Elements in Base Oils by Inductively Coupled Plasma Atomic Emission Spectrometry, is being used by many laboratories for the determination of elements such as lead and phosphorus in crude oils. ICP has an advantage over AA because a number of elements can be determined simultaneously; although detection limits by AA are often better. X-ray fluorescence spectrophotometry is also sometimes used, although matrix effects can be a problem. The method to be used is generally a matter of individual preference.

Many advances have been made in techniques for trace and ultra-trace sample preparation and elemental analysis including atomic absorption spectrometry, inductively coupled plasma emission and mass spectrometry, isotope dilution mass spectrometry, and other multihyphenated methods. A number of these are discussed in ASTM STP 1468 [31].

Vapor Pressure

Vapor pressure is an important physical property of crude oils impacting shipping, storage, and refinery-handling practices. The greater the vapor pressure of a crude oil, the greater is the potential for atmospheric emission of hydrocarbons and other volatile compounds such as hydrogen sulfide. With the increasingly more stringent environmental limitations on emission of these compounds, it is important that the vapor pressure be known so that crude oil stocks can be stored and handled in an appropriate manner. ASTM D323, Test Method for Vapor Pressure of Petroleum Products (Reid Method), and ASTM D5191, Test Method for Vapor Pressure of Petroleum Products (Mini Method), have been used for determining this property. Although the latter method is not scoped for crude oil, it is used by many laboratories for this determination. The Reid vapor pressure differs from the true vapor pressure of the sample under test due to some small sample vaporization and the presence of water vapor and air in the sample chamber used in the test. The latter technique also does not take into account dissolved water in the sample in determining total pressure. The measured total vapor pressure can be converted to a dry vapor pressure equivalent by use of a correlation equation. A newer test method–ASTM D6377, Test Method for Determination of Vapor Pressure of Crude Oil: $VPCR_x$ (Expansion Method)–covers determination for vapor-liquid ratios of 4:1 to 0.02:1. The former ratio can be related to values determined by ASTM D323. The latter mimics closely the situation of an oil tanker and approaches the true vapor pressure.

ASTM Manual 51 *Distillation and Vapor Pressure Measurement in Petroleum Products* includes detailed discussion of the several ASTM test methods used to measure vapor pressure of crude oil, that is, ASTM D323, D5191, and D6377 [32]. This will provide the analyst a better understanding of the details of each method and how they apply to determination of this parameter. A separate, complementary chapter provides a more in-depth discussion of the importance of crude oil vapor pressure measurements as they relate to determining regulatory compliance.

Total Acid Number

Total acid number, as determined by ASTM D664, Test Method for Acid Number of Petroleum Products by Potentiometric Titration, provides an indication of the organic acid content of a crude oil. It will also indicate the presence of remnant inorganic acids such as hydrochloric and hydrofluoric that may have been used in production well workover operations. Collectively, these acids contribute to increased rates of corrosion in the refinery and can contribute to instability in refined products. The method does not differentiate acid species, such as carboxylic, naphthenic, or inorganic, and does not provide any indication of relative strength. While no general correlation is known between acid number and the corrosive tendency of oils toward metals, knowledge of the acid number is important in planning for injection of neutralizing agents in refinery streams or reducing the acid content to an acceptable level by other means.

Carbon Residue

Carbon residue is a useful measure of the amount of material left after evaporation and pyrolysis and provides some indication of the relative coke-forming propensity of crude oil. The residue formed is not composed entirely of carbon but is a coke, the composition of which can be changed by further pyrolysis. Two methods have historically been used for determination of carbon residue. These are ASTM D189, Test Method for Conradson Carbon Residue of Petroleum Products, and ASTM D524, Test Method for Ramsbottom

Carbon Residue of Petroleum Products. No exact correlation of the results obtained by these two test methods exists, although an approximate correlation has been derived. A newer test method, ASTM D4530, Test Method for Determination of Carbon Residue (Micro Method), can be correlated to Test Method D189 and offers the advantages of better control of test conditions, smaller samples, and less operator attention.

Other Tests

Other properties that are generally determined on a more limited basis include those listed in the sections that follow.

TOTAL NITROGEN CONTENT

Nitrogen compounds can contaminate refinery catalysts and, increasingly, are of concern to refiners. They also tend to be the most difficult class of compounds to hydrogenate. The nitrogen content remaining in the product from a hydrotreater is a measure of the effectiveness of the hydro-treating process. Three test methods are available for the determination of total nitrogen. They are ASTM D3228, Test Method for Total Nitrogen in Lubricating Oils and Fuel Oils by Modified Kjeldahl Method; ASTM D4629, Test Method for Trace Nitrogen in Liquid Petroleum Hydrocarbons by Syringe/Inlet Oxidative Combustion and Chemiluminescence Detection; and ASTM D5762, Test Method for Nitrogen in Petroleum and Petroleum Products by Boat-Inlet Chemiluminescence.

ORGANIC HALIDES

Organic halide compounds are not known to occur naturally in crude oils, and their presence commonly results from contamination by solvents used in cleaning operations at production sites and in pipelines and tanks. These compounds are potentially damaging to refinery processes. For example, hydrochloric acid can be produced in hydrotreating or reforming reactors, following which the highly corrosive acid can accumulate in condensing regions of the refinery. Large or unexpected concentrations of the resulting acids cannot be effectively neutralized, and damage can result. Organic halide species can also poison catalysts in reformers and adversely affect gasoline yields. Total organic halide content of the naphtha fraction can be effectively determined using ASTM D4929, Test Method for Determination of Organic Chloride Content in Crude Oil. It is imperative that the sample be distilled as described in the test method to eliminate potential interference by inorganic salts.

ASPHALTENES

Asphaltenes are the organic molecules of highest molecular mass and carbon-hydrogen ratio normally occurring in crude oil. They may give problems during storage and handling if the suspension of these molecules is disturbed through excess stress or incompatibility. Their composition normally includes a disproportionately high quantity of the sulfur, nitrogen, and metals present in crude oil. ASTM D6560, Test Method for Determination of Asphaltenes (Heptane Insolubles) in Crude Petroleum and Petroleum Products, covers a procedure for their determination. ASTM D3279, Test Method for *n*-Heptane Insolubles, is similar in scope to D6560 and is useful in quantifying the asphaltene content of crude oils.

ASH

Ash present in crude oil results from the presence of noncombustible extraneous solids such as dirt, pipeline scale, and rust. Normally there is a close correlation between a crude oil's ash content and its sediment content. In the use of crude oil as a fuel, it is important to know its ash content, as this can be related directly to particulate emissions. ASTM D482, Test Method for Ash from Petroleum Products, covers the determination of this property.

METHANOL

Methanol, as with organic halides, does not occur naturally in crude oils but is introduced artificially to prevent formation of gas hydrates. These large matrixes of natural gas and water can block or impede flow in gas pipelines. Use of methanol as a hydrate inhibitor occurs mostly in production of crude oils from deep waters such as the Outer Continental Shelf of the Gulf of Mexico, offshore West Africa, and in areas of the North Sea. For crude oil produced from the Gulf of Mexico, methanol contamination commonly occurs after production restarts following hurricanes. As the methanol is water miscible, it gets carried with water present in the crude oil to the refinery where it comes out in the water effluent from the desalter unit. When it reaches the wastewater treatment system, it can drastically upset the balance of the system. The bacteria used in the plant to digest oily components prefer the methanol, leaving hydrocarbons and some other toxic substances untreated. Large incursions of methanol can lead to a "bug kill" effectively deactivating the system. Either of these situations can result in discharge of pollutants and environmental excursions that exceed permitted levels.

Currently, there is no standard test method for determining methanol in crude oils containing water. ASTM D7059, Test Method for Determination of Methanol in Crude Oil by Multidimensional Gas Chromatography, is applicable only to crude oils containing a maximum of 0.1 % (v/v) water. As such, it is not applicable to analysis of most production quality crude oil streams that commonly contain 0.25 to 1.0 % (v/v) water. Several instrument manufacturers have worked on development of suitable analytical methods. A prototype, on-line gas chromatographic system for real-time measurement of methanol in a crude oil stream was tested by a crude oil pipeline company at one of its onshore Gulf of Mexico terminals, but the system did not have the necessary capabilities or ruggedness. In the absence of a standard test method, methanol can be determined by washing a sample with water, then analyzing the eluate by gas chromatography. This method is time consuming and does not allow for continuous monitoring of a stream. The latter is an important consideration, as methanol is usually disposed of in batches from offshore operations rather than on a continuous basis.

Boiling Point Distribution

Boiling point distribution provides an insight into the composition of crude oil and an estimation of the quantity of products likely to be yielded in refinery processes. Simulated distillation using gas chromatography can be used to rapidly determine this parameter without the need for a conventional potstill distillation, which is a lengthy process and requires a relatively large volume of sample. ASTM D2887, Test Method for Boiling Range Distribution of Petroleum Fractions by Gas Chromatography, originally approved in 1973, was the first standardized gas chromatographic method for determining boiling range distribution of petroleum. It is, however, restricted to petroleum products and fractions in the range of 55 to 538°C, which limits its application to crude oils. ASTM D5307, Test Method for Determination of the Boiling Range Distribution of Crude Petroleum by Gas Chromatography, covers determination of the boiling range distribution of water-free crude oil, but still only through 538°C, which corresponds to n-C_{43}. Material boiling above 538°C is reported as residue. ASTM D7169, Test Method for Boiling Point Distribution of Samples with Residues Such as Crude Oils and Atmospheric and Vacuum Residues by High Temperature Gas Chromatography, a method commonly abbreviated HTSD, extends the boiling range distribution through a temperature of 720°C. This temperature corresponds to the elution of n-C_{100}. This extended range is important to the refinery engineer, because a number of heavy crude oils available in today's market have a substantial amount of residue boiling well beyond 538°C.

Further discussion on application of gas chromatography to determination of boiling point distribution can be found in ASTM Manual 51, cited earlier [32]. This includes important considerations such as instrument requirements and column selection and a comparison to physical distillation.

Determination of boiling point distribution by HTSD is useful for rapidly obtaining information on the potential mass percent yield of products. These data provide refiners the ability to quickly evaluate crude oils and to select those with economic advantages and more favorable refining margins [33]. The information it provides can be input to LP models and used in establishing operations conditions in the refinery. Data on the boiling point distribution also serve as a rapid method for screening for the presence of diluents or residuum, constituting what is commonly referred to as "dumb-bell crude."

Gas chromatographic simulated distillation does not, however, provide any material for quality assessments. This requires that samples be fractionated by conventional pot-still distillation methods described later in this chapter.

The inspection assay tests discussed earlier are unquestionably not exhaustive but are the ones most commonly used. These tests will provide the refiner with data on a crude oil's handling characteristics, some of the impurities that are present, evidence of "spiking," and a general idea of the products that may be recoverable.

A summary of these inspection test methods is provided in Table 1. These tests will not, however, provide the data essential to determining whether a given crude oil or blend of crude oils will yield an economically attractive product slate. This requires that a comprehensive assay be performed.

TABLE 1—Crude Oil Inspection Assay Properties

Property	ASTM Referee Test Method
API, relative density, or specific gravity	D1298 or D5002
Pour point	D97 or D5853
Nitrogen, total, mass percent	D5762
Sulfur, total, mass percent	D4294
Viscosity, cSt @ 25°C and 40°C	D445
Water, volume percent	D4928
Sediment, mass percent	D473
Salt, ppm	D6470
Total acid number, mg KOH/g	D664
Organic chlorides, ppm	D4929
Hydrogen sulfide and mercaptans, ppm	UOP 163
Vapor pressure, kPa @ 37.8°C	D323 or D6377
Trace metals, ppm	D5708 or D5863
Carbon residue, mass percent	D4530
Asphaltenes, mass percent	D6560
Methanol, ppm	Gas chromatography
Boiling range distribution	D7169

COMPREHENSIVE ASSAY

In addition to the whole crude oil tests performed as part of the inspection assay, a comprehensive or full assay requires that the crude be fractionally distilled and the fractions characterized by appropriate tests. This is necessary so that the refiner can assess the quantity and quality of products recoverable from a given crude oil and determine if that product slate economically satisfies the market requirements of a particular refinery. Refiners tailor a comprehensive assay to their individual needs, and the number of cuts or fractions taken may vary from as few as 4 to 24 or more. The following ten fractions provide the basis for a moderately thorough evaluation:

C_2–C_4	Gas
C_5–79°C	Light naphtha
79–121°C	Medium naphtha
121–191°C	Heavy naphtha
191–277°C	Kerosene
277–343°C	Distillate fuel oil
343–455°C	Light vacuum gas oil (LVGO)
455–566°C	Heavy vacuum gas oil (HVGO)
343°C+	Atmospheric (long) residuum
566°C+	Vacuum (short) residuum

Commonly, from 5 to 50 L of crude oil will be needed for a comprehensive assay, depending on the number of cuts to be taken and the tests to be performed on the fractions. Fractionation of the crude oil begins with a true boiling point (TBP) distillation using a fractionating column having an efficiency of 14 to 18 theoretical plates and operated at a reflux ratio of 5:1 [ASTM D2892, Test Method for Distillation of Crude Petroleum (15-Theoretical Plate Column)]. The TBP distillation may be used for all fractions up to a maximum cut point of about 350°C atmospheric equivalent temperature (AET) provided reduced pressure is used to avoid cracking. Beyond an AET of 350°C, it is necessary to continue the distillation at further reduced pressures under conditions that provide approximately a one-theoretical plate fractionation. ASTM D5236, Test Method for Distillation of Heavy Hydrocarbon Mixtures (Vacuum Potstill Method), enables the distillation to be continued to a temperature of up to about 565°C AET at a pressure of 0.013 kPa, provided significant cracking does not occur. The maximum achievable AET is dependent on the heat tolerance of the charge. An older method, ASTM D1160, Test Method for Distillation of Petroleum Products at Reduced Pressure, is useful only up to a maximum liquid temperature of approximately 400°C at a pressure of 0.13 kPa (640°C AET). Figure 1 graphically depicts typical TBP distillation curves for a heavy (22°API) and a light (38°API) crude oil.

A detailed discussion of the three ASTM methods commonly used for a true boiling point distillation is provided in the ASTM MNL 51 cited earlier in the discussion on vapor pressure [32]. This includes field of application, important parameters, and precision and accuracy, as well as a comparison of ASTM D2892 and D5236. ASTM D1160 has now largely been replaced by ASTM D5236 for high vacuum distillation of crude oil residues.

Wiped-wall or thin-film molecular stills can also be used to separate the higher boiling fractions under conditions that minimize cracking. In these units, however, cut points cannot be directly selected, because vapor temperature in the distillation column cannot be measured accurately under operating conditions. Instead, the wall (film) temperature, pressure, and feed rate that will produce a cut equivalent to a D1160 or D5236 fraction with a given end point are determined from in-house correlations developed by matching yields between the wiped-wall distillation and the D1160 or D5236 distillation. ASTM D7169 should be useful in determining cut points of the higher boiling fraction material recovered by wiped-wall distillation. Despite this indirect approach, wiped-wall stills are often used because they allow higher end points to be attained than with either D1160 or D5236 and can easily provide large quantities of material for characterization.

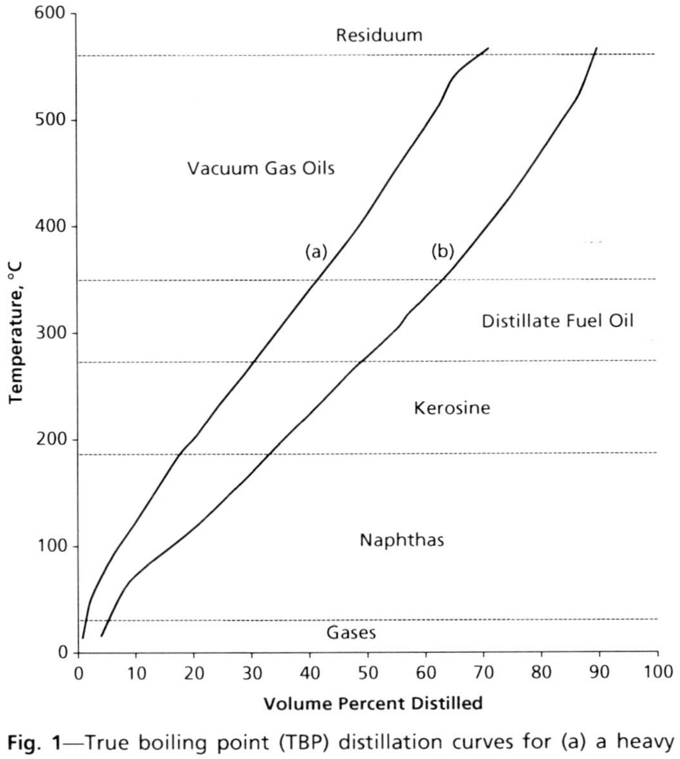

Fig. 1—True boiling point (TBP) distillation curves for (a) a heavy (22°API) crude oil and (b) a light (38°API) crude oil.

Following fractionation of the crude oil, each of the fractions is analyzed to determine one or more of its physical or chemical characteristics depending on the needs of the refiner. In the following discussion, the properties or constituents generally measured in a detailed analysis of each of the given ten fractions are listed. All of the various tests that could be performed on each of the fractions are too numerous to be included here. Table 2 summarizes this comprehensive assay format and indicates representative test methods for determining the properties. As with Table 1, the methods listed are those generally accepted as referee methods for determination of the property.

Numerous standard test methods are available for the determination of the properties and constituents of the distillate and residual fractions described. These test methods will not be enumerated in the following discussion as they are described in detail in the other chapters of this manual and elsewhere [34–36]. Although not listed in the table or succeeding discussion, volume and mass percent yields are an integral part of the analysis. These provide critical information on the quantity of product yields, allow calculation of mass balances, and permit the analyst or refiner to reformat data using LP models and empirically derived correlations to obtain characteristics of fractions suitable to their individual needs.

Gas

Typically, the gas or debutanization fraction is analyzed by high-resolution gas chromatography for quantitative determination of individual C_2 to C_4 and total C_5+ hydrocarbons. Relative density (specific gravity) can be calculated from the compositional analysis.

Naphtha Fractions

Density or specific gravity, total sulfur, mercaptan sulfur, hydrogen sulfide, and organic halides are typically determined on these fractions. Because these fractions, and especially the light naphtha fraction, are important both as a petrochemical feedstock and as a gasoline blending component or reformer feedstock, it is likely that they would also be analyzed by high-resolution gas chromatography for quantitative determination of their paraffin, isoparaffin, aromatic, and naphthene (cycloparaffin) components (PIAN analysis) (ASTM D5134, Test Method for Detailed Analysis of Petroleum Naphthas through *n*-Nonane by Capillary Gas Chromatography). Through judicious selection of columns and operating conditions, some laboratories have extended this method to *n*-dodecane (bp 216°C).

Octane numbers would also be determined for these fractions if they were to be included as a gasoline-blending component. Historically, octane numbers are determined using special engines that require relatively large volumes of sample. Today, many companies are now using semi-micro methods that require considerably less sample than the engine test methods for determination of octane numbers [37]. Other laboratories use PIAN data to calculate octane numbers [5]. For the heavy naphtha fraction, aniline point would also normally be determined.

Included in the information that can be derived from the PIAN analysis are the concentrations of benzene–benzene precursors (compounds that ultimately form benzene in a refinery's reforming unit), ethyl benzene, toluene, and xylene (B-E-T-X). These data are important because of environmental regulations limiting the maximum concentration of benzene in

TABLE 2—Crude Oil Comprehensive Assay Format

Property	ASTM Representative Test Methods	Whole Crude	C_2–C_4 Gases	C_5–79°C Light Naphtha	79°–121°C Medium Naphtha	121–191°C Heavy Naphtha	191–277°C Kerosine	277–343°C Distillate Fuel Oil	343–455°C Light VGO	455–566°C Heavy VGO	343 °C + Atmospheric Residuum	566°C + Vacuum Residuum
Volume and mass percent yields	D2892 and D5236		X	X	X	X	X	X	X	X	X	X
°API, density, specific gravity	D5002	X		X	X	X	X	X	X	X	X	X
Sulfur, total, mass percent	D4294	X		X	X	X	X	X	X	X	X	X
Sediment, mass percent	D473	X										
Water, volume percent	D4928	X										
Salt, mass percent	D6470	X										
Nitrogen, total, mass percent	D5762	X					X	X	X	X	X	X
Nitrogen, basic, mass percent	D2896								X	X	X	X
Carbon residue, mass percent	D4530	X							X	X	X	X
Pour point	D5853	X					X	X	X	X	X	
Metals: Ni, V, Fe, Cu	D5708	X									X	X
Organic chlorides, total, ppm	D4929	X		X	X	X	X					
UOP "K" factor	UOP 375	X										
Vapor Pressure, kPa @ 37.8°C	D323 or D6377	X										
Acid number, mg KOH/g	D664	X					X	X	X	X	X	X
H_2S and mercaptans, ppm	UOP 163	X		X	X	X	X					
Paraffins, isoparaffins, aromatics, naphthenes (PIAN)	D5134	B-T-E-X	X	X	X	X						

Viscosity, cSt, @ 25°C	D445	X					X					
40°C		X					X	X				
55°C								X	X	X	X	
80°C									X	X	X	X
100°C												X
120°C												
High temp. sim. distillation	D7169	X										
Hydrogen and carbon, mass percent	D5291								X	X	X	X
Refractive index @ 20°C	D1218						X	X				
Research and motor octane numbers	Calcula-tion from PIAN data			X	X	X						
Asphaltenes, mass percent	D6560									X	X	X
Aniline point	D611					X	X	X	X	X		
Cetane index	D976						X	X	X			
Naphthalenes, volume percent	D1840						X	X				
Smoke point, mm	D1322						X	X				
Freezing point	D2386						X					
Cloud point	D5773						X	X	X	X		
Penetration	D5											X
Softening point	D36											X

reformulated gasoline and because of the importance of these compounds as petrochemical feedstocks and intermediates.

Kerosene

Typically, density or specific gravity, total sulfur, mercaptan sulfur, hydrogen sulfide, organic halides, aniline point, total acid or neutralization number, naphthalene content, smoke point, total nitrogen, viscosity, and pour, cloud, and freezing points would be determined for this fraction and a cetane index calculated. Other tests that might be performed, depending on the intended end use of the fraction, are flash point, corrosiveness, and thermal stability.

As discussed earlier in "Sulfur Content," thermally reactive sulfur compounds such as mercaptans may be present in crude oils. On heating or distillation, these can decompose to form hydrogen sulfide, giving rise to its presence in the naphtha and kerosene fractions.

Distillate Fuel Oil

Tests of the distillate fuel oil fraction, which includes material used to produce aviation turbine fuel, normally include determination of density or specific gravity, total sulfur, aniline point, total acid number, naphthalene content, smoke point, total nitrogen, viscosity, cloud, freeze, and pour points, and calculation of cetane index. Thermal stability and corrosiveness may also be determined in more thorough evaluations. Measurement of refractive index is also useful in correlative methods, especially in LP models, for determination of the gross composition of this fraction.

Vacuum Gas Oil Fractions

Density or specific gravity, total sulfur, total nitrogen, aniline point, viscosity, acid number, cloud and pour points, and carbon residue would normally be determined on these fractions. Cetane index would be determined on the light vacuum gas oil if the material is to be used as diesel fuel oil blending stock. If the heavy gas oil fraction is to be used as catalytic cracker feedstock, asphaltenes would also be determined. Hydrogen and carbon content would also be determined on fractions to be used as catalytic cracker feedstock. Lube stock evaluations would include determination of wax content. Basic nitrogen is also typically determined on these fractions. In general, however, the ratio of basic to total nitrogen is on the order of 0.3:1 for many crude oils and virgin stocks. It also appears that the types of nitrogen compounds present in various crude oils are essentially the same, although the actual amounts may vary considerably [38]. Consequently, for most assays it is sufficient to determine total nitrogen.

Residuum

Tests of the residuum fractions usually include density or specific gravity, total sulfur, total and basic nitrogen, acid number, viscosity, trace metals, asphaltenes, and carbon residue. Hydrogen/carbon ratio and pour point determinations on the atmospheric (long) and vacuum (short) residua are also important. Increasingly, refiners are minimizing production of material that formerly went into heavy fuels such as No. 6 and bunker oil as markets for these diminish.

Determination of the properties of asphalt, such as penetration (ASTM D5, Test Method for Penetration of Bituminous Materials), softening point [ASTM D36, Test Method for Softening Point of Bitumen (Ring-and-Ball Apparatus)], and viscosity, would also be included in some assays. These tests were, however, developed in an era of less traffic and significantly lower pavement loadings and are no longer suitable for evaluating asphalt to be used as a binder. Newer "Performance Grade" tests such as Dynamic Shear Rheometer, Rolling Thin Film Oven, and Direct Tension Test have been developed [39].

Assay Summary

In the preceding discussion, the tests listed for each fraction and for the whole crude oil assay are not exhaustive but are illustrative of those used to evaluate quality. As noted earlier, refiners tailor their analytical scheme to their particular crude oil and product slates, although one refiner is reported to have said "The best crude oil assay is a 100,000 bbl run through my refinery" [40]. While this opinion carries some validity, the assay tests presented here provide data that are sufficient for most refiners to evaluate crude oil streams.

With the proliferation of computer LP models and their associated assay libraries [9], many refiners no longer need to perform comprehensive assays as frequently as in the past. Often, an inspection assay is all that is required for them to plan for changes in processing that will be necessitated by varying levels of impurities or small changes in crude oil composition resulting from changes in the production commingled to make up the crude oil stream. Most important, an inspection assay can be completed much more rapidly than a comprehensive assay and requires considerably less material. Collectively, these can result in savings in analyst's time and in shipping and handling expenses.

The quality of cargoes arriving at refineries may not conform to what is known or expected for that stream. This can be the result, among other causes, of new production coming on-stream and being commingled with existing production, or field maintenance. The analyst or refiner can use the "Flash Assay Tool" [41] to update an existing comprehensive assay on the stream in question with a minimum of data. The tool uses HTSD and various whole crude properties such as API gravity or density and total sulfur content to adjust the older assay. The "new" assay, together with a refinery LP model, can then be used to make quick decisions on purchasing and running a challenged (opportunity) crude, for example. The flash assay can also be used to indicate when another full assay should be run. With the proliferation of new streams and with new production being added into old, a flash assay can result in considerable cost and time savings.

Moreover, today's trend is toward automated, real-time analysis using on-line detectors to the extent permitted by available instrumentation. Some of this instrumentation has the capability to provide data suitable for custody transfer determination of crude oil properties, but this is currently quite limited. It seems likely, however, that significant advances in this area will be accomplished in the near future, further removing the laboratory analyst from characterization of crude oil feedstocks.

REFEREE TEST METHODS

When two or more test methods are available for determining a property, one is customarily designated as the referee or primary method in testing protocols. This provides for resolving disputes in cases where two methods yield different results on the same material. The methods listed in Table 1 are those generally accepted as referee methods for determination of the property. In some cases, two test methods are listed for the same property, as their respective scopes differ or the methods provide equivalent results. In these cases, it is important that the purpose of the analysis and the nature of the material are clearly understood in selecting a suitable referee method.

SAMPLING

The importance of adhering to a rigorous sampling protocol to ensure that samples are representative of the bulk material cannot be overemphasized. Representative samples are required for the determination of chemical and physical properties used to establish standard volumes and compliance with contractual specifications. Maintaining compositional integrity of these samples from the time of collection until they are analyzed requires care and effort.

Moreover, it is critically important that the sampling procedure does not introduce any contaminant into the sample or otherwise alter the sample so that subsequent test results are affected. Procedures for collection and handling of samples for hydrogen sulfide (H_2S) determination are especially critical because of the highly reactive nature and volatility of this compound. Appendix A provides recommended procedures suitable for collection and handling of samples for determination of H_2S in crude oil. These were developed by the U.S. Department of Energy's Strategic Petroleum Reserve in support of its crude oil assay program and underwent rigorous field and laboratory testing [42]. With proper handling, samples do not exhibit loss of their H_2S for a minimum of 10 days.

ASTM D4057, Practice for Manual Sampling of Petroleum and Petroleum Products, provides procedures for manually obtaining samples from tanks, pipelines, drums, barrels, and other containers. This practice addresses, in detail, the various factors that need to be considered in obtaining a representative sample. It must be kept in mind that, in many liquid manual sampling applications, the material to be sampled contains a heavy component, such as free water, which tends to separate from the main component. Unless certain conditions can be met to allow for this, an automatic sampling system as described in ASTM D4177, Practice for Automatic Sampling of Petroleum and Petroleum Products, is recommended.

ASTM D5854, Practice for Mixing and Handling of Liquid Samples of Petroleum and Petroleum Products, covers the handling, mixing, and conditioning procedures that are required to ensure that a representative sample is delivered from the primary sample container or receiver into the analytical test apparatus or into intermediate containers. This practice also provides a guide for selecting suitable containers for crude oil samples for various analyses.

ASTM D5842, Practice for Sampling and Handling of Fuels for Volatility Measurements, covers procedures and equipment for obtaining, mixing, and handling representative samples of volatile fuels. While directed to products such as gasoline and reformulated fuels, the guidance provided is useful in sampling and handling of crude oils.

Chain-of-custody procedures are a necessary element in a program to ensure one's ability to support data and conclusions adequately in a legal or regulatory situation. ASTM D4840, Guide for Sample Chain-of-Custody Procedures, contains a comprehensive discussion of potential requirements for a sample chain-of-custody program and describes the procedures involved. The purpose of these procedures is to provide accountability for and documentation of sample integrity from the time samples are collected until their disposal.

CRUDE OIL AS FUEL

Crude oil has been used directly as a fuel for power generation in a number of applications for more than 100 years. Among the earliest documented uses was as fuel for the Cleveland City Cable Railway power plant in Cleveland, Ohio, in 1891 [43]. In 1892, it was reported "there are some such progressive brick manufacturers in Chicago, who use neither coal nor wood in the drying or burning of their clay products. Crude oil is the fuel which they employ" [44]. In 1900, crude oil was used to power a railroad locomotive in a trial run in Corsicana, Texas. A history of Anheuser Busch, the American brewing giant, mentions that Adolphus Busch obtained the American rights to build diesel engines in 1897 and that the company's first engine was installed in the Anheuser Busch power plant in 1898. While not stating explicitly that crude oil was used to fuel this engine, the history states "as now is well known, the Diesel engine operates very economically, using crude oil as fuel" [45].

In 1915, the *New International Encyclopedia* stated, "nearly all crude oil can be used by fuel (for ships). If it contains very little sulphur or asphalt, it is suitable for heavy oil engines of the Diesel type." The article went on to state that "crude oil is not, however, much used for either of these purposes and the reasons are ... the presence of the more volatile oils renders storage dangerous" [46].

In 1975, crude oils with a wide range of properties were investigated for direct use as fuel in U.S. Army high-speed diesel engines [47]. Results of this study indicated that, with proper selection and pretreatment, crude oils are feasible for emergency use.

In recent years, crude oil has been used as fuel under boilers, in gas turbines used for standby power generation, and in diesel engines worldwide. Since 1993, crude oil has replaced other sources of energy to fire cement kilns in Cuba. One of two heating systems in the Republic of Macedonia uses crude oil as fuel. In 2004, China purchased three diesel engines from a Finnish company for installation on one of its offshore production platforms specifically to generate electricity using crude oil from the oilfield's production as fuel [48]. In 2006, Japan used on the order of 90,000–100,000 barrels per day of crude oil to fire some of its thermal power plants [49]. And, contrary to the admonition in the 1915 *New International Encyclopedia* article about crude oil's dangers, a number of tanker operators have, at times, surreptitiously and unlawfully diverted some of their cargo to power their ships. These incidents have not been without disastrous consequences, and fires and explosions have occurred.

Cost and availability—especially in remote operational areas—are certainly factors influencing the selection of crude oil over conventional burner, diesel, and gas turbine fuels. Other factors must also be considered, among them safety, environmental emissions, and the impact use of crude oil will have on the performance and maintenance of engines, turbines, and boilers.

No specifications have been established for crude oil to be used in these applications, but certain properties critical to the operation of burners, engines, and turbines must be taken into consideration. Among these are flash point, sulfur, ash content, pour point and viscosity, wax content, carbon residue, and metals content—especially vanadium, nickel, and sodium. As crude oils usually have relatively low flash points, their use as a fuel has an associated fire hazard and inherent risk of explosion necessitating that storage facilities be appropriately configured. Furthermore, crude oils selected for use as burner fuels must have suitable fluidity properties to ensure they are pumpable and can be aspirated in the burner or turbine. Moreover, the crude oil should have low ash and sulfur contents. These are important environmental considerations from the standpoints of fly ash (particulates) and sulfur dioxide emissions. The asphaltic residuum from burning of crude oils can build up on heat transfer surfaces, reducing their efficiency, and can clog burners. Crude oils to be used for gas turbines must also have low concentrations of trace metals to ensure that turbine blades are not adversely affected.

The standard test methods described elsewhere in this chapter for characterizing crude oils as refinery feedstocks are equally applicable to determining its use as a fuel.

CRUDE OIL COMPATIBILITY

The blending of compositionally different crude oils and crude oil with condensate can result in significant problems in transportation, storage, and refining. This is usually manifested by agglomeration or flocculation of waxes or asphaltenes, or both. These can accumulate and clog tubular components such as pipelines and heat exchangers or contribute to sludge buildup in tanks, requiring costly and time-consuming cleanup and equipment downtime.

The problem has been recognized for many years, but no standard method has been developed that reliably predicts whether particular blends will be incompatible, quantify the extent to which waxes and asphaltenes will accumulate, or determine when onset of incompatibility will begin once components are mixed. This latter is complicated by the fact that considerable empirical evidence indicates there can be an induction period of 3 or more days prior to onset.

Historically, two test methods used for testing compatibility of residual fuel oil mixtures have been applied to crude oil blends: ASTM D4740, Test Method for Stability and Compatibility of Residual Fuels by Spot Test, and ASTM D4870, Test Method for Determination of Total Sediment in Residual Fuels. Other test methods, which range from simple to complex, include "modified Shell Spot test" [50], "Asphaltene Stability Index" [51], and "Oil Compatibility Model" [52].

Two automated instrumental test methods seem promising for determining stability and compatibility of crude oils. ASTM D7112, Test Method for Determining Stability and Compatibility of Heavy Fuel Oils and Crude Oils by Heavy Fuel Oil Stability Analyzer (Optical Detection), is a procedure involving titration and optical detection of precipitated asphaltenes. ASTM D7157, Test Method for Determination of Intrinsic Stability of Asphaltene-Containing Residues, Heavy Fuel Oils, and Crude Oils (*n*-Heptane Phase Separation; Optical Detection), is a procedure for quantifying the intrinsic stability of the asphaltenes in an oil using an optical device.

FUTURE NEEDS IN CRUDE OIL CHARACTERIZATION

The average quality of crude oil being processed in refineries is becoming higher in sulfur and heavier [53]—that is, a greater content of heavy ends or residuum. With the growing demand for transportation fuels, the refiner is faced with mounting pressure to make better use of the "bottom of the barrel"—the residuum that formerly went into low-quality products such as No. 6 and bunker fuel oils. Diminishing worldwide demand for these latter fuels is exacerbating the need to use this material to produce other products, especially lighter fuels. Moreover, as environmental restrictions increasingly limit sulfur and aromatics in transportation and burner fuel oils, refiners are facing new challenges to economically produce a marketable slate of products from heavier, higher-sulfur feedstocks. These challenges generally require new or expanded processing and treatment technology at the refinery. This, in turn, translates into the need for new analytical test methods in the laboratory to adequately evaluate feedstocks and monitor product quality.

Existing analytical methods may not, however, be suitable for characterizing many of today's crude oil streams. The scope of these methods may not be sufficiently broad to cover the range in quality of some of the heavier and higher sulfur crude oils and nonconventional streams such as bitumen and "synthetic crude oil" derived from Canadian oil sands and Venezuelan heavy oils. Consequently, at a minimum, it may be necessary to conduct new interlaboratory round robin studies to validate the precision statements. With the consolidations within the industry and the growing demands being placed on the analyst, this may not be practicable.

The Canadian Crude Quality Technical Association has used the extensive experience of its members to compile a manual of *Heavy Oil and Bitumen Analytical Methods* [54]. At the end of 2008, this manual provided a review of existing test methods for eight parameters commonly used in assessing quality of heavy oils and bitumen and discussed their capabilities and limitations. When available, precision data obtained from round robin studies on heavy oils and bitumen are included. The analytical sections in this manual are not a substitute for test methods specific to heavy oils and bitumen, but they do help users to make informed decisions on methods' selection, sample preparation, test modifications, and interpretation of resultant data.

Sulfur reduction processes are sensitive to both amount and structure of the sulfur compounds being removed. Tests that can provide information about both are becoming increasingly important. A number of laboratories have combined gas chromatography with sulfur-selective detectors to provide data on the boiling range distribution of the sulfur compounds and probable molecular types. In addition, gas chromatography detectors that provide selectivity for other constituents of interest, such as nitrogen and organometallics, are also commercially available and being increasingly used for characterization.

The supply of high-TAN crude oils, that is those with a TAN >1.0, is increasing substantially on a global basis. To adequately assess and plan the effective refinery processing of these, it is important to know the acid species present. Naphthenic acids in particular can be especially detrimental throughout a refinery. A number of crude oils today contain high concentrations of calcium that will contribute to formation of egregious naphthenic acid soaps. These soaps can cause formation of highly stable emulsions that can be carried through to wastewater treatment plants. Naphthenic acids can be very corrosive, but it is important to know their species as they behave differently depending on the various conditions of temperature, flow velocity, and metallurgy existing within a refinery. ASTM D664, which is used almost universally for determining the acid content of a crude oil, provides no information on either the acid species present or their relative strength. A rapid and reliable method for characterizing the components contributing to the acidity of a crude oil as well as their relative strength is needed.

Upgrading the "bottom of the barrel" involves converting more–ideally all–of the residuum into a more salable, higher-valued product (100 % residuum conversion). Whatever the means to this end, improved characterization methods are necessary for process design, crude oil evaluation, and operational control. Characterization methods that define the boiling range and the hydrocarbon type distribution of heavy residuum are of growing importance. High-temperature, gas chromatography simulated distillation has been used to obtain cut point data up to an atmospheric equivalent boiling point temperature of 740°C [55].

The distributions of hydrocarbon types in gas oil and heavier materials are important in evaluating them as feedstocks for further processing. Some older mass spectrometric methods for determining hydrocarbon types are being updated for use with modern quadrupole mass spectrometers, either with batch inlets or with gas chromatographic inlets (GC/MS). Another technique that has been successfully applied for determining hydrocarbon types in these materials involves the use of high-performance liquid chromatography (HPLC), which can provide information comparable to the mass spectrometric methods. Standardized HPLC techniques are now being used to determine aromatic hydrocarbon types in middle distillate and aviation fuels.

With the large volumes of crude oil being moved in today's markets, and with the growing availability of opportunity or challenged crude oils, comes increasing pressure to determine key properties in less time and with a high degree of confidence in the results. While opportunity crudes may be financially attractive to the crude oil buyer, they can be a problem to the refiner. They may be high-TAN or contain excessive levels of certain contaminants. Their product yields and qualities may be unknown or undesirable. For example, excessive residuum yields can overload a coker, and the coke produced may not be of the quality needed to satisfy the existing market. With the need for rapid turnaround of stocks and the relatively limited storage available in most refineries, these streams may be processed before conventional analyses are complete.

Rapid, automated instrumental methods of analysis continue to be the option of choice in responding to these demands. Laboratories are constantly seeking to reduce analysis time, improve the quality of test results, and eliminate dependency on time-consuming manual methods of analysis. Advances in technology are helping to meet the analytical challenges of the industry.

Predictive methods that can rapidly provide accurate physiochemical and boiling range data are the "holy grail" of the crude oil analyst. A number of systems have been proposed but all have limitations that constrain their usefulness. Molecular modeling can be used to provide the degree of characterizations necessary, but it involves tedious sample preparation, lengthy component separation, use of advanced analytical instrumentation, and development of a large database containing reference material [15].

In the absence of these predictive methods, refiners and analysts are fortunate to have at their disposal LP programs with their built-in correlations and ability to recut data to their specific needs. Ultimately, however, they still must rely on a comprehensive assay to validate output from computer models.

APPENDIX A: PROCEDURES FOR COLLECTION OF SAMPLES FOR H_2S DETERMINATION

These procedures have been found suitable for collection of samples for determination of H_2S in crude oil [41]. Prior to collecting either high density polyethylene (HDPE) or floating piston cylinder samples, thoroughly flush the sampling point and all connections.

HDPE bottles—1 L capacity. Prepare the bottles by placing about 10 g of dry ice into each. Place the cap on the bottle and tighten loosely. Shake the bottle vigorously, and periodically loosen–but do not remove the cap–to relieve excess pressure. Continue this process until the dry ice has evaporated. Once the dry ice has evaporated, tighten the cap and set aside until the bottle is needed for sampling. Do not overpressure the bottles. If the bottles are not relieved of pressure buildup, they may explode.

Bottles may be prepared up to 2 days in advance of when they will be needed. It is advisable to prepare at least one extra bottle in case one leaks.

When ready to collect the sample, remove the bottle cap. There must be an audible hiss indicating the presence of carbon dioxide (CO_2) overpressure. If not, use another bottle. Slowly fill the bottle using a polytetrafluoroethylene (PTFE) tube extending to the bottom. When the bottle is full to the top of the shoulder, that is, just below the threads, squeeze the bottle at the center just enough to cause a small amount (a few drops) of oil to spill over the lip of the bottle. Screw the cap tightly onto the bottle and seal with plastic tape. Keep the bottle refrigerated or on ice. If shipping is necessary, package samples with dry ice and in accordance with International Air Transport Association (IATA) regulations.

Samples collected in this manner, when kept cold, may be used for determination of H_2S for up to 10 days following their collection.

Floating piston cylinders (Sulfinert®-treated). Prior to use, cylinders should have a back-pressure at least 100 psi greater than that of the pipeline from which samples are to be collected. Argon should be used as back-pressure gas, and not nitrogen or helium. No further preparation is necessary. Make connections to the pipeline with Sulfinert®-treated stainless steel or high pressure PTFE tubing. Slowly open the valve nearest the pipeline and check for leaks. Next, slowly open the bleed valve on the cylinder and bleed at least 250 mL to waste to purge the system and displace air. Then, slowly open the third valve and gradually reduce the back pressure until it approaches that of the pipeline. Once the indicator rod begins to move, continue to slowly bleed the back-pressure until the tip of the indicator rod is within approximately 1 cm of the red end cap. Several minutes should be allowed for this process in order to maintain a single phase in the cylinder. Tighten all valves, and then disconnect from the pipeline. Replace all plugs using PTFE tape.

Floating piston cylinders do not need to be refrigerated.

ASTM and UOP Standards

ASTM	UOP	Title
D5		Penetration of Bituminous Materials
D36		Softening Point of Bitumen (Ring-and-Ball Apparatus)
D97		Pour Point of Petroleum Products

ASTM	UOP	Title
D129		Sulfur in Petroleum Products (General Bomb Method)
D189		Conradson Carbon Residue of Petroleum Products
D287		API Gravity of Crude Petroleum and Petroleum Products (Hydrometer Method)
D323		Vapor Pressure of Petroleum Products (Reid Method)
D341		Viscosity-Temperature Charts for Liquid Petroleum Products
D445		Kinematic Viscosity of Transparent and Opaque Liquids (and the Calculation of Dynamic Viscosity)
D473		Sediment in Crude Oils and Fuel Oils by the Extraction Method
D482		Ash from Petroleum Products
D524		Ramsbottom Carbon Residue of Petroleum Products
D664		Acid Number of Petroleum Products by Potentiometric Method
D1160		Distillation of Petroleum Products at Reduced Pressure
D1250		Petroleum Measurement Tables
D1298		Density, Relative Density (Specific Gravity), or API Gravity of Crude Petroleum and Liquid Petroleum Products by Hydrometer Method
D1552		Sulfur in Petroleum Products (High-Temperature Method)
D2161		Conversion of Kinematic Viscosity to Saybolt Universal Viscosity or to Saybolt Furol Viscosity
D2622		Sulfur in Petroleum Products by X-Ray Spectrometry
D2887		Boiling Range Distribution of Petroleum Fractions by Gas Chromatography
D2892		Distillation of Crude Petroleum (15-Theoretical Plate Column)
D3228		Total Nitrogen in Lubricating Oils and Fuel Oils by Modified Kjeldahl Method
D3230		Salts in Crude Oil (Electrometric Method)
D3279		*n*-Heptane Insolubles
D4006		Water in Crude Oil by Distillation
D4007		Water and Sediment in Crude Oil by the Centrifuge Method (Laboratory Procedure)
D4057		Manual Sampling of Petroleum and Petroleum Products
D4177		Automatic Sampling of Petroleum and Petroleum Products

ASTM	UOP	Title
D4294		Sulfur in Petroleum Products by Energy-Dispersive X-Ray Fluorescence Spectrometry
D4377		Water in Crude Oil by Potentiometric Karl Fischer Titration
D4530		Determination of Carbon Residue (Micro Method)
D4629		Trace Nitrogen in Liquid Petroleum Hydrocarbons by Syringe/Inlet Oxidative Combustion and Chemiluminescence
D4740		Cleanliness and Compatibility of Residual Fuels by Spot Test
D4807		Sediment in Crude Oil by Membrane Filtration
D4840		Sampling Chain-of-Custody Procedures
D4870		Determination of Total Sediment in Residual Fuels
D4928		Water in Crude Oil by Coulometric Karl Fischer Titration
D4929		Determination of Organic Chloride Content in Crude Oil
D5002		Density and Relative Density of Crude Oils by Digital Density Analyzer
D5134		Detailed Analysis of Petroleum Naphthas through *n*-Nonane by Capillary Gas Chromatography
D5185		Determination of Additive Elements, Wear Metals, and Contaminants in Used Lubricating Oils and Determination of Selected Elements in Base Oils by Inductively-Coupled Plasma Atomic Emission Spectrometry (ICP-AES)
D5191		Vapor Pressure of Petroleum Products (Mini Method)
D5236		Distillation of Heavy Hydrocarbon Mixtures (Vacuum Pot Still Method)
D5291		Instrumental Determination of Carbon, Hydrogen, and Nitrogen in Petroleum Products and Lubricants
D5307		Determination of the Boiling Range Distribution of Crude Petroleum by Gas Chromatography
D5708		Determination of Nickel, Vanadium, and Iron in Crude Oils and Residual Fuels by Inductively-Coupled Plasma (ICP) Atomic Emission Spectrometry
D5762		Nitrogen in Petroleum and Petroleum Products by Boat-Inlet Chemiluminescence
D5842		Sampling and Handling of Fuels for Volatility Measurement
D5853		Pour Point of Crude Oils

ASTM	UOP	Title
D5854		Practice for Mixing and Handling of Liquid Samples of Petroleum and Petroleum Products
D5863		Determination of Nickel, Vanadium, Iron, and Sodium in Crude Oils and Residual Fuels by Flame Atomic Absorption Spectrometry
D6377		Determination of Vapor Pressure of Crude Oil: $VPCR_X$ (Expansion Method)
D6470		Salt in Crude Oils (Potentiometric Method)
D6560		Determination of Asphaltenes (Heptane Insolubles) in Crude Petroleum and Petroleum Products
D6822		Density, Relative Density, and API Gravity of Crude Petroleum and Liquid Petroleum Products by Thermohydrometer Method
D7059		Determination of Methanol in Crude Oils by Multidimensional Gas Chromatography
D7112		Determining Stability and Compatibility of Heavy Fuel Oils and Crude Oils by Heavy Fuel Oil Stability Analyzer (Optical Detection)
D7157		Determination of Intrinsic Stability of Asphaltene-Containing Residues, Heavy Fuel Oils, and Crude Oils (*n*-Heptane Phase Separation; Optical Detection)
D7169		Boiling Point Distribution of Samples with Residues Such as Crude Oils and Atmospheric and Vacuum Residues by High Temperature Gas Chromatography
	163	Hydrogen Sulfide and Mercaptan Sulfur in Liquid Hydrocarbons by Potentiometric Titration
	375	Calculation of UOP Characterization Factor and Estimation of Molecular Weight of Petroleum Oil

References

[1] Rossini, F. D., "Hydrocarbons in Petroleum," *J. Chem. Educ.*, Vol. 37, 1960, pp. 554–561.

[2] Mair, B. J., "Annual Report for the Year Ending, June 30, 1967," American Petroleum Institute Research Project 6, Carnegie Institute of Technology, Pittsburgh, PA, 1967.

[3] Rail, H. T., Thompson, C. J., Coleman, H. J., and Hopkins, R. L., "Sulfur Compounds in Oil," Bulletin 659, U.S. Department of the Interior, Bureau of Mines, 1972.

[4] Csoklich, C., Ebner, B., and Schenz, R., "Modern Crude Oil Practices–Austria's OEMV," *Oil Gas J.*, March 21, 1983, pp. 86–90.

[5] O'Donnell, R. J., "Modern Crude Oil Practices–Standard Oil of California Companies," *Oil Gas J.*, March 21, 1983, pp. 90–93.

[6] McNelis, F. B., "Modern Crude Oil Practices–Exxon Organizations," *Oil Gas J.*, March 21, 1983, pp. 94–97.

[7] Wampler, R. J., and Kirk, E. L., "Modern Crude Oil Practices–Gulf Companies," *Oil Gas J.*, March 21, 1983, pp. 98–104.

[8] Nelson, G. V., Schierberg, G. R., and Sequeira, A., "Modern Crude Oil Practices–The Texaco System," *Oil Gas J.*, March 21, 1983, pp. 108, 112, 116, 118–120.

[9] McCleskey, G., and Joffe, B. L., "Modern Crude Oil Practices–Phillips Petroleum Co.," *Oil Gas J.*, March 21, 1983, pp. 124, 126–127.

[10] Aalund, L. R., "Guide to Export Crudes for the '80s–1 to 13," *Oil Gas J.*, April 11, May 2, 23, June 6, 20, July 4, 25, Aug. 22, Sept. 5, Oct. 24, Nov. 7, 21, Dec. 12, 19, 1983.

[11] "Crude oil assay," *Wikipedia, The Free Encyclopedia,* 12 September 2008, http://en.wikipedia.org/wiki/Crude_oil_assay.

[12] O'Donnell, J., "Crude Oils," *Criteria for Quality of Petroleum Products*, J. P. Allison, Ed., John Wiley, New York, 1973, pp. 10–21.

[13] Smith, N. A. C., Smith, H. M., Blade, O. C., and Garton, E. L., "The Bureau of Mines Routine Method for the Analysis of Crude Petroleum 1. The Analytical Method," Bulletin 490, U.S. Department of the Interior, Bureau of Mines, 1951.

[14] Ali, M. A., "Full Range Crudes, Analytical Methodology of," *Encyclopedia of Analytical Chemistry*, R. A. Meyers, Ed., Vol. 8, pp. 6709–6726, John Wiley and Sons Ltd., Chichester, 2000.

[15] Watts, M. R., and Roussis, S. G., "Crude Assay," *Practical Advances in Petroleum Processing,* C. S. Hsu and P. R. Robinson, Eds., pp. 103–116, Springer, New York, 2006.

[16] "Protecting Crude Oil Quality," a report of the API Ad Hoc Crude Oil Quality Task Force, American Petroleum Institute, Washington, DC, February 1993.

[17] Neihof, R. A., Hydrogen Sulfide Analyzer with Protective Barrier. U.S. Patent No. 5,529,841, U. S. Patent and Trademark Office, Washington, DC, June 25, 1996.

[18] *Impurities in Petroleum*, Petrolite Corporation, Houston, 1958.

[19] "Manual of Petroleum Measurement Standards Chapter 17.3–Guidelines for Identification of the Source of Free Waters Associated with Marine Petroleum Cargo Movements," First Edition, American Petroleum Institute, Washington, DC, 1992.

[20] Watson, K. M., Nelson, E. F., and Murphy, G. B., "Characterization of Petroleum Fractions," *Ind. Eng. Chem.*, Vol. 27, 1935, pp. 1460–1464.

[21] Nelson, W. L., "Which Base of Crude Oil Is Best?" *Oil Gas J.*, Jan. 8, 1979, pp. 112–113.

[22] Valkovic, V., *Trace Elements in Petroleum,* The Petroleum Publishing Co., Tulsa, OK, 1978.

[23] Yen, T. F., Ed., *The Role of Trace Metals in Petroleum,* Ann Arbor Science Publishers, Inc., Ann Arbor, MI, 1975.

[24] Jones, P., "Trace Elements and Other Elements in Crude Oil: A Literature Review," unpublished manuscript, The British Petroleum Co., Ltd., BP Research Center, Sunbury, UK, September 1975.

[25] Jones, M. C. K., and Hardy, R. L., "Petroleum Ash Components and Their Effect on Refractories," *Ind. Eng. Chem.*, Vol. 44, 1952, pp. 2615–2619.

[26] Woodle, R. A., and Chandler, W. B., Jr., "Mechanisms of Occurrence of Metals in Petroleum Distillates," *Ind. Eng. Chem.*, Vol. 44, 1952, pp. 2591–2596.

[27] Dreyfus, S., "Origin and Nature of Trace Metals in Crude Oil," Crude Oil Quality Group, March 6, 2008, New Orleans, LA, www.coqg.org Meeting Archives

[28] Dekkers, C., and Daane, R., "Metal Contents in Crudes Much Lower than Expected," *Oil Gas J.* March 1, 1999, pp. 44–51.

[29] Wilhelm, S. M., Kirchgessner, D. A., Liang, L., and Kariher, P. H. "Sampling and Analysis of Mercury in Crude Oil," R. A. K. Nadkarni, Ed., *Elemental Analysis of Fuels and Lubricants: Recent Advances and Future Prospects,* STP 1468, ASTM International, West Conshohocken, PA, 2005.

[30] Long, S. E., Kelly, W. R., Mann, J. L. and Giles, H. N., "Determination of Mercury in Crude Oil Using a Novel Method," *Proceedings 10th International Conference on Stability,*

Handling, and Use of Liquid Fuels, Tucson, AZ, October 7–11, 2007.

[31] Nadkarni, R. A. K., Ed., *Elemental Analysis of Fuels and Lubricants: Recent Advances and Future Prospects*, STP 1468, ASTM International, West Conshohocken, PA, 2005.

[32] Montemayor, R. G., Ed., *Distillation and Vapor Pressure Measurement in Petroleum Products*, MNL 51, ASTM International, West Conshohocken, PA, 2008.

[33] Villalanti, D. C., Raia, J. C., and Maynard, J. B., "High-Temperature Simulated Distillation Applications in Petroleum Characterization," *Encyclopedia of Analytical Chemistry*, R. A. Meyers, Ed., Vol. 8, pp. 6726–6741, John Wiley and Sons Ltd., Chichester, 2000.

[34] Drews, A. W., Ed., *Manual on Hydrocarbon Analysis*, 6th ed., ASTM International, West Conshohocken, PA, 1998.

[35] Nadkarni, R. A. K, *Guide to ASTM Test Methods for the Analysis of Petroleum Products and Lubricants*, ASTM International, West Conshohocken, PA, 2000.

[36] Totten, G. E., Ed., *Fuels and Lubes Handbook: Technology, Properties, Performance, and Testing*, MNL 37, ASTM International, West Conshohocken, PA, 2003.

[37] Childs, W. V., and Vickery, E. H., "The Phillips Small Sample Octane Number Methods, Automation of a Knock-Test Engine," *Symposium on Laboratory and Pilot Plant Automation*, Washington, DC, Aug. 28–Sept. 2, 1983, American Chemical Society, Washington, DC, 1983, pp. 979–990.

[38] Richter, F. P., Caesar, P. D., Meisel, S. L., and Offenhauer, R. D., "Distribution of Nitrogen in Petroleum According to Basicity," *Ind. Eng. Chem.*, Vol. 44, 1952, pp. 2601–2605.

[39] *Asphalt Binder Testing Manual*, MS-25, Asphalt Institute, Lexington, KY, 2007.

[40] Aalund, L. R., "Guide to Export Crudes for the '80s–1," *Oil Gas J.*, April 1, 1983, p. 71.

[41] Alexander, D., "How to Update an Old Assay," Crude Oil Quality Group, New Orleans, LA, January 2003. www.coqg.org Meeting Archives.

[42] *U.S. Department of Energy, Strategic Petroleum Reserve Crude Assay Manual*, 3rd edition, Appendix B, p. 17, U.S. Department of Energy, Washington, DC, August 2008. www.spr.doe.gov/reports/docs/CrudeOilAssayManual.pdf

[43] "Cable City Cable Railway," *Manufact. Builder*, Vol. 23, No. 11, November 1891.

[44] "Burning Brick with Crude Oil Fuel–The Use of Petroleum in Brick Kilns–Its Advantages, Cleanliness, and Cheapness," *Sci. Am.* Suppl., No. 841, February 13, 1892.

[45] Krebs, R., in collaboration with P. J. Orthwein, *Making Friends Is Our Business: 100 Years of Anheuser Busch*, Anheuser Busch, St. Louis, MO, 1953.

[46] "Fuel for Ships," *The New International Encyclopedia*, Vol. IX, p. 329, Dodd, Mead and Company, New York, 1915.

[47] Owens, E. C., and Frame, E. A., "Direct Utilization of Crude Oils as Fuels in U. S. Army Diesel Engines," Interim Report AFLRL No. 66, Southwest Research Institute, San Antonio, TX, June 1975.

[48] "Crude-Oil-Burning Wärtsilä Engines for Offshore China," *Marine News*, Vol. 3, 2004, p. 27.

[49] "Japan Studies Effect of Possible Sudan Crude Oil Ban," *Sudan Tribune*, November 17, 2007.

[50] Schermer, W. E. M., Melein, P. M. J., and van den Berg, F. G. A., "Simple Techniques for Evaluation of Crude Oil Compatibility," *Petrol. Sci. Technol.*, Vol. 22, 2004, Nos. 7 and 8, pp. 1045–1054.

[51] Stark, J., and Falkler, T. J. "Index Correlates Asphaltene Stability to Coker Furnace Fouling," *Oil Gas J.*, September 13, 2004, pp. 72–74.

[52] Wiehe, I. A., and Kennedy, R. J., "The Oil Compatibility Model and Crude Oil Incompatibility," *Energy Fuels*, Vol. 14, 2000, pp. 56–59.

[53] Swain, E. "U.S. Refiners Continue to Process Crudes with Lower Gravity, Higher Sulfur," *Oil Gas J.*, January 3, 2005, pp. 51–55.

[54] *Heavy Oil and Bitumen Analytical Methods: Understanding Their Capabilities and Limitations*, Canadian Crude Quality Technical Association, Edmonton, AB, 2008.

[55] Boduszynski, M. M., Grudoski, D. A., Rechsteiner, C. E., and Iwamoto, J. D., "Deep-Cut Assay Reveals Additional Yields of High-Value VGO," *Oil Gas J.*, September 11, 1995, pp. 39–44.

9

Properties of Petroleum Coke, Pitch, and Manufactured Carbon and Graphite

C. O. Mills[1] and F. A. Iannuzzi[2]

TEST METHODS USED TO CHARACTERIZE THE properties of petroleum coke, pitch, carbon artifacts (anodes and cathodes), and manufactured carbon products and graphite are presented in this chapter. This includes the raw materials and artifacts used in the electrolytic production of aluminum as well as graphite and its artifacts. The information presented here was extracted from ASTM standards and from *Introduction to Carbon Technologies, Sciences of Carbon Materials, Carbon and Graphite Handbook*, and *Chemistry and Physics of Carbon* [1–4]. ASTM Subcommittee D02.05 on Properties of Fuels, Petroleum Coke and Carbon Material has developed the test methods for coke, pitch, and for electrode carbon products. ASTM Subcommittee D02.F0 on Manufactured Carbon and Graphite has developed the test methods for manufactured carbon and graphite. This subcommittee was originally ASTM Committee C05 on Manufactured Carbon and Graphite, and it has recently been assimilated into ASTM Committee D02 on Petroleum and Lubricants as Subcommittee D02.F0.

Carbon is the chemical element that is the basis of organic chemistry. The two common, crystalline allotropes of carbon are diamond and graphite. Diamond is one of the hardest substances known, and graphite is one of the softest substances known.

Artifacts for the aluminum industry, prebaked amorphous carbon electrodes, are prepared from pitch and petroleum coke that has not been graphitized.

Graphite is resistant to chemical attack under harsh conditions. For this reason, the electrolytic industry may use graphite electrodes since no other material can withstand this aggressive environment. Its excellent high-temperature properties have allowed its use in nuclear, aerospace, and propulsion applications. Melting crucibles, pencils, and lubricants are examples of applications where graphite has been used throughout the centuries.

Carbon in its powder form is used for printing, paints, coatings, and in the tire industry. In its porous char form, it is used in filtering, decolorizing, and adsorbing applications. As a component in searchlights, electric arcs, sustained by carbon vapor, function as a high-intensity white light source. Its extensive use in the electrical industries facilitates the operation of electrical motors and other devices.

ATOMIC STRUCTURE OF VARIOUS FORMS

The international atomic weight of carbon is 12.01. There are seven isotopes that differ in atomic weight. The three "common" naturally occurring isotopes are carbon-12, carbon-13, and carbon-14. There are a number of allotropes that differ in crystal structure and form. The allotropes of carbon are classified as amorphous, semigraphitic, and crystalline. Diamond, graphite, and buckminsterfullerene represent three of the best-known allotropes. Certain amorphous carbons may be transformed into crystalline graphite at high temperatures in an electric furnace and into diamond at both high temperature and high pressure.

The structure of a diamond, Fig. 1a, is face-centered cubic. Solid carbons and graphite, excluding diamond, are, in general, made from platelets of carbon atoms arranged in a hexagonal pattern, with these platelets being capable of a variety of degrees of ordering. The most highly ordered structure is that in which platelets are of considerable size, parallel to each other, and oriented with respect to the "c" crystallographic axis of the hexagonal structure (Fig. 1b). This is the structure of crystalline graphite. The most disordered structure is the one found in "amorphous" carbon, in which a small number of layer planes, having a mean diameter of 25 Å, lie parallel to each other but are otherwise randomly oriented in the "c" direction. The small parallel packets thus formed are randomly oriented with respect to each other. Between these two extremes, there exists a range of intermediate structures in which the parallel packets are essentially parallel to each other and in which the size and number of layer planes in a parallel packet may continuously increase. In the late stages of this ordering process, some layer planes within a parallel pocket may orient themselves with respect to their neighbors to form a true ordered graphitic structure.

The carbons produced by pyrolysis of hydrocarbons have the intermediate type of structure and encompass most of the intermediate structures. The deposits formed at low temperatures are composed of groups of layer planes all lying parallel to each other. As the temperature of deposition increases, the ordering of structure becomes more developed; at the highest temperature of deposition, limited numbers of layer planes within the parallel structures are oriented with respect to each other as in true graphite.

PETROLEUM COKE AND PITCH

Petroleum coke and coal tar pitch provide the basic raw materials for the production of synthetic graphite as well as for the carbon anodes and lining materials used in the electrolytic production of aluminum.

Raw or green petroleum coke is manufactured from the heaviest portions of crude oil. In the past, petroleum coke

[1] *Consultant, Ponca City, OK*
[2] *Southern Research Institute, Birmingham, AL*

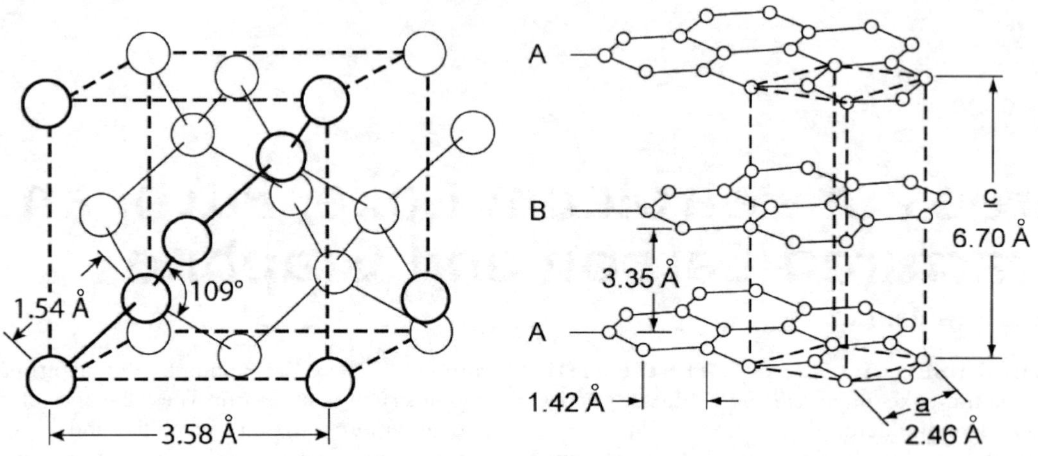

Fig. 1—(a) The diamond cubic structure. (b) The graphite crystal structure.

was produced as a by-product of a cracking process used to obtain additional amounts of gasoline and distillate from the residual portion of the crude oils. The raw petroleum coke currently produced is used for several purposes, which depend on its properties.

Many factors impact the quality and value of raw petroleum coke. Most of these factors are dependent on the feed material used in production of the coke and the actual process variables used during production.

Some petroleum coke produced today, usually of high sulfur and metals content, is used as fuel in electric generating facilities.

As improved processes have developed, markets have developed for higher-quality petroleum coke for use in preparation of electrodes used in the steel and aluminum industries as well as other applications. These applications require further processing of the raw petroleum coke (as well as closer control of the petroleum feed stock, etc.). Petroleum coke used as feedstock for the production of graphite articles and prebaked anodes for the production of aluminum is typically low in sulfur and metallic impurities. Petroleum coke must be calcined prior to being processed into materials used in production of carbon artifacts or graphite.

Calcining is a high-temperature heating process used to eliminate the volatile materials left during production of raw petroleum coke. This volatile fraction is composed of water and high-molecular-weight hydrocarbons. Calcining also promotes further structural changes to the remaining carbon material. In addition, some impurities, including sulfur (some of which may be contained in the volatile components) may be reduced while most metallic elements will be correspondingly increased during the calcining process.

Carbon artifacts such as electrodes, anodes, and specialty shapes are generally made by mixing a carbon aggregate (usually petroleum coke) with a binder and then forming, baking (at about 750 to 1,200°C), and in some cases graphitizing (about 2,800°C). Binders are usually manufactured from coal tar by distillation of the coal tar to a pitch. Binders can also be manufactured from petroleum-based materials (petroleum pitch) or a combination of coal tar and petroleum materials. Binder pitches are usually designated as carbon pitch, coal-tar pitch, petroleum pitch, industrial pitch, or, simply, pitch.

Carbon anodes used in electrolytic cells for the production of aluminum are manufactured using a variety of production techniques and raw materials. Therefore, it is important to test the anodes in order to optimize anode performance and to minimize the cost of aluminum production.

Figure 2 is a cursory flow diagram of how petroleum coke is utilized both as fuel and in production of carbon and carbon artifacts.

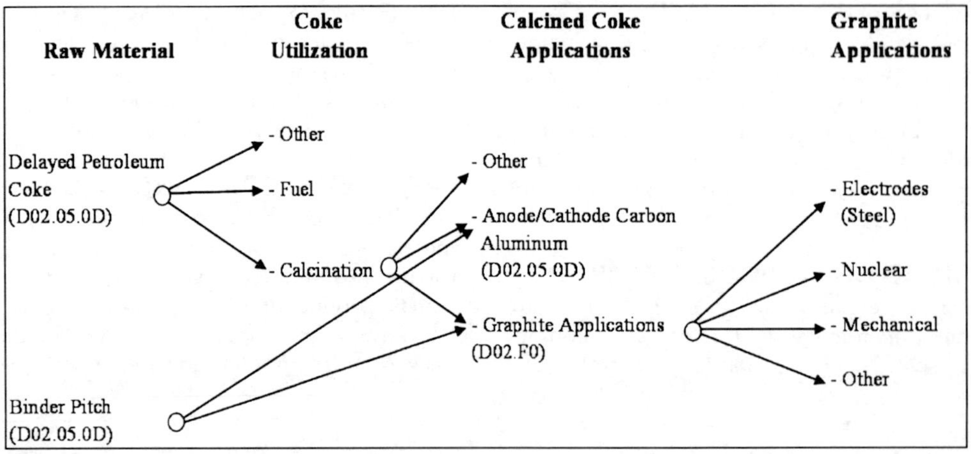

Fig. 2—Flow diagram for carbon and graphite production.

THE GRAPHITIZATION PROCESS

The transformation of amorphous carbon into artificial (synthetic) graphite by heat treatment is known as the "graphitization process." Graphitization can be described in a series of steps that occur as the temperature is raised from approximately 1,000 up to about 3,000°C. A typical graphitization process involves the graphitization heat treatment of a composite material composed of filler carbon held in a pitch-based binder matrix.

As the filler carbon's calcining temperature (1,000 to 1,300°C) is surpassed, the incipient graphitic structure slowly develops. The binder evolves hydrogen, sulfur, and other heteroatoms in the range of 1,500 to 2,000°C and undergoes a volume expansion (0.2 to 0.6 %). As 1,800°C is surpassed, the graphitic structure grows more rapidly and grain size increases as temperatures exceeds 2,200°C. At approximately 2,600°C, a volume contraction takes place and crystallite growth predominates.

One can follow the development of crystallites by X-ray diffraction analysis. Crystallite size increases from 50 to about 1,000 Å, while interlayer spacing decreases from 3.44 Å, typical of amorphous carbon, to 3.35 Å, typical for graphitic structure.

Graphitization involves a displacement and rearrangement of layer planes and small groups of planes to achieve a three-dimensional ordering. Movements of individual atoms or single carbon rings to fill vacancies or improve perfection in existing crystallites (annealing) may supplement the growth of such planes. The probability of rearrangement is related to the existing degree of disorientation and the degree of carbon-to-carbon bonding between layer planes (cross-linking). A high degree of interplanar cross-linking is known to inhibit graphitization.

In addition to the structure of the precursor, maximum temperature and residence time at temperature are important factors in the achievement of desirable properties. The presence of oxygen and carbon dioxide enhances the beneficial effects of graphitization. The crystallites interlayer spacing reaches a limiting value at each temperature, even after long residence periods. When the temperature is increased, the rate approach to the limiting value is also increased. Where graphitization temperatures are higher than 2,500°C and the times are longer than 12 h, the effects of kinetics are minimal.

Mechanical properties follow a similar behavior with respect to the ultimate graphitization temperature. As temperatures are increased from 2000 to 2,500–2,600°C, the tensile strength remains constant or increases slightly to a maximum. From 2,600 to 2,900°C, a decrease in tensile strength is observed, followed by a flat region from 3,000 to 3,200°C. Electrical resistivity typically decreases as calcining temperature is raised to about 1,000°C and then may stay constant or increase as temperatures are increased. Not all physical properties achieve optimum values at any one temperature.

At ordinary temperatures, carbon is one of the most inert of the elements, but at high temperatures, it becomes one of the most reactive. Mantell [3] describes the reactions with the various compounds and chemical elements. Mantell also gives information on other properties: contact resistance, creep, crystal structure, elasticity, electrical conductivity, electrochemical equivalent, entropy, hardness, heat capacity, heat of combustion, heat of vaporization, magnetic susceptibility, melting and boiling points, and radiation constants.

MANUFACTURING METHODS

Carbon and graphite parts can be manufactured from the thermal decomposition of hydrocarbons or by mechanical, chemical, and thermal processing of carbon-base materials.

Pyrolytic Graphite

Pyrolytic graphite is formed by the thermal decomposition of hydrocarbon vapor/gas upon a hot surface. The density of the graphite deposited depends primarily on the deposition temperature. Minimum density is 1.24 g/cm^3 for deposition at 1,700°C and maximum 2.22 g/cm^3 at 2,100°C. The density of the deposit in a closed system is also a function of the initial pressure of the gas. Increases in pressure lead to increased density. The deposited carbon consists of layered planes of hexagonally arranged carbon atoms lying parallel to the surface of the original substrate but otherwise somewhat disordered.

The electrical resistivity and thermal conductivity of the deposits are dependent on temperature of deposition. For material produced at 1,800°C, the electrical resistivity is 2,200 μÓ/cm, the thermal conductivity 0.06 W/m-K; for that deposited at 2,100°C, the corresponding values are 200 μÓ/cm and 1.3 W/m-K.

Artificial Carbon and Graphite

Artificial carbon and graphite can be manufactured to achieve a wide variety of properties. Among the variables that influence the resulting properties are starting materials, conditioning processes, fineness, forming methods, carbonization parameters, impregnation conditions, and the graphitization process.

Carbon and graphite can be made from nearly any organic matter that in heating produces a large amount of carbon (coke). Petroleum coke is the most used starting material. Lampblack and pitch cokes are generally used for specialty applications.

To begin the process, the coke is calcined to stabilize its morphology and remove volatiles. Ultimately it is processed to a size that fits the particular application. The size and shape of the particle are important factors in the determination of the final properties.

The coke powder is next combined with coal tar pitch and heated to make it formable. This coke-pitch "dough" is next formed. There are many methods used to form the blend, which include isostatic and uniaxial molding and extrusion. The forming method is another important parameter that influences the final properties of the finished product.

Next, the formed coke-pitch mixture is carbonized at temperatures ranging from 750 to 900°C. At this stage, the material is carbon, a porous, hard, amorphous material with random crystallinity. After the primary carbonization process (baking), the carbon may be reimpregnated with pitch by a vacuum/pressure method. Pitch impregnation impacts the final density of the carbon article being manufactured. Impregnation results in higher density carbon and graphite, which influences the properties of the final material.

The final step of the production of graphite is the graphitization process, which is performed at heat treatment temperatures ranging from 2,600 to 3,000°C as discussed earlier, and the method and parameters during graphitization have a paramount influence on the final properties of the graphite made.

TEST METHODS

The test methods, practices, and specifications under the jurisdiction of ASTM Subcommittee D02.05 on Properties of Fuels, Petroleum Coke and Carbon Material applicable to petroleum coke, pitch, and carbon material and Subcommittee D02.F0 on Manufactured Carbon and Graphite applicable to manufactured carbon and graphite are summarized in Table 1 and indexed below. Here the properties measured

TABLE 1—Summary of Test Methods for Petroleum Coke, Pitch, and Manufactured Carbon and Graphite

ASTM Designation (Use Latest Version)	Title	Analysis Type	Application				
			Raw Coke	Calcined Coke	Anode/ Pitch	Carbon	Graphite
C561	Ash in a Graphite Sample	Ash					X
D2415	Ash in Coal Tar and Pitch	Ash			X		
D4422	Ash in Analysis of Petroleum Coke	Ash	X	X		X	?
D2416	Coking Value of Tar and Pitch (Modified Conradson)	Coking value			X		
D4715	Coking Value of Tar and Pitch (Alcan)	Coking value			X		
D5187	Determination of Crystallite Size (L_C) of Calcined Coke by X-Ray Diffraction	Crystallite size		X		X	
C1039	Apparent Porosity, Apparent Specific Gravity, and Bulk Density of Graphite Electrodes	Density				?	X
C559	Bulk Density by Physical Measurements of Manufactured Carbon and Graphite Articles	Density					X
C838	Bulk Density of As-Manufactured Carbon and Graphite Shapes	Density				?	X
D71	Relative Density of Solid Pitch and Asphalt (Displacement Method)	Density			X		
D2320	Density (Relative Density) of Solid Pitch	Density			X		
D2638	Real Density of Calcined Petroleum Coke by Helium Pycnometer	Density		X		X	?
D4292	Determination of Vibrated Bulk Density of Calcined Petroleum Coke	Density		X			
D4892	Density of Solid Pitch (Helium Pycnometer Method)	Density			X		
D5004	Real Density of Calcined Petroleum Coke by Xylene Displacement	Density		X		X	?
D5502	Apparent Density by Physical Measurements of Manufactured Anode and Cathode Carbon Used by the Aluminum Industry	Density				X	?
D7454	Standard Test Method for Determination of Vibrated Bulk Density of Calcined Petroleum Coke using a Semi-Automated Apparatus	Density		X			
D4930	Dust Control Material on Calcined Petroleum Coke	Dust Control		X			

TABLE 1—Summary of Test Methods for Petroleum Coke, Pitch, and Manufactured Carbon and Graphite (Continued)

ASTM Designation (Use Latest Version)	Title	Analysis Type	Application				
			Raw Coke	Calcined Coke	Anode/ Pitch	Carbon	Graphite
D5003	Hargrove Grindability Index (HGI) of Petroleum Coke	Friability	X	X			
D6791	Grain Stability of Petroleum Coke	Friability		X			
C748	Rockwell Hardness of Graphite Materials	Hardness					X
C886	Scleroscope Hardness Testing of Carbon and Graphite Materials	Hardness				?	X
D2318	Quinoline-Insoluble (QI) Content of Tar and Pitch	Insolubles			X		
D2764	Dimethylformamide-Insoluble (DMF-I) Content of Tar and Pitch	Insolubles			X		
D4072	Toluene-Insoluble (TI) Content of Tar and Pitch	Insolubles			X		
D4312	Toluene-Insoluble (TI) Content of Tar and Pitch (Short Method)	Insolubles			X		
D4746	Determination of Quinoline Insolubles (QI) in Tar and Pitch by Pressure Filtration	Insolubles			X		
D7280	Standard Test Method for Quinoline-Insoluble (QI) Content of Tar and Pitch by Stainless Steel Crucible Filtration	Insolubles			X		
C565	Tension Testing of Carbon and Graphite Mechanical Materials	Mechanical				?	X
C651	Flexural Strength of Manufactured Carbon and Graphite Articles Using Four-Point Loading at Room Temperature	Mechanical				?	X
C695	Compressive Strength of Carbon and Graphite	Mechanical				?	X
C749	Tensile Stress-Strain of Carbon and Graphite	Mechanical					X
C808	Reporting Friction and Wear Test Results of Manufactured Carbon and Graphite Bearing and Seal Materials	Mechanical				?	X
C560	Chemical Analysis of Graphite	Metals					X
D5056	Metals in Petroleum Coke by Atomic Absorption	Metals	X	X	?	?	?

(Continued)

TABLE 1—Summary of Test Methods for Petroleum Coke, Pitch, and Manufactured Carbon and Graphite (Continued)

ASTM Designation (Use Latest Version)	Title	Analysis Type	Application				
			Raw Coke	Calcined Coke	Anode/ Pitch	Carbon	Graphite
D5600	Metals in Petroleum Coke by ICP-AES	Metals	X	X	?	?	?
D6376	Determination of Trace Metals in Petroleum Coke by Wavelength Dispersive X-Ray Fluorescence Spectroscopy	Metals	X	X	?	?	?
In Development	New Standard Test Method for Determination of Trace Metals in Pitch by Wavelength Dispersive X-Ray Fluorescence Spectroscopy	Metals			X		
C1025	Modulus of Rupture in Bending of Electrode Graphite	Modulus					X
C747	Moduli of Elasticity and Fundamental Frequencies of Carbon and Graphite Materials by Sonic Resonance	Modulus					X
C769	Sonic Velocity in Manufactured Carbon and Graphite Materials for Use in Obtaining an Approximate Young's Modulus	Modulus					X
C562	Moisture in a Graphite Sample	Moisture				?	X
D4931	Gross Moisture in Green Petroleum Coke	Moisture	X				
C625	Reporting Irradiation Results on Graphite	Nuclear					X
C781	Testing Graphite and Boronated Graphite Components for High-Temperature Gas-Cooled Nuclear Reactors	Nuclear					X
C1179	Oxidation Mass Loss of Manufactured Carbon and Graphite Materials in Air	Reactivity				?	X
D6558	Determination of TGA C02 Reactivity of Baked Carbon Anodes and Cathode Blocks	Reactivity				X	?
D6559	Determination of TGA Air Reactivity of Baked Carbon Anodes and Cathode Blocks	Reactivity				X	?
C611	Electrical Resistivity of Manufactured Carbon and Graphite Articles at Room Temperature	Resistivity				?	X

TABLE 1—Summary of Test Methods for Petroleum Coke, Pitch, and Manufactured Carbon and Graphite (Continued)

ASTM Designation (Use Latest Version)	Title	Analysis Type	Application				
			Raw Coke	Calcined Coke	Anode/ Pitch	Carbon	Graphite
D6120	Electrical Resistivity of Anode and Cathode Carbon Material at Room Temperature	Resistivity				X	?
C783	Core Sampling of Graphite Electrodes	Sampling				?	X
D4296	Standard Practice for Sampling Pitch	Sampling			X		
D6353	Guide for Sampling Plan and Core Sampling or Prebaked Anodes Used in Aluminum Production	Sampling				X	?
D6354	Guide for Sampling Plan and Core Sampling of Carbon Cathode Blocks Used in Aluminum Production	Sampling				X	?
D6969	Standard Practice for the Preparation of Calcined Petroleum Coke Samples for Analysis	Sampling			X		
D6970	Standard Practice for Collection of Calcined Petroleum Coke Samples for Analysis	Sampling			X		
D5709	Sieve Analysis of Petroleum Coke	Sieve	X	X			
D61	Softening Point of Pitches (Cube-in-Water Method)	Softening			X		
D2319	Softening Point of Pitch (Cube-in-Air Method)	Softening			X		
D3104	Softening Point of Pitches (Mettler Softening Point Method)	Softening			X		
D3461	Softening Point of Asphalt and Pitch (Mettler Cup-and-Ball Method)	Softening			X		
D4616	Microscopical Analysis by Reflected Light and Determination of Mesophase in a Pitch	Structure			X		
C816	Sulfur in Graphite by Combustion-Iodometric Titration Method	Sulfur	?	?		?	X
D1552	Sulfur in Petroleum Products (High Temperature Combustion)	Sulfur	X	X	X	X	?
C709	Standard Terminology Relating to Manufactured Carbon and Graphite	Terminology	?	?		?	X

(Continued)

TABLE 1—Summary of Test Methods for Petroleum Coke, Pitch, and Manufactured Carbon and Graphite (Continued)

ASTM Designation (Use Latest Version)	Title	Analysis Type	Application				
			Raw Coke	Calcined Coke	Anode/ Pitch	Carbon	Graphite
C714	Thermal Diffusivity of Carbon and Graphite by a Thermal Pulse Method	Thermal				?	X
D6744	Determination of the Thermal Conductivitiy of Anode Carbons by Guarded Heat Flow Meter Technique	Thermal				X	?
D6745	Linear Thermal Expansion of Electrode Thermal Carbons	Thermal				X	?
D5018	Shear Viscosity of Coal-Tar Petroleum Pitches	Viscosity			X		
D4421 (withdrawn)	Volatile Matter in Petroleum Coke	Volatiles	X				
D6374	Volatile Matter in Green Petroleum Coke, Quartz Crucible Procedure	Volatiles	X				
D2569	Distillation of Pitch	Volatility			X		
D4893	Determination of Pitch Volatility	Volatility			X		
D7219	Standard Specification for Nuclear Graphite Suitable for Components Subjected to Low Neutron Irradiation Dose	Specification					X
D7301	Standard Specification for Nuclear Graphite Suitable for Components Subjected to Low Neutron Irradiation Dose	Specification					X

X = applicable.
? = Not within scope, but possible application with modification.

and the significance of test methods, practices, and specifications are given.

Petroleum Coke Materials

ASH, D4422

Ash content is one of the properties used to evaluate petroleum coke and indicates the amount of undesirable residue present. Acceptable ash content varies with the intended use of the petroleum coke. Generally speaking, higher ash content reflects lower value of the coke.

DENSITY, D2638, D4292, D5004, D7454

Density of calcined petroleum coke directly influences the physical and chemical properties of the manufactured carbon and graphite artifacts for which it is used. Density is a major quality specification of calcined petroleum coke and is used as a control of the calcining process. Density is a measure of the porosity of the calcined petroleum coke.

GRAIN STABILITY, D6791

The grain stability of calcined petroleum coke determines the resistance to breakdown of 4- to 8-mm particles used in the manufacture of carbon anodes for use in the reduction process of aluminum.

Cokes have to be relatively easy to grind for fine production but strong enough to withstand forming pressures and thermal stresses occurring when the anodes are used in the reduction process.

METALS, D5056, D5600, D6376

The presence and concentrations of various metallic elements in a petroleum coke are major factors in the suitability of a coke for various uses. The concentration of these elements is one factor in determining the economic value of the coke. Coke used for production of electrodes will have different specification requirements dependent on what service the electrode will be used in. Generally, fuel cokes are highest in metallic element concentration and have the least economic value.

In addition to measuring the concentration of metallic elements in the coke, D6376 has a further advantage in that it determines the concentration of sulfur in the coke.

SULFUR, D1552, D6376

The concentration of sulfur present in petroleum coke is a major factor in the suitability of a coke for various uses. Sulfur contents are used to evaluate the potential formation of sulfur oxides (SO_x), a source of atmospheric pollution. In some cases, sulfur is detrimental to production and the use of artifacts produced from the petroleum coke.

VOLATILE MATTER, D6374

Volatile matter testing is only applicable to petroleum coke having a thermal history of less than 600°C (green petroleum coke). The volatile matter of green petroleum coke affects the density of the coke particles produced during calcining, which can affect the properties of artifacts produced from the coke. Volatile matter also provides an estimate of materials lost during the calcining process.

Volatile matter can also be used in estimating calorific value of coke used for fuels.

Pitch Materials

SOFTENING POINT OF PITCH, D61, D2319, D3104, D3461

Softening point is found in practically all customer specifications and is easily one of the most important pitch tests. Coal tar can be distilled to various softening points from about 50 to over 175°C; the softening point for most binder applications is about 110°C. Pitches do not have a discreet melting point (such as ice does) but gradually soften as temperature increases. Since pitch softens gradually, it is important that the softening point procedures specified be carefully followed.

The softening point of the pitch determines the mixing temperature of the mix of petroleum coke and pitch. Mixing temperature is typically about 50°C higher than the softening point of the pitch. The quality of the final carbon product is strongly influenced by the conditions in the mixer (temperature and mixing time); therefore, it is important that the softening point of the pitch be manufactured to a consistent softening point to ensure a quality finished carbon product.

QUINOLINE INSOLUBLES, D2318, D4756, D7280

Coal-tar pitch consists of carbonaceous particles dispersed in a liquid phase, which consists of a mixture of aromatic organic chemicals. The liquid phase is soluble in quinoline but the carbonaceous particles are not. The quinoline insolubles (QI) test determines the quantity of these particles (solids). The QI consist mainly of normal QI (about 80 to 90 % of the total solids); the balance is mostly carryover QI. Thermal cracking of coal volatiles in the coke oven forms normal QI; carryover QI consists of carbonized coal particles that were "carried over" during the charging of coal into the coke oven. Photomicrographs of these two types of QI are given in ASTM D4616 (mesophase).

The QI content of coal-tar pitch can vary from about 5 to 20 %. Petroleum pitch has a very low QI content (<1 %) because the feedstock for petroleum pitch did not encounter cracking conditions found in a coke oven. The QI value in the binder is important because the amount of binder required is proportional to the QI content. Also, QI alters the structure of pitch coke formed during the baking of the carbon artifact; the pitch coke holds the aggregate coke particles together, and the structure of the pitch coke bridges can affect the properties of the carbon artifact.

TOLUENE INSOLUBLES, D4072, D4312

The liquid portion (solids free) of pitch is completely soluble in quinoline but only partially soluble in toluene. Consequently, the toluene insolubles (TI) are higher than the QI because the TI includes the QI material and the portion of the quinoline soluble material insoluble in toluene.

The TI of pitch can be affected by the tar distillation conditions. The amount of TI can influence the conditions required to prepare a carbon artifact; therefore, the TI needs to be known by the user.

MESOPHASE, D4616

Mesophase is formed in coal tar and petroleum pitches by condensation and polymerization reactions of aromatic hydrocarbons initiated by thermal treatment at temperatures of about 360 to 430°C. The presence of mesophase in pitch may alter the way the pitch is used; therefore, the user must know the amount and size distribution of the mesophase spheres. Mesophase is a liquid crystal that can be seen in pitch using reflected-polarized-light microscopy.

COKING VALUE, D2416, D4715

During the baking operation of a green carbon artifact, the pitch binder forms pitch coke bridges between the petroleum coke aggregate particles to form a rigid carbon form. The amount of the pitch binder that is converted to pitch coke can be estimated by the coking value tests. The coking value is an important test that helps the user optimize the carbon-forming process to produce a quality artifact.

ASH, D2415

Ash is an estimate of minerals and metal impurities that are present due to carryover of coal, coke, or mineral particles in coal-tar pitch or catalyst in petroleum pitch. A carbon artifact that has an abnormally high concentration of mineral or metal contaminants can adversely affect the product that is made using the carbon material. For example, if a carbon anode used in the production of aluminum has a high concentration of silicon, the silicon will be transferred to the aluminum and may affect the extrusion of the aluminum.

Ash testing is used to ensure that pitch feed stocks will meet customer specifications.

Major Anode/Cathode Tests

The following ASTM tests are currently being utilized to determining anode quality; other ASTM tests are being developed.

APPARENT DENSITY, D5502

High density equates to longer life in pot electrode operation. Baked apparent density (B.A.D.) is calculated by taking the weight of a core piece divided by its measured volume. The results are reported in grams per cubic centimeter and vary from approximately 1.52 to 1.62 g/cm^3. Using this method of determination means any internal pores or holes are included in the density. Low density is indicative of an anode that will not make it through normal shift life with a large enough butt, resulting in high current density and lower pot operating efficiencies. Too high a density can be a problem in plants with large anodes because a very dense anode is prone to damage and breakup due to thermal shock.

ELECTRICAL RESISTIVITY OF ANODES AND CATHODE CARBON MATERIALS, D6120

Efficient power use in the pot room means that materials with the lowest electrical resistance to current flow are required. In an operating pot the resistance losses are measured as voltage loss (volts = amps × resistance). The electrical resistance of an anode is determined by measuring core "resistivity." This is obtained by applying set amperage through the core and measuring the voltage drop across a 5-cm distance of that core. The resistance is measured and "resistivity" is then calculated in micro ohmmeter.

AIR REACTIVITY AT 525°C, D6559

The anode is consumed by oxygen from the air above the bath line wherever it is not adequately protected against air-burn (oxidation) by either the ore/bath cover or by the aluminum spray coating. How well an anode can resist air-burn is determined in the lab by performing an air reactivity rate test. This test is performed on a 2-in.-long core segment by heating the core to 525°C for 3.5 h and then determining the weight lost during the test. The results are reported as milligrams consumed per square centimeter of original core surface per hour.

CARBON DIOXIDE REACTIVITY AT 970°C, D6558

Below the bath line, carbon dioxide (CO_2) gas, formed during aluminum smelting, reacts with the anode. This CO_2 reactivity rate is measured in the lab by maintaining a core sample at 970°C in a carbon dioxide atmosphere for 13.5 h. The weight lost from the core during the test is the "total loss" and is reported as milligrams consumed per square centimeter surface area per hour. In addition, another important CO_2 reactivity property, "dust loss," is reported. The coke material that has fallen to the bottom of the sample tray during the test represents the "dust loss." The "dust loss" is filler coke that is lost when pitch binder coke is preferentially attacked by CO_2. This dust loss should be reflective of the coke dusting levels that the pot room may experience during operation. Normally, binder coke is slightly more reactive than filler coke, but this can be greatly aggravated by overcalcining the filler coke by underbaking the anodes or from sodium contamination.

LINEAR THERMAL EXPANSION OF ELECTRODE CARBON, D6745

The method covers determination of the coefficient of linear thermal expansion (CTE) for baked carbon anodes and carbon lining materials using a vitreous silica dilatometer. Linear thermal expansion is the change in length per unit length resulting from a temperature change. The application temperature range for this method is ambient to 1,000°C. The petroleum cokes used in anode production influence the anode CTE. A lower CTE is preferred when looking at anode thermal shock resistance.

THERMAL CONDUCTIVITY, D6744

The test method covers a steady state technique for the determination of thermal conductivity of carbon materials in thickness of less than 25 and 50 mm in diameter by guarded heat flowmeter technique. The test is designed to measure and compare the thermal conductivity properties of material under controlled conditions and their ability to maintain required thermal conductance levels. Thermal conductivity is affected by changes in density of the anodes. The more dense the anode, the more heat characteristics will change at the anode surface, resulting in carbon loss due to air burn.

Manufactured Carbon and Graphite Materials

TERMINOLOGY, C709

The terms used in the standards on manufactured carbon or graphite are given in the sections that follow.

BULK DENSITY, C838, C559

ASTM C838 covers the determination of the density of as-manufactured carbon and graphite from measurements of mass and dimensions at room temperature. This test method is also applicable to boronated carbon and graphite.

This test method is not applicable to carbon- and graphite-containing materials that are thermally unstable at or below temperatures of 473 K. This test method provides a means of determining bulk density on as-manufactured logs, blocks, or shaped articles. This test method is suitable for manufacturing control and acceptance specifications.

ASTM C559 covers the determination of the bulk density of manufactured articles of carbon and graphite of at least 500 mm^3 volume. The bulk density is calculated to an accuracy of 0.25 % using measurements of mass and dimensions in air at 25 ± 5°C. "Bulk density as determined by this test method is a basic material property of importance in manufacturing and application of carbon and graphite. This test method can be used for quality and process control, material characterization and description, and other purposes."

APPARENT POROSITY, APPARENT SPECIFIC GRAVITY, AND BULK DENSITY, C1039

These test methods cover the determination of apparent porosity, apparent specific gravity, and bulk density of cores taken from graphite electrodes manufactured for use in electric arc furnaces.

The results of these test methods can be used as a quality control or quality assurance check of electrodes either during their manufacture or at the user's location. The results of these methods tend to be operator sensitive; therefore, care must be taken in the execution of the test in order to obtain reproducible results.

ASH, C561

This test method provides a practical determination for the ash content in a graphite sample.

This test method provides a practical estimate of nonburnable residues in commercially available graphite materials. The ash values determined by this test method are of use in comparing the relative purity of various grades of graphite. To facilitate use, this test method institutes simplifications that preclude the ability to determine absolutely the ash values of the test graphite material due to uncontrolled sources of trace contamination.

This test method is not intended for use in determining the ash content of purified graphites, for example, nuclear materials. The relationship between the mineral content of a graphite sample and the ash content of that sample is unknown and is not determined by the application of this test method.

ELECTRICAL RESISTIVITY, C611

This test method covers the determination of the electrical resistivity of manufactured carbon and graphite articles at room temperature.

This test method provides a means of determining the electrical resistivity of carbon or graphite specimens. The use of specimens that do not conform to the specimen size limitations described in the test method may result in an alteration of test method accuracy.

SCLEROSCOPE HARDNESS, C886

This test method covers the apparatus and procedures for determining the hardness of fine-grained carbon and graphite materials using the Model C-2 scleroscope with the hammer calibrated for use on fine-grained (particles smaller than 0.8 mm [0.032 in.]) carbon and graphite materials.

The scleroscope is a rebound hardness tester with a scale divided into 140 equal parts. For carbon and graphite materials, there is no established correlation between the scleroscope hardness scale and other hardness scales. The test is useful in the evaluation and the manufacturing control of carbon and graphite materials.

ROCKWELL HARDNESS, C748

This test method covers Rockwell L scale hardness testing of fine-grained graphite materials with a grain size of less than 0.8 mm ($^1/_{32}$ in.) and a hardness range between 0 and 120.

A Rockwell L scale hardness number is a number derived from the net increase in depth of impression as the load on a 6.3500 ± 0.0025 mm (0.2500 ± 0.0001 in.) diameter steel-ball indenter is increased from a fixed 10-kg minor load to a 60-kg major load and then returned to the minor load.

A Rockwell hardness number is directly related to the indentation hardness of a material; the higher the Rockwell L scale reading, the harder the material. The test is useful in the evaluation and the manufacturing control of carbon and graphite materials.

CHEMICAL ANALYSIS, C560

These test methods cover the chemical analysis of graphite. The analytical procedures appear in the following order: Silicon by the Molybdenum Blue (Colorimetric) Test Method (Sections 8 to 14); Iron by the o-Phenanthroline (Colorimetric) Test Method (Sections 15 to 21); Calcium by the Permanganate (Colorimetric) Test Method (Sections 22 to 28); Aluminum by the 2-Quinizarin Sulfonic Acid (photometric) Test Method (Sections 29 to 35); Titanium by the Peroxide (Colorimetric) Test Method (Sections 36 to 43); Vanadium by the 3,3′-Dimethylnaphthidine (Colorimetric) Test Method Oxalic Acid (Colorimetric) Test Method (Sections 52 to 59).

These test methods provide a practical way to measure the concentration of certain trace elements in graphite. Many end uses of graphite require that it be free of elements that may be incompatible with certain nuclear applications. Other elemental contamination can affect the rate of oxidative degradation.

These test methods allow measurement of trace amounts of contaminants with a minimal amount of costly equipment. The colorimetric procedures used are accessible to most laboratories.

MOISTURE, C562

This test method provides a method for the determination of the percentage of volatile moisture in a graphite sample.

This test method is feasible only for the volatile moisture present on the graphite due to adsorption from the atmosphere and is not intended to give representative moisture data for graphite that has been exposed to liquid water contamination.

TENSILE PROPERTIES, C565

These test methods, and the figures that are part of the test methods, cover the apparatus, specimen, and procedures for the tension testing of carbon and graphite mechanical materials with a grain size smaller than 0.79 mm ($^1/_{32}$ in.).

These test methods may be used for quality control testing of established grades of carbon and graphite materials, in the development of new grades, and for other purposes where relative strength levels are the primary quantities of interest. This test method may be applicable only if the ratio of specimen diameter to grain size, or flaw size, is greater than 5. These test methods do not substitute for that described in Test Method C749, but are useful where less sophisticated data and less expensive techniques are sufficient.

Carbon and graphite materials exhibit significant physical property differences within parent materials. Exact sampling patterns and grain orientations must be specified in order to make meaningful tensile strength comparisons.

REPORTING IRRADIATION RESULTS, C625

This practice covers information recommended for inclusion in reports giving graphite irradiation results.

The purpose of this practice is to identify sample and test parameters that may influence graphite irradiation test results. This practice should not be construed as a requirement or recommendation that proprietary information be disclosed.

Irradiation results on graphite include dimensional changes and changes in properties that are used in reactor design. The irradiation data are reported in government documents and open literature publications and are assembled into data manuals for use by reactor designers.

IMPERVIOUS GRAPHITE PIPE AND THREADING, C662

The purpose of this specification is to standardize the pipe sizes and types of threads used to join impervious graphite pipe and fittings. The thread standards may also be applied to impervious carbon pipe and fittings. It is limited to physical dimensions.

COMPRESSIVE STRENGTH, C695

This test method covers the determination of the compressive strength of carbon and graphite at room temperature.

Carbon and graphite can usually support higher loads in compression than in any other mode of stress. This test, therefore, provides a measure of the maximum load-bearing capability of carbon and graphite objects.

MODULI OF ELASTICITY AND FUNDAMENTAL FREQUENCIES, C747

This test method covers the measurement of the fundamental transverse, longitudinal, and torsional frequencies of isotropic and anisotropic carbon and graphite materials. These measured resonant frequencies are used to calculate dynamic elastic moduli for any grain orientations.

This test method is primarily concerned with the room temperature determination of the dynamic moduli of elasticity and rigidity of slender rods or bars. It can be adapted for other materials that are elastic in their initial stress-strain behavior and could be modified to determine elastic moduli behavior at temperatures from −75 to +2,500°C.

TENSILE STRESS-STRAIN, C749

This test method covers the testing of carbon and graphite in tension to obtain the tensile stress-strain behavior, to failure, from which the ultimate strength, the strain to failure, and the elastic moduli may be calculated as required for engineering applications. Suggested sizes of specimens that can be used in the tests are given.

This test method is intended to be used for both carbons and graphites whose particle sizes are of the order of 1 mL to 0.25 in. (0.0254 to 6.4 mm) and larger. This wide range of carbons and graphites can be tested with uniform gage diameters with minimum parasitic stresses to provide quality data for use in engineering applications rather than simply for quality control. This test method can be easily adapted to elevated temperature testing of carbons and graphites without changing the specimen size or configuration by simply using elevated temperature materials for the load train. This test method has been used for temperatures as high as 4,352 T (2,400°C). The design of the fixtures and description of the procedures are intended to bring about, on the average, parasitic stresses of less than 5 %. The specimens for the different graphites have been designed to ensure fracture within the gage section commensurate with experienced variability in machining and testing care at different facilities. The constant gage diameter permits rigorous analytical treatment.

FLEXURAL STRENGTH USING FOUR-POINT LOADING, C651

This test method covers determination of the flexural strength of manufactured carbon and graphite articles using a simple beam in four-point loading at room temperature. The values stated in inch-pound units are to be regarded as the standard. The metric equivalents of inch-pound values may be approximate.

THERMAL DIFFUSIVITY BY THERMAL PULSE, C714

This test method covers the determination of the thermal diffusivity of carbons and graphite to ±5 % at temperatures up to 500°C. It requires only a small easily fabricated specimen. Thermal diffusivity values in the range from 0.04 to 2.0 cm/s are readily measurable by this test method; however, for materials outside this range this test method may require modification.

SONIC VELOCITY YOUNG'S MODULUS, C769

This test method describes a procedure for measuring the sonic velocity in manufactured carbon and graphite materials having a grain size less than 0.80 mm ($^{1}/_{32}$ in.). The sonic velocity can be used to obtain an approximate value for Young's modulus.

Sonic velocity measurements are useful for comparing materials, and a value for Young's modulus can be obtained for many applications, which will generally be within 10 % of the value obtained by other methods, such as in Test Method C747.

HIGH-TEMPERATURE GAS-COOLED NUCLEAR REACTORS, C781

This practice covers the test methods for measuring those properties of graphite and boronated graphite materials that may be used for the design and evaluation of high-temperature gas-cooled reactors. The test methods referenced in C781 are applicable to materials used for replaceable and permanent components and include fuel and removable reflector elements; target elements and insulators; permanent side reflector elements; core support pedestals and elements; control rod, reserve shutdown, and burnable poison compacts, and neutron shield material.

This practice includes test methods that have been selected from existing ASTM standards, ASTM standards that have been modified, and new ASTM standards that are specific to the testing of materials noted above.

Property data obtained with the recommended test methods identified in C781 may be used for research and development, design, manufacturing control, specifications, performance evaluation, and regulatory statutes pertaining to high-temperature gas-cooled reactors.

The test methods are applicable primarily to specimens in the unirradiated and unoxidized state. Many are also applicable to specimens in the irradiated or oxidized state, or both, provided the specimens meet all requirements of the test method. The user is cautioned to consider the instructions given in the test methods.

CORE SAMPLING, C783

This practice was developed for electric-arc furnace graphite electrodes and covers a procedure and equipment for obtaining core samples from electrodes in a manner that does not destroy the electrode or prevent its subsequent use as originally intended. However, the minimum electrode diameter, for which extraction of a core sample using this practice does not influence subsequent use, is influenced by the particular application and must be determined by the user. Graphite electrodes for use in electric arc furnaces are usually solid cylinders of graphite with threaded sockets machined in each end.

Core sampling is an acceptable way of obtaining a test specimen without destroying the usefulness of the electrode. Specimens obtained by this practice can be used by producers and users of graphite electrodes for the purpose of conducting the tests to obtain comparative physical properties. This practice may not provide a specimen of the appropriate size (with respect to particle size/sample dimension ratios) to allow the determination of absolute property values.

FRICTION AND WEAR, C808

C808 is a guideline for reporting friction and wear test results of manufactured carbon and graphite bearing and seal materials. The guidelines cover the following test parameters:

1. Description of test device and techniques
2. Description of carbon and graphite material test specimen
3. Description of mating member test specimen
4. Report of friction and wear test results

The purpose of this guideline is twofold. First, it is a research tool that will aid in the analysis and correlation of test results obtained on various test devices by different investigators. Second, it serves to identify important considerations that must be made in testing to make the results easily understood and comparable with the results of other investigators.

SULFUR CONTENT, C816

This test method covers the determination of sulfur in graphite in the concentration range from 1 to 200 μg/g in a 1-g sample or 5 to 1,000 μg/g in a 0.2-g sample.

Sulfur, even in very low concentrations, is of concern in a nuclear reactor because of potential corrosion of metallic components. This test method has the sensitivity to analyze very low sulfur contents in graphite using very small samples. This test method can be used to characterize graphite for design purposes.

MODULUS OF RUPTURE IN FOUR-POINT BENDING, C1025

This test method covers determination of the modulus of rupture in bending. The specimens are cut from graphite electrodes using a simple square cross section beam and loaded in four-point loading at room temperature.

This test method provides a means for determining the modulus of rupture of a square cross section graphite specimen machined from the electrode core sample obtained according to Practice C783, with a minimum core diameter of 57 mm (2.25 in.). This test method is recommended for quality control or quality assurance purposes but should not be relied on to compare materials of radically different particle sizes or orientation characteristics.

The maximum particle size and maximum pore size vary greatly for manufactured graphite electrodes, generally increasing with electrode diameter. The test is on a rather short stubby beam; therefore, the shear stress is not insignificant compared to the flexural stress, and the test results may not agree when a different ratio or specimen size.

OXIDATION MASS LOSS, C1179

This test method provides a comparative oxidation mass loss of manufactured carbon and graphite materials in air.

This test method is primarily concerned with the oxidation mass loss of manufactured carbon and graphite materials in air at temperatures from 371 to 677°C (700 to 1,250°F). The test method will provide acceptable results at preselected test temperatures that yield less than 10 % mass loss in 100 h. These results can be used to determine relative service temperatures.

STANDARD SPECIFICATION FOR ISOTROPIC AND NEAR-ISOTROPIC NUCLEAR GRAPHITES, D7219

This specification defines the properties that a graphite type must have to be eligible for use in regions of a nuclear reactor where the radiation doses are "high."

This specification covers the classification, processing, and properties of nuclear-grade graphite billets with dimensions sufficient to meet the designer's requirements for fuel elements, moderator or reflector blocks, in a high-temperature gas-cooled reactor. The graphite classes specified here would be suitable for reactor core applications where neutron irradiation–induced dimensional changes are a significant design consideration.

The values stated in SI units are to be regarded as standard. No other units of measurement are included in this standard. (See IEEE/ASTM SI 10.)

STANDARD SPECIFICATION FOR NUCLEAR GRAPHITE SUITABLE FOR COMPONENTS SUBJECTED TO LOW NEUTRON IRRADIATION DOSE, D7301

This specification defines the properties that a graphite type must have to be eligible to be used in regions of a nuclear reactor where the radiation doses are relatively "low."

This specification covers the classification, processing, and properties of nuclear-grade graphite billets with dimensions sufficient to meet the designer's requirements for reflector blocks and core support structures, in a high-temperature gas-cooled reactor. The graphite classes specified here would be suitable for reactor core applications where neutron irradiation induced dimensional changes are not a significant design consideration.

The purpose of this specification is to document the minimum acceptable properties and levels of quality assurance and traceability for nuclear-grade graphite suitable for components subjected to low irradiation dose. Nuclear graphites meeting the requirements of ASTM D7219 are also suitable for components subjected to low neutron irradiation dose.

The values stated in SI units are to be regarded as standard. No other units of measurement are included in this standard.

ACKNOWLEDGMENTS

Important contributions to the writing and editing of this chapter have been made by John T. Baron, Koppers Industries, Ralph E. Gehlbach, Kenneth C. Krupinski, Koppers Industries, and Albert V. Tamashausky.

References

[1] *Introduction to Carbon Technologies*, H. Marsh, E. A. Heintz, and F. Rodriguez-Reinoso, Eds., University of Alicante, 1997.
[2] *Sciences of Carbon Materials*, H. Marsh and F. Rodriguez-Reinoso, Eds., University of Alicante, 2000.
[3] Mantell, C. L., *Carbon and Graphite Handbook*, Interscience Publishers, John Wiley & Sons Inc., New York, 1968.
[4] Walker, P. L., Jr., and Thrower, P. A., *Chemistry and Physics of Carbon*, Marcel Dekker, Inc., New York [this series has several volumes].

10

Sampling Techniques

Peter W. Kosewicz,[1] Del J. Major,[2] and Dan Comstock[3]

TO COMPLETE THE HYDROCARBON CUSTODY transfer transaction, one needs to determine the quality of the hydrocarbons in addition to the volume. Samples of petroleum and petroleum products are examined by various methods of test for the determination of physical and chemical characteristics. The test results are used for quality, quantity, and price determinations that are part of the hydrocarbon custody transfer transaction. Because of the way the test results are used, it is important to note that sampling becomes the most important part of any analysis performed. Sampling is also required for inventory quality validation at many points in addition to the custody transfer points. The same care in sampling is required when sampling for regulatory compliance, process control, or internal quality validations. The samples should be representative of the petroleum or petroleum products in question. The automatic sampling method provides the most representative sample, but it is not always available because it is a dynamic method that requires the product be moving through a pipeline. When automatic sampling is not available then manual sampling becomes the best available method. Manual sampling is performed while the fluid is at rest in a tank or vessel. When manual sampling is used, there is no way to prove the samples are truly representative so they are usually deemed as representative because automatic sampling is not available and manual sampling is the only method of sampling that can be used. We will cover the procedures and equipment necessary to obtain a representative sample on which quantitative and qualitative analyses can be performed in the sections titled "Manual Sampling" and "Automatic Sampling." We will also cover how sampling affects volume when performing static measurements in "Static Measurements."

MANUAL SAMPLING

General

The objective of manual sampling (ASTM D4057) is to obtain a small portion of material from a selected area within a container that is representative of the material in the area or, in the case of running or all-levels samples, a sample whose composition is representative of the column of material in the container from which the sample is taken. Manual sampling is the least preferred method of sampling behind automatic sampling but is often used because it is the only available method to obtain samples. The precautions required to ensure the sample can be deemed as representative are numerous and depend on the type of material sampled; the tank, carrier, container, or line from which the sample is obtained; the type and cleanliness of the sample container; and the sampling procedure used. A summary of the sampling procedures and their application is presented in Table 1. Each procedure is suitable for sampling a number of specific materials under definite storage, transportation, or container conditions. The basic principle of each procedure is to obtain a sample or a composite of several samples in such a manner and from such locations in the tank or other container that the sample or composite can be deemed as being representative of the petroleum or petroleum product upon analysis.

Definitions

All-level sample—This is obtained by submerging a stoppered beaker or bottle to a point as near as possible to the draw-off level and then opening the sampler and raising it at a rate such that it is about three-fourths full (maximum 85 %) as it emerges from the liquid. Alternately, all-levels samples may be taken with samplers designed for filling as they pass downward through the liquid, closing when raised. In either case they must be between 70 and 85 % full when exiting the liquid.

Boring sample—This is obtained from the material contained in a barrel, case, bag, or cake that is obtained from the chip created by boring holes into the material with a ship auger.

Bottom sample—This is obtained from the material on the bottom of the tank, container, or line at its lowest point. In practice, the term *bottom sample* has a variety of meanings. Therefore, the exact sampling location should be specified when using this term.

Clearance sample—This is a spot sample taken with the inlet opening of the sampling apparatus 10 cm (4 in.) below the bottom of the tank outlet. This term is normally used when referring to small tanks, such as lease tanks.

Composite spot sample—This is a blend of spot samples mixed volumetrically proportional for testing. Some tests may also be made on the spot samples before blending and the results averaged. Spot samples from crude oil tanks are collected in the following ways:

Three-way—On tanks larger than 1,000-bbl capacity, which contain in excess of 15 ft of oil, equal volume samples should be taken at the upper, middle, and lower or outlet connection of the merchantable oil in the order named. This method may also be used on tanks up to and including a capacity of 1,000 bbl.

[1] *Assistant Director, The University of Texas–Petroleum Extension Service, Houston, TX 77073.*
[2] *Senior Measurement and Allocation Specialist, Shell Exploration and Production Company, Houston, TX, 79079.*
[3] *Training Specialist, The University of Texas–Petroleum Extension Service, Houston, TX 77073.*

TABLE 1—Summary of Sampling Procedures and Applicability

Application	Type of Container	Procedure
Liquids of not more than 101 kPa (14.7 psia) RVP	Storage tanks, ship and barge tanks, tank cars, tank trucks	Bottle sampling
		Core sampling
Liquids of 101 kPa (14.7 psia) RVP or less	Storage tanks with taps	Tap sampling
Bottom sampling of liquids of 13 kPa (2 psia) RVP or less	Storage tanks with taps	Tap sampling
Liquids of 101 kPa (14.7 psia) RVP or less	Pipes or lines	Manual pipeline sampling
Liquids of 13 kPa (2 psia) RVP or less	Free or open discharge streams	Dipper sampling
Liquids of 13 kPa (2 psia) RVP or less	Drums, barrels, cans	Tube sampling
Bottom or core sampling of liquids of 13 kPa (2 psia) RVP or less	Tank cars, storage tanks	Core sampling
Liquids and semiliquids of 13 kPa (2 psia) RVP or less	Free or open-discharge streams; open tanks or kettles with open heads; tank cars, tank trucks; drums	Dipper sampling
Crude petroleum	Storage tanks, ship and barge tanks, tank cars, tank trucks, pipelines	Automatic sampling
		Core sampling
		Bottle sampling
		Tap sampling
Industrial aromatic hydrocarbons, waxes, solids, bitumen, other soft solids	Storage tanks, ship and barge tanks	Bottle sampling
	Barrels, cases, bags, cakes	Boring sampling
Petroleum coke, lumpy solids	Freight cars, conveyors, bags, barrels, boxes	Grab sampling
Greases, soft waxes, asphalts	Kettles, drums, cans, tubes	Grease sampling
Asphaltic materials	Storage tanks, tank cars, lines, packages	
Emulsified asphalts	Storage tanks, tank cars, lines, packages	

Two-way–On tanks larger than 1,000-bbl capacity that contain in excess of 10 ft and up to 15 ft of oil, equal volume samples should be taken at the upper and lower or outlet connection of the merchantable oil in the order named. This method may also be used on tanks up to and including a capacity of 1,000 bbl.

Core sample–This is a sample of a uniform cross-sectional area taken at a given height in a tank.

Dipper sample–This is a sample obtained by placing the dipper or other collecting vessel in the path of a free-flowing stream to collect a definite volume from the full cross section of the stream at regular time intervals for a constant time rate of flow or at time intervals varied in proportion to the flow rate.

Dissolved water–This is water in solution in the oil at the defined temperature and pressure.

Drain sample–This is obtained from the water draw-off valve. Occasionally, a drain sample may be the same as a bottom sample, as in the case of a tank car.

Emulsion–This is an oil/water mixture that does not readily separate.

Entrained water–This is water suspended in the oil. Entrained water includes emulsions but does not include dissolved water.

Floating roof sample–This is a spot sample taken just below the surface to determine the density of the liquid on which the roof is floating.

Flow proportional sample–This is taken from a pipe during the entire transfer. The rate of sampling is proportional to the flow of the liquid in the pipe at any point in time during the transfer.

Free water–This is the measured volume of water present in a container that is not in suspension in a contained liquid at observed temperature.

Grab sample–This is a sample obtained by collecting equal quantities from parts or packages of a shipment of loose solids such that the sample is representative of the entire shipment. "Grab sampling" is also used as a description for the automatic sampling process.

Grease sample–This is obtained by scooping or dipping a quantity of soft or semi-liquid material contained in a package in a representative manner.

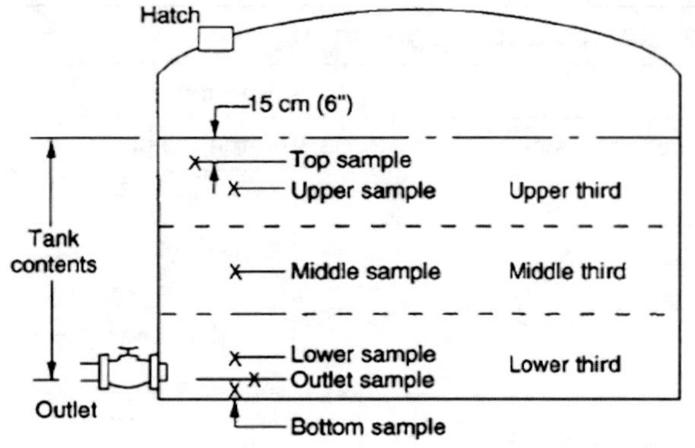

Notes:

1. The location shown for the outlet sample applies only to tanks with side outlets. It does not apply when the outlet comes from the floor of the tank or turns down into a sump. Bottom sample location must be specified.

2. Samples should not be obtained from within solid stand pipes as the materials normally are not representative of the material in the tank at that point.

Fig. 1—Spot sampling locations.

Intermediate container–This is the vessel into which all or part of the sample from a primary container or receiver is transferred for transport, storage, or ease of handling.

Lower sample–This is a spot sample obtained at the midpoint of the lower third of the tank contents (Fig. 1).

Middle sample–This is a spot sample obtained from the middle of the tank contents (a point halfway between the upper and lower sample points) (Fig. 1).

Middle spot sample–This is taken on tanks of a larger than 1,000-bbl capacity containing ten feet or less of crude oil. One middle spot sample should be taken as near the center of the vertical column of oil as possible.

Multiple tank composite sample (ships, barges, and so forth)–This is a mixture of individual samples from several compartments, each of which contains the same grade of petroleum material. The mixture is blended in proportion to the volume of material in each compartment.

Official sample–This is taken at the point of custody transfer and is used for custody transfer properties determination.

Outlet sample–This is a spot sample taken at the level of the bottom of the tank outlet (either fixed or swing pipe) but not higher than 1 m above the bottom of the tank (see Fig. 1).

Primary manual sample container–This is the container in which a manual sample is initially collected. These containers can include glass and plastic bottles, cans, or any manual sampling apparatus.

Representative sample–For manual sampling, this is a portion extracted from the total volume that is deemed to contain the constituents in the same proportions that are present in that total volume. The true representative nature of any sample may not be determinable but when automatic sampling is not available, then manual sampling becomes the best practice. Automatic sampling provides the most representative sample.

Running sample–This is one obtained by lowering an unstoppered beaker or bottle from the top of the oil to the level of the bottom of the outlet connection or swing line and returning it to the top of the oil at a uniform rate of speed such that the beaker or bottle is about three-fourths full when withdrawn from the oil. A running sample is not necessarily a representative sample because the tank volume may not be proportional to the depth and because the operator may not be able to raise the sampler at the variable rate required for proportionate filling. The rate of filling is proportional to the square root of the depth of immersion.

Sample–This is a portion extracted from the total volume that may or may not contain the constituents in the same proportions that are present in that total volume.

Sampling–These are all the steps required to obtain a sample that is representative of the contents of any pipe, tank, or other vessel and to place that sample in a container from which a representative specimen can be taken for analysis.

Spot sample–This is a sample taken at a specific location in a tank or from a pipe at a specific time during a pumping operation.

Stand pipes–These are vertical sections of pipe or tubing used for gaging that extend from the gaging platform to near the bottom of the tanks that are equipped with external or internal floating roofs. These stand pipes shall be continuously slotted from bottom to top, with overlapping/offset slots; otherwise a representative sample cannot be obtained.

Surface sample–This is a spot sample skimmed from the surface of a liquid in a tank.

Tank composite sample–This is a blend of the upper, middle, and lower samples. For a tank of uniform cross section, such as an upright cylindrical tank, the blend consists of equal parts of the three samples. For a horizontal cylindrical tank, the blend consists of samples in various proportions, which are dependent on liquid depth, sampling level, and number of proportional parts of the upper, middle, and lower levels (Table 2).

Tank-side sample–This is a spot sample taken from a sample connection in the side of a tank.

Tap sample–This is a spot sample taken from a sample tap on the side of a tank. It may also be referred to as a tank-side sample.

Test specimen–This is a representative sub-sample taken from the primary or intermediate sample container for analysis.

Top sample–This is a spot sample obtained 15 cm (6 in.) below the top surface of the liquid (Fig. 1).

Tube and thief sample–This is obtained with a sampling tube or special thief, either as a core sample or spot sample from a specified point in the tank or container.

Uniform petroleum product–This is one in which spot samples from top, upper, middle, lower, and outlet agree within the precision of the laboratory tests. Similarly, in pipeline transfers, spot samples taken at 1, 20, 50, 80, and 99 % of the total volume agree within the precision of the laboratory tests.

Upper sample–This is a spot sample taken at the midpoint of the upper third of the tank contents (Fig. 1).

Sample Containers

Sample containers come in a variety of shapes, sizes, and materials. The selection of the correct container for the application requires knowledge of the material to be sampled for compatibility, remixing requirements, and analytical tests to be performed. ASTM D5854 provides detailed procedures for matching the container to the analytical test. Additionally,

TABLE 2—Sampling Instructions for Horizontal Cylindrical Tanks

Liquid Depth (% of Diameter)	Sampling Level (% of Diameter Above Bottom)			Composite Sample (Proportionate Parts)		
	Upper	Middle	Lower	Upper	Middle	Lower
100	80	50	20	3	4	3
90	75	50	20	3	4	3
80	70	50	20	2	5	3
70		50	20		6	4
60		50	20		5	5
50		40	20		4	6
40			20			10
30			15			10
20			10			10
10			5			10

ASTM D5842 provides the specific container requirements for volatile products, and ASTM D4306 is specific to aviation fuel samples.

The use of intermediate containers is especially prevalent in the manual sampling of finished products. When using intermediate containers for finished products, the same requirements must be met for their application and preparation as for the primary container.

In general, sample containers should contain no internal pockets or dead spots; internal surfaces that minimize corrosion, encrustation, and water or sediment clingage; closure of a size to facilitate filling, inspection, and cleaning; ability to allow preparation of a homogeneous mixture and not affect the loss of any constituents; and the ability to allow the transfer of samples from the container to the analytical apparatus without loss of their representative nature.

Sample containers may be clear or brown glass bottles. The clear glass bottle may be examined visually for cleanliness and allows visual inspection of the sample for free water or solid impurities. The brown glass bottle affords some protection from light.

Plastic bottles made of suitable unpigmented linear polyethylene may be used for the handling and storage of gas oil, diesel oil, fuel oil, and lubricating oil. They should not be used for gasoline, aviation jet fuel, kerosene, crude oil, white spirit, medicinal white oil, and special boiling point products unless testing indicates there is no problem with solubility, contamination, or loss of light ends. (*Note.* In no circumstances shall nonlinear (conventional) polyethylene containers be used to store samples of liquid hydrocarbons. This is to avoid sample contamination or sample bottle failure.)

Container Closure

Cork or glass stoppers, screw caps of plastic or metal, may be used for glass bottles. Corks must be of good quality, clean, and free from holes and loose bits of cork. Never use rubber stoppers, as they may not be compatible with the hydrocarbon sample and will contaminate it. Wrapping tin or aluminum foil around the cork before forcing it into the bottle may prevent contact of the sample with the cork. Glass stoppers must be a perfect fit.

A disk faced with tin, aluminum foil, or other material that will not affect petroleum or petroleum products must protect screw caps.

Cleaning Procedure

All sample containers must be absolutely clean and free of water, dirt, lint, washing compounds, naphtha, other solvents, soldering fluxes, or acids and corrosion, rust, or oil. The same cleaning requirements exist for both primary and intermediate containers, especially when sampling finished products.

Crude Petroleum and Heavy Fuel Oils

Crude petroleum and heavy fuel oils usually are nonhomogeneous. Automatic samplers are recommended whenever representative samples are required for sediment and water and density measurements.

Tank samples may not be representative because of the following reasons:

1. The concentration of entrained water is higher near the bottom. The running sample or the composite of the upper, middle, and lower samples may not represent the concentration of entrained water.
2. The interface between oil and free water is difficult to measure, especially in the presence of emulsion, layers, or sludge.
3. The determination of the volume of free water is difficult because the free water level may vary across the tank bottom surface. The bottom is often covered by pools of free water or water emulsion impounded by layers of sludge or wax.

Sample Handling of Crude Petroleum and Nonuniform Products

The transfer of crude oil samples from the sample receiver to the laboratory glassware in which they will be analyzed requires special care to maintain their representative nature. Generally, the samples of nonuniform products must be rehomogenized prior to transfer. The number of transfers should be minimized.

Gasoline and Distillate Products

Gasoline and distillate products are usually homogeneous, but they are often shipped from tanks that have clearly separated water on the bottom. Tank sampling, if performed in strict accordance with the indicated procedures, is acceptable under the following conditions:

- Sufficient time must have elapsed for the heavy component, such as free water, to adequately separate and settle.
- It must be possible to measure the level of the settled component to stay well above that level when drawing representative samples.
- If one or more of these conditions cannot be met, then it is recommended that sampling be performed by means of an automatic sampling system (see Practice ASTM D4177).

Extreme care must be exercised in any product transfer, and transfers of finished products should be minimized. Unnecessary transfers of finished products can result in the degradation of the representivity of the sample. Normally, transfers of finished product samples have very explicit handling requirements that must be followed.

Marine Custody Transfer

By mutual agreement, samples can be taken from shore tanks, from ships' tanks, or from pipelines. Pipeline samples may be taken automatically or manually. Properly taken automatic pipeline samples are the most representative.

Manual pipeline samples are less representative than automatic pipeline samples.

Tank samples will usually not be representative unless the tank is completely homogeneous and contains no free water.

STATIONARY OR SHORE TANKS

Crude petroleum tanks may be sampled by mutual agreement in the following ways: by composite spot, middle spot, all-levels, running samples, or sample cocks. Additional samples may be taken as necessary.

SHIP OR BARGE TRANSFERS

Samples of ship cargoes of crude petroleum may be taken by mutual agreement by the following methods:

1. From the shore tanks before loading and both before and after discharging.
2. From the pipeline during discharging or loading. Pipeline samples may be taken either manually or with an automatic sampler. If the pipeline requires displacement or flushing, care must be taken that the pipeline sample includes the entire cargo and none of the displacement. Separate samples may be required to cover the effect of the line displacement on the prior or following transfer. Pipeline samples taken with an automatic sampler is the preferred method.
3. From the ship's tanks after loading or before discharging. An all-levels sample or a running sample shall be obtained from each compartment of the ship's cargo tanks.

When loading a ship, the shore tank sample or the pipeline sample taken from the loading line is normally the official sample. However, ship's tank samples may also be tested for sediment and water and for other quality aspects when required. The results of these ship's tank sample tests, together with the shore tank sample tests, may be shown on the cargo certificate.

When discharging a ship, the pipeline sample taken from a properly designed and operated automatic line sampler in the discharge line is the preferred method. However, where no proper line sample is available, the ship's tank sample is normally used as the official sample except where specifically exempted.

Obtaining Samples

Extreme care and good judgment are necessary to ensure that samples are obtained that represent the general character and average condition of the material.

Sampling Procedures

The standard sampling procedures described in this method are summarized in Table 1. Alternate sampling procedures may be used if the parties involved have reached a mutually satisfactory agreement. Such agreement should be put in writing and signed by authorized officials.

Bottle or Beaker Sampling

APPLICATION

The bottle or beaker sampling procedure is applicable for sampling liquids of 101 kPa (14.7 psia) RVP or less in tank cars, tank trucks, shore tanks, ship tanks, and barge tanks. Solids or semiliquids that can be liquefied by heat may be sampled by this procedure provided they are true liquids at the time of sampling.

APPARATUS

A suitable sampling bottle or beaker, as shown in Fig. 2, is required. Recommended uses and diameter of openings in the bottle or beaker are given in Table 3.

PROCEDURE

1. *All-Levels Sample.* Lower the weighted, stoppered bottle or beaker as near as possible to the draw-off point or, if higher, to a point just above the free water or other heavy material, pull out the stopper with a sharp jerk of the line, and raise the bottle at a uniform rate so that it is about three-fourths full as it emerges from the liquid. For light products or deep tanks, a restricted opening may be needed to avoid filling the bottle. Care should be taken with an all-levels sample because the tank volume may not be proportional to the depth and because the operator may not be able to raise the sampler at the variable rate required for proportionate filling. If this occurs, the all-levels sample may not necessarily produce a representative sample. The rate of filling is proportional to the square root of the depth of immersion.
2. *Running Sample.* Lower the unstoppered bottle or beaker as near as possible to the level of the bottom of the outlet connection or swing line and then raise the bottle or beaker to the top of the oil at a uniform rate of speed such that it is about three-fourths full when withdrawn from the oil. For light products or deep tanks, a notched cork or other restricted opening may be needed to avoid filling the bottle.

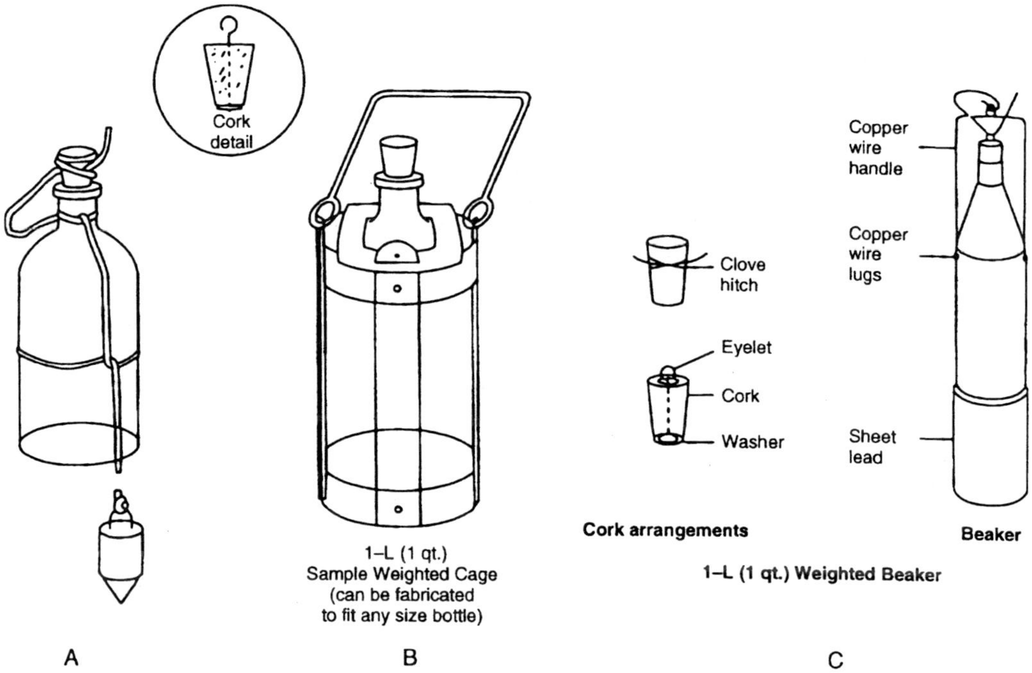

Fig. 2—Typical assemblies for bottle/beaker sampling.

3. *Top, Upper, Middle, Lower, and Outlet Samples.* Lower the weighted, stoppered bottle to the proper top, upper, middle, lower, or outlet depths (see Fig. 1). At the selected level, pull out the stopper with a sharp jerk of the line and allow the bottle or beaker to fill completely. When judged full, raise the bottle or beaker, pour off a small amount, and stopper immediately.
4. *Multiple Tank Composite Samples.* Prepare a composite sample by mixing portions of the individual tank samples. The volume of product in each tank is recorded and the composite is made by volume proportions. For nonhomogeneous products (e.g., crude oil, heavy heating oils, etc.), the composite is to be prepared in the laboratory or other location where the individual samples may be homogenized before extracting out the aliquot for the composite sample. In the field, multiple tank composites are usually made, especially for clean products for ease of handling and transport.
5. *Composite Spot Sample.* Prepare a composite spot sample by mixing spot samples in equal proportions using either the three-way or the two-way procedure, whichever applies.
6. *Middle Spot Sample.* Obtain this sample in the manner specified.
7. *Handling.* Stopper and label bottle samples immediately after taking them and deliver the samples to the laboratory in the original sampling bottles. These labels should include as a minimum:
   - Date and time
   - Name of the sample
   - Product type
   - Name and number and owner of the vessel, car, or container
   - Location of sample
   - Sampling procedure and apparatus used to obtain the sample
   - Type of and reason for sample
   - Sampler's name

There is an advantage to using a bottle and a sampling cage instead of a weighted beaker for volatile products. The loss of light ends is likely to occur when transferring the sample from a weighted beaker to another container.

TABLE 3—Weighted Sampling Bottle or Beaker

Material	Diameter of Opening	
	cm	in.
Light lubricating oils, kerosene, gasolines, transparent gas oils, diesel fuels, distillates	2	3/4
Heavy lubricating oils, nontransparent gas oils	4	1 1/2
Light crude oils (less than 44 cSt at 37.8°C)	2	3/4
Heavy crude and fuel oils	4	1 1/2

Manual Sampling of Pipeline Streams

APPLICATION

The manual pipeline sampling procedure is applicable to liquids of 101 kPa (14.7 psia) RVP or less and semi-liquids in pipelines, filling lines, and transfer lines.

When a custody transfer is involved, continual automatic sampling is the preferred method. In the event of automatic sampler failure, manual samples may be needed. Such manual samples should be taken as representatively as possible.

APPARATUS

The function of the sampling probe is to withdraw from the flowing stream a portion that will be representative of the

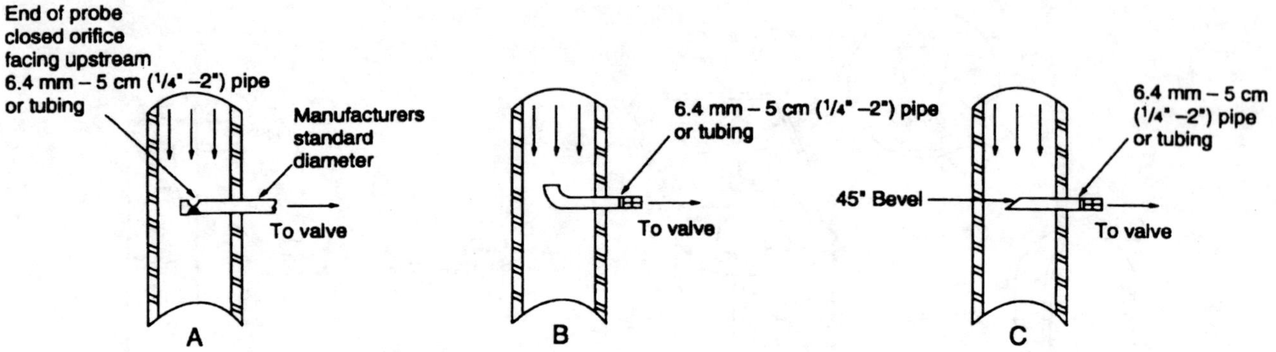

NOTE 1—Probes may be fitted with valves or plug cocks. The probe should be oriented horizontally.

Fig. 3—Probes for spot manual samples.

entire stream. Probe designs that are commonly used are shown in Fig. 3 and described here.

- A closed-end tube with a round orifice spaced near the closed end that should be positioned in such a way that the orifice is in the center third of the pipeline and is facing the stream as shown in Fig. 3a.
- A long-radius forged elbow or pipe bend extending to the center third of the pipe and facing upstream. The end of the probe should be externally beveled to give a sharp entrance edge (Fig. 3b).
- A tube extending to the center third of the line and beveled at a 45° angle facing upstream (Fig. 3c).

Probe Location

Since the fluid to be sampled may not always be homogeneous, the location, position, and size of the sampling probe should be such as to minimize any separation of water and heavier particles that would make their concentration different in the gathered sample than in the main stream.

The probe should always be in a horizontal position to prevent drainback of any part of the sample to the main stream.

The sampling probe should be located in a vertical run of pipe. The probe may also be located in a horizontal run of pipe, provided the flowing velocity is high enough to provide adequate turbulent mixing. While adequate flowing velocity may not eliminate a concentration difference between the bottom of the pipe and the top of the pipe, it may provide an average concentration at the mid-pipe probe location that will be representative of the entire stream at the sampling station. A flow stream conditioner may be required upstream of the probe to provide adequate mixing for some sampling applications.

The sampling point should be as near as is practicable to the point where the fluid passes to the recipient.

The sampling lines should be as short as is practicable and should be cleared before any samples are taken.

To control the rate at which the sample is withdrawn, the probe should be fitted with valves or plug cocks.

Procedure

Adjust the valve or plug cock from the sampling probe so that a steady stream is drawn from the probe. Whenever possible, the rate of sample withdrawal should be such that the velocity of liquid flowing through the probe is approximately equal to the average linear velocity of the stream flowing through the pipeline. Measure and record the rate of sample withdrawal as gallons per hour. Divert the sample stream to the sampling container continuously or intermittently to provide a quantity of sample that will be of sufficient size for analysis.

In sampling crude petroleum, samples of one-half pint or more should be taken every hour or at increments less than an hour, as necessary. By mutual agreement, the sample period or sample size, or both, may be varied to accommodate the parcel size. It is important that the size of the samples and the intervals between the sampling operations be uniform for a uniform flow rate. When the main stream flow rate is variable, the sampling rate must be varied accordingly so that the amount of sample is always withdrawn from any given amount of fluid passing the sampling point in the main stream. In practice, this is difficult to accomplish.

The sample of the petroleum or petroleum product should be placed in a closed container, and, at the end of the agreed upon time period, the combined samples should be mixed and a composite sample taken for test purposes, mixing, and handling. The sample container should be stored in a cool, dry place; exposure to direct sunlight should be avoided.

Alternatively, line samples may be taken at regular intervals and individually tested. The individual test results may be arithmetically averaged, adjusting for variations in flow rate during the agreed upon time period.

Either composite or arithmetically averaged results are acceptable by mutual agreement.

Core Sampling

APPLICATION

The core-type thief is used to obtain spot samples of liquids in storage tanks, tank cars, tank trucks, and ship and barge tanks. The core sampler is widely used in sampling crude petroleum in storage tanks. In this application it may be used for taking samples at different levels, as well as for bottom samples of non-merchantable oil and water at the bottom of the tank. The core thief can be used in some cases to obtain a quantitative estimate of the water at the bottom of the tank.

APPARATUS

The thief should be designed so that a sample can be obtained within $1/2$ in. of the bottom of the car or tank. Two types of thiefs are illustrated in Figs. 4 and 5. One type is lowered into the tank with valves open to permit the fluid to flush through

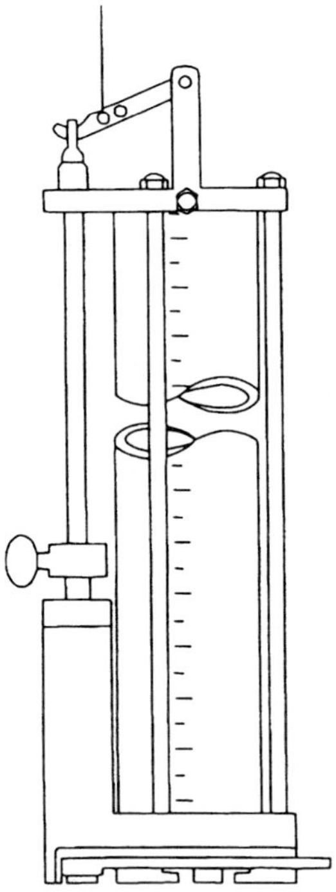

Fig. 4—Core-type sampling thief.

the container. When the thief strikes the bottom of the tank, the valves shut automatically to trap a bottom sample. The other type has a projecting stem on the valve rod, which opens the valves automatically as the stem strikes the bottom of the tank. The sample enters the container through the bottom valve and air is released simultaneously through the top. The valves snap shut when the thief is withdrawn.

A core-type thief similar to that shown in Fig. 4, with a uniform cross section and bottom closure and with a capacity depending upon the size of the sample required, should be used for sampling crude petroleum. Additionally, it can be used to sample other petroleum products. The thief should be capable of penetrating the fluid in the tank to the required level, mechanically equipped to permit filling at any desired level and also be capable of being withdrawn without undue contamination of the contents. The thief may be equipped with the following:

1. Extension rods for use in obtaining samples at levels corresponding with requirements for high connections or for samples to determine high settled water and sediment levels.
2. A thief cord marked so that the sample can be taken at any depth in the vertical cross section of the tank.
3. A hook to hang the thief in the hatch vertically.
4. Sample cocks for obtaining samples at specific levels. Use clean, dry glass bottles as sample containers.

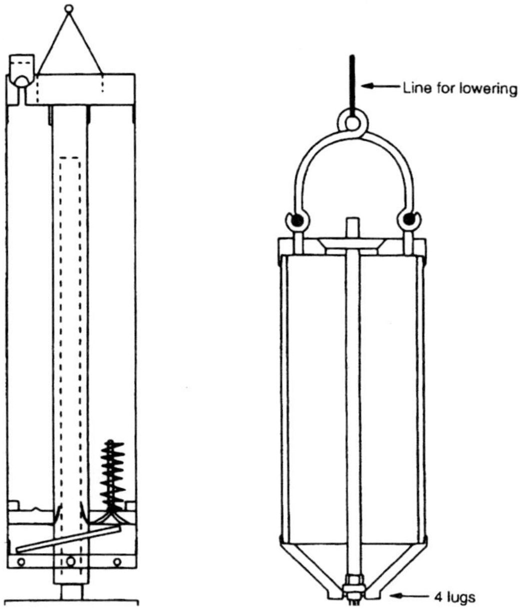

Fig. 5—Closed-core type sampling thief.

Procedure

Lower the clean, dry thief through the dome of the tank car or tank hatch until it strikes the bottom. When full, remove the thief and transfer the contents to the sample container. Close and label the container immediately.

In sampling of crude petroleum or petroleum products, the following procedure should be used:

1. Obtain top sample by lowering and tripping the open thief when the top of the thief is about 6 in. below the hydrocarbons surface, and after pouring off 2 or 3 in. of the hydrocarbons, partly fill a clean bottle with hydrocarbons from the thief.
2. Obtain bottom sample by lowering and tripping the open thief when the bottom of the thief is on a level with or slightly above the bottom of the outlet, and after pouring off about two thirds of the hydrocarbons, partly fill a second clean bottle with hydrocarbons from the thief.
3. For closed-core type sampling thiefs, after sampling, place the bottom outlet of the thief over the intermediate container opening and release the liquid into the intermediate container. Allow sufficient time for the sample to drain to capture all contaminants.

AUTOMATIC SAMPLING

General

Automated sampling systems are designed to extract a representative sample of petroleum and petroleum products from a flowing stream and storing them in a sampler receiver. Careful consideration must be given the variables when sizing a sample system. It is essential to determine the rate of flow through the meter, the total volume to be delivered, and the volume of composite sample required. After the composite sample is obtained, it is essential that proper mixing and handling procedures of the sample at the field location and during transportation to the laboratory be followed. It does not make sense to spend large sums of money to collect representative samples only to lose the representativeness in transit and in laboratory techniques. Automatic sampling is addressed in ASTM D4177: Standard Practice for Automatic Sampling of Petroleum and Petroleum Products.

Samples of petroleum and petroleum products are examined by various methods of test for the determination of physical and chemical characteristics. Sediment and water content, sulfur content, and density are generally included in the testing of crude oil. The test results are used for custody transfer and pricing determinations. The samples must be representative of the petroleum or petroleum products in question. As petroleum products are increasingly sampled using automatic sampling systems, different procedures may become necessary to ensure the proper operation of the systems. However, though the use of the water injection method might be used, it introduces an unwanted contaminant that then must be dealt with. The requirements to ensure the representative character of the samples can be summarized in four distinct steps for determining the sediment and water content accurately:

1. Free water must be adequately distributed at the sample point to allow representative sampling.
2. The parcel must be sampled representatively.
3. Samples must be handled and mixed during preparation to ensure dispersion of free water to eliminate density stratification and ensure homogeneity.
4. The sample must be accurately analyzed.

Definitions

Automatic sampler–This is used to retrieve a representative sample from the liquid flowing in a pipe. The automatic sampler generally consists of a probe, an extracting mechanism, an associated controller, and a sample receiver. The probe and extracting mechanism may or may not be individual components.

Automatic sampling system–This consists of stream conditioning, an automatic sampler, and sample handling.

Fast sample loop (or slip stream)–This is a bypass from the main pipeline being sampled, through which a representative portion of the total flow is brought to the automatic sampler.

Flow proportional sample–This is taken from a pipe during the entire transfer. The rate of sample collection is proportional to the rate of flow of the liquid in the pipe over any time period.

Isokinetic sampling–This is sampling in such a manner that the linear velocity through the opening of the sample probe is equal to the linear velocity in the pipeline at the sampling location and is in the same direction as the bulk of the liquid approaching the sample probe.

Minimum pipeline velocity–This is a velocity that exists at the lowest operating flow rate, excluding those rates, which occur infrequently (for example, one of ten cargoes) or for short time periods (less than 5 min).

Pipeline–This is any section of pipe used for the transfer of fluids.

Powered mixer–This depends on an external source of power for the energy required to mix the fluid.

Probe–This is the sampler element that extends into the pipeline.

Profile testing–This is a technique for simultaneously sampling at several points across the diameter of a pipe to identify the extent of stratification at a proposed probe location.

Rangeability–This is the capability of a meter or flow-measuring device to operate between the minimum and maximum flow range, within an acceptable tolerance. Rangeability is generally expressed as ratio of a maximum flow to the minimum flow (for example, 5:1, 10:1, and so forth).

Representative sample–For automatic sampling, this is a portion extracted from the total volume that contains the constituents in the same proportions that are present in that total volume. The true representative nature of any sample may not be determinable but an automatic sampling system provides the most representative sample if the system is properly designed, installed, and operated.

Sample conditioning–This is the mixing required to prepare the sample prior to sample transfer for analysis.

Sample controller–This governs the operation of the extracting mechanism either in proportion to time or to flow.

Sample extractor–This extracts the grabs from the pipeline or from the sample loop.

Sample grab. This is the volume of liquid extracted from the pipe by a single actuation of the sample extractor. The sum of all grabs results in the total sample.

Sample handling and mixing–This is the extracting, conditioning, transferring, and transporting of the representative sample from the container to the analytical glassware.

Sample proving–This is a technique used to validate an automatic sampling system. Injecting a known quantity of water ahead of the sampling system, taking samples, analyzing the samples, and comparing those results with the known water injection can make an acceptable validation.

Primary sample receiver/container (receptacle)–This is a container into which all sample grabs are initially collected. A receiver may be fixed or portable.

Sediment water (S&W)–This is a material coexisting with, yet foreign to, petroleum liquid that requires separate measurement for reasons that include sales accounting. These foreign materials include free water and sediment and emulsified or suspended water and sediment.

Static mixer–This has no moving parts. The kinetic energy of the moving fluid provides the power required for mixing.

Stream conditioning–This is the mixing of the pipeline contents, upstream of the sampling location, necessary for delivery of a representative sample.

Time proportional sample–This is composed of equal sample grabs taken from a pipeline at uniform time intervals during the entire transfer.

Automatic Sampler Probe Location

Before selecting the system to be used, a number of factors concerning sample probe location should be considered. The automatic sampler probe must be located at a point in the pipeline where the flowing stream is well mixed. Generally, an oil sample extracted from a pipeline for sediment and water analysis will only be representative if the sediment and water and the oil in the line are mixed sufficiently–that is, to have distribution and dispersion of the water droplets evenly over the pipe cross section.

A vertical line with adequate upstream mixing is an acceptable probe location. Based on tests with low viscosity crudes, locate the probe in the downstream section of the vertical piping loop to obtain the benefit of the additional mixing from the two 90° elbows. Locate the probe a minimum of one-half pipe diameter before the final exiting elbow or fitting in the loop (Fig. 6).

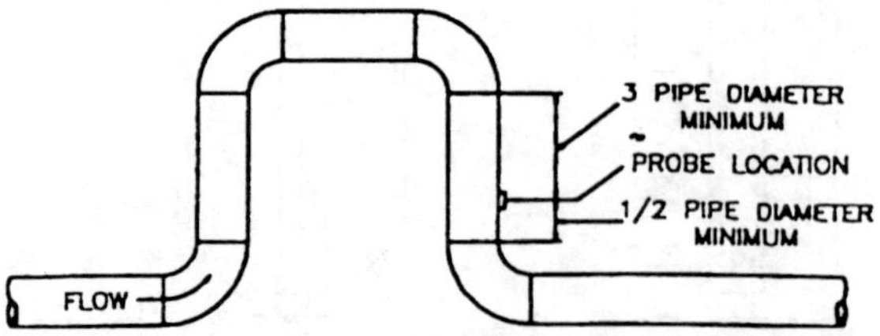

Fig. 6—Sample probe location (vertical).

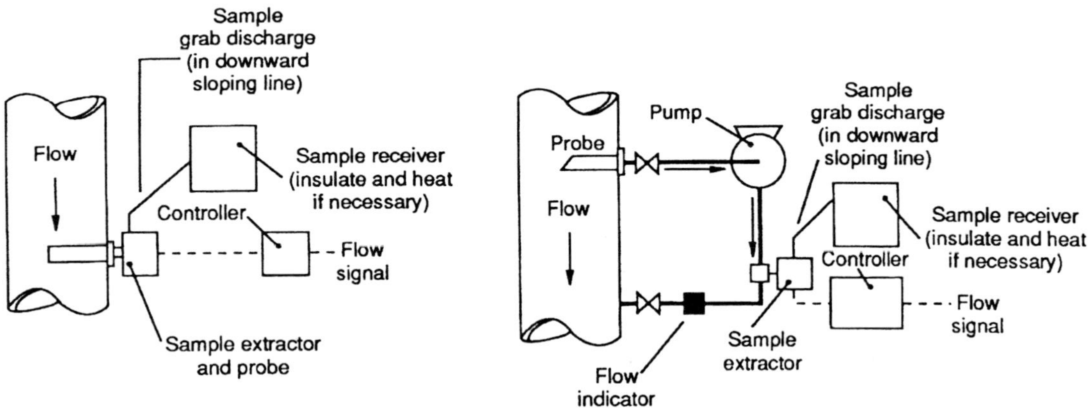

Automatic Sampling-in-Line **Automatic Sampling With a Fast Loop**

Fig. 7—Typical automatic sampling systems. *Note.* Arrow does not indicate piping orientation.

The probe shall be installed in a horizontal position regardless of where the sample point is located.

The best sampling point is the center third of the pipeline cross section.

Minimum Velocities

The minimum velocity for a sampling system, using piping elements as the mixing element and the piping installed in the vertical loop configuration, is 5 ft/s.

Automatic Sampling Systems

An acceptable automatic sampling system consists of stream conditioning upstream of the sampling location (necessary for a representative sample), a device to physically extract fluid from a flowing stream, a sample receiver to contain the sample grabs from the stream, and a means to control the amount of sample taken (by varying the sample frequency or grab size).

There are two general types of automatic sampling systems (Fig. 7): in one system, the sample extractor and probe are located directly in the main line; in the other system, the sample extractor is located in a fast sample loop. Both types require the same minimum pipeline velocities.

The use of single and multiple receivers is shown in Fig. 8.

Sample Extractors

Sample extractors are devices that physically grab a representative sample from the pipeline. Marine operation or pipeline operation may require different sampling frequency.

Sample Frequency

Sample frequency can be expressed in different ways. For example:

1. One grab per unit of volume (grab/barrel, grab/gallon, or grab/cubic meter).
2. Grabs per unit of time (1/min, $^1/_2$ min, 900/h, 10,000/day).
3. Sufficient grabs such that a unit of volume (1, 5, or 10 gal [4, 20, or 40 L] of sample is collected per cargo, tender, day, or week).
4. One grab per multiple lineal feet (meters) or pipeline volume (particularly for marine and large-volume pipeline operation).

Sample frequency will be determined so that a desired volume of sample is collected during a cargo tender, day, week, or some other period of time. This frequency can be expressed as a grab per volume pumped, a grab per unit time, or a grab per lineal meters (feet) of pipeline volume.

Lease automatic custody transfer (LACT) operation and other small-volume movements are typically expressed as grab per volume or grab per unit time.

Sample Frequency in Large-Volume Pipeline Operations

Large-volume pipeline operations typically present a less severe sampling problem than marine operations because there are more sediment and water peaks per parcel and the duration of the peaks is longer.

Typical marine cargo sampling practice has been to take one sample grab per 80 lineal ft of pipeline volume. However, for pipeline operations, it is suggested that one grab per 320 ft of pipeline volume is suitable for samplers located near the supply tanks (worst case), and 800 ft of pipeline volume is suitable for samplers far away from the supply tanks.

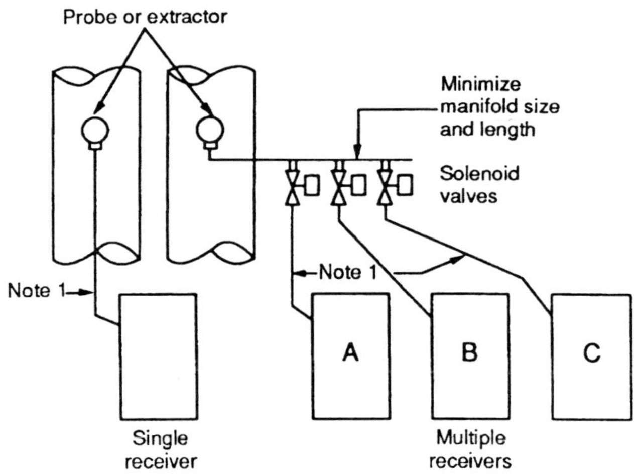

Fig. 8—Single/multiple fixed receivers. *Note 1.* Tubing of 6.4 or 9.5 mm ($^1/_4$ or $^3/_8$ in.) tubing, as short as possible and sloping continuously toward the sample receiver, should be used. Tubing of 9.5 mm ($^3/_8$ in.) should be used where long sampling lines cannot be avoided or in crude oil service. Heat trace and insulate these lines when necessary.

For severe pipeline service (for example, stratified tanks), sampling frequency should be increased.

Example–Illustrative Calculation for Determining Sample Frequency in Pipeline Operations

A. Given

Flow rate, barrels per hour	5,000
Sample period, days per parcel	30
Parcel size, barrels per parcel	3,600,000
Line diameter (*D*), inches	16
Maximum sampler frequency (R_{max}) grabs per hour	900
Grab size, milliliters	1.5
Lineal feet of line volume per grab (LFV)	320

B. Calculate

1. Minimum sample frequency, barrels per grab

$$\text{Frequency} = 0.00097\ D^2 \times \text{LFV}$$

where
$0.00097\ D^2$ = volume, in barrels, of 1 ft of pipe with a diameter of D inches, and LFV = lineal feet of line volume per grab (given as 80).
Thus:

$$\begin{aligned}\text{Frequency} &= 0.00097(16)^2\ (320)\\ &= 79.46 \text{ barrels per grab}\end{aligned}$$

2. Number of grabs per parcel

$$\text{Grabs} = \frac{\text{parcel size (given)}}{\text{minimum sampling frequency (see Step 1)}}$$

Thus:

$$\begin{aligned}\text{Grabs} &= \frac{3{,}600{,}000}{79}\\ &= 45{,}570 \text{ grabs per parcel}\end{aligned}$$

3. Sample volume size, gallons per parcel

$$\text{Volume size} = \frac{\text{(grabs per parcel)(grab size)}}{3{,}785 \text{ mL per gallon}}$$

Thus:

$$\begin{aligned}\text{Sample volume size} &= \frac{(45{,}570)(1.5)}{3{,}785}\\ &= 18.05 \text{ gal per parcel}\end{aligned}$$

4. Actual sampler operating rate, grabs per hour

$$\text{Operating rate} = \frac{\text{grabs per parcel (Step 2)}}{\text{sample period (given)}}$$

Thus:

$$\begin{aligned}\text{Operating rate} &= \frac{45{,}570}{(30)(24 \text{ h per day})}\\ &= 62.93 \text{ grabs per hour}\end{aligned}$$

C. Final Check

Since the maximum sampler frequency (given) is 900 grabs per hour, the maximum sampler-operating rate has not been exceeded.

Design Considerations for Sampling Systems

SAMPLE PROBE

The sampler should be the true proportional type paced by the meter. The probe should be located on a down-comer a minimum of three pipe diameters from any bends or elbows. The probe must be mounted horizontally with the sample point located in the center of the pipe and the tip opening shall face upstream. The sample probe should be located as near as possible to the sample container. Typical sample probes are shown in Fig. 9.

SAMPLER PACING

The meter, proportional to flow, will pace the sampler.

SAMPLE LINES

Sample lines should be as short as possible and slope continuously downward from the sample probe to the sample receiver or container. Sample lines should connect to the sample container through a nipple extending at least 1 in. into the interior of the container. Sample lines shall be a minimum of 3/8-in. metal tubing.

SAMPLE CONTAINER

The sample container should be between 5 and 15 gal and be designed to hold the sample under sufficient pressure to prevent the escape of vapors. The container should be internally coated with a plastic or epoxy coating with a bottom continuously sloped toward the center drain to facilitate mixing and complete withdrawal of the contents. Container must be equipped with a means for the withdrawal of the representative sample. The previous sizing example demonstrates some extremes, in time interval (30 days) and volume (3,600,000 bbl). Normally, for large batch sizes the sampling interval (days) is substantially reduced.

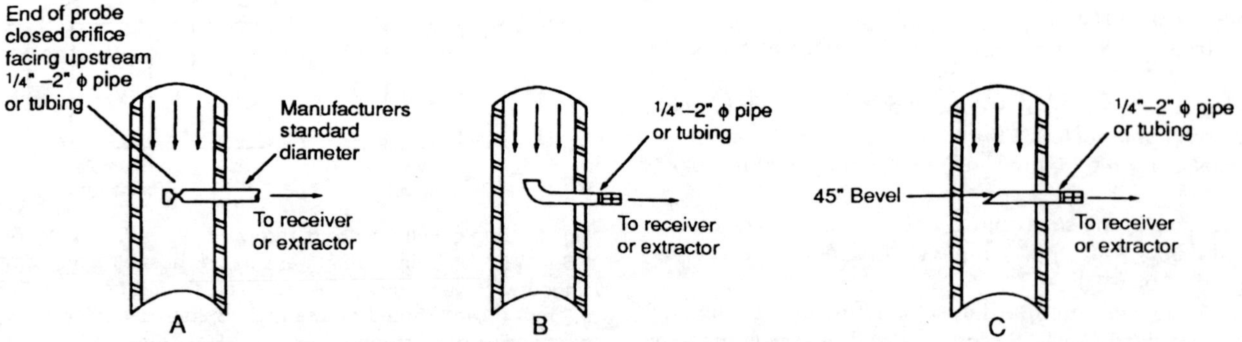

Fig. 9—Typical sample probes.

The receiver or container can be either portable or stationary. However, systems with portable receivers or containers are becoming the system of choice for crude oil marine systems, petroleum product, and some crude oil pipeline systems. The portable receiver system allows the integrity of the sample to be maintained while the sample is transported to the lab for analysis. Each transfer of a sample from one container to another may cause degradation in the quality of the sample analysis. Therefore, sample transfers should be minimized to the absolute minimum.

MIXING APPARATUS

The mixing apparatus must be arranged in such a way as to avoid trapping sediment and water in low spots. Normally, the system is capable of displacing the sample volume once or twice per minute, which is normally sufficient to mix the sample. Mixing elements such as static mixers or orifice plates are sometimes used in the mixing system. Testing should be conducted to validate the performance of the system.

PUMP OUT

Sample container pump out piping should be arranged in such a way as to permit complete withdrawal of the contents of the sample container.

SLIP STREAM SAMPLING

Slip stream sampling occurs when a smaller volume is diverted from the main pipeline, where sampling occurs. The fluid velocity in both the main pipeline and the fast loop should be the same at all times.

STATIC MEASUREMENT

General

Static measurement involves several processes to determine the quantity and quality of a product such as crude oil at rest in an upright cylindrical tank or marine vessel. Those processes are:

- Tank calibration
- Tank gaging
- Temperature determination
- Sampling
- Density determination
- Sediment and water determination
- Calculation of petroleum quantities

Sampling plays a key role in the process of determining both the quantity and quality of a product. Density (American Petroleum Institute [API] gravity or relative density) and sediment and water content are used in custody transfer quantity calculations and are also often used as quality indicators for crude oil. Accurate determination of temperature and density of liquid petroleum products is required for conversion of measured volumes to volumes at standard temperatures of 15°C, 20°C, or 60°F. For this reason, the manual sampling process is as important as the analysis itself because both can affect the quantity and quality determination. An overview of static measurements is provided in the following section to draw attention to the importance of sampling and its affect on the quantity and quality of a custody transfer transaction.

Definitions

Capacity table or tank capacity table–The capacities of, or volumes in a tank for various liquid levels measured from the reference gage point.

Floating roof adjustment–The adjustment made to offset the effect of the displacement of the floating roof.

Gage tape and bob (innage)–Graduated tapes in feet, inches and fractions of an inch, or metric versions of the same with an attached bob that is graduated to have the tip of the bob as the zero point.

Gage tape and bob (outage/ullage)–Graduated tapes in feet, inches and fractions of an inch, or metric versions of the same with an attached bob that is graduated to have the top of the bob as the zero point.

Reference gage height–The vertical distance between the reference point on the gage hatch and the striking point on the tank floor; the distance should be clearly marked on the tank top near the gage hatch.

Reference gage point–A point marked on the gage hatch to indicate the position at which gaging shall be carried out.

Tank increments–Factors used to generate the capacity tables in a condensed form for use in computer systems.

Tank strapping–The term commonly applied to the procedure for measuring tanks to provide the dimensions necessary for computing capacity tables that will reflect the quantity of product in a tank at any given depth or level.

Abbreviations

CSW–Correction for sediment and water.

CTSh–Correction for temperature of the shell: The correction factor for the effect of the temperature, both ambient and liquid, on the shell of the tank.

GOV–Gross observed volume: The total volume of all petroleum liquids and sediment and water, excluding free water, at observed temperature and pressure.

GSV–Gross standard volume: The total volume of all petroleum liquids and sediment and water, excluding free water, corrected by the appropriate volume correction factor (correction for the temperature of liquid [CTL]) for observed tank temperature and density to a standard temperature such as 60°F.

NSV–Net standard volume: The total volume of all petroleum liquids, excluding sediment and water and free water, corrected by the appropriate volume CTL for the observed tank temperature and density to a standard temperature such as 60°F.

TOV–Total observed volume: Total volume of all petroleum liquids, sediment and water, free water, and bottom sediments at observed temperature as taken from the tank capacity table prior to any corrections.

VCF–Volume correction factor: This is the same as CTL. The two symbols are interchangeable. This factor compensates for the effect of temperature on a liquid. It corrects a volume at an observed temperature to a standard temperature.

Calculation of Petroleum Quantities

There is a standard method for calculating static petroleum and petroleum liquid quantities, at atmospheric conditions, in upright cylindrical tanks or marine vessels. The standard method provides a uniform approach to volumetric calculations by specifying equations, calculation sequences, rules for rounding, and discrimination levels to be used with the aim that different operators can arrive at identical results using the same standardized input data. Fundamental to this calculation process is the understanding that in order for different parties to be able to reconcile volumes, they must start with the same basic information listed next:

- Tank capacity tales
- Liquid levels (tank gages)
- Temperatures
- Free water volume (should remain constant for opening and closing)
- Tank shell temperature correction
- Floating roof adjustment (if applicable)
- Sediment and water content

Because of the rigor in the calculation method, it is important to note that, if the input data are not representative of the total volume, the volume calculations will be in error. Most of the input or test data derived for computing a tank or vessel volume begin with the sampling process. This is why the sampling techniques used must be carefully followed to ensure that the most representative sample is used for determination of results in calculation of quantity and quality.

Tank Calibration

Accurate tank measurements are critical in the determination of liquid volume and the development of capacity tables or tank increments used for custody transfer and inventory control measurements. Sound engineering principles are used to obtain data necessary for the calculation of tank capacity tables or increments. The density, the temperature of the tank's contents, and the ambient temperature at the time of strapping or calibration are necessary for computing the tank's capacity. Additional information on tank calibration can be found in the *API Manual of Petroleum Measurement Standards,* chapter 2, "Tank Calibration."

Tank Gaging

Manual gaging is the term used to define the process for determining the liquid level of petroleum or petroleum products in nonpressurized upright cylindrical tanks or marine vessels. Gaging can also be used to determine the level of free water, as a separate phase, that coexists in a tank with the petroleum or petroleum product. Gage readings of petroleum or petroleum products and free water are used with tank capacity tables or increments to determine the total observed volume (TOV) of petroleum or petroleum products contained in the tank. The TOV is then used with various correction factors to calculate the gross standard volume (GSV) and the net standard volume (NSV). There are two basic procedures for obtaining gage readings: innage and outage. For the innage method, the gage reading is defined as the measure of linear distance along a vertical path from the datum plate or tank bottom to the surface of the liquid being gaged. For the outage method, the gage reading is defined as the measure of the linear distance along a vertical path from the surface of the liquid being gaged to the tank reference gage point. The outage gage is an indirect measurement of liquid level in a tank or vessel. Both measurements are taken with a gage line with the appropriate bob attached. All equipment used should be traceable to a National Metrology Institute such as NIST.

Additional information on tank gaging can be found in the *API Manual of Petroleum Measurement Standards,* chapter 3, "Tank Gauging."

Temperature Determination

Temperature has the most significant effect on the accurate determination of liquid quantities when correcting to standard conditions for custody transfer and inventory control purposes. The average temperature of a liquid is required to calculate its volume at a standard temperature. Although taking manual temperature readings from a tank does not require removing a sample, it is a form of sampling in that the temperature is read directly from a device inserted into the tank at different levels or using fixed automatic tank thermometers. Depending on the level in the tank and the size of the tank, several temperatures may be taken at various levels and then averaged to determine the average temperature of the liquid in the vessel. The average temperature along with the product density will allow the volume to be corrected to

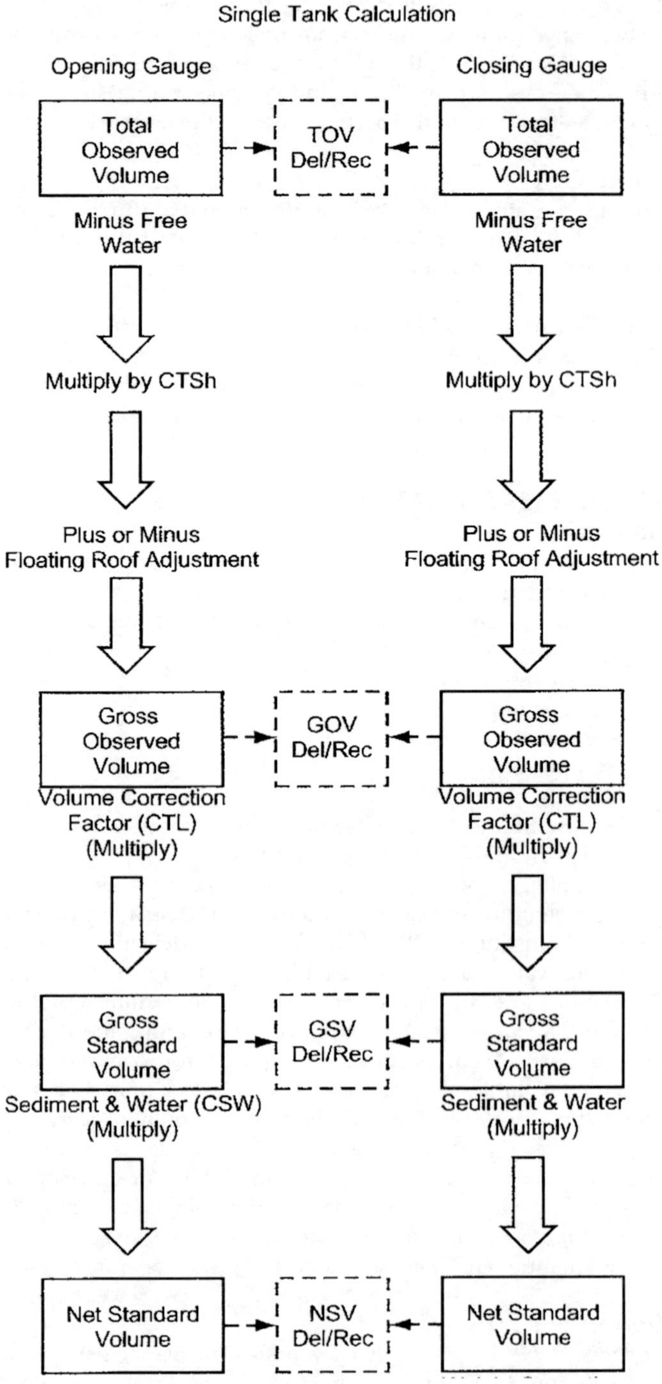

Fig. 10—Typical static measurement volume calculation sequence.

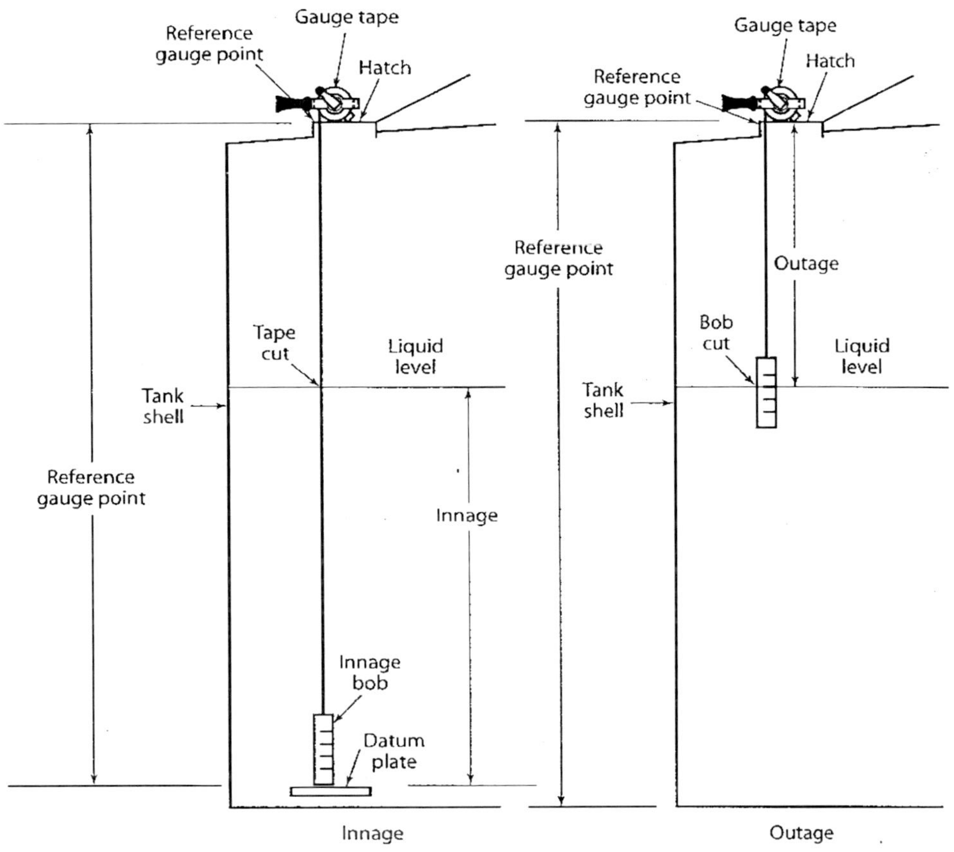

Fig. 11—Gauging diagram for innage and outage methods

what it would be at a standard temperature such as 60°F. Additional information on temperature determination can be found in the *API Manual of Petroleum Measurement Standards*, chapter 7, "Temperature Determination."

Sampling

Automatic sampling is the most preferred method of sampling and provides the most representative sample if it is properly designed, installed, and operated. However, automatic sampling is often not available when trying to determine the chemical and physical properties of a product in a tank or marine vessel, so manual sampling must be used to establish compliance with commercial and regulatory specifications as well as volume determination at standard conditions. As discussed earlier in "Manual Sampling," there are many different procedures and sampling apparatuses that can be used to obtain samples that are deemed to be representative. Care must be taken to ensure that the samples are handled and mixed properly so that they are not altered prior to testing.

Density Determination

Accurate determination of density of crude petroleum or petroleum products is required for the conversion of measured volumes to volumes at a standard temperature such as 60°F. Because a thermohydrometer incorporates both the hydrometer and thermometer in one device, it is more applicable in field operations for determining density of crude petroleum or petroleum products; however, there are several other methods to obtain the density of a product. The density is determined by lowering a thermohydrometer into a hydrometer cylinder or sample thief and allowing the temperature to reach equilibrium. The device is then read by reading its vertical scale nearest to the point of intersection of the horizontal plane surface of the liquid. The observed density reading and temperature reading are then used to determine the density at a standard temperature using Petroleum Measurement Tables. This corrected value will be used in conjunction with an average tank temperature to arrive at a volume correction factor (VCF) that will correct the gross observed volume (GOV) to a GSV.

Sediment and Water Determination

Sediment and water are often found in crude petroleum and sometimes in petroleum products. A determination of the quantity is required to accurately determine the net volumes of crude petroleum or petroleum products. There are many different means of determining the quantity of either sediment or water, but there is only one method that can determine the quantity of both with one test and that is the centrifuge test method. The procedure involves placing known quantities of crude petroleum, solvent, and demulsifier, if required, into a centrifuge tube and heating the mixture to at least 140°F. After centrifuging the samples, the percentage of sediment and water is read at the bottom of the tubes. The samples taken at various locations in a tank or marine vessel are tested for suspended sediment and water that has not separated and fallen to the bottom of the tank. Suspended sediment and water is very hard to determine accurately unless the tank or vessel has internal mixers to uniformly disperse the sediment and water throughout the volume. The percentage of sediment and water read directly from a centrifuge tube or the average of several tubes is deducted directly from the GSV to arrive at the NSV for a custody transfer transaction.

Applicable ASTM/API/ISO Standards

ASTM	API	ISO	
D4057			Standard Practice for Manual Sampling of Petroleum and Petroleum Products
D4177			Standard Practice for Automatic Sampling of Petroleum and Petroleum Products
D5854			Standard Practice for Mixing and Handling Liquid Samples of Petroleum and Petroleum Products
D5842			Standard Practice for Sampling and Handling of Fuels for Volatility Measurement
D4306			Practice for Sampling Aviation Fuel for Tests Affected by Trace Contamination
D6822			Test Method for Density, Relative Density (Specific Gravity), or API Gravity of Crude Petroleum and Liquid Petroleum Products by Thermohydrometer Method
	MPMS 2		Tank Calibration
	MPMS 3		Tank Gauging
	MPMS 7		Temperature Determination
	MPMS 8.1		Standard Practice for Manual Sampling of Petroleum and Petroleum Products
	MPMS 8.2		Standard Practice for Automatic Sampling of Petroleum and Petroleum Products
	MPMS 8.3		Standard Practice for Mixing and Handling Liquid Samples of Petroleum and Petroleum Products
	MPMS 8.4		Standard Practice for Sampling and Handling of Fuels for Volatility Measurement
	MPMS 9		Density Determination
	MPMS 10		Sediment and Water
	MPMS 12		Calculation of Quantities
		ISO 3170	Petroleum Products—Liquid Hydrocarbons—Manual Sampling
		ISO 3171	Petroleum Liquids—Automatic Pipeline Sampling

11

Methods for Assessing Stability and Cleanliness of Liquid Fuels

David R. Forester[1] and Harry N. Giles[2]

STABILITY GENERALLY REFERS TO THE RESIST-ance of a fuel to changes in its chemical quality, while cleanliness is a parameter relating to the physical presence of contaminants: most notably water and sediment, but also degradation products such as gums, asphaltenes, or polymerized products that can form following production of a fuel. There are significant differences in cleanliness and stability between light fuels (distillates) and residual fuels, and very different test methods are required. Also, while these would appear to be unrelated properties, the presence of contaminants can exacerbate reactions resulting in a loss of storage stability. In like manner, products such as gum, asphaltenes, or polymerized products produced when a fuel degrades can adversely affect fuel performance and can contribute to other reactions seriously affecting its usefulness.

Two types of materials can make a fuel less suitable or unsuitable for the intended use. *Fuel contaminants* include materials such as water and rust or other particulate matter introduced subsequent to fuel manufacture, and *fuel degradation products* form in fuel during extended storage as a manifestation of fuel instability. Whether degradation products are soluble or insoluble depends on many factors, which include the chemistry of the degradation products and also the chemistry and solubility of the fuel itself (e.g., aromatics or oxygenate content). In addition, instability of heavy, residual fuels can result from mixing different heavy fuels or diluents (cutter stocks) designed to reduce viscosity or improve handling. Insoluble degradation products can combine with other fuel contaminants to reinforce deleterious effects. Soluble degradation products, such as gums, can form due to complex interactions and oxidation of small amounts of olefins, and sulfur, nitrogen, and oxygen compounds present in some fuels. These soluble gums are less volatile than fuel and can carbonize during combustion. The formation of degradation products can be catalyzed by dissolved metals such as iron, zinc, and copper.

The increasing diversity of products in today's market, ranging from highly refined gasoline, through aviation turbine fuel and ultra-low sulfur diesel fuel, to residual fuel oil as well as newer biofuels, such as ethanol and alkyl ester biodiesel (fatty acid methyl esters [FAME]), demands an equally diverse array of test methods. Consequently, many test methods have been developed over the years, with quite a few new methods having been introduced within the past decade for assessing the stability of petroleum products, biofuels, and blends of the two and determining their cleanliness.

Stability is determined on the basis of thermal and oxidative changes fuels may undergo during use and that affect performance. Stability is also defined in terms of time a product is stored. Operational or bulk storage can be for a period of up to about 12 months, long term is for 12 to 24 months, and strategic storage is generally for periods greater than 24 months. A number of accelerated stability methods are available for assessing the storage potential of conventional fuels, biofuels, and blends of the two. However, there is a lack of good data correlating accelerated stability test results with storage stability or service performance in actual use.

It must be kept in mind that stability properties of distillate products are highly dependent on the crude oil source, severity of refinery processing, and whether additional finishing or treating processes such as caustic washing or Merox treating have been carried out. However, with gasoline now less than 30 ppm sulphur, and most diesel fuel becoming S15 grade, even off-road, both requiring fairly severe processing, the impact of crude oil source is less and less important. Also, the stability properties of comparable fuels produced from the same crude oil source but at different refineries may be different. Moreover, blends of fuels may result in stability properties worse than expected and may create problems depending on the characteristics of the individual fuels. In addition, the introduction of biofuels into some petroleum fuels can significantly change solubility parameters and impact relative instability.

From the standpoint of contaminants, water and sediments are those most frequently found in detailed product requirements and identified in workmanship statements. And in terms of product quality complaints, dirt and water are the most common sources of customer problems. Sediments can accumulate in storage tanks and on filter screens, the latter obstructing flow from the tank to the burner or gas turbine combustor. Particulate contaminants, especially abrasive materials, can also shorten the life of fuel system components (fuel pumps and fuel injectors) through increased wear rates. High levels of particulate matter also lead to short cycle times in the operation of filters, filter separators, centrifuges, and electrostatic purifiers. Some water will always be present in fuels in dissolved form. Excessive contamination by water will exceed the solubility of water in fuel, resulting in free (undissolved) water. This free water can cause corrosion of tanks and equipment, can promote microbiological growth, and can lead to emulsions in heavier grades of fuels. Corrosion products can plug filters and flow

[1] *Director of Technical Services, Power Service Products, Weatherford, TX 76086.*
[2] *PetroStorTech LLC, Arlington, VA 22207.*

lines and lead to increased wear rate in pumps and fuel injectors. Corrosion can also lead to failure of tanks and lines, leading to fuels leaks and losses.

Incompatibility, a form of instability, usually manifests itself when different fuels are mixed, including biofuels. This can lead to deposition of solids, which can result in handling or use problems (e.g., plugged fuel dispensers or vehicle filters).

Microorganisms are another contaminant that can have a profound impact on fuel quality. Microbial action can produce slimes that can plug filters and clog fuel handling systems. Microbes growing in slime films (biofilms) on tank and pipe walls create a physicochemical environment that is conducive to aggressive corrosion. Sulfate reducing bacteria, living under anoxic (oxygen-free) conditions within biofilms, produce toxic, corrosive hydrogen sulfide.

In assessing quality and cleanliness of stored fuels, the importance of proper sampling technique and the appropriate sample containers themselves cannot be overemphasized. Sampling is especially important in assessing quality of stored fuel. Fuel contaminants and insoluble degradation products will usually settle to the bottom of a quiescent tank. Microbial growth can occur at the fuel–water interface or wherever water is found in the tank. Soluble degradation products can be unevenly distributed in the stored fuel and may be relatively concentrated near the air–fuel interface and near the walls of the storage tank. All these factors must be taken into consideration when choosing a sampling technique (e.g., bottom sample, water–interface sample, or all-levels' sample).

Although crude oils have been used as a fuel in some applications, no limits on cleanliness have been generally established in this respect. Sellers, buyers, and shippers of crude oil usually set limits on sediment and water content (typically $<$ 0.5 % sediment and water), but this is to maximize transfer of mercantile commodity. Crude oils generally exhibit excellent storage stability, but commingling of different crude oils can result in incompatibility reactions [1].

This chapter is intended to provide an overview of the test methods available for assessing stability and cleanliness of liquid fuels. Products such as petroleum waxes, gaseous fuels, light hydrocarbons, and lubricating oils are not addressed. More detailed discussion of the significance of these test methods, as they relate to specific products, is found in other chapters of this manual.

WORKMANSHIP AND GENERAL REQUIREMENTS

Many product specifications include a workmanship or general requirements statement limiting the quantity of water and sediment that may be present, without including a maximum quantity in the detailed requirements. These statements usually stipulate that "the finished fuel shall be visually free of water (or undissolved water), sediment, and suspended matter" and shall be "clear and bright." Variations on this include "free from excessive amounts of solid or fibrous foreign matter," "free of solid or fibrous matter that could cause system handling or maintenance problems," and "essentially free of water ... and excessive amounts of particulate contaminants." While the phraseology differs somewhat between specifications, the intent is effectively the same. Requiring that the fuel be "clear and bright" limits the quantity of water and sediment that may be present to very low levels. The trained observer can usually detect as little as 40 ppm sediment and water in a fuel.

Some specifications include not only a workmanship statement but also detailed requirements limiting the quantity of sediment and water present.

While the above examples specifically require that a product be "clear and bright," the specification for Mineral Lubricating Oil Used in Steam and Gas Turbines (ASTM D4304) refers to this in an appendix describing significance of functional properties of turbine oils. The nonmandatory information indicates as follows: "XI. 1.5 *Visual Examination*–in the manufacture, distribution, and use of turbine oils, fresh oils should be examined for appearance and clarity as a check against contamination."

TEST METHODS FOR STABILITY

From the time a fuel is manufactured, it is susceptible to a decrease in its inherent stability. Inevitably, the fuel comes into contact with air, which can lead to the formation of gums and sediments through reactions with unstable compounds such as olefins, and trace amounts of sulfur-, nitrogen-, and oxygen-containing compounds present in some fuels. In service, diesel and aviation turbine fuels can be used as a heat transfer fluid, subjecting them to high temperatures that promote formation of deposits in the fuel system. During long-term storage, prolonged contact with air can result in increased formation of gums and sediments through further oxidation of olefins, and sulfur-, nitrogen-, and oxygen-containing compounds.

OXIDATION STABILITY

"Stability" can be differentiated between thermal or oxidative stability. Different accelerated test methods may tend to evaluate just one mechanism of either thermal or oxidative stability. The storage stability properties of distillate fuels depend on complex oxidative and nonoxidative interactions of olefins, dienes, and nitrogen-, sulfur-, and oxygen-containing compounds present in some fuel. These reactions can also be promoted by other contaminants, such as dissolved metal salts. A number of methods have been developed for determining the oxidation stability of petroleum products, including gasoline, middle distillate fuels such as diesel fuel and aviation turbine fuels, and more recently for biodiesels as well as for miscellaneous products including lubricating greases and steam-turbine oils. Many of these methods can be used to estimate long-term storage stability, while others are most appropriately restricted to assessing service life of the product. Some stability test methods will identify *very good* and *very bad* fuels—but most fuels fall in the "intermediate" range, and the correlation is not always clear. Appropriate methods for measuring the oxidation stability of biodiesel and its correlation to long-term storage are continuing to be evaluated.

ASTM D525/IP 40, Test Method for Oxidation Stability of Gasoline (Induction Period Method), is used for determining the stability of finished gasolines under accelerated oxidation conditions. The induction period is useful as an indication of the resistance of fuel to gum formation in storage.

ASTM D873/IP 138, Test Method for Oxidation Stability of Aviation Fuels (Potential Residue Method), covers the determination of the tendency of aviation reciprocating, turbine, and jet engine fuels to form gum and deposits under accelerated aging conditions. Potential gum is a significant factor affecting cleanliness, handling, and storage stability of these products. The results of this test method can be used to indicate storage stability of the tested fuels. The tendency of these fuels to form gum and deposits has not, however, been correlated with field performance. The formation of gum and deposits can vary markedly under different storage conditions.

ASTM D942/IP 142, Test Method for Oxidation Stability of Lubricating Greases by the Oxygen Bomb Method, is used to determine resistance of lubricating greases to oxidation when they are stored statically in an oxygen atmosphere in a sealed system at an elevated temperature. This test method may be used for quality control purposes to indicate batch-to-batch uniformity. The method does not, however, predict the stability of greases under dynamic service conditions, when stored in containers for long periods, or as films on bearings and motor parts.

ASTM D943, Test Method for Oxidation Characteristics of Inhibited Mineral Oils, is used to evaluate the oxidation stability of inhibited steam-turbine oils in the presence of oxygen, water, and copper and iron metals at an elevated temperature. The method can also be used for testing other products, such as hydraulic oils and circulating oils containing rust and oxidation inhibitors. This test method is used for specification purposes and is considered of value in estimating the oxidation stability of lubricants, especially those that are prone to water contamination. It should be recognized, however, that correlation between results of this method and the oxidation stability of a lubricant in field service may vary markedly with field service conditions and with various lubricants. ASTM D4310, Test Method for Determination of the Sludging and Corrosion Tendencies of Inhibited Mineral Oils, is a modification of D943 and involves gravimetric determination of the mass of insoluble material formed.

ASTM D2272, Test Method for Oxidation Stability of Steam Turbine Oils by Rotating Pressure Vessel, uses an oxygen-pressured vessel to evaluate the oxidation stability of new and in-service turbine oils having the same composition. The estimate of oxidation stability is useful in controlling the continuity of this property for batch acceptance of production lots having the same operation. This test method is also used to assess the remaining oxidation test life of in-service oils. This method is not intended for comparing the service lives of new oils of different compositions. A modification of this test method, which uses the same procedure and apparatus but at a lower bath temperature, has been published as ASTM D2112, Test Method for Oxidation Stability of Inhibited Mineral Insulating Oil by Pressure Vessel.

Oxidation stability of turbine oils (ASTM Specification D4304) may be determined by either Test Method ASTM D943 or ASTM D2272. Test Method ASTM D943 is the accepted method for determining oxidation stability of new steam turbine oils. As this is a lengthy procedure, ASTM D2272, a shorter test, is useful for quality control.

ASTM D2274/IP 388, Test Method for Oxidation Stability of Distillate Fuel Oil (Accelerated Method), covers the measurement of inherent stability of middle distillate petroleum fuel under accelerated oxidizing conditions. This test method provides a basis for the estimation of the storage stability of middle distillate fuels, such as No. 2 fuel oil. It is not applicable to fuels containing residual oil or any significant component derived from a nonpetroleum source. This test method may not provide a prediction of the quantity of insolubles that will form in field storage over any given period of time. The amount of insolubles formed in such field storage is subject to specific conditions, which are too variable for this method to predict accurately. This test method yields results more rapidly than ASTM D4625 Test Method for Distillate Fuel Storage Stability at 43°C. However, as a result of the significantly elevated temperature and the pure oxygen atmosphere, the nature and amount of insolubles can deviate from those formed in field storage to a greater extent than indicated by Test Method ASTM D4625. This method has been found to be unsuitable for evaluation of biodiesel B100 and many biodiesel blends containing more than 5 % biodiesel. A similar method, ASTM D7462-08, Standard Test Method for Oxidation Stability of Biodiesel (B100) and Blends of Biodiesel with Middle Distillate Petroleum Fuel (Accelerated Method), has been developed that is suitable for these fuels. It has been shown to provide results that suggest the potential instability of biodiesels and biodiesel containing blends to degrade and form deposits under accelerated test conditions. However, it has not been shown to correlate with real world storage stability conditions or a variety of field stability issues in tanks or equipment.

ASTM D4636, Test Method for Corrosiveness and Oxidation Stability of Hydraulic Oils, Aircraft Turbine Engine Lubricants, and Other Highly Refined Oils, is used to determine resistance to oxidation and corrosion degradation and tendency to corrode various metals. This test simulates the environment encountered in actual service and uses an accelerated oxidation rate to permit measurable results to be obtained in reasonable time.

ASTM D4742, Test Method for Oxidation Stability of Gasoline Automotive Engine Oils by Thin-Film Oxygen Uptake (TFOUT), evaluates the oxidation stability of engine oils under conditions partially simulating those to which oil may be subjected in a gasoline combustion engine. This test method is intended to be used as a bench screening test and quality control tool for lubricating base oil manufacturing, especially for re-refined lubricating base oils, and is useful for quality control of oxidation stability of re-refined oils from batch to batch. This test method is also useful for screening formulated oils prior to engine tests.

Two new methods have been developed and recently approved by ASTM: D7525, Standard Test Method for Oxidation Stability of Spark Ignition Fuel–Rapid Small Scale Oxidation Test (RSSOT), and D7545, Oxidation Stability of Middle Distillate Fuels–Rapid Small Scale Oxidation Test (RSSOT). ASTM D7525 covers the quantitative determination of the stability of spark ignition fuel, including those containing alcohols or other oxygenates, under accelerated oxidation conditions, as determined by a prescribed drop in oxygen pressure in the small-scale, automated RSSOT. The method has been shown to give similar results to ASTM D525 except in much shorter test times (1 to 2 h) using a much smaller sample size, only 5 mL of fuel.

The second test method ASTM D7545 covers the determination of the stability of middle distillate fuels under accelerated oxidation conditions (as determined by a prescribed drop in oxygen pressure) in the small-scale, automated RSSOT. The apparatus specified in this new test method has been internationally evaluated with diesels, biodiesels (FAME), and blends of middle distillate fuels and biodiesels in the range between B0 and B100 with good precision. It is anticipated that this new test method can be used for the following materials: ASTM D6751 for biodiesel (as a possible alternative method to EN 14 112 because it has a good correlation); ASTM D975 for diesel fuel oils (up to 5 % biodiesel); ASTM D396 for burner fuel oils (up to 5 % biodiesel); and ASTM D7467 for diesel fuel oil (6 to 20 % biodiesel). The method also uses a much smaller sample size, only 5 mL of fuel, in much shorter test times (1 to 2 h) than EN 14 112.

THERMAL STABILITY

For aviation turbine fuels, stability to oxidation and polymerization at the operating temperatures encountered in certain jet aircraft is an important performance requirement. The thermal stresses experienced by aviation fuel in modern jet engines can lead to the formation of undesirable and possibly harmful insoluble materials, such as lacquers on heat exchangers and control surfaces. These reduce efficiency and necessitate extra maintenance. Thermal stability measurements are related to the amount of deposits formed in the engine fuel system on heating the fuel. Also, there is usually a direct relationship between thermal stability and long-term storage stability.

Products other than aviation turbine fuels may have a thermal stability requirement. The most important of these is diesel fuels because, like aviation fuels, they may be used as a heat transfer fluid, which subjects them to high temperatures that promote formation of deposits in the fuel system.

Arguably, more study has been devoted to the thermal stability of middle distillate fuels, such as diesel fuel and aviation turbine fuel, than any other group of petroleum products. This is likely because of the requirement that these fuels function effectively as a heat transfer fluid and not produce degradation products, which could foul injectors and other fuel system components. Bacha and others [2] is an excellent source of information on all aspects of S500 and S5000 diesel fuels, including their refining chemistry, specifications, and additive treatment. However, most diesel fuels used today are much lower sulfur, particularly in North America and Europe with sulfur levels of 10 to 50 ppm.

Originally, thermal stability of aviation and gas turbine fuels was measured by Test Method ASTM D1660, known as the ASTM Coker. This test method has been discontinued and replaced by ASTM D3241/IP 323, Test Method for Thermal Oxidation Stability of Aviation Turbine Fuels (JFTOT Procedure). This latter test method covers a procedure widely used for rating the tendency of gas turbine fuels to deposit decomposition products within the fuel system. The test results are indicative of fuel performance during gas turbine operation and can be used to assess the level of deposits that form when liquid fuel contacts a heated surface.

ASTM D6811, Test Method for Measurement of Thermal Stability of Aviation Turbine Fuels Under Turbulent Flow Conditions (HiReTS Method), has recently been developed. This test method describes a procedure for measuring the tendency of aviation turbine fuels to deposit insoluble materials and decomposition products within a fuel system. The test method provides a quantitative result for fuel under turbulent flow conditions. While most large-scale test rigs operate in the turbulent flow regime, fuel volumes are very large and test times are very long. Test Method ASTM D6811 provides continuous analysis of results and allows performance of the fuel to be monitored in real time, thus enabling the test time to be shortened.

The HiReTS test aims to match more closely real-life operating conditions. Correlation between this test method and ASTM D3241, the JFTOT procedure, is unlikely due to differences between the two methods. The HiReTS test operates in the turbulent flow regime using stainless steel tubes, while the JFTOT operates in the laminar flow regime using aluminum tubes.

ASTM D6468, Test Method for High Temperature Stability of Distillate Fuels, covers relative stability of middle distillate fuels under high temperature aging conditions with limited air exposure, leading to insoluble degradation products. This test method is suitable for all Grades Nos. 1 and 2, in ASTM Specifications D396, D975, D2880, and D3699, and for grades DMX and DMA in ASTM Specification D2069. This test method is based on the widely used Du Pont/Octel F 21 149°C (300°F) Accelerated Stability Test (also known as the "pad test" because it collects insoluble degradation products on a filter pad and evaluates the amount of insolubles based on light reflection; higher levels of insolubles reduce the light reflectance of the white filter pad). It provides an indication of thermal oxidative stability of distillate fuels when heated to high temperatures that simulate those that may occur in some types of recirculating engine or burner fuel delivery systems. Although results have not been substantially correlated to engine or burner operation, the test method can be useful for investigation of operational problems related to fuel thermal stability. When the test method is used to monitor manufacture or storage of fuels, changes in filter rating values can indicate a relative change in inherent stability. Storage stability predictions are more reliable when correlated to longer-term storage tests such as ASTM D4625. Note that ASTM D6468 does not detect soluble degradation products.

ASTM D2070, Test Method for Thermal Stability of Hydraulic Oils, is designed primarily to evaluate the thermal stability of hydrocarbon-based hydraulic oils, although oxidation may occur during the test. No correlation of the test to field service has been made.

Storage stability varies enormously, due to refinery feedstock source (crude oil or otherwise) and the type of processing used to produce components in the finished fuel. Because the chemical reactions leading to formation of sediment (and color changes) vary depending on the type and amount of unstable materials present, the effects of degradation-accelerating conditions will also vary. This was especially true of higher sulfur S500 and S5000 diesel fuels and higher gsulfur-containing gasoline. However, most diesel fuels used today are much lower sulfur, particularly in North America and Europe with sulfur levels of 10 to 50 ppm, and gasoline is typically 20 to 40 ppm sulfur in most developed nations.

Guidance on means of storage of distillate fuels for extended periods to minimize product degradation is provided in Appendix X2 of Specification ASTM D975. The guidance provided includes suggestions on the operation and maintenance of existing fuel storage and handling facilities, and identifying where, when, and how fuel quality should be monitored. This appendix has not been updated to reflect the instability of B5 blends. Conventional, hydrocarbon diesel fuel typically had very good stability (hence no specification requirement for stability), but biodiesel blends, even B1–B5, should be used within 6 months due to serious concerns about poor stability of alkyl ester biodiesel.

ASTM D4418, Practice for Receipt, Storage, and Handling of Fuels for Gas Turbines, provides detailed guidance for the control of substances in a fuel that could cause deterioration of either the fuel system, or the gas turbine, or both. This guidance is generally applicable to extended storage of any distillate fuel.

ASTM D4625/IP 378, Test Method for Distillate Fuel Storage Stability at 43°C (110°F), is a method for evaluating the inherent storage stability of distillate fuels conforming to Specification D396 Grades Nos. 1 and 2, Specification D975 Grades 1-D and 2-D, and Specification D2880 Grades 1-GT and 2-GT. Fuel oxidation and other degradative reactions leading to formation of sediment are mildly accelerated by the test conditions, compared to typical storage conditions. Test results have been shown to predict storage stability more reliably than other more accelerated tests. Because the storage periods are long (4 to 24 weeks), the test method is not suitable for quality control testing. Moreover, because environmental effects and the materials and nature of tank construction affect storage stability, the results obtained by this test are not necessarily the same as those obtained during storage in a specific field storage situation.

ASTM D5304, Test Method for Assessing Distillate Fuel Storage Stability by Oxygen Overpressure, covers a procedure for assessing the potential storage stability of middle distillate fuels such as Specification D975 Grades No. 1-D and 2-D. The test method is applicable to either freshly refined fuels or fuels already in storage. The results of this test method are useful in ranking a specific fuel sample against other specific fuel samples or standards with or without stabilizer additives when tested under identical conditions. This test method is not meant to relate a specific fuel to specific field handling and storage conditions.

ASTM D6748, Test Method for the Determination of Potential Instability of Middle Distillate Fuels Caused by the Presence of Phenalenes and Phenalenones (Rapid Method by Portable Spectrophotometer), has recently been developed. This new method is useful for marine, automotive, heating, and gas turbine fuels, such as those in ASTM Specifications D396, D975, D2069, and D2880. It is applicable to both dyed and undyed fuels at all points in the distribution chain from refinery to end user. It is not applicable to fuels containing residual components. The phenalenes and phenalenones detected by this test method are unstable reactive compounds that can be present in fuels containing catalytically cracked or straight-run materials. These compounds can affect the storage stability of fuels, leading to formation of degradation products, which might later cause performance problems.

In addition to the method just described, ASTM Test Methods D873 and D2274, discussed earlier in the section on "Oxidation Stability," are useful in assessing storage stability of aviation fuels and middle distillate fuels, respectively.

TEST METHODS FOR CLEANLINESS AND CONTAMINATION

Fuel contaminants, in principle, are any fuel components other than hydrocarbons that are not designed to be combusted in engines or equipment. These contaminants make the fuel less suitable or even unsuitable for its intended use. The contaminants of primary interest and concern are foreign materials introduced subsequent to the manufacture of specification quality fuel. They can be introduced in the distribution system (storage tanks, refinery, terminals, pipelines, trucks, barges, and elsewhere), in the user's storage and handling systems, or generated within these systems (for example, rust produced in pipelines and tanks by moisture-laden fuel). Contaminants can be soluble or insoluble in the fuel. They may be grouped into four broad categories: water, particulate solids, fuel degradation products, and microbial slimes.

Water will be present in the fuel in dissolved form and can be present as "free" (undissolved) water. This water can be "fresh" or saline. It can enter fuel from leaking steam coils in storage tanks, from condensation out of moisture-laden air as tanks "breath," or from water bottoms in tankers and barges. Water will cause corrosion of tanks, piping, flow dividers, and pumps. Corrosion or corrosion products can cause plugging and can stop flow to injectors, combustors, and burners. Free water is potentially corrosive and, in sulfur-containing fuels, it can be particularly corrosive. Free water will also encourage microbiological growth and formation of microbial slimes (discussed in more detail later).

Particulate solids can enter a fuel from the air as suspended dirt or aerosols or be picked up from distribution and storage systems. These solids can shorten the life of fuel system components, such as fuel pumps and various close-tolerance devices. Over the past 25 years, fuel pressures in diesel fuel systems have increased tenfold (from 200 bar/atm -2×10^4 kPa to >2,000 ± bar/atm -2×10^5 kPa), with a concurrent decrease in clearances in fuel pumps and injectors, thus requiring much cleaner fuels than previously. High levels of particulate matter can lead to short cycle times in the operation of filters, filter/separators, centrifuges, and electrostatic purifiers. Also, as separation devices do not remove all the particulate matter, certain quantities will be present in the down-stream fuel.

Microbial slimes (biofilms) can result from the growth of microorganisms that are always present in air, fuel, and water, and most fuels are sensitive to microbial contamination, especially in warmer and more humid conditions. The presence of free water is essential to the growth of these microorganisms. Most commonly, they grow at the tank fuel–water interface and feed on the fuel as "food," using the water for trace nutrients and metabolic purposes. Microbial slimes can plug filters and other close-tolerance openings. Some organisms can cause corrosion as well as produce slimes. Under anoxic (oxygen-free) conditions–created within biofilms that form on infrastructure surfaces and at fuel–water interfaces–sulfate-reducing bacteria can produce hydrogen sulfide, which can be very corrosive and is also highly toxic. Sulfate-reducing bacteria and several other taxa of obligate anaerobes use the enzyme *hydrogenase*, which scavenges hydrogen ions (protons) from the cathodic end of electrolytic cells that form on metallic surfaces beneath biofilms. Consequently, hydrogenase activity depassivates and thereby dramatically accelerates galvanic corrosion. More importantly, other facultatively and obligately anaerobic bacteria growing under these anoxic conditions produce weak organic acids. The weak organic acids react readily with chloride, nitrate, nitrite, and sulfate anions to form strong inorganic acids, which attack infrastructure surfaces. Moreover, these acids contribute to the fuel's acid number and increase fuel corrosivity. ASTM Manual 47 and Guide D6469 provide excellent overviews of fuel microbiology and the process of diagnosing fuel systems for microbial contamination. ASTM Practice D7464 addresses the unique considerations for sampling from fuel systems when the samples are for microbiological testing. Microbiological testing by culture methods is explained in ASTM Practice D6974, and the detection of the biological marker-molecule adenosine triphosphate (ATP) is detailed in ASTM Method D7463.

Fuel degradation products are those materials that are formed in fuel during extended storage. Oxidation, polymerization, and other degradation reactions in the fuel can produce relatively significant quantities of particulate solids. Insoluble degradation products can combine with other fuel contaminants to reinforce deleterious effects. Soluble degradation products (soluble gums) are less volatile than fuel and can carbonize during combustion. They usually form in fuels due to complex interactions and oxidation of small amounts of olefins and oxygen-, sulfur-, and nitrogen-containing compounds. Dissolved metals, especially copper, zinc, and iron, can catalyze the formation of degradation products.

Cleanliness of a fuel is generally assessed in terms of its content of water and sediment, which will affect the general workmanship of the fuel for its intended purpose. This may be determined by whether a fuel is "clear and bright," or whether it contains water and sediment as determined by any one of several standard test methods. The color and clarity of a fuel will usually dictate which of the many methods available will be used in assessing these contaminants (Table 1).

Automotive spark-ignition engine fuels (ASTM Specification D4814) include a requirement limiting its solvent-washed gum content (ASTM Test Method D381). Solvent-washed gum consists of both fuel-soluble and fuel-insoluble gums. Both can be deposited on surfaces of carburetors, fuel injectors, and intake manifolds, ports, valves, and valve guides when the fuel evaporates. The fuel-insoluble portion can clog fuel filters. ASTM Specification 4814 also requires that the finished fuel be free of any adulterant or contaminant that may render the fuel unacceptable for its commonly used applications.

For unleaded aviation gasoline (ASTM Specification D6227), existent gum is a significant factor affecting its cleanliness, handling, and storage stability.

The cleanliness of aviation turbine fuels is an essential performance requirement to minimize problems such as

TABLE 1—Summary of Stability and Cleanliness Requirements and Test Methods in Petroleum Product Specifications

Product Specification	Water and Sediment	Workmanship (Clear and Bright)	Thermal Stability	Oxidation Stability	Storage Stability	Cleanliness Contaminants
D396 Fuel Oils	D2709					
	D95 and D473					
D910 Aviation Gasolines[A]		X		D873		D381
D975 Diesel Fuel	D2709	X				
	D1796					
D1655 Aviation Turbine Fuels[B]		X	D3241		X[B]	D381, D1094
D2880 Gas Turbine Fuel Oils[C]	D1796, D2709					X[D]
D3699 Kerosene						X[E]
D4304 Mineral Lube				D943		X[F]
Oil Used in Turbines				D2272		
D4814 Automotive Spark-Ignition Engine Fuel[G]		X		D525		D381
D6227 Grade 82 UL Aviation Gasoline		X		D873		D381
D6448 Industrial Burner Fuels from Used Lube Oils	D95, D1796, D2709, D4377, and D473	X				X[H]
D6751 Biodiesel (B100) Blend Stock	D2709 (D1796)	X		D7427, other methods being evaluated	Methods being evaluated	

[A] Specification includes a requirement for water reaction (D1094).
[B] Specification includes requirements for a number of performance characteristics defined as fuel storage stability and fuel cleanliness and handling.
[C] While not specifying limits on contaminants other than water and sediment, the specification provides a detailed discussion on sources and significance of these in the operation of gas turbines and gas turbine fuel systems.
[D] The fuels shall be free of excessive amounts of solid or fibrous foreign matter likely to make frequent cleaning of suitable strainers necessary.
[E] Contains a general requirement that the product shall be essentially free of water and excessive amounts of particulate contaminants.
[F] Cleanliness may be described by either the microscopic particle count or the electronic particle count methods.
[G] Includes a water tolerance requirement (D6422).
[H] Has a general requirement that the fuel oil shall be free of solid or fibrous matter that could cause system handling or maintenance problems.
Methods in parentheses may be used, but are not the referee method.

wear, corrosion, or plugging of filters or orifices. It can be defined as the relative absence of free water and solid particulate matter.

Gas turbine fuels are required to be free of solid or fibrous foreign matter likely to make frequent cleaning of strainers necessary.

ASTM D6728, Test Method for Determination of Contaminants in Gas Turbine and Diesel Engine Fuel by Rotating Disc Electrode Atomic Emission Spectrometry, provides a rapid on-site means of determining contamination and corrosive elements in gas turbine and diesel engine fuels. Ash-forming substances present in these fuels can lead to high-temperature corrosion, ash deposition, and fuel system fouling. Ash-forming materials can be in a fuel as oil-soluble organometallic compounds, as water-soluble salts, or as solid foreign contamination. Although distillate fuel oils are typically contaminant-free when leaving a refinery, ash-forming materials can be introduced during transportation and storage.

For mineral lubricating oil used in steam or gas turbines, cleanliness of the oil may be described by either microscopic particle count or electronic particle count methods.

Filtration procedures provide a means of evaluating fuel cleanliness. A change in filtration performance after storage, pretreatment, or commingling can be indicative of changes in fuel condition. Causes of poor filterability include fuel degradation products, contaminants picked up during storage or transfer, incompatibility of commingled fuels, or interaction of the fuel with the filter media.

The procedures in ASTM Test Method D4176 provide a rapid but nonquantitative means for detecting free water and particulate contamination by determining whether a distillate fuel is *clear and bright*. Quantitative determination of particulate contaminants such as dirt and rust may be accomplished by filtration using ASTM Test Methods D2276 or D5452. However, gravimetric methods for determining particulate contamination have little relevance to in-service problems. Particle counting is a new test method for fuel cleanliness, and this will likely be the test method used in the future, when correlations of particle counts with operability are determined. Undissolved water in aviation fuels may be quantitatively determined by means of ASTM Test Method D3240.

ASTM D381/IP 131, Test Method for Gum Content in Fuels by Jet Evaporation, provides a means of detecting quality deterioration (formation of gum) or contamination with heavier products, or both. This method covers the determination of the existent gum content of aviation fuels and the gum content of motor gasolines or other volatile distillates in their finished form. High gum content can cause induction-system deposits and sticking of intake valves. In most instances, it can be assumed that low gum will ensure absence of induction-system difficulties. It should be realized, however, that the test method is not correlative to induction-system deposits. When applied to motor gasoline, the primary purpose of the test method is the measurement of the oxidation products formed in the sample prior to or during the comparatively mild conditions of the test procedure. With respect to aviation turbine fuels, large quantities of gum are indicative of contamination of fuel by higher boiling oils or particulate matter and generally reflect poor handling practices in distribution downstream of the refinery.

ASTM D1094, Test Method for Water Reaction of Aviation Fuels, covers the determination of the presence of water-miscible components in aviation gasoline and turbine fuels. When applied to aviation turbine fuels, the technique reveals the presence of relatively large quantities of partially soluble contaminants such as the presence of alcohols in aviation gasoline, but is much less useful for determining surfactants in trace amounts. ASTM D1094 has now been deleted from the ASTM D1655 jet fuel standard. Surfactant contaminants are apt to disarm filter-coalescers quickly and allow free water and particulate matter to pass. A similar version of this test method, ASTM D7451-08a, Standard Test Method for Water Separation Properties of Light and Middle Distillate, and Compression and Spark Ignition Fuels, has been developed. This test method covers the evaluation of the tendency of water and fuels with a final boiling point of less than 390°C, as measured in Test Method D86, to separate cleanly rather than create emulsions when they may contain potential emulsion forming additives or components, or have been additized with potential emulsion-forming additives or components. This test method applies primarily to gasoline, diesel, kerosene, and distillate grades of gas turbine, marine, home heating oils, and furnace fuels. It is common for gasoline to contain alcohol as an oxygen-enriching agents for combustion improvement. When testing gasoline known to contain alcohols such as ethanol, for instance, one can expect to see the amount of recovered water increase due to the extraction of such species from the fuel by the aqueous phase. Since the amount of ethanol in the gasoline is usually known, it is possible to anticipate the volume increase to expect assuming that the aqueous phase will extract nearly all of the ethanol. This increase will not change how the fuel and interface are rated. The presence of biodiesel in distillate fuel can cause water to become emulsified in the fuel, resulting in haze in much the same manner as detergents and other surfactant additives. One can anticipate that as the level of biodiesel in the distillate fuel is increased from B5 (5 % biodiesel in petroleum distillate fuel) to B20 (20 % biodiesel in petroleum distillate fuel), the severity of the emulsion formed at the interface will increase and, in some cases, can become quite stable. In any case, the water separation performance of a biodiesel tested under this method would be rated no differently than a traditional petroleum distillate fuel containing typical additives.

ASTM D2068, Test Method for Filter Plugging Tendency of Distillate Fuel Oils, describes a procedure for determining the filter-plugging tendency (FPT) of distillate fuel oils, where the end use demands an exceptional degree of cleanliness. This method is applicable to Specification D396 Grades Nos. 1 and 2, Specification D975 all grades except Grade 4, Specification D2880 Grades 1-GT and 2-GT, and Specification D3699 kerosene. This test method is intended for use in evaluating distillate fuel cleanliness in those applications that demand a high throughput per installed filter. Causes of poor filterability include fuel degradation products, contaminants picked up during storage or transfer, or interaction of the fuel with the filter media. This test method has been recently revised to allow its use to evaluate biodiesel and biodiesel containing blends.

ASTM D3240, Test Method for Undissolved Water in Aviation Turbine Fuels, covers the measurement of undissolved (free) water in aviation turbine fuels in flowing fuel systems without exposing the fuel sample to the atmosphere or to a sample container. The use of sample containers for the temporary storage of samples will result in large errors in the determination. Undissolved water in aviation fuel can encourage the growth of microorganisms and subsequent corrosion in the tanks of aircraft. It can also lead to icing of filters in the fuel system.

ASTM D3948, Test Method for Determining Water Separation Characteristics of Aviation Turbine Fuels by Portable Separometer, commonly referred to as MSEP, provides a rapid means for rating the ability of aviation turbine fuels to release entrained or emulsified water when passed through fiberglass coalescing material. This test method provides a measure of the presence of strong surfactants in these fuels and can also detect traces of treating residues carried over from the refinery. Surfactants, which may be contaminants or deliberately added materials, affect the ability of filter separators to remove free water and particulate matter from the fuel. This test method is commonly used in the field to detect the presence of strong surfactants, which would disarm filter-coalescers and allow water to pass if the fuel became wet. Thus, MSEP is used to determine if the fuel needs to be clay treated to remove strong surfactants. ASTM D7224, Determining Water Separation Characteristics of Kerosine-Type Aviation Turbine Fuels Containing Additives by Portable Separometer, is the updated jet fuel MSEP with current filter media, and ASTM D7261, Standard Test Method for Determining Water Separation Characteristics of Diesel Fuels by Portable Separometer, is a very similar method more specific to diesel fuels.

ASTM D6426, Test Method for Determining Filterability of Distillate Fuel Oils, covers all Grades 1 and 2 fuels in Specifications D396, D975, and D2880. This test method is intended for use in the laboratory or field in evaluating distillate fuel cleanliness. This test method is similar to D2068; however, the instrumentation used represents a significant improvement over the other method in terms of ruggedness and reliability. These improvements make the instrument sufficiently rugged, so that it can be used in the field, and eliminate the need for a calibration fluid inherent in D2068.

ASTM D6824, Test Method for Determining Filterability of Aviation Turbine Fuel, based largely on D6426, has recently been developed. This test method is applicable to fuels falling within Specification D1655 and to military grades JP4, JP5, JP7, and JP8. ASTM D7261, Standard Test Method for Determining Water Separation Characteristics of Diesel Fuels by Portable Separometer, is similar to ASTM D3948 and ASTM D7224 but intended for use with nonaviation middle distillates. It uses a special filter (labeled "D") appropriate for diesel fuels. Historically, diesel fuels have contained few if any additives and have allowed very good separation of water, if the fuels became wet in distribution or use. However, with the introduction of ultra-low sulfur diesel (ULSD) fuel, more additives are being used (e.g., conductivity and lubricity), which can interfere with the separation of water should the fuel become wet.

ASTM D2276/IP 216, Test Method for Particulate Contaminant in Aviation Fuel by Line Sampling, covers the determination of particulate contaminant in aviation turbine fuel using a field monitor (Millipore filter). This test method provides a gravimetric measurement of the particulate matter present in a sample of aviation turbine fuel by line sampling using a field monitor. The objective is to minimize these contaminants to avoid filter plugging and other operational problems. Two methods are described in the test. The basic test method is used to evaluate the level of contamination gravimetrically. The second test method describes a color rating technique that is used for rapid qualitative assessment of changes in contamination level without the time delay required for the gravimetric determinations by stringent laboratory procedures. Membrane color is a simple means of detecting changes in the fuel and may be used for qualitative assessment of contaminant level in fuels or of changes in other visual characteristics. Changes in membrane color may be indicative of changes in fuel contaminant level, contaminant type, the fuel handling system, or refinery process conditions. No quantitative relationship exists between gravimetric and color rating test results.

ASTM D4176, Test Method for Free Water and Particulate Contamination in Distillate Fuels (Visual Inspection Procedures), covers two procedures for estimating the presence of suspended free water and solid particulate contamination in distillate fuels. Both procedures are intended for use as field tests at storage temperatures. One procedure provides a rapid pass/fail method for contamination, while the second provides a gross numerical rating of haze appearance. Many fuel specifications include a requirement that the fuel be *clear and bright* and *free of visible particulate matter.* This test method provides procedures for making that determination. Other Test Methods, including ASTM D1796, ASTM D2276, ASTM D2709, and ASTM D4860, permit quantitative determination of contaminant levels.

A new Test Method for Free Water, Particulate and Other Contamination in Aviation Fuels (Visual Inspection Procedures), based in part on ASTM D4176, is currently under development. This test method covers two procedures for establishing the presence of suspended free water, solid particulate, and other contaminants in aviation gasoline and aviation turbine fuels. Both procedures are intended primarily for use as field tests at ambient temperatures and provide rapid methods for detection of contamination and include numerical ratings of haze appearance and particulate presence.

ASTM D4860, Test Method for Free Water and Particulate Contamination in Mid-Distillate Fuels (Clear and Bright Numerical Rating), provides a rapid, portable means for field and laboratory use to visually inspect for particulate matter and free water, similar to ASTM Test Method D4176, and numerically rate free water in aviation turbine and distillate fuels. Other test methods, such as ASTM D2276 and ASTM D2709, can be used for quantitative determination of contaminant levels.

ASTM D5452/IP 423, Test Method for Particulate Contamination in Aviation Fuels by Laboratory Filtration, covers the gravimetric determination by filtration of particulate contaminants in aviation turbine fuel. The measurement of total contaminants in a fuel is normally of interest, even though tolerable levels of particulate matter have not yet been established for all points in fuel distribution systems. This test method also provides a practice that describes how to determine the filter membrane color rating of a sample of aviation fuel delivered to a laboratory. The filter membrane color rating may be used for the qualitative assessment of contaminant levels in the fuel. No quantitative relationship exists between gravimetric results and membrane color rating. Nevertheless, the filter membrane color rating provides a simple means of detecting changes in the fuel. Changes in membrane color may be indicative of changes in fuel contaminant level, contaminant type, the fuel handling system, or refinery process conditions. As such, this practice provides a rapid means of screening fuels.

ASTM D6217/IP 415, Test Method for Particulate Contamination in Middle Distillate Fuels by Laboratory Filtration, covers the determination of the mass of particulate contamination in a middle distillate fuel. This test method is suitable for all Grades Nos. 1 and 2 fuel in Specifications D396, D975, D2880, and D3699, and for Grades DMA and DMB in the international marine fuel oil specification ISO 8217. This test method can be used in specifications and purchase documents as a means of controlling particulate contamination levels in fuels purchased.

D6217 is the first ASTM test method developed for assessing the mass quantity of particulate matter in middle distillate fuels. Test Methods ASTM D5452 and ASTM D2276, which were developed for aviation fuels, employ gravimetric procedures for determining the level of particulate contamination, but require considerably greater quantities of sample. To assess middle distillate fuels, which can contain greater particulate levels than aviation fuels, using these latter methods would likely require excessive time to complete the filtration using the larger volumes. This test method was found unsuitable for use with some high cloud (FAME) biodiesels, so ASTM D7321-08, Standard Test Method for Test Method for Particulate Contamination of Biodiesel B100 Blend Stock Biodiesel Esters and Biodiesel Blends by Laboratory Filtration, was developed.

A new test method, Counting and Sizing Particles in Light and Middle Distillate Fuels, and Liquid Bio Fuels, by Automatic Particle Counter, is being developed to assist with rapid measurement of fuel cleanliness. This test method uses an automatic particle counter (APC) to count and measure the size of dispersed dirt particles, water droplets, and other particulates in light and middle distillate fuel and biofuels such as FAME biodiesel and biodiesel blends, in the overall range of 4 $\mu m_{(c)}$ to >30 $\mu m_{(c)}$ and in the size bands ≥ 4 $\mu m_{(c)}$, ≥ 6 $\mu m_{(c)}$, and ≥ 14 $\mu m_{(c)}$. ASTM specification fuels falling within the scope of this test method include ASTM Specifications D975, D3699, D4814, D6751, and D6985 and distillate grades of D396 and D2880. For the purposes of this method, water droplets are considered as particles, and agglomerated particles are detected and reported as a single larger particle. Dirt includes biological particulates. Although the projected area of a particle is measured, this is expressed as the diameter of a sphere for the purposes of this method.

ASTM D7501, Determination of Fuel Filter Blocking Potential of Biodiesel (B100) Blend Stock by Cold Soak Laboratory Filtration, has been developed to cover the determination by filtration time after cold soak of the suitability for a biodiesel (B100) blend stock that meets all other requirements of ASTM D6751 and has cloud point below 20°C to provide adequate low temperature operability performance to at least the cloud point of the finished blend. Some substances that are soluble or appear to be soluble in biodiesel (B100) at room temperature will, upon cooling to temperatures above the cloud point or standing at room temperature for extended periods of time, come out of the solution. This phenomenon has been observed in both B100 and Bxx blends. These substances can cause filter plugging. This method provides an accelerated means of assessing the presence of these substances in B100 and their propensity to plug filters. B100 biodiesel fuels that give short filtration times are expected to give satisfactory operation of Bxx blends *at least* down to the cloud point of the biodiesel blends. The test method can be used in specifications as a means of controlling levels of minor filter plugging components in biodiesel and biodiesel blends.

In addition to the test methods described above, ASTM Test Method D873, discussed earlier in the section on "Oxidation Stability," is useful in assessing fuel cleanliness.

MICROBIAL CONTAMINATION

Until relatively recently, microbial contamination in fuels and fuel systems was not widely recognized or understood. Yet, this means of contamination has been responsible for innumerable incidents of fuel quality degradation, some resulting in tragic operational consequences. Uncontrolled microbial contamination in fuels and fuel systems remains a largely unrecognized but costly problem at all stages of the petroleum industry from crude oil production, through fleet operations, to consumer end use. Microbial contamination in fuel systems can cause or contribute to a variety of problems, including corrosion, odor, filter plugging, decreased stability, and deterioration of fuel/water separation characteristics. In addition to system component damage, off-specification fuel can result.

Microbes can be introduced into fuels as products cool in refinery tanks. Bacteria and fungi are carried along with dust particles and water droplets through tank vents. In seawater-ballasted tanks on marine vessels, whether used for product transportation or propulsion, microbes are carried with the ballast.

After arriving in fuel tanks, microbes can either stick to overhead surfaces or settle through the product. Some microbes will adhere to tank walls, whereas others will settle to the fuel–water interface where most growth and activity takes place. Typically, fuel and system deterioration is caused by the net activity of complex microbial communities living within slimy layers called *biofilms*. These biofilms have been found on tank roofs, shells, at the fuel–water interface, and within bottom sludge or sediment.

ASTM D6469, Guide for Microbial Contamination in Fuels and Fuel Systems, provides personnel who have a limited microbiological background with an understanding of the symptoms, occurrence, and consequences of chronic microbial contamination and focuses on microbial contamination in refined petroleum products and product handling systems. The guide also suggests means for detection and control of microbial contamination in fuels and fuel systems. This guide applies primarily to gasoline, aviation, boiler, industrial gas turbine, diesel, marine, and furnace fuels. Even so, the principles discussed therein also apply generally to crude oil and all liquid petroleum fuels.

This guide provides information addressing the conditions that lead to fuel microbial contamination and biodegradation and the general characteristics of and strategies for controlling microbial contamination. It complements information provided in ASTM Practice D4418 and D2880 on handling gas-turbine fuels.

Additional methods developed to assist with microbial contamination determination include ASTM D6974-04, Standard Practice for Enumeration of Viable Bacteria and Fungi in Liquid Fuels–Filtration and Culture Procedures; ASTM D7463-08, Standard Test Method for Adenosine Triphosphate (ATP) Content of Microorganisms in Fuel, Fuel/Water Mixtures and Fuel Associated Water; and ASTM D7464-08, Standard Practice for Manual Sampling of Liquid Fuels, Associated Materials and Fuel System Components for

Microbiological Testing. Other methods are in process of development. ASTM Manual 47 *Fuel and Fuel System Microbiology* includes chapters on fuel microbiology fundamentals, sampling and treatment of microbial contaminated fuel systems. Manual 47 also includes a compilation of ASTM guides, practices, and methods that can be used to facilitate the microbial contamination detection and biodeterioration diagnosis.

WATER AND SEDIMENT

Knowledge of the water and sediment content of crude oils and petroleum products is important in the refining, purchase, sale, and transfer of these commodities. Moreover, the water and sediment content must be known to measure accurately net volumes of actual fuel oil.

Appreciable amounts of water and sediment in a fuel oil tend to cause fouling of fuel handling facilities and trouble in burner mechanisms. Sediment may accumulate in storage tanks and on filter screens or burner parts, resulting in obstruction to flow of oil from the tank to the burner or combustor. Water in distillate fuel can cause corrosion of tanks and equipment. It can impart a hazy appearance to fuels and may contribute to formation of emulsions, especially with residual fuels. The presence of water in a burner fuel can also cause sputtering in a burner flame and lead to damage of burner nozzles (erosion or explosive damage) due to the rapid expansion of water vapor at a hot nozzle tip. Excessive water in burner fuel could lead to *flame out* or extinguishing the flame. The presence of water in lubricating oils can lead to premature corrosion and wear, an increase in the debris load resulting in diminished lubrication, premature plugging of filters, and an impedance in the effect of additives. The presence of water in these products can also support deleterious microbial growth.

Insoluble contaminants comprising sediment can include metallic and nonmetallic materials that can cause abrasive wear of bearings, pumps, and seals; faulty control functioning; plugged oil lines; and reduced filter life. The nature of these insoluble materials may be evaluated by different techniques, such as microscopic particle analysis and counting or electronic particle counting. McCrone et al [3] is an excellent reference for use in microscopic particle analysis. Development of electronic particle counters and relevant test methods, as discussed previously, has greatly facilitated fuel evaluation.

The methods specified for determination of water and sediment varies according to the nature of the fuel. The transparency or opacity of the product will generally govern which of a number of different test methods are appropriate. In lighter products, such as diesel and heating oils, ASTM D2709 is specified. For heavier and darker products, ASTM D1796 is specified.

For marine fuels and other products, where water and sediment are to be determined separately, ASTM D95 is specified for water and ASTM D473 is specified for sediment. Similarly, industrial burner fuels made in whole or in part with hydrocarbon-based or used or reprocessed lubricating fluids specify water and sediment are to be determined by ASTM D95 and ASTM D473, respectively.

ASTM D6751 specification for biodiesel blend stock (B100), a more recent addition to the distillate fuel oil pool, allows for the use of either D1796 or D2709 for determination of water and sediment, but stipulates that D2709 shall be the referee method. ASTM D6304 Karl Fischer Method for Total Water and D7321 are under discussion as D2709 is known to be indiscriminatory for biodiesel and biodiesel blends as a measure of water and sediment contamination, but representative sampling is critical. D6304 measures total, dissolved, and free water. It is critical that in any test measuring free water that a truly representative sample must be obtained, and this can be difficult.

Water can also support microbial growth and fuel degradation. ASTM D6469, discussed earlier in the section on "Microbial Contamination," provides detailed information on the conditions that lead to fuel microbial contamination and biodegradation, and the general characteristics of and strategies for its control.

ASTM D95/IP 74, Test Method for Water in Petroleum Products and Bituminous Materials by Distillation, is applicable to all liquid petroleum products, tars, and other bituminous materials.

ASTM D473/IP 53, Sediment in Crude Oils and Fuels Oils by the Extraction Method, is applicable to crude oils and fuel oils. These two test methods are applicable to all petroleum products and crude oils, regardless of their opacity, and are usually specified as the referee methods for determination of water and sediment.

ASTM D1796, Test Method for Water and Sediment in Fuel Oils by the Centrifuge Method (Laboratory Procedure), covers the laboratory determination of the water and sediment content of fuel oils using a centrifuge. With some types of fuel oils, such as residual fuel oils or distillate fuel oils containing residual components, it is difficult to obtain water or sediment contents with this test method. In these cases, ASTM Test Method D95 or D473 should be used.

ASTM D2709, Test Method for Water and Sediment in Middle Distillate Fuels by Centrifuge, covers the determination of the volume of free water and sediment in middle distillate fuels, corresponding to Specification D975 Grades 1D and 2D, Specification D2880 Grades 0-GT, 1-GT, and 2-GT, and Specification D3699 Grades 1-K and 2-K. Test Method D1796 is intended for higher viscosity fuels.

ASTM D4377/IP 356, Test Method for Water in Crude Oils by Potentiometric Karl Fischer Titration, and ASTM D4928/IP, 386 Test Methods for Water in Crude Oils by Coulo metric Karl Fischer Titration, are specific to crude oils. The latter is the more widely used of the two.

ASTM D4870/IP 375, Test Method for Determination of Total Sediment in Residual Fuels, covers the determination of total sediment in distillate fuel oils containing residual components and in residual fuel oils. This test method can be used for the assessment of total sediment after regimes of fuel pretreatment designed to accelerate the aging process.

ASTM D6304, Test Method for Determination of Water in Petroleum Products, Lubricating Oils, and Additives by Coulometric Karl Fischer Titration, covers the direct determination of entrained water using automated instrumentation.

TEST METHODS FOR COMPATIBILITY OF HEAVY AND RESIDUAL FUELS

Blends of fuels, particularly heavy and residual fuels, from various sources can interact to give stability properties worse than expected, based on the characteristics of the individual fuels. The stability properties of fuels are highly dependent on the crude oil sources, severity of processing, and other refinery treatment. These result partially from the fact that the composition and stability properties of heavy fuels produced at specific refineries may be different. If fuels having different fundamental compositions–paraffinic, aromatic, and naphthenic–are blended, incompatibility reactions can be expected. This can be a precipitation of asphaltenes or flocculation of waxes. Both reactions can lead to problems in handling and end use of the fuel, including blockage of fuel transfer lines, accumulation of tank bottom deposits, and burner fouling or flame out. Hence, expertise is required to produce a good fuel.

ASTM D4740, Test Method for Cleanliness and Compatibility of Residual Fuels by Spot Test, describes two procedures for determining the cleanliness of residual fuel oil and the compatibility of a residual fuel oil with a blend stock. These procedures may be used alone or in combination to identify fuels or blends that could result in excessive centrifuge loading, strainer plugging, tank sludge formation, or similar operating problems. Although a fuel may test clean when subjected to the cleanliness procedure, suspended solids can precipitate when the fuel is mixed with a blend stock. Hence, it is recommended to check blending in a lab on a small scale first.

ASTM D6560, Test Method for Determination of Asphaltenes (Heptane Insolubles) in Crude Petroleum and Petroleum Products, is applicable to gas oil, diesel fuel, residual fuel oils, lubricating oil, bitumen, and crude petroleum. Asphaltenes in these products may give problems during storage and handling if the suspension of these molecules is disturbed through excess stress or incompatibility.

Four other "instrument specific" methods to measure storage and potential instability of heavy oils include ASTM D7060, Standard Test Method for Determination of the Maximum Flocculation Ratio and Peptizing Power in Residual and Heavy Fuel Oils (Optical Detection Method); ASTM D7061, Standard Test Method for Measuring n-Heptane Induced Phase Separation of Asphaltene-Containing Heavy Fuel Oils as Separability Number by an Optical Scanning Device; ASTM D7112, Standard Test Method for Determining Stability and Compatibility of Heavy Fuel Oils and Crude Oils by Heavy Fuel Oil Stability Analyzer (Optical Detection); and ASTM D7157, Standard Test Method for Determination of Intrinsic Stability of Asphaltene-Containing Residues, Heavy Fuel Oils, and Crude Oils (*n*-Heptane Phase Separation; Optical Detection). More recently, new precision statements for these methods have been developed during an interlaboratory study. The general description of these methods is as follows: Three test specimens are dissolved in three different quantities of very light aromatic solvent (VLAS) (toluene). The three specimen/VLAS solutions are automatically and simultaneously titrated with light paraffinic solvent LPS (*n*-heptane) to *cause* precipitation of the asphaltenes. The optical probe monitors the formation of flocculated asphaltenes during the titration. Flocculated asphaltenes will alter the detected light intensity. Start of flocculation is interpreted when the optical probe detects a significant and sustained decrease in rate-of-change of the light intensity. A computer routine calculates stability parameters and subsequently the intrinsic stability of the oil from the added LPS at the inversion point, the mass of specimen, and the volume of VLAS, for the three specimen/VLAS solutions.

MISCELLANEOUS PROPERTIES RELATED TO STABILITY AND CLEANLINESS

Water Reaction Interface Rating

This may be applied to aviation or motor gasoline fuels to reveal the presence of relatively large quantities of partially soluble contaminants, such as alcohols. Contaminants such as surfactants that affect the interface are apt to disarm filter separators quickly and allow free water and particulate matter to pass. The interface rating may be determined using ASTM D1094, Test Method for Water Reaction of Aviation Fuels. ASTM D7451-08a, Standard Test Method for Water Separation Properties of Light and Middle Distillate, and Compression and Spark Ignition Fuels, may also be used.

To ensure that low levels of polar compounds are present, ASTM D1401, Test Method for Water Separability of Petroleum Oils and Synthetic Fluids, may be used. This test method provides a guide for determining the water separation characteristics of oils subject to water contamination and covers measurement of the ability of oils to separate from water. Although specifically developed for steam-turbine oils, the test method can be used to test oils of other types.

Color

Color as such is not normally a significant indicator of fuel quality. However, a *change* in fuel color can be very important. Darkening of fuel or a change in fuel color can be a useful indicator of the overall purity of products such as kerosene and diesel fuel and, as such, is a useful parameter to ensure freedom from trace contamination with heavier products, which may render the products unsuitable for designated applications.

Visible Lead Precipitate (ASTM D910 only)

The formation of a lead precipitate during the aging period of the potential gum test (ASTM D873) indicates a potential instability. Small amounts of insoluble material may foul the induction system and plug filters. Consequently, a limit on the amount of precipitate formed in this determination is included in specifications such as ASTM D910.

SAMPLING

The importance of adhering to a rigorous sampling protocol to ensure that samples are representative of the bulk product, or appropriate to the specific requirement, cannot be overemphasized. Moreover, it is critically important that the sampling procedure does not introduce any contaminant into the sample or otherwise alter the sample so that subsequent test results are affected.

A number of aviation gasoline and jet fuel properties are very sensitive to trace contamination, which can originate from sample containers. Because of the importance of proper sampling procedures in establishing fuel quality, it is imperative that appropriate procedures be used.

ASTM D4057, Practice for Manual Sampling of Petroleum and Petroleum Products, provides procedures for manually obtaining samples from tanks, pipelines, drums, barrels, and other containers. It addresses, in detail, the various factors that need to be considered in obtaining a representative sample. It must be kept in mind that, in many liquid manual sampling applications, the material to be sampled contains a heavy component, such as free water, which tends to separate from the main component. Unless certain conditions can be met to allow for this, an automatic sampling system, as described in ASTM D4177, Practice for Automatic Sampling of Petroleum and Petroleum Products, is recommended.

ASTM D4306, Practice for Aviation Fuel Sample Containers for Tests Affected by Trace Contamination, describes the types and preparation of containers found most suitable for the handling of aviation fuel samples for the determination of critical properties. A number of aviation fuel properties are established or affected by trace levels of polar or other compounds. Measurement significance, therefore, requires that the sample containers not add or adsorb any materials. Several properties, including particulate contamination, free water content, and the presence of corrosive species (e.g., hydrogen sulfide or elemental sulfur), involve materials easily removed by any sampling container.

In ASTM D4418, Practice for Receipt, Storage, and Handling of Fuels for Gas Turbines, two of the annexes provide guidance on where and when samples should be collected in distillate and residual turbine fuel storage, treatment, and handling systems. Sampling and inspection are important steps in identifying contaminated, improperly inhibited, or otherwise unsatisfactory fuel before it reaches the combustor(s).

ASTM D7464, Standard Practice for Manual Sampling of Liquid Fuels, Associated Materials and Fuel System Components for Microbiological Testing, is also applicable for collecting samples intended to determine microbial contamination.

Applicable ASTM/IP Standards

Stability Methods

ASTM	IP	Title
D525	40	Oxidation Stability of Gasoline (Induction Period Method)
D873	138	Oxidation Stability of Aviation Fuels (Potential Residue Method)
D942	142	Oxidation Stability of Lubricating Greases by the Oxygen Bomb Method
D943		Oxidation Characteristics of Inhibited Mineral Oils
D2070		Thermal Stability of Hydraulic Oils
D2112		Oxidation Stability of Inhibited Mineral Insulating Oil by Pressure Vessel
D2272		Oxidation Stability of Steam Turbine Oils by Rotating Pressure Vessel
D2274	388	Oxidation Stability of Distillate Fuel Oil (Accelerated Method)
D3241	323	Thermal Oxidation Stability of Aviation Turbine Fuels (JFTOT Procedure)
D4310		Determination of the Sludging and Corrosion Tendencies of Inhibited Mineral Oils
D4625	378	Distillate Fuel Storage Stability at 43°C (110°F)
D4636		Corrosiveness and Oxidation Stability of Hydraulic Oils, Aircraft Turbine Engine Lubricants, and Other Highly Refined Oils
D4742		Oxidation Stability of Gasoline Automotive Engine Oils by Thin-Film Oxygen Uptake (TFOUT)
D5304		Assessing Distillate Fuel Storage Stability by Oxygen Overpressure
D6468		High Temperature Stability of Distillate Fuels
D6748		Determination of Potential Instability of Middle Distillate Fuels Caused by the Presence of Phenalenes and Phenalenones (Rapid Method by Portable Spectrophotometer)
D6811		Measurement of Thermal Stability of Aviation Turbine Fuels Under Turbulent Flow Conditions (HiReTS Method)
D7462		Oxidation Stability of Biodiesel (B100) and Blends of Biodiesel with Middle Distillate Petroleum Fuel (Accelerated Method)
D7525		Oxidation Stability of Spark Ignition Fuel—Rapid Small Scale Oxidation Test (RSSOT)
D7545		Oxidation Stability of Middle Distillate Fuels—Rapid Small Scale Oxidation Test (RSSOT)

Cleanliness and Contamination Methods

ASTM	IP	Title
D381	131	Gum Content in Fuels by Jet Evaporation
D1094		Water Reaction of Aviation Fuels
D2068		Filter Plugging Tendency of Distillate Fuel Oils
D2276	216	Particulate Contaminant in Aviation Fuel by Line Sampling
D3240		Undissolved Water in Aviation Turbine Fuels
D3948		Determining Water Separation Characteristics of Aviation Turbine Fuels by Portable Separometer
D4176		Free Water and Particulate Contamination in Distillate Fuels (Visual Inspection Procedures)
D4740		Cleanliness and Compatibility of Residual Fuels by Spot Test
D4860		Free Water and Particulate Contamination in Mid-Distillate Fuels (Clear and Bright Numerical Rating)
D5452	423	Particulate Contamination in Aviation Fuels by Laboratory Filtration
D6217	415	Particulate Contamination in Middle Distillate Fuels by Laboratory Filtration
D6426		Determining Filterability of Distillate Fuel Oils
D6469		Microbial Contamination in Fuels and Fuel Systems

ASTM	IP	Title
D6728		Determination of Contaminants in Gas Turbine and Diesel Engine Fuel by Rotating Disc Electrode Atomic Emission Spectrometry
D6824	131	Determining Filterability of Aviation Turbine Fuel
D6974		Standard Practice for Enumeration of Viable Bacteria and Fungi in Liquid Fuels—Filtration and Culture Procedures
D7224		Determining Water Separation Characteristics of Kerosine-Type Aviation Turbine Fuels Containing Additives by Portable Separometer
D7261		Determining Water Separation Characteristics of Diesel Fuels by Portable Separometer
D7321		Particulate Contamination of Biodiesel B100 Blend Stock Biodiesel Esters and Biodiesel Blends by Laboratory Filtration
D7451		Water Separation Properties of Light and Middle Distillate, and Compression and Spark Ignition Fuels
D7463		Adenosine Triphosphate (ATP) Content of Microorganisms in Fuel, Fuel/Water Mixtures and Fuel Associated Water
D7501		Determination of Fuel Filter Blocking Potential of Biodiesel (B100) Blend Stock by Cold Soak Laboratory Filtration
New method		Test Method for Counting and Sizing Particles in Light and Middle Distillate Fuels, and Liquid Bio Fuels, by Automatic Particle Counter

Water and Sediment Methods

ASTM	IP	Title
D95	74	Water in Petroleum Products and Bituminous Materials by Distillation
D473	53	Sediment in Crude Oils and Fuel Oils by the Extraction Method
D1796		Water and Sediment in Fuel Oils by the Centrifuge Method (Laboratory Procedure)
D2709		Water and Sediment in Middle Distillate Fuels by Centrifuge
D4377	356	Water in Crude Oils by Potentiometric Karl Fischer Titration
D4870	375	Determination of Total Sediment in Residual Fuels
D4928	386	Water in Crude Oils by Coulometric Karl Fischer Titration
D6304		Determination of Water in Petroleum Products, Lubricating Oils, and Additives by Coulometric Karl Fischer Titration

Compatibility Methods

ASTM	IP	Title
D4740		Cleanliness and Compatibility of Residual Fuels by Spot Test
D6560	143	Determination of Asphaltenes (Heptane Insolubles) in Crude Petroleum and Petroleum Products
D7060		Determination of the Maximum Flocculation Ratio and Peptizing Power in Residual and Heavy Fuel Oils (Optical Detection Method)
D7061		Measuring n-Heptane Induced Phase Separation of Asphaltene-Containing Heavy Fuel Oils as Separability Number by an Optical Scanning Device
D7112		Determining Stability and Compatibility of Heavy Fuel Oils and Crude Oils by Heavy Fuel Oil Stability Analyzer (Optical Detection)
D7157		Determination of Intrinsic Stability of Asphaltene-Containing Residues, Heavy Fuel Oils, and Crude Oils (n-Heptane Phase Separation; Optical Detection)

Sampling Methods

ASTM	IP	Title
D4057		Manual Sampling of Petroleum and Petroleum Products
D4177		Automatic Sampling of Petroleum and Petroleum Products
D4306		Aviation Fuel Sample Containers for Tests Affected by Trace Contamination
D4418		Receipt, Storage, and Handling of Fuels for Gas Turbines
D7464		Standard Practice for Manual Sampling of Liquid Fuels, Associated Materials and Fuel System Components for Microbiological Testing

References

[1] Giles, H. N., Koenig, J. W. J., Neihof, R. A., Shay, J. Y., and Woodward, P. W., "Stability of Refined Products and Crude Oil Stored in Large Cavities in Salt Deposits: Biogeochemical Aspects," *Energy & Fuels*, Vol. 5, 1991, pp. 602–608.

[2] Bacha, J. D. and others, *Diesel Fuel Technical Review*, Chevron Products Company, Richmond, CA, 1998.

[3] McCrone, W. C, Delly, J. G., and Palenik, S. J., *The Particle Atlas. An Encyclopedia of Techniques for Small Particle Identification*, 2nd ed., Ann Arbor Publishers, Ann Arbor, MI, 1973.

12

Gaseous Fuels and Light Hydrocarbons[1] [Methane through Butanes, Natural Gasoline, and Light Olefins]

Andy Pickard[2]

THE MATERIALS AND THEIR ASSOCIATED properties and test methods in this chapter will be considered in two groups–"raw" materials and finished products. The raw materials are produced or "raw" natural gas, raw natural gas liquids, crude oil, and liquefied refinery (LR) gases. Finished products include sales quality natural gas, high purity ethane, ethane/propane mixes, liquefied petroleum (LP) gases (propane, *n*-butane, isobutane, and mixtures), natural gasoline, and light olefins.

"RAW MATERIALS"

Note that the names of the raw materials and properties can differ by geographic area around the world.

Natural Gas

Natural gas, as produced from wells, is the primary source of most of the materials and products in this chapter. "Raw" natural gas is a mixture of hydrocarbon and nonhydrocarbon gases found in porous formations beneath the earth's surface, often in association with crude petroleum. The principal constituent of natural gas is typically methane (CH_4) with varying amounts of heavier hydrocarbons and certain nonhydrocarbon gases such as nitrogen, carbon dioxide, hydrogen sulfide, and helium. Indeed, some produced "natural gases" can contain over 50 % hydrogen sulfide or carbon dioxide. Other contaminants include water and dissolved salts, mercaptans, carbonyl sulfide (COS), and light disulfides.

Raw natural gas can be subclassified as follows:

NONASSOCIATED GAS

Free gas *not* in contact with significant amounts of crude oil in the reservoir.

ASSOCIATED GAS

Free gas in contact with crude oil in the reservoir.

"SWEET" NATURAL GAS

Natural gas that is low in sulfur compounds including hydrogen sulfide (H_2S).

"SOUR" NATURAL GAS

Natural gas that contains H_2S, sometimes in very high concentrations (exceeding 65 %). Some natural gas streams contain significant amounts of carbon dioxide (CO_2) and these are also considered "sour." "Sour" natural gas is very corrosive toward metals in the presence of moisture.

Raw Natural Gas Liquids (NGLs)

Raw natural gas liquids (NGLs) are mixtures of hydrocarbons (from ethane and propane to heavier hydrocarbons) that are extracted in liquid form from produced natural gas in lease separators, field facilities, gas processing plants, and "straddle plants" associated with major natural gas pipelines. Such facilities typically remove over 80 % of ethane and virtually 100 % of propane, mixed butanes, and heavier hydrocarbons that were in the produced natural gas. Note that hydrocarbons associated with raw natural gas are virtually all *saturated* hydrocarbons with no unsaturated species: ethane but no ethylene; propane but no propylene; and butanes but no butylenes.

In some areas, NGLs are referred to as condensate. NGLs or condensates are similar in characteristics and application to a very light crude oil, and consist primarily of four-carbon to ten-carbon (C_4 to C_{10}) hydrocarbons.

Raw NGLs may be sold as a "raw mix" or unfractionated stream or fractionated into various commercial mixtures including LP gases and natural gasoline, and sometimes further fractionated into individual compounds including ethane, propane, *n*-butane, and isobutane.

Crude Oil

Light crude oils can also be a source of propane, butanes, and light hydrocarbons in the range of C_5 to C_{10}. These materials come directly from the atmospheric distillation tower in refineries–the first separation step in a refinery after the crude oil is cleaned and dried in a desalting unit.

Liquefied Refinery Gas

LR gases are mixtures of hydrocarbons that are recovered during the refining of crude oil–especially from conversion processes like cokers and catalytic cracking units. These materials are composed principally of propane, propylene, *n*-butanes and isobutanes, and butylenes, and they can be stored and handled at ambient temperatures and moderate pressures.

[1] In preparing this chapter, the author acknowledges the content and contributions of the author of the seventh edition, the late Ron Brunner; the authors of the sixth edition, Nicholas Schroeder and Mark Sutton; and earlier authors.

[2] *Consultant, Fuels and Technology Support, Qualicum Beach, B.C., Canada; Member of Committee D02 and chairman of Sub-committee H on LP Gases.*

FINISHED PRODUCTS AND QUALITY CRITERIA

Quality of a product may be defined as its fitness for a purpose. Once the required quality is determined by the application, it is controlled by appropriate processing, testing, and analysis. This section will outline quality criteria for significant finished product gaseous fuels and light hydrocarbons. Appropriate ASTM International and Energy Institute (formerly the Institute of Petroleum) methods for testing and analysis will be listed in parentheses (ASTM/IP). The full titles of the methods can be found at the end of the chapter under the heading "Applicable ASTM/IP Standards."

Natural Gas

The primary uses for natural gas are as a fuel for the production of heat (home and commercial/industrial heating) and power generation. Significant quantities of natural gas are also used to produce hydrogen and in the production of ammonia.

Following processing (especially removal of moisture, H_2S, CO_2, and gases that are not condensable during ordinary processing) and conditioning into a merchantable or finished product, natural gas is composed primarily of CH_4 and can contain a little ethane (C_2H_6), propane (C_3H_8), butanes (C_4H_{10}), and trace amounts of heavier hydrocarbons, depending on the source and extent of extraction of ethane and heavier hydrocarbons. Natural gas pipelines set the quality requirements for natural gas, which usually include restrictions on total sulfur, hydrogen sulfide, oxygen, carbon dioxide, nitrogen, solids, water, and minimum energy content in Btu (or MJ).

The principal quality criterion of natural gas is its *heating value* (ASTM D1826 and D3588). It is common practice during periods of very high natural gas demand (during periods of low temperatures) to boost the energy content (Btu content) of natural gas by the addition of small quantities of propane and air (also called "peak shaving"). Most gas companies sell natural gas based on the energy (Btu) content, not the volume of natural gas or methane, because of constant variability in composition.

Natural gas must be readily transportable through high-pressure pipelines. Therefore, the *water content*, as defined by the water dew point (ASTM D1142, D4888, and D5454), must be considered to prevent the formation of ice or hydrates in the pipeline and to protect pipelines against corrosion, which can be severe if moisture is present as well as H_2S or CO_2–even at just parts per million concentrations. Likewise, the amount of *entrained hydrocarbons heavier than ethane*, as defined by the hydrocarbon dew point, should be considered to prevent accumulation of condensable liquids that could block the pipeline or interfere with process instrumentation and measurement.

Natural gas and its products of combustion must not be unduly corrosive to the materials with which they come in contact. Thus, the detection, measurement, and reduction of *hydrogen sulfide* (ASTM D4084 or D4810) and *total sulfur* (ASTM D1072 or D6228 or D5504) are important. When natural gas is used for fuel purposes (heating or engine fuel), it is normal practice to add one or more odorants (mercaptans or sulfides) to the natural gas so that leaks can be readily detected.

If the gas is to be liquefied by cryogenic processing and stored in liquid form (LN gas [LNG]), carbon dioxide will separate out of the cold liquid as a solid and interfere with the refrigeration system. *Carbon dioxide* can be determined by ASTM D4984 or ASTM D1945. If the gas is to be used as a feedstock for the production of hydrogen or petrochemicals, a *complete analysis* (ASTM D1945) must be made, because some gases contain materials that are deleterious to these processes.

Ethane

This normally gaseous paraffinic compound (C_2H_6) has a boiling point of approximately −88°C (−127°F). Ethane can be handled as a liquid at very high pressures (typically above 1,800 psig) and at temperatures below 32°C (90°F).

Virtually all ethane produced (that is, separated from raw natural gas) is used as a feedstock for the production of high purity *ethylene*–a basic petrochemical whose applications are covered more fully in the section on olefins and diolefins.

Quality criteria for ethane as a petrochemical feedstock are specific to the particular process used for petrochemical manufacture. In general, however, such feedstocks must be very low in *water content* (ASTM D1142), *oxygen* (ASTM D1945), *carbon dioxide* (ASTM D1945), *sulfur compounds* (ASTM D1072), and *other elements* that will interfere with the process or the catalysts used in petrochemical manufacture.

Liquefied Petroleum Gases

A mixture of hydrocarbons, principally propane and butanes, obtained from either natural gas liquids or refinery sources, LP gas can be stored and handled as liquids at ambient temperatures and moderate pressures.

Propane

A normally gaseous paraffinic hydrocarbon (C_3H_8) having a boiling point of approximately −42°C (−44°F), propane is handled as a liquid at ambient temperatures and moderate pressures. It may include some ethane (maximum is typically about 6 % by volume), and up to 2.5 liquid volume (LV) percent of butanes and heavier hydrocarbons. Vapor pressure at 38°C (100°F) is never to exceed 1,434 kPa (208 psig). Commercial propane can include varying amounts of *propylene*, and *commercial propane–butane* mixes can include varying amounts of propane, propylene, butanes, and butylenes.

Butane

A light paraffinic hydrocarbon (C_4H_{10}), butane is usually handled as a liquid at ambient temperatures and moderate pressures. Butane can be fractionated further into normal butane (*n*-butane), which has a boiling point of approximately 0°C (32°F), and isobutane, which has a boiling point of approximately −12°C (11°F).

LP Gases

The original use for LP gas was as a fuel gas in areas where pipelined natural gas was not available. Later, most LP gases were fractionated into their separate components, principally propane, *n*-butane, and isobutane, as preferred feedstock for the petrochemical industry.

There are many uses for LP gases, among the most important being the following:

1. Residential, commercial, and construction fuel for heating
2. Engine fuels
3. Farm uses such as crop drying, tobacco curing, and powering irrigation pumps, and railroad uses to prevent icing of track switching gear
4. Chemical feedstock for production into ethylene and propylene
5. Manufacture of gasoline components such as alkylate

6. Gasoline blending stocks (butanes)–isobutane is a particularly good, high octane blending component, with a motor octane number of 97

Special applications such as *alkylate* and *petrochemical feedstock* are not covered by ASTM D1835 because they are usually purchased and manufactured to meet specific quality requirements that differ from one process application to another.

The ASTM specifications for LP gases are designed to properly define acceptable products for historical residential and industrial uses. In many cases, it has been found that the products meeting these specifications will be usable in applications other than the ones for which they were designed. However, some new applications such as fuel cells and microturbines can have much more stringent requirements (such as sulfur in the low ppm concentration range and essentially no heavy ends or residues).

The following descriptions will serve as a general guide to the applications of the four LP gas fuel types:

Special Duty Propane (HD-5)
This fuel type is tailored to meet the needs of internal combustion engines operating under moderate-to-severe conditions (modern automotive applications). Special duty propane will be less variable in composition and combustion characteristics than the other fuels described in ASTM D1835. Note that the high concentration of propane and limitations of allowable other hydrocarbons means that HD-5 propane will have consistent *octane quality*; thus, it is not necessary to measure octane quality on each batch of HD-5 propane. For information, the "5" in "HD-5" refers to a maximum of 5 % propylene. Similarly, an "HD-10" propane, allowed in some states, may contain up to 10 % propylene.

Since most distributors of propane handle only a single grade of propane, where HD-5 propane is available, it is also used for less demanding applications that could use commercial propane, such as heating.

Commercial Propane
This fuel type, sometimes called commercial LP gas, can contain small quantities of ethane (limited by vapor pressure restrictions) and varying quantities of propylene from refinery gases. It is preferred for residential and commercial uses (mostly heating) in geographic areas and in seasons where low ambient temperatures are encountered. Commercial propane, having a lower octane quality than HD-5 propane, is only suitable for low severity internal combustion engine applications (such as some agricultural irrigation pumps).

Commercial Propane/Butane Mixtures
This fuel type covers a wide range of mixtures and permits the tailoring of fuels to specific needs. This fuel finds applications in warmer areas and seasons where low ambient temperatures are not frequently encountered. This fuel is not common in the North American market, but may be used in other, more tropical, parts of the world.

COMMERCIAL BUTANE
This fuel type has limited use as a heating fuel and may only be used in warm climates and in industrial applications where fuel vaporization problems are not present. High purity butane, not described in ASTM D1835, has been used as a propellant in pressurized personal care products.

LP gases are essentially propane or butane or mixtures of the two. As a result, the important characteristics of these products can be defined and controlled by relatively few measurements. The *vaporization* and *combustion characteristics* of specification LP gases are defined for normal applications by *vapor pressure, volatility* and *(relative) density*. The significance of the tests is summarized here.

Sampling LP Gases
Before discussing the important properties and test methods of products in this section, a discussion of sampling is vital. Correct sampling of LP gases, especially propane, presents different challenges from sampling other fuels. Most fuels, like gasoline, diesel fuel, and aviation fuels, are liquids and are used by the application as liquids. Thus, the importance of most conventional fuel sampling is to obtain a representative sample of the liquid fuel. In the case of LP gases, the products are stored as liquids under moderate pressure, with a vapor phase above the liquid. However, in most applications, LP gases are actually used as gases. Improper sampling, sample storage, even removal of an aliquot for analysis, and the way the test specimen is introduced to an analyzer can alter the composition of the test specimen and give a false result for certain properties, especially vapor pressure. Thus, correct sampling according to ASTM D1265 or D3700 is critical. Further, some properties such as corrosion toward copper alloys, caused by aggressive sulfur species in the LP gas, can be changed by the sampling procedure or sample container, again giving a false test result. Sample containers with an inert internal coating must be used for LP gases for certain tests when determining reactive materials.

Test Methods
All the test methods cited in this chapter are ASTM test methods. Note, however, that the Gas Processors Association (GPA; http://gpaglobal.org/) has many test methods for properties described here, and some laboratories in North America will use GPA methods. In Europe, EN or ISO methods may be used.

VAPOR PRESSURE
For most purposes, LP gases will be stored, handled, and transported as a liquid (under moderate pressure) but used as a gas. To store, handle, and transport it safely, the *vapor pressure* must be known (ASTM D1267 and recently ASTM D6897). Vapor pressure gives an indication of the low temperature conditions under which initial vaporization will take place. It can also be considered a semiquantitative measure of the amount of the most volatile material present in the product. The predominant volatile component in propane is ethane, and the predominant volatile component in butane is propane. Thus, the vapor pressure limit in HD-5 propane is an effective limit on the maximum ethane (and methane) content in the propane. Vapor pressure is also used as a means for predicting the maximum pressures that can be experienced in storage and ensures trouble-free performance in commercial equipment.

Vapor pressure can also be *calculated* from a compositional analysis using ASTM D2598. The accuracy of the vapor pressure calculation is entirely dependent on the accuracy and completeness of the compositional analysis.

VOLATILITY
Volatility in LP gases (ASTM D1837) is expressed in terms of the *95 % evaporated temperature* and is a measure of the

amount of the *least* volatile fuel components present in the product. This specification controls the heavy ends in the fuel and is, in effect, a restriction on higher boiling fractions that might not vaporize for use at system temperatures.

DENSITY AND RELATIVE DENSITY

Density and relative density are important in measurement and custody transfer calculations. It is not a specification property but is usually determined for commercial or transport reasons. ASTM D1657 is the test method to actually measure the density or relative density of LP gases. Relative density of liquefied LP gases is the density of the LP gas relative to the density of water (density of the LP gas/density of water) at a stated temperature, usually 15°C (60°F). More commonly, density or relative density is calculated from a compositional analysis using ASTM D2598. In addition to vapor pressure, volatility, and density, there are other limits imposed on the LP gas that provide assurance that the product will perform dependably under actual conditions of use.

RESIDUE

Residue (ASTM D2158) is a measure of the concentration of contaminants boiling above 38°C (100°F) that can be present in the LP gas, usually as a result of contamination during distribution, downstream of the original production. Contaminants that have frequently been seen and identified are compressor oils, lubricants from valves, plasticizers from hoses, corrosion inhibitors, and other petroleum products from pumps, pipelines, and storage vessels that are used in multiple service applications. Contaminants in the gasoline boiling range are usually not a problem, but contaminants of diesel fuel (from transfer of LP gas in a multiproduct pipeline) can create low volatility liquids in vaporization units, depending on the temperature of vaporization.

Contaminants, whatever their source, can be particularly troublesome in liquid withdrawal systems followed by vaporization, such as those used in internal combustion engine fuel systems where the materials accumulate in the vaporizer or at low points in a fuel gas transfer line and will ultimately plug the fuel system. The oil stain observation portion of D2158 can provide some insight into the nature of the residue and show the presence of oily contaminants that might not be detected visually in the first step of the test (evaporation of 100 mL of propane in a centrifuge tube).

Note that the temperature of the residue separation step has a huge impact on the quantity of residue collected. Thus, a residue test performed at 100°C can show a very much smaller amount of residue than a test performed at 40°C on the same LP gas fuel.

COPPER CORROSION

Copper strip corrosion (ASTM D1838) limits provide assurance that difficulties will not be experienced in the deterioration of copper and copper alloy (brass) fittings and connections commonly used in LP gas systems. Reactive sulfur compounds such as *elemental sulfur, hydrogen sulfide*, and some other sulfur species or decomposition products of those species are corrosive to copper. The copper strip corrosion test is an extremely sensitive test that will detect virtually all species of corrosive sulfur, including minute traces of hydrogen sulfide. As little as 0.5 to 2 ppm of H_2S can cause a copper strip test failure. Mixtures of H_2S and elemental sulfur below 1 ppm can be very corrosive to copper. It is important that the product being tested does not contain any additives (such as "anticorrosion" additives) that can diminish the reaction with the copper strip, or a "false pass" can be obtained. Note that batches of LP gases can pass the copper strip corrosion test at point of manufacture, but through mixing with other batches of LP gases, reactions of sulfur species during transit (COS to H_2S, for example), or contamination during transportation and distribution can become corrosive to copper in the field.

ASTM D2420 is a simple pass/fail test for *hydrogen sulfide* and is based on the discoloration of a *lead acetate solution* on a filter paper exposed to propane vapors. The test method is required by ASTM D1835 as an additional safeguard that hydrogen sulfide is not present.

SULFUR CONTENT

Total sulfur content is determined by ASTM D2784, D6667, and other test methods. Sulfur content is limited to ensure that the sulfur oxides formed during combustion are low enough that there will not be acidic condensates in exhaust gases. It should be noted that ASTM D2784 does not include the usual precision statement. This was in recognition of the difficulty of shipping stable samples of LP gas containing various concentrations of sulfur to a number of cooperating laboratories. Studies have confirmed that the sulfur species can change during transit. Recent developments in shipping LP gas samples for laboratory proficiency testing are expected to allow the development of precision data for many LP gas test methods that lack such data.

Note that for decades, LP gases had much lower sulfur content than gasoline or diesel fuel. However, most gasolines now contain less than 30 ppm sulfur, and many diesel fuels now contain less than 15 ppm sulfur, making LP gases higher in sulfur content than other major fuels. Thus, HD-5 propane, with up to 123 ppm sulfur by mass, could not be used with current gasoline or diesel engine emission control technologies, which cannot tolerate such higher sulfur fuels. If LP gas were to be used in engines with current emission control technologies, the sulfur content would have to be reduced significantly, and a new warning mechanism would be needed to replace the current sulfur-containing odorants.

WATER CONTENT

Moisture content (ASTM D2713) is a measure of the approximate water content of the product. Liquid fuels like diesel fuel, jet fuel, gasoline, and even butane must contain no free water, but may be saturated with dissolved water. Only commercial and special duty propane types of LP gas must actually be *subsaturated with water* at normal ambient temperature. In fact, the allowable water content in propane is equivalent to the water saturation content of propane at about −25°C. Water content is controlled to provide assurance that pressure reducing regulators and similar equipment will operate consistently without freeze-ups caused by autorefrigeration at points of pressure reduction in LP gas systems, leading to separation of dissolved water from the product and freezing.

Water content must also be controlled to low concentrations to avoid hydrate formation in pipelines. *Gas hydrates* are formed as a granular solid or slushy substance when liquid water is intimately mixed with some of the light hydrocarbons, principally methane, ethane, and propane under certain conditions of temperature and pressure. Hydrates can, and do, form in pipelines well above the freezing point of water, if sufficient moisture is present in the stream and the pressure is high enough.

COMPOSITIONAL ANALYSIS

A component analysis can be performed using ASTM D2163. A gas chromatograph is used to obtain a component distribution of the LP gas. Precise compositional data are often needed for chemical feedstock applications and specification analysis. The component distribution data may be used to calculate certain physical properties, such as relative density, vapor pressure, and octane number (ASTM D2598).

Odorization

LP gases are odorless, and leaks cannot be detected by smell. Thus, for safety, LP gases to be used for fuel purposes are usually odorized prior to distribution to end users so that leaks in residential and commercial distribution systems can readily be detected "at one fifth the lower explosive limit (LEL) in air" by most people. The only two odorants currently used in North America for LP gases are ethyl mercaptan and tetrahydrothiophene (thiophane). To ensure that the LP gas is properly odorized with ethyl mercaptan, ASTM D5305, a "stain tube" test method, may be performed, either in the field or in a laboratory.

Note that LP gas is not odorized when intended for feedstock use or further processing. When users require unodorized products, suppliers generally require written instructions from the end user to document that they do not want odorized product and that the product will not be used for fuel purposes.

Percentage Fill of LP Gas Cylinders

While not a specification property, regulations pertaining to safe and proper handling of cylinders of LP gases, such as propane barbecue cylinders, propane storage tanks, and even sampling cylinders, typically restrict the fill level of LP gas cylinders and tanks to 80 % of the total capacity. This restriction is a safety requirement because light hydrocarbons like propane and butane have a very high coefficient of expansion as a function of temperature. If a cylinder were filled above 80 % at a cool temperature, and the cylinder was then exposed to a high ambient temperature, the liquid LP gas could expand to totally fill the cylinder, leaving no vapor space. Any further expansion would result in product release through overpressure protection devices (pressure relief valves). Should these devices fail or the heat input be so great that the relief devices cannot relieve pressure within safe limits, the container itself could rupture, releasing all remaining product at once–a very serious fire hazard. Thus, consideration should always be given to maximum temperatures that could be experienced during transportation and storage of LP gases and initial fill percentages adjusted accordingly. For example, if a rail car is likely to sit in a rail yard in the desert, it could experience extremely high summertime temperatures and fill levels may need to be restricted to 70 % maximum.

Natural Gasoline

Natural gasoline is a mixture of hydrocarbons (extracted from natural gas) that consists mostly of pentanes (C_5H_{12}) and heavier hydrocarbons (up to about C_{10}).

In the beginning of the natural gasoline industry, the only use for natural gasoline was as low octane motor fuel or as a blending agent in the production of automotive gasoline. Even today, natural gasoline is still used as a blending component in automotive gasoline and is also used as a denaturant in ethanol used for fuel purposes. Some of the individual components of natural gasoline, namely *isobutanes, butanes, pentanes,* and *isopentanes,* may be separated as feed stocks for reforming, alkylation, and production of synthetic rubber and other petrochemicals.

For use as motor fuel or as a component of motor fuel, the primary criteria for the quality of natural gasoline are its *volatility and knock performance (combustion characteristics).* The basic measures of volatility are *vapor pressure* (ASTM D4953, D5190, and D5191) and *distillation.* ASTM D216/IP 191 was originally used for distillation of natural gasoline, but D216 has been withdrawn and replaced by ASTM D86 (IP 123). *Knock performance* is measured by rating in knock test engines by both the *motor octane number* (ASTM D2700/IP 236) and *research octane number* (ASTM D2699/IP 237) methods.

Other considerations for natural gasoline used in motor fuels are *copper corrosion* (ASTM D130/IP 154), *density* or *API gravity* (ASTM D1298/IP 160), and *sulfur content* (ASTM D1266/IP 191 or D6667). Although density/API gravity has only a minor relationship with quality, it is necessary to determine density or gravity for measurement (purchase–sale agreements) and shipping.

When natural gasoline is used as a feedstock for further processing or petrochemicals, the list of quality criteria is tied to the processes to be used in the end application. For nearly all petrochemical uses, *composition by hydrocarbon types* is needed, and, frequently, a *complete analysis* of specific components is made (ASTM D2427). If a catalytic process is involved, total *sulfur* (ASTM D1266/IP 191 or D6667) and *nitrogen* are very important, even down to sub-part per million concentrations, because sulfur and nitrogen will destroy some catalyst activity.

Olefins and Diolefins

Many light olefins and diolefins that are produced by refinery processes are isolated for petrochemical use. Some saturated light hydrocarbons are specifically processed to generate olefins. The individual products are as follows:

ETHYLENE (C_2H_4)

This is a normally gaseous olefinic compound having a boiling point of approximately −104°C (−155°F). It can be handled as a liquid at very high pressures and low temperatures. Ethylene is normally made by cracking an *ethane* or *naphtha* feedstock in a high-temperature furnace (thermal cracking) and subsequent separation from other components by distillation.

The major uses of ethylene are in the production of *ethylene oxide, ethylene dichloride,* and *polyethylenes.* Other uses include the coloring of fruit, EP rubbers, ethyl alcohol, and medicine (anesthetic).

Since ethylene is a high-purity product (normally supplied at 99.5 mol percent purity or higher), the quality criteria of interest are *trace components (or contaminants).* The components of greatest concern are *hydrogen, carbon monoxide, carbon dioxide, oxygen,* and *acetylene. Water* and *sulfur content* are also critical to ethylene-based processes. All of the above impurities are generally catalyst poisons to polymerization processes, even in the low ppm concentration ranges on a mol basis. Therefore, the analytical problems are magnified greatly. To date, ASTM International does not have a standard method for moisture at these low levels, but a trace sulfur method, based on microcoulometry, has been adopted (ASTM D3246). Ethylene purity is determined by quantitative analysis of the *trace impurities* and subtracting the sum of their concentrations from 100 % (ASTM D2504 and D2505).

PROPYLENE CONCENTRATES (C_3H_6)

These are mixtures of propylene and other hydrocarbons, principally propane and trace quantities of ethylene, butylenes, and butanes. Propylene concentrates can vary in propylene content from 70 % mol up to over 95 % mol and can be handled as a liquid at normal temperatures and moderate pressures. Propylene concentrates are isolated from refinery thermal cracking processes. Higher purity propylene streams are produced by distillation and extractive techniques. Propylene concentrates are used in the production of propylene oxide, isopropyl alcohol, polypropylene, and the synthesis of isoprene.

Propylene concentrate streams generally require a *component analysis*, depending on their final use. The best method is by gas chromatography (ASTM D2427). Another gas chromatography method is used to identify *major impurities* (ASTM D2163/IP 264). *Sulfur* may be determined by the oxidative microcoulometric technique (ASTM D3246) or by the Wickbold combustion method (ASTM D2784). As is the case for ethylene, *water* in propylene is critical. Several field tests and a few laboratory tests are in use by individual firms, but no standard method for moisture has been adopted to date. The problems in sampling for moisture content, especially in the less than 10 ppm range, are hard to overcome.

The trace impurities in 90 % or better propylene, which is used in polymerization processes, become quite critical. *Hydrogen, oxygen,* and *carbon monoxide* are determined by one technique (ASTM D2504), and *acetylene, ethylene, butenes, butadiene, methyl acetylene,* and *propadiene* are determined by using another very sensitive analytical method (ASTM D2712).

BUTYLENE CONCENTRATES (C_4H_8)

These are mixtures of butene-1, *cis-* and *trans-*butene-2, and sometimes isobutene (2-methylpropene). These products are stored as liquids at ambient temperatures and moderate pressures. Various impurities such as butane, butadiene, and C_5S can be present in butylene concentrates.

Virtually all of the butylene concentrates are used as a feedstock for either:

1. An *alkylation plant,* where isobutane and butylenes are reacted in the presence of either sulfuric acid or hydrofluoric acid to form a mixture of *high octane C_7 to C_9 paraffins (alkylate)* used in gasoline blending, or
2. Butylene dehydrogenation reactors at *butadiene* production facilities.

The major quality criterion for butylene concentrates is the *distribution of butylenes*, which is measured, along with other components, by gas chromatography. *Trace impurities* generally checked are *sulfur* (ASTM D2784), *chlorides* (ASTM D2384), and *acetylenes*. These impurities are sometimes catalyst poisons or become unwanted impurities or both in the final product. When butylene concentrates are used as an alkylation unit feedstock, the *diolefin* content becomes important.

BUTADIENE (C_4H_6)

A reactive, gaseous hydrocarbon having a boiling point of −4.4°C (24°F), butadiene can be handled as a liquid at moderate pressure. Subambient temperatures are generally used for long-term storage due to the easy formation of butadiene dimer (4-vinyl cyclohexene-1).

Butadiene is produced by two methods:

1. The catalytic dehydrogenation of butane or butylenes
2. As a by-product from the production of ethylene

In either case, the butadiene must be isolated from other components by extractive distillation techniques and subsequent purification to polymerization-grade specifications by fractional distillation.

The largest end use of butadiene is as a monomer for production of GR-S synthetic rubber. Butadiene is also chlorinated to form 2-chloro-butadiene (chloroprene), which is used to produce the polychloroprene rubber known as neoprene.

The major quality criteria for butadiene are the various *impurities* that can affect the polymerization reactions for which butadiene is used. The gas chromatographic examination (ASTM D2593/IP 194) of butadiene determines the gross *purity* as well as *C_3, C_4, and C_5 impurities*. Most of these hydrocarbons are innocuous to polymerizations, but some, such as *1,2-butadiene and 1,4-pentadiene*, are capable of polymer cross-linking, which is harmful to the process.

Acetylenes are determined by using a chemical test method, while *carbonyls* are determined by the classical hydroxylamine hydrochloride reaction with carbonyls (ASTM D1089). *Total residue* is determined to check for nonvolatile matter (ASTM D1025).

ASTM D1550, ASTM Butadiene Measurement Tables, provides the data necessary for proper calculation of commercial quantities of butadiene.

Associated Products

While the thrust of this chapter is on light hydrocarbons, two significant non-hydrocarbons are associated with the processing of raw natural gas–sulfur and helium.

SULFUR

Much of the sulfur used in world commerce now comes from gas plants cleaning up raw natural gas to make pipeline or sales quality natural gas. This involves removing H_2S and converting it (by the Claus process) to elemental sulfur, typically of very high purity.

During times of high worldwide demand for sulfur, an important industrial chemical, gas plants and refineries with sulfur plants ship all the sulfur they produce to distribution terminals around the world. In fact, during periods of high sulfur prices, some gas plants will process very "sour" (high H_2S content) natural gas streams more for the sulfur content than for the natural gas. Conversely, during periods of low demand and low prices for sulfur, gas plants may need to stockpile the sulfur they produce in large, rectangular yellow "mountains" beside the plants, until sulfur markets improve.

Testing of sulfur for quality generally involves determination of moisture (water) content, hydrocarbons, acidity, and measurement of trace (ppm) contaminants including arsenic and ash.

HELIUM

The primary sources of helium are certain natural gas reservoirs in Canada and the United States. Quality is determined primarily by gas chromatography for trace impurities.

TEST METHODS USED FOR GASEOUS AND LIGHT HYDROCARBONS

This section describes some general comments about the test methods used for gaseous fuels and light hydrocarbons.

- Many test methods in this section are "old" methods–ASTM test methods with numbers under D2000 were published prior to about 1962, and standards with numbers below D5000 were generally first published before 1990. While the equipment for many older methods has been updated, the basic test method or approach remains unchanged.
- While the "older" test methods have been suitable for many historic purposes, such as fuels for heating, they are often manual methods that do not lend themselves to automation. Further, the older test methods may not be precise enough for some new product applications.
- Some test methods for light hydrocarbon gases have limited or no precision because of the historical difficulty of preparing multiple, identical samples and distributing them to multiple laboratories both in a timely manner and without degradation or change to the samples prior to analysis. Fortunately, developments on inert sample containers and acceptable shipping practices in the past few years, in connection with a laboratory proficiency testing program operated by the Alberta Innovates–Technology Futures (formerly the Alberta Research Council), have overcome these problems, and it may now be possible to develop proper precision data and statements for test methods for LP gases.
- ASTM test methods are commonly specified for characterizing finished products in this section. However, since there are generally no ASTM specifications for the "raw materials," many gas plants use test methods based on their process suppliers' or instrument suppliers' recommendations.
- There is a distinct need for new analytical test methods for many properties for characterizing products in this section. There is a need for test methods with higher accuracy and precision and a reduced dependence on "operator skill and training," which allow more automation and require far less product.
  - As an example, D2158 for residue in propane requires release of a lot of propane to chill the apparatus, then 100 mL of propane has to be evaporated from a centrifuge tube without "bumping" and losing much of the sample. The result is just a general indication of potential contaminants, which has not proven to be very useful in protecting LP gas customers. Use of a new analytical technique such as direct liquid injection into a high-pressure liquid chromatograph (HPLC) would reduce the amount of sample dramatically and allow identification of many residual contaminants such as gasoline, diesel fuel, lubricating oil, and phthalate plasticerizers.
  - ASTM D2713, the Valve Freeze Method for moisture in LP gases, requires a high degree of skill to get an indication of approximate water content. However, the released propane vapors present a serious safety hazard. Electronic dew point analyzers, routinely used for in-line water measurement in process control, give precise measurement of water content in real time, with continuous measurement.
- The LP gas industry faces a challenging future that will require new test methods for new requirements. However, unlike other areas such as spark ignition (gasoline) and compression ignition (diesel) fuels, the LP gas industry is quite small, and there are limited technical resources to develop new methods and limited facilities to introduce new grades of products, which will be required for new applications such as fuel cells and microturbines.

Applicable ASTM Specification

ASTM D1835 Specifications for Liquefied Petroleum (LP) Gases

ASTM	IP	Title
Natural Gas and Ethane		
D1070		Relative Density of Gaseous Fuels
D1072		Total Sulfur in Fuel Gases by Combustion and Barium Chloride Titration
D1142		Water Vapor Content of Gaseous Fuels by Measurement of Dew-Point Temperature
D1826		Calorific Value of Gases in Natural Gas Range by Continuous Recording Calorimeter
D1945		Analysis of Natural Gas by Gas Chromatography
D1988		Mercaptans in Natural Gas Using Length of Stain Detector Tubes
D3246		Sulfur in Petroleum Gas by Oxidative Microcoulometry
D3588		Practice for Calculating Heat Value, Compressibility Factor, and Relative Density of Gaseous Fuels
D4084		Analysis of Hydrogen Sulfide in Gaseous Fuels (Lead Acetate Reaction Rate Method)
D4810		Hydrogen Sulfide in Natural Gas Using Length of Stain Detector Tubes
D4888		Test Method for Water Vapor in Natural Gas Using Length of Stain Detector Tubes
D4984		Test Method for Carbon Dioxide in Natural Gas Using Length of Stain Detector Tubes
D5454		Test Method for Water Vapor Content of Gaseous Fuels Using Electronic Moisture Analyzers
D5504		Test Method for Determination of Sulfur Compounds in Natural Gas and Gaseous Fuels by Gas Chromatography and Chemiluminescence
D6228		Test Method for Determination of Sulfur Compounds in Natural Gass and Gaseous Fuels by Gas Chromatography and Flame Photometric Detection

ASTM	IP	Title
Liquefied Petroleum Gas		
D1265		Practice for Sampling Liquefied Petroleum (LP) Gases, Manual Method
D1267		Gage Vapor Pressure of Liquefied Petroleum (LP) Gases (LP-Gas Method)
D1657		Density or Relative Density of Light Hydrocarbons by Pressure Hydrometer.
D1837		Volatility of Liquefied Petroleum (LP) Gases
D1838		Copper Strip Corrosion by Liquefied Petroleum (LP) Gases
D2158		Residues in Liquefied Petroleum (LP) Gases
D2163	264	Analysis of Liquefied Petroleum (LP) Gases and Propene Concentrates by Gas Chromatography
D2420		Hydrogen Sulfide in Liquefied Petroleum (LP) Gases (Lead Acetate Method)
D2421		Practice for Interconversion of Analysis of C_5 and Lighter Hydrocarbons to Gas-Volume, Liquid Volume or Mass Basis
D2497		Test Method for Demethanized Hydrocarbon Liquid Mixtures Containing Nitrogen and Carbon Dioxide by Gas Chromatography
D2597		Analysis of Demethanized Hydrocarbon Liquid Mixtures Containing Nitrogen and Carbon Dioxide by Gas Chromatography
D2598		Practice for Calculation of Certain Physical Properties of Liquefied Petroleum (LP) Gases from Compositional Analysis
D2713		Dryness of Propane (Valve Freeze Method)
D2784		Sulfur in Liquefied Petroleum Gases
D3700		Practice for Obtaining LPG Samples Using a Floating Piston Cylinder
D5305		Determination of Ethyl Mercaptan in LP-Gas Vapor
D6667		Determination of Total Volatile Sulfur in Gaseous Hydrocarbons and Liquefied Petroleum Gases by Ultraviolet Fluorescence
D6849		Practice for Storage and Use of Liquefied Petroleum Gases (LPG) in Sample Cylinders for LPG Test Methods
D6897		Test Method for Vapor Pressure of Liquefied Petroleum Gases (LPG) (Expansion Method)
Natural Gasoline		
D86	123	Distillation of Petroleum Products at Atmospheric Pressure
D130	154	Corrosiveness to Copper from Petroleum Products by Copper Strip Test

ASTM	IP	Title
D216 Withdrawn	191	Distillation of Natural Gasoline Replaced by D86
D323		Vapor Pressure of Petroleum Products (Reid Method)
D1266	101	Sulfur in Petroleum Products (Lamp Method)
D1298	160	Density, Relative Density (Specific Gravity), or API Gravity of Crude Petroleum and Liquid Petroleum Products by Hydrometer Method
D2427		Determination of C_2 through C_5 Hydrocarbons in Gasolines by Gas Chromatography
D2699	237	Knock Characteristics of Motor Fuels by the Research Method
D2700	236	Knock Characteristics of Motor and Aviation Type Fuels by the Motor Method
D4953		Test Method for Vapor Pressure of Gasoline-Oxygenate Blends (Dry Method)
D5190		Test Method for Vapor Pressure of Petroleum Products (Automatic Method)
D5191		Test Method for Vapor Pressure of Petroleum Products (Mini Method)
Ethylene		
D2504		Non-condensable Gases in C_3 and Lighter Hydrocarbon Products by Gas Chromatography
D2505		Ethylene, Other Hydrocarbons, and Carbon Dioxide in High-Purity Ethylene by Gas Chromatography
D3246		Sulfur in Petroleum Gases by Oxidative Microcoulometry
D5234		Guide for Analysis of Ethylene Products
D6159		Determination of Hydrocarbon Impurities in Ethylene by Gas Chromatography
Propylene Concentrates		
D1265		Sampling Liquefied Petroleum (LP) Gases
D2163	264	Analysis of Liquefied Petroleum (LP) Gases and Propane Concentrates by GC
D2504		Non-condensable Gases in C_2 and Lighter Hydrocarbon Products by Gas Chromatography
D2712		Hydrocarbon Traces in Propylene Concentrates by Gas Chromatography
D4864		Determination of Traces of Methanol in Propylene Concentrates by Gas Chromatography
D5273		Guide for Analysis of Propylene Concentrates
D5303		Trace Carbonyl Sulfide in Propylene by Gas Chromatography

ASTM	IP	Title
Butylene Concentrates		
D1265		Sampling Liquefied Petroleum (LP) Gases
D2384		Traces of Volatile Chlorides in Butane-Butene Mixtures
D2784		Sulfur in Liquefied Petroleum Gases (Oxy-Hydrogen Burner or Lamp)
D4423		Determination of Carbonyls in C_4 Hydrocarbons
D4424		Butylene Analysis by Gas Chromatography
Butadiene		
D1025		Nonvolatile Residue of Polymerization-Grade Butadiene
D1089		Carbonyl Content of Butadiene
D1157		Total Inhibitor Content (TBC) of Light Hydrocarbons
D1550		Standard ASTM Butadiene Measurement Tables
D2426		Butadiene Dimer and Styrene in Butadiene Concentrates by Gas Chromatography
D2593	194	Butadiene Purity and Hydrocarbon Impurities by Gas Chromatography
D5274		Guide for Analysis of 1,3-Butadiene Product
D5799		Determination of Peroxides in Butadiene

Bibliography

Katz, D. L., *Handbook of Natural Gas Engineering,* McGraw-Hill, New York, 1969.

13

Petroleum Solvents

R. G. Montemayor[1]

THE TERM "PETROLEUM SOLVENTS" DESCRIBES the special liquid hydrocarbon fractions obtained from petroleum for use in industrial processes and commercial formulation to dissolve, suspend, or transport the other ingredients of the process or formulation. In recent years, the variety of petroleum solvents has increased considerably due to the development of refinery processes designed primarily for the transformation of low-octane feedstocks to high-octane fuels, as well as due to environmental regulations and concerns. These new developments led to the production of a number of important solvents and intermediates, which were obtained previously only from pine trees or from the coal carbonizing industry. This chapter deals mainly with petroleum hydrocarbon solvents consisting only of hydrogen and carbon atoms. Solvents that contain chemical species other than hydrogen and carbon such as oxygen (alcohols, ethers, aldehydes, esters, etc.), nitrogen (amines, nitriles, etc.), and sulfur (mercaptans, sulfolanes, etc.), are outside the scope of this chapter. Such solvents are obtained from various petrochemical industries.

Petroleum solvents are available for a variety of applications. Modern refining techniques produce solvents with a high degree of purity and stability in a wide range of grades and boiling points. Petroleum solvents are particularly attractive because they are readily available in large volumes at reasonably low costs compared to solvents from other sources.

GENERAL USES

Solvents are used extensively by industry in manufacturing processes for such diverse products as paint, printing ink, polish, adhesives, perfumes, vegetable oils, varnish, etc. Further uses are found in the dry cleaning, leather, fur, and tire industries. Solvents are important ingredients of pesticide formulations and are used in large quantities in cleaning machine parts and military hardwares. Solvents, in highly purified conditions, are also becoming increasingly important for use as reaction media in certain catalytic processes.

PRODUCT REQUIREMENTS

The variety of applications emphasizes the versatility of petroleum solvents. The main characteristics, which determine the suitability of a petroleum fraction for a particular use, are its solvency properties, polarity, volatility, purity, density, odor, toxicity, and air pollution control/limitation.

Solvency

In many applications, petroleum solvents are used as a vehicle to dissolve resins, oils, gums, or waxes. In other cases, the solvent merely suspends such items as pigments, fillers, and water. Examples of these are their use in the printing ink, rubber coating and dipping, paint, lacquer, adhesive, and polish industries.

Solvent properties are determined by the types of hydrocarbon species making up the solvent. Among hydrocarbon solvents in general, aromatic hydrocarbons have the highest solvency or solvent power, and straight-chain and branched-chain aliphatic hydrocarbons exhibit the lowest solvency.

A primary characteristic of solvent power is the ability of the solvent to dissolve a resin or other film-forming materials. In general, the more polar resins require the more polar solvents such as the oxygen-containing alcohols, esters, and ketones.

The amount of film former that the solvent will dissolve determines the solid content (or non-volatile content) of the system and ultimately the viscosity and flow properties of the finished product. This is the most common basis for rating solvents as to solvency for a particular resin.

Hydrocarbon solvents are frequently used in conjunction with other solvents to reduce the overall cost of the resin-solvent system. As a hydrocarbon solvent is added to a solvent-resin system in increasing amounts, there is a point at which the resin precipitates or "kicks out." This "dilution limit" is expressed as the percent age nonvolatile when precipitation occurs. The dilution limit is used frequently as an indication of the relative solvency between solvents.

Volatility

Next to solvency, volatility may be the most important property of a solvent since volatility largely governs evaporation rate or drying time. A measure of volatility is the distillation temperature or distillation temperature range of the solvent. Pure hydrocarbon solvents such as pentane, hexane, heptane, benzene, toluene, and xylene are characterized by a narrow boiling range of 1 to 2°C (2 to 4°F). Other petroleum solvents that are mixtures of many hydrocarbons have wide boiling ranges that vary with their composition.

Another measure of the volatility of solvents is vapor pressure. The higher the vapor pressure of a solvent, the higher is its volatility. Conversely, the lower the vapor pressure of a solvent, the lower is its volatility. Lower vapor pressure indicates a solvent composition with higher initial boiling point components, while higher vapor pressure is consistent with solvent components with lower initial boiling points. For applications that involve the removal of the solvent after use, the lower the initial boiling point is, the better for the application. Vapor pressure is also related to evaporation rate. The higher the vapor pressure of a solvent, the faster is the evaporation rate.

An excellent example of the importance of evaporation rate can be found in the factors considered in the selection

[1] *Imperial Oil Ltd., Sarnia, Ontario, Canada.*

of a solvent for coating applications. In coating applications, the solvent evaporation rate influences leveling, flowing, sagging, wet-edge time, and gloss. The optimum evaporation rate varies with the method of application; from fastest for spraying, intermediate for brushing, and slowest for flow coating and conveyor.

The choice of a solvent for a given application is also dictated by the temperature of use. A high-boiling solvent will be required for a heat-set ink where the operating temperature may be as high as 316°C (601°F), but a low-boiling solvent would be selected for fast drying in a foil printing press. Fortunately, solvents are available to satisfy the wide range of requirements for evaporation rates and temperatures.

Purity

The purity of the various hydrocarbon solvents pertains not only to the minor concentrations of non-hydrocarbon contaminants, but in some cases, particularly with the highly purified lower boiling aromatic solvents, to the presence of other undesirable hydrocarbons.

Petroleum solvents are often mixtures of straight-chain, branched-chain, and cyclic paraffins and aromatic hydrocarbons with a possible trace of olefinic material. With this composition, they are, in the refined state, inert materials. As inert materials, solvents may be used in the preparation of surface coatings and adhesives and in a multitude of applications without risk of side effects due to reaction with other substances in the formulation or with the application surface. Abnormally high aromatic or olefinic material in a solvent could present a problem to some applications, due to the potential reactivity of the aromatic or olefinic species.

The importance of any particular impurity will be a function of the solvent's intended application. As a result, purchasers generally set solvent specifications. Normally, part of these specifications will contain limits on the maximum level of specific contaminants critical to their processes.

Specific Gravity

Specific gravity, sometime referred to as relative density, is directly related to the density of a solvent. Specific gravity will be defined later in the chapter. Since cost calculations are based on density, this solvent property becomes very important in the marketplace for solvent applications.

Odor

The odor of a solvent generally refers to the odor of the vapor emitted during and shortly following application. Occasionally, there may be a persistent residual odor contributed by trace contaminants in the solvent.

Odor is most critical for interior trade sales products such as paints used in homes. Even the mild odor of the typical aliphatic hydrocarbon solvents frequently is considered unpleasant. This has contributed to the widespread acceptance and use of latex and other water-based paints, which by comparison, have subdued odors. Odor of solvents used in the vegetable oil industry is also critical since any undesirable residual odor of the solvent could find its way to the final vegetable oil product.

In the case of paints, there is very little leeway regarding odor, since the solvents must be chosen primarily for solvency and evaporation rates. A partial solution is the use of mineral spirits, some grades of which are almost odor free or odorless.

Toxicity

The toxicological properties of the various hydrocarbon solvents are quite important. The threshold limit values (TLVs) adopted by the American Conference of Governmental Industrial Hygienists (ACGIH) limit the use of solvents in high concentration in workroom air. In some cases, the choice between different solvent systems is based solely on TLVs. TLVs for individual solvents are available from ACGIH.

Toxicity of solvents pertains to the potential injury to health that could occur if the solvent gets into the bloodstream through inhalation, skin contact, or ingestion. Hydrocarbon solvent manufacturers provide detailed toxicological and health information to the solvent user through the material safety data sheet (MSDS). MSDSs should be consulted whenever solvent systems are considered. They provide a wealth of information not only helpful for solvent selection, but also for potential health effects, symptoms of exposure, and hazard control measures.

Typical outcomes of excessive exposure to hydrocarbon solvents include respiratory irritation and central nervous system effects. Depending on the solvent, target organ effects may also be encountered. Solvents used in oil-based paints for home user are relatively low in toxicity. However, prolonged breathing of concentrated solvent vapors should be avoided. Always consult the label and the MSDS before using any hydrocarbon solvent product.

Air Pollution Requirements

The ability of a solvent to meet air pollution requirements has become an increasingly important consideration. Most air pollution regulations prescribe the quantity of a given solvent that may discharged into the air over a given time interval and the composition of allowable solvent blends.

Ground level ozone and smog are important air pollution concerns. Under certain conditions, hydrocarbons released into the air can react with nitrogen oxides in the atmosphere producing ozone. Hydrocarbons vary in their capability to produce ozone. As a general rule, hydrocarbon solvents containing aromatics are more reactive than aliphatic hydrocarbon solvents. In the 1960s, hydrocarbons containing aromatics were regulated in California [Los Angeles Rule 102 (formerly Rule 66)]. Rule 102 (formerly Rule 66) limited the use of toluene and ethylbenzene to 20 volume percent of the solvent, and C8 and higher aromatics like xylene and its isomers to 8 volume percent.

Volatile organic compounds (VOCs) values of solvents have become increasingly important in recent years due to increased awareness of the environmental impact of these materials. A number of regulations deal with VOC parameters.

A new concept has been adopted for regulation of certain products in California. Rather than restrict the total mass of VOCs in a product, the regulation addresses the ozone formation capability of the product. This ozone formation capability is expressed in terms of maximum incremental reactivity (MIR). MIR values refer to the maximum mass of ozone produced under worst-case condition per unit mass of VOC released. This MIR numeric ratio can guide the product formulator to select solvents having the lowest potential impact to the environment.

TYPES OF HYDROCARBONS

Hydrocarbons may be divided into subgroups in different ways based on chemical composition. The classification that promotes the easiest understanding is:

1. Aliphatic hydrocarbons (aliphatics)
   a. Normal or straight-chain paraffins, C_nH_{2n+2}
   b. Iso or branched-chain paraffins, C_nH_{2n+2}
2. Naphthenic hydrocarbons (naphthenes) or cycloparaffins, C_nH_{2n}
3. Aromatic hydrocarbons (aromatics), C_nH_{2n-6}
4. Olefinic hydrocarbons (olefins), C_nH_{2n}

Each of these classifications will be discussed in subsequent paragraphs. Corresponding structural formulas, based on compounds having six carbon atoms, are shown in Fig. 1.

Aliphatics

There are two types of open-chain or aliphatic hydrocarbon solvents: normal paraffins and isoparaffins. The normal paraffins have a straight chain of carbon atoms, while the isoparaffins have side chains. Commercial solvents are in the range of compounds with 6 to 15 carbons with the most widely used solvents being compounds with 7 to 12 carbons.

Both types of paraffins conform to the general principle that increasing molecular weight (i.e., the larger the number of carbon atoms) results in lower solvency and lower volatility. The normal paraffins are characterized by low solvency and mild odor. As implied by the name, the isoparaffins are isomers (same molecular formula but different structural formula) of the normal paraffins, which they resemble in terms of solvency. A distinctive and significant property of the isoparaffins is their extremely low odor. They are almost odorless in the carbon-9 (C_9) and higher range and are the solvent choice where low odor is a critical parameter.

Naphthenes or Cycloparaffins

Naphthenes are cyclic paraffins. The only naphthenes that are present in paint solvents in significant quantity are cyclopentane, cyclohexane, and their alkyl-substituted homologues. The amount of naphthenes in nonaromatic hydrocarbon solvents varies greatly–from 0 to 90 volume percent. Because of their cyclic structure, the naphthenes are less volatile than paraffins with the same number of carbons and have stronger solvency.

Aromatics

Most of the aromatic species that are found in petroleum solvents are based on the unsaturated, six carbon ring structure as shown in Fig. 2. However, in the solvents that contain heavier aromatics components like those used for pesticide applications, some naphthalene (two fused or condensed aromatic ring structures) or alkyl-substituted naphthalene species, or both, can be found. Alkyl-substituted benzenes are commonly referred to as alkyl benzenes because they are compounds that contain the benzene ring with one or more of the hydrogen atoms replaced by alkyl groups such as methyl, ethyl, or butyl groups. Fig. 2 also shows the structural formulas for toluene and *ortho*-xylene, one of the three isomers of xylene.

The more important products that are of higher molecular weight than xylene are *n*-propylbenzene, iso-propylbenzene (commonly known as cumene), *n*-butylbenzene, and

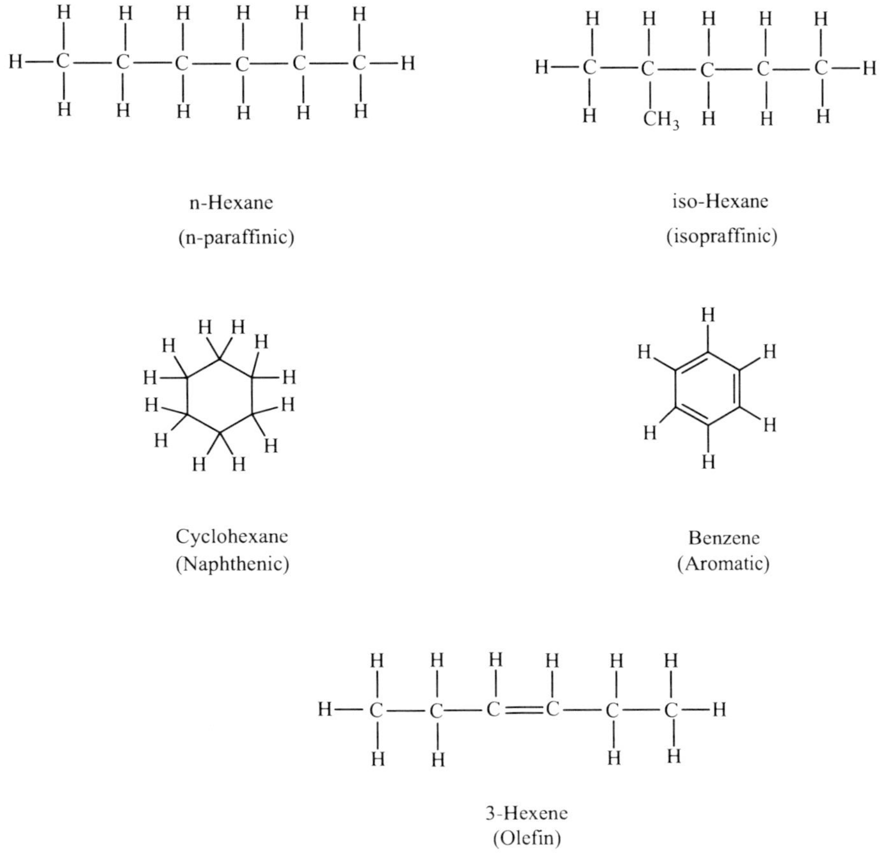

Fig. 1—Chemical structure of some C_6 hydrocarbons.

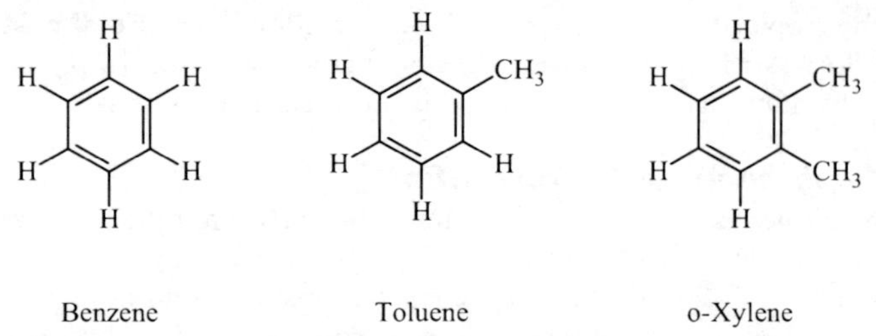

Fig. 2—Chemical structure of lower-boiling aromatic hydrocarbons.

iso-butylbenzene. Mixtures of these higher-molecular-weight compounds together with some naphthalene species are sold as solvents under the name high-flash aromatic naphthas.

Characteristics of the aromatic hydrocarbon solvents, in comparison with aliphatic hydrocarbons, are much higher solvency and stronger odor. Although naphthenic solvents are a minor factor, it is not incorrect to say that the solvent power of commercial hydrocarbon solvents is roughly proportional to their aromatic content.

Examples of very high purity aromatic hydrocarbons (+99 volume percent purity) are benzene, toluene, and xylene. Commercial grades of toluene and xylene with distillation ranges up to 10°C are equally satisfactory for use in coatings. Benzene is not used in coatings due to its toxicity. These materials were referred to in the past as benzol, toluol, and xylol, although the use of these terms has declined over the years.

Olefins

There are no petroleum-derived olefinic hydrocarbon solvents in commercial use. Small amounts may be found in some petroleum solvents. They are undesirable components of solvents since they can contribute to discoloration and color formation during storage. Nevertheless, there are several important olefinic hydrocarbon solvents, derived from pine trees in commercial use. These are terpenes, examples of which are turpentine, dipentenes, and pine oil. Another example is limonene, which is obtained from citrus oils.

Although their use as a solvent is limited, olefinic hydrocarbons are used as raw material for the manufacture of certain petrochemicals because of the reactivity of the olefinic carbon-to-carbon double bond.

COMMERCIAL HYDROCARBON SOLVENTS

General Solvent Types

The foregoing discussion of hydrocarbon solvents by chemical types provides a background for the commercial products and their usage. While a paint chemist should have some interest in the chemical constituents of solvents, his primary concern must be with commercial solvents as purchased. Of necessity, his selection and use of solvents are based solely on the properties of commercial products as determined by tests. Solvent users are guided mainly by the following commercial classification of hydrocarbon solvents:

1. Narrow-cut aliphatics
2. Special boiling range solvents
3. Mineral spirits types
4. Aromatics
5. Kerosene

Typical properties of these classes of hydrocarbon solvents are shown in Table 1.

Special Types of Solvents

Although industrial users of solvents are guided by the commercial classification just outlined, there are specific types of commercial solvents that merit a brief description.

PETROLEUM ETHER

Petroleum ether is a misnomer that has been erroneously used in the past and still used today in some industries. It is not really an ether, which by definition should have the C-O-C functional group, but it is a type of hydrocarbon solvent consisting generally of mixtures of lower-boiling aliphatic species, primarily pentane and hexane. It is one of the fastest evaporating of the commonly used solvents and therefore has obvious advantages. However, its flash point is extremely low. It is used as a solvent for oils, fats, waxes, in paints and varnishes, and as a fuel. The use of the term petroleum ether is misleading and should be discontinued.

HEXANE AND HEPTANE

Hexane and heptane are used frequently in multistage processes where the solvent must still be present in one step but be gone completely before a subsequent step. They are also used as solvents and suspending agents in the petrochemical industry where they are recovered and reused. Heptane is used in place of hexane when a higher flash point or higher boiling range solvent is needed.

"TEXTILE" SPIRITS

This narrow-cut solvent, sometimes referred to as "rubber solvent," is predominantly hexane. It is used only when very fast setting or drying of lacquers, inks, or adhesives is desired. Since it has a wider boiling range than hexane, evaporation occurs over a longer time interval.

LACQUER DILUENT

Lacquer diluent is designed primarily for use in nitrocellulose lacquer as a replacement, or partial replacement, for toluene. Lacquer diluent is used in place of toluene because it is generally cheaper. The quality of lacquer diluent is directly proportional to aromatic content: the more aromatics, the better the quality. Lacquer diluents are used also as solvents in gravure inks for printing.

VM&P NAPHTHA

Large quantities of VM&P (varnish makers and painters) naphtha are used. These solvents resemble regular mineral spirits in hydrocarbon composition and exhibit the same solvency. However, they have a lower boiling point and lower flash point, and they evaporate faster. The fast evaporation rates and lower flash points of these type of solvents make them generally unsuitable for interior architectural finishes, but they are well adapted for application by spraying, and in some cases, by dipping. The grades with an initial boiling point of about 138°C (280°F) or higher are differentiated frequently as high-flash VM&P naphtha.

MINERAL SPIRITS

Mineral spirits constitute about 75 % of all hydrocarbon solvents used in the paint industry, and 90 % of the solvents used in architectural finishes. There are currently four

TABLE 1—Approximate Properties of Typical Hydrocarbon Solvents

	Distillation Range °C (°F)		Evaporation Rate N = BuAc = 100	Flash Point. Tag Closed Cup °C (°F)		Kauri. Butanol Value	Aniline Point. °C (°F)	Aromatics Volume Percent	Sp Gr 60/60
	Initial	Dry Points							
Narrow Cut Aliphatics									
Hexane	66 (151)	69 (157)	>1000	<−15	(<0)	31	63 (146)	<1	0.680
Heptane	94 (201)	98 (209)	600	−5	(23)	38	53 (127)	3	0.728
Special Boiling Point Solvents									
Petroleum ether	35 (95)	60 (140)	>1000	15	(20)	26	66 (150)	0	0.645
Textile spirits	63 (145)	79 (175)	850	−5	(20)	34	57 (134)	5	0.690
Rubber solvent	62 (144)	120 (248)	700	<−18	(<0)	36	55 (131)	5	0.712
Lacquer diluent	93 (200)	104 (220)	400	−7	(20)	40	49 (120)	15	0.728
VM&P naphtha	113 (235)	177 (350)	150	7	(45)	38	54 (130)	10	0.752
Ink oil	257 (495)	293 (560)	<1	110	(240)	26	80 (176)	<5	0.820
Mineral Spirits									
Regular (Stoddard)	160 (320)	193 (380)	10	42	(108)	39	54 (130)	15	0.789
140-flash point	182 (360)	204 (400)	5	60	(140)	35	62 (143)	15	0.792
Odorless	177 (350)	193 (380)	7	52	(125)	27	84 (184)	<1	0.759
High solvency	160 (320)	193 (380)	8	43	(110)	44	41 (105)	30	0.808
Aromatics									
Benzene	79 (175)	81 (177)	500	<−12	(<10)	115	14 (57)[A]	99+	0.885
Toluene	110 (230)	111 (232)	200	4	(40)	105	8 (46)[A]	99+	0.870
Xylene	139 (282)	141 (285)	60	28	(83)	98	11 (52)[A]	99+	0.869
High-flash aromatic naphtha	154 (310)	177 (350)	19	42	(107)	92	13 (56)[A]	98+	0.874
High-flash aromatic naphtha	182 (360)	210 (410)	4	66	(151)	90	18 (64)[A]	97+	0.891
Heavy aromatic naphtha	182 (360)	282 (540)	1	67	(152)	105	24 (76)[A]	86+	0.933
Kerosine									
Regular	177 (359)	268 (515)	1	60	(140)	30	71 (160)	<25	0.804
Odorless	204 (400)	249 (480)	1	77	(170)	27	88 (190)	<1	0.782

[A] Mixed aniline point, °C (°F).

general types of mineral spirits: (1) Type I: Full Range, (2) Type II: High Flash Point, (3) Type III: Odorless, and (4) Type IV: Low Dry Point. Within each of these types, there are products that offer a choice in boiling range, aromatic content, and volatility, and thus permit some control of "setting" rate, brushability, and wet-edge time.

BENZENE

Benzene is no longer used in coatings industry because of its toxicity. But in the past, it was used in some solvent-type paint and varnish removers where it is more effective than the higher-boiling aromatics.

TOLUENE

Toluene is used widely when maximum solvency and relatively fast evaporation rate are desired. It is the best fast diluent for nitrocellulose lacquers and is also used in synthetic enamels of several types. The industrial pure grades with distillation ranges of a few degrees centigrade give the same results as the 1°C boiling range nitration-grade toluene.

XYLENES

Xylene is essentially a mixture of the *ortho-*, *meta-*, and *para-* isomers of xylene and ethyl benzene. It is used instead of toluene where slower evaporation is sought. In both lacquers and synthetic enamels, xylene is used frequently in admixture with toluene. It is the sole or main hydrocarbon solvents in many synthetic industrial enamels that are applied by spraying.

HI-FLASH SOLVENT NAPHTHA

Formerly this term referred only to a high-boiling fraction obtained from coal tar. Today, it also covers an equivalent product that is derived from petroleum. Many industrial paints require at least a portion of a strong hydrocarbon solvent that evaporates slower than xylene. Hi-flash solvent naphtha fills this need.

KEROSENE

Kerosene is the only hydrocarbon solvent that is rated as having a very wide distillation range. It is used when extremely slow setting and rather low solvency are desired and the kerosene odor is permissible. Two such purposes are wood fillers and putties. In most applications, kerosene is combined with a large amount of mineral spirits.

INK OILS

Ink oils are very high-boiling hydrocarbons, typically ranging from 232 to 315°C (450 to 600°F). Ink oils are used in place of kerosene where narrow-cut ink solvents are desired. The narrow cut provides better drying control of the ink. There was a time when lower-boiling, odorless hydrocarbon solvents were required for liquid toners. However, dry powder toners have essentially replaced the need for liquid toners, especially in photocopy applications.

TEST METHODS FOR SOLVENTS

The most important tests conducted on solvents are shown in Table 2. These tests are of two general types. Quality control tests are those used to test shipments for uniformity and compliance with specifications, and evaluation tests are those conducted on new materials to determine their properties and suitability for use. As a rule, the quality control tests are simpler. Some tests may be used for both purposes.

The following paragraphs address sampling techniques for petroleum solvents and discuss the purposes and principles of the tests listed in Table 2. Specific details on the various tests can be found in the *Annual Book of ASTM Standards*, volumes 5.01, 5.02, 5.03, 5.04, 5.05, 6.01, 6.02, 6.04, 10.03, and 15.05.

TABLE 2—ASTM Tests Commonly Conducted on Solvents

Specific gravity (relative density)	D891 A or B or D4052
Distillation	
(a) petroleum type solvents	D86, D7344, or D7345
(b) aromatic solvents	D850
(c) other solvents	D1078
Flash point	
(a) Tag closed-cup tester	D56
(b) Cleveland open-cup tester	D92
(c) Pensky-Martens closed-cup tester	D93
(d) Tag open-cup tester	D1310
Evaporation rate	D3539
Copper corrosion, aromatic hydrocarbon	D849
Sulfur content	
Trace sulfur by oxidative microcoulometry	D3120
Sulfur by hydrogenolysis and rateometric colorimetry	D4045
Sulfur by GC and selective sulfur detection	D5623
Sulfur in aromatic solvents	D4045
Sulfur in light hydrocarbons by ultraviolet (UV) fluorescence	D5453
Dilution ratio in nitrocellulose solutions	D1720
Viscosity of nitrocellulose solutions	D1343
Color	D156, D1209, D848, D5386, or D6045
Kauri-butanol value	D1133
Aniline point and mixed-aniline point	D611
Bromine index	D1492, D2710, or D5776
Water	D6304, D7375, or E 1064

Sampling

Because of the standards set for petroleum solvents, it is essential to employ the correct techniques when taking samples for tests. Mishandling, or the slightest trace of contaminant, can give rise to misleading results. Sampling methods are described in ASTM D4057, Standard Practice for Manual Sampling of Petroleum and Petroleum Products. Special care is necessary to ensure that containers are scrupulously clean and free from odor. Samples should be taken with minimum disturbance, so as to avoid loss of volatile components. In the case of the lightest solvents, it may be necessary to chill the sample. While waiting to be tested, samples should be kept in a cool, dark place to ensure that they do not discolor or develop odors.

Specific Gravity (Relative Density)

The determination of specific gravity (relative density) serves two purposes: it provides a check on the uniformity of shipments, and it permits calculation of the weight per gallon.

Specific gravity is defined as the ratio of a weight of a given volume of material to the weight of an equal volume of water under specified conditions. Manual determination of the specific gravity of petroleum solvents can be done using a glass hydrometer or pycnometer by ASTM D891, Standard Test Method for Specific Gravity of Liquid Industrial Chemicals. An instrumental method of determining specific gravity, ASTM D4052, Standard Test Method for Density and Relative Density by Digital Density Meter, has gained popularity because of its ease of operation and digital display of results.

In the past, producers of petroleum solvents used American Petroleum Institute (API) gravity instead of specific gravity. However, this practice has virtually been discontinued in favor of measured specific gravity.

For figuring formula yields and important compositional relationships, paint laboratories must have a list of raw materials that includes the weight per gallon of all liquid materials at normal operating temperature. When the specific gravity of solvents is determined at the customary 15.6/15.6°C (60/60°F), a temperature conversion factor must be used to obtain the specific gravity at 20°C (68°F). This temperature conversion factor differs from solvent to solvent. The corrected specific gravity must then be converted to pounds per gallon.

Distillation

The significance of distillation results is based on the close relationship to volatility properties, which, in turn, largely governs evaporation rate. ASTM D86, Standard Test Method for Distillation at Atmospheric Pressure, is used for solvents with a wide distillation range (this includes the common aliphatic and aromatic naphtha solvents). Distillation of pure aromatic solvents like toluene and xylene is done using ASTM D850, Standard Test Method for Distillation of Industrial Aromatic Hydrocarbons and Related Materials. For other solvents with a narrow distillation range (alcohols, esters, ketones, etc.), ASTM D1078, Standard Test Method for Distillation Range of Volatile Organic Liquids, is used. Regardless of solvent type, a very narrow distillation range has no direct advantage in coatings. However, it may have significance as an indication of the degree of purity, especially for oxygenated solvents. Two recently developed distillation methods that are applicable to solvents are: ASTM D7344, Standard Test Method for Petroleum Products (Mini Method), and ASTM D7345, Standard Test Method for Petroleum Products (Micro Method). These two new distillation methods make use of less than 10 mL sample and test parameters quite distinct from D86, D1078, and D850.

Flash Point

The flash point of a liquid is the lowest temperature corrected for atmospheric pressure at which application of an ignition source causes the vapor above the specimen to ignite. Flash point is indicative of potential fire hazard. The most common method for determining flash point of solvents is using ASTM D56, Flash Point by Tag Closed-Cup Tester, which confines the vapor until the instant an ignition source is applied. However, Department of Transportation (DOT) regulations on the shipping of flammable solvents are based on ASTM D56, ASTM D3828, Standard Test Method for Flash Point by Small Scale Closed Cup Tester, or ASTM D3278, Standard Test Method for Flash Point of Liquids by Small Scale Closed Cup Tester. There are other test methods of flash point determination for other types of materials. In the past, the determination of flash point was based on the use of a test flame and visual observation of the occurrence of the flash point. The advent of automatic flash point instruments made other sources of ignition possible such as electric ignitors and arcs, in addition to automatic detection of flash point using changes in ionization current, thermal properties, and pressure.

In mixtures of solvents and in coatings, the flash point is substantially that of the lowest flash point component that is present in appreciable amount.

Evaporation Rate

Although there is a correlation between distillation range and evaporation rate, the relationship is not direct. Because it is important, several methods of measuring evaporation rates of straight solvents (not the solvent in a film coating) have been developed by various commercial and governmental organizations. In the past, simple gravimetric or volumetric procedures were used to measure evaporation rate. However, the determination of solvent evaporation rate is currently described in ASTM D3539, Standard Test Method for Evaporation Rates of Volatile Liquids by Shell Thin-Film Evaporometer. In this method, a measured volume of liquid solvent is spread on a known area of filter paper that is suspended from a sensitive analytical balance under a controlled environment cabinet. The weight loss of the filter paper/solvent liquid is measured as a function of time as the solvent evaporates.

Evaporation rates are usually expressed relative to *n*-butyl acetate, which is assigned a value of one (or 100). A value of 0.4 (or 40) means that the solvent evaporates 40 % as fast as
n-butyl acetate, and a value of 2.5 (or 250) means 2.5 times faster.

There is a formula for calculating evaporation rates from distillation temperatures. However, in view of the complexity of the calculations, direct determination would probably involve a shorter time and be more accurate.

Color

Most solvents are "water-white" or clear and essentially colorless. Color is an important specification requirement. If a solvent has a color worse than the specification limit, it may be an indication of (1) inadequate processing; (2) contamination that might have occurred during storage or distribution (e.g., pick-up of rust from tanks, pipes, or extracted from improper tank linings or hoses); or (3) oxidative degradation as a result of aging.

There are two visual test methods used for measuring the color of solvents: ASTM Method D156, Saybolt Color of Petroleum Products (Saybolt Chromometer Method), and D1209, Color of Clear Liquids (Platinum-Cobalt Scale). D156 is used only for hydrocarbon solvents. D1209 is used mainly for oxygenated solvents, but is gaining acceptance for hydrocarbon solvents as well. More recently, color measurement by automatic tristimulus colorimetry has gained acceptance and popularity with petroleum solvent manufacturers because of better precision and less subjectivity. ASTM D5386, Color of Liquids Using Tristimulus Colorimetry, is used by the aromatic solvent industry to report correlated or equivalent D1209 color values. ASTM D6045, Color of Petroleum Products by the Automatic Tristimulus Method, can report correlated or equivalent D156 color.

Kauri Butanol Value

ASTM D1133, Standard Test Method for Kauri Butanol Value of Hydrocarbon Solvent, is used to determine relative solvent power. The kauri-butanol (K-B) value is the number of milliliters of the solvent at 25°C (77°F) required to produce a defined degree of turbidity when added to a 20 g sample of a standard solution of kauri gum resin in *n*-butyl alcohol. For K-B values of 60 and over, the reference standard is toluene, which has an assigned value of 105. For K-B values under 60, the reference standard is a blend of 75 %

n-heptane and 25 % toluene. This blend has an assigned value of 40. Higher values indicate greater solvent power. The K-B value of products that are classified as regular mineral spirits normally varies between 34 and 44; xylene is 93; and aromatic naphtha solvents range all the way from 55 to 108.

Aniline Point and Mixed Aniline Point

ASTM D611, Standard Test Method for Aniline Point and Mixed Aniline Point of Petroleum Products and Hydrocarbon Solvents, is another test method for determining the solvent power of hydrocarbon solvents. It is a more precise technique than the test method for K-B value, and the results have better correlation with performance in some coatings. The two methods may indicate a reversal of solvent power when applied to solvents that are in the same range. Therefore, in evaluation testing, it is best to employ both methods. For solvents with very high aromatic content, such as those of the high-flash aromatic naphthas, the mixed aniline point is determined instead of aniline point. Mixed aniline point uses a sample made up of 5 mL of the solvent being tested and 5 mL of *n*-heptane. Automatic instruments are becoming more prevalent these days in the determination of aniline and mixed aniline points of solvents.

Bromine Index

Olefinic and other unsaturated hydrocarbon impurities in solvents could have some deleterious adverse effect on certain solvent applications. Because of the reactivity of the olefinic bond, the color of solvents could be negatively impacted by unsaturated chemical species. When solvents are used as reaction solvents, unsaturated hydrocarbon impurities have to be monitored accurately to prevent deterioration of catalyst materials. ASTM D2710, Test Method for Bromine Index of Petroleum Hydrocarbons by Electrometric Titration, is commonly used. For aromatic hydrocarbons, similar test methods, D1492, Test Method for Bromine Index of Aromatic Hydrocarbons by Coulometric Titration, and D5776, Test Method for Bromine Index for Aromatic Hydrocarbons by Electrometric Titration, are used. All these test methods for determining bromine index do not distinguish the type of unsaturated chemical species in the solvent.

Sulfur

Previous to the advent of accurate quantitative determination of sulfur content of solvents, copper strip corrosion tests were used to detect the presence and determine the relative amounts of sulfur compounds in solvents. Sulfur compounds in the form of mercaptans and di-sulfides are objectionable because they impart odor and may cause darkening of cooked varnishes and resins.

ASTM D849, Standard Test Method for Copper Corrosion of Industrial Aromatic Hydrocarbons, is used for the determination of sulfur in aromatic hydrocarbons, while ASTM D130, Standard Test Method for Detection of Copper Corrosion from Petroleum Products by the Copper Strip Tarnish Test, is applicable for sulfur species in aliphatic hydrocarbons. With modern refining technology, copper corrosion tests have assumed less importance, although some solvent specifications still call for it occasionally.

Doctor's Test is a qualitative test for the presence of sulfur. It is described in ASTM D4952, Test Method for Qualitative Analysis for Active Sulfur Species in Fuels and Solvents (Doctor's Test). In recent years, quantitative instrumental analysis for sulfur content in petroleum solvents has improved tremendously, and it has gained widespread acceptance in the petroleum and solvent industries. Among the many ASTM quantitative sulfur test methods applicable to solvents are: D3120, Test Method for Trace Quantities of Sulfur in Light Petroleum Hydrocarbons by Oxidative Microcoulometry; D3961, Trace Quantities of Sulfur in Liquid Aromatic Hydrocarbons by Oxidative Microcoulometry; D4045, Test Method for Sulfur in Petroleum Products by Hydrogenolysis and Rateometric Colorimetry; D5453, Test Method for Determination of Total Sulfur in Light Hydrocarbons, Spark-Ignition Engine Fuel, Diesel Engine Fuel, and Engine Oil by Ultraviolet Fluorescence; D5623, Test Method for Sulfur Compounds in Light Petroleum Liquids by Gas Chromatography and Sulfur Selective Detection; D6212, Test Method for Total Sulfur in Aromatic Compounds by Hydrogenolysis and Rateometric Colorimetry; D6313, Test Method for Total Sulfur in Aromatic Compounds by Hydrogenolysis and Sulfur Specific Difference Photometry; and D6428, Test Method for Total Sulfur in Liquid Aromatic Hydrocarbons and Their Derivatives by Oxidative Combustion and Electrochemical Detection.

Dilution Ratio in Nitrocellulose Solutions

Solvent mixtures for nitrocellulose lacquers contain three classes of volatile liquids: active solvents, latent solvents (e.g., alcohols), and diluents. The active and latent solvents may be grouped together under the term "oxygenated solvents." The diluents are hydrocarbon solvents, which have no solvency for nitrocellulose and are employed mainly to reduce cost without sacrifice of quality. The proportion of hydrocarbon solvents that can be used without precipitation of the nitrocellulose differs with the kind of hydrocarbon. The dilution ratio test determines this proportion for particular hydrocarbon ratio and is described in ASTM D1720, Test Method for Dilution Ratio of Active Solvents in Cellulose Nitrate Solutions.

$$\text{Dilution ratio} = \frac{\text{Volume of hydrocarbon solvent}}{\text{Volume of butyl acetate}} \quad (1)$$

at the end point. This ratio is roughly proportional to the aromatic content of the hydrocarbon solvent.

The dilution ratio test is also applied to oxygenated solvents. Oxygenated solvents differ in the amount of diluent that they will tolerate. This affects the cost of formulating with a particular oxygenated solvent. For this purpose, a single diluent, toluene, is generally employed. The amount varies with each oxygenated solvent being tested.

Viscosity of Nitrocellulose Solutions

One of the disadvantages of nitrocellulose in coatings is that the more durable grades produce solutions that have high viscosity and, therefore, low solids at a viscosity suitable for application. The viscosity of a solution is affected by the kind of solvent or mixture of solvents. ASTM D1343, Test Method for Viscosity of Cellulose Derivatives by Ball-Drop Method, describes the test for viscosity. The test is designed to determine the inherent viscosity characteristic of various grades of nitrocellulose. The test is equally suitable for comparing the viscosity effect of solvents by varying the solvent while holding the kind and amount of nitrocellulose constant.

Applicable ASTM Standard Specifications and Test Methods

The following boxes summarize the various ASTM standard specifications applicable to petroleum solvents and give the various ASTM standard test methods that can be used for testing petroleum solvents, including cross-reference to Institute of Petroleum (IP) test methods applicable to solvents.

Applicable ASTM Standard Specifications

ASTM	Title
Aliphatic Hydrocarbons	
D235	Mineral Spirits (Petroleum Spirits, Hydrocabon Dry Cleaning Solvent)
D1836	Commercial Hexanes
D3735	VM&P Naphthas
Aromatic Hydrocarbons	
D841	Toluene, Nitration Grade
D843	Xylene, Nitration Grade
D2359	Benzene, Refined 535
D3193	Ethylbenzene
D3734	High-Flash Aromatic Naphthas
D4077	Cumene (Isopropyl Benzene)
D4734	Benzene, Refined 545
D5136	High Purity *p*-Xylene

Applicable ASTM/IP Standard Test Methods

ASTM	IP	Title
Physical Properties		
Volatility		
Distillation		
D86	123	Distillation of Petroleum Products at Atmospheric Pressure
D850		Distillation of Industrial Aromatic Hydrocarbons and Related Materials
D1078		Distillation Range of Volatile Organic Liquids
D7344		Distillation of Petroleum Products at Atmospheric Pressure (Mini Method)
D7345		Distillation of Petroleum Products at Atmospheric Pressure (Micro Method)
Flash Point		
D56		Flash Point by Tag Closed-Cup Tester
D92	36	Flash Point by Cleveland Open-Cup Tester
D93	34	Flash Point by Pensky-Martens Closed-Cup Tester
D2887		Flash Point of Liquids by Small Scale Closed-Cup Appartus
D3828		Flash Point by Small Scale Closed-Cup Tester
D6450		Flash Point by Continuously Closed-Cup Tester
Autoignition Temperature		
E659		Autoignition Temperature of Liquid Chemicals
Residue		
D1353		Non-Volatile Matter in Volatile Solvents for Use in Paint, Varnish, Lacquer, and Related Products
Density and Specific Gravity		
D891		Specific Gravity of Liquid Industrial Chemicals
D1217		Density and Relative Density by Bingham Pycnometer
D1298	160	Density, Relative Density (Specific Gravity) by Hydrometer Method
D1555		Calculation of Volume and Weight of Industrial Aromatic Hydrocarbons
D3505		Density and Relative Density of Pure Liquid Chemicals
D4052		Density and Relative Density of Liquids by Digital Density Meter
Freezing Point		
D852		Solidification Point of Benzene
D1015		Freezing Points of High-Purity Hydrocarbons
D1493		Solidification Point of Industrial Aromatic Hydrocarbons
D2500		Cloud Point (Manual Method)
D5771		Cloud Point (Optical Detection Stepped Cooling Method)
D5772		Cloud Point (Linear Cooling Rate Method)
D5773		Cloud Point (Constant Cooling Rate Method)
Viscosity		
D445		Kinematic Viscosity of Transparent and Opaque Liquids
D1343		Viscosity of Cellulose Derivative by Ball Drop Method

ASTM	IP	Title
Color		
D156		Saybolt Color of Petroleum Products (Saybolt Chromometer Method)
D848		Acid Wash Color of Industrial Aromatic Hydrocarbons
D1209		Color of Clear Liquids (Platinum-Cobalt Scale)
D5386		Color of Liquids Using Tristimulus Colorimetry
D6045		Color of Petroleum Products by the Automatic Tristimulus Method
Refractive Index		
D1218		Refractive Index and Refractive Dispersion of Hydrocarbon Liquids
Solvency		
D611		Aniline Point and Mixed Aniline Point of Petroleum Products
D1133		Kauri-Butanol Value of Hydrocarbon Solvents
D1720		Dilution Ratio of Active Solvents in Cellulose Nitrate Solutions
Odor		
D1296		Odor of Volatile Solvents and Diluents
Definitions and Handling		
D268		Sampling and Testing Volatile Solvents and Chemical Intermediates
D1250	200	Petroleum Measurement Tables
D4057		Manual Sampling of Petroleum and Petroleum Products
D4177		Automatic Sampling of Petroleum and Petroleum Products
Chemical Properties		
Bulk Composition		
D1016		Purity of Hydrocarbons from Freezing Points
D1319	156	Hydrocarbon Types in Liquid Petroleum Products by FIA
D2008		Ultraviolet Absorbance and Absorptivity of Petroleum Products
D2268		Analysis of High-Purity n-Heptane and *Iso*octane by Capillary GC
D2308		C_8 Aromatic Hydrocarbon Analysis by GC
D3797		Analysis of *o*-Xylene by GC

ASTM	IP	Title
D3798		Analysis of *p*-Xylene by GC
D4492		Analysis of Benzene by GC
D5399		Boiling Point Distribution of Hydrocarbon Solvents by GC
Impurity Determination		
Hydrocarbon Types		
D2360		Trace Impurities in Monocylic Aromatic Hydrocarbons
D3257		Aromatics in Mineral Spirits by GC
D4367		Benzene in Mineral Spirits
D5060		Determining Impurities in High-Purity Ethylbenzene by GC
D5917		Trace Impurities in Monocyclic Aromatic Hydrocarbons by GC and External Calibration
D1492		Bromine Index of Aromatic Hydrocarbon by Coulometric Titration
D2710		Bromine Index of Petroleum Hydrocarbon by Electrometric Titration
D5776		Bromine Index of Aromatic Hydrocarbon by Electrometric Titration
D6229		Trace Benzene in Hydrocarbon Solvent By Capillary GC
Sulfur		
D130	154	Copper Corrosion from Petroleum Products by Copper-Strip Tarnish Test
D849		Copper Corrosion of Industrial Aromatic Hydrocarbons
D853		Hydrogen Sulfide and Sulfur dioxide Content (Qualitative) of Industrial Aromatic Hydrocarbon
D1266	107	Sulfur in Petroleum Products (Lamp Method)
D1685		Traces of Thiophene in Benzene by Spectrophotometry
D3120		Trace Quantities of Sulfur in Light Petroleum Hydrocarbon by Oxidative Microcoulometry
D4045		Sulfur in Petroleum Products by Hydrogenolysis and Rateometric Colorimetry
D4735		Traces of Thiophene in Refined Benzene by GC
D4952		Qualitative Analysis of Active Sulfur Species in Fuels and Solvents (Doctor's Test)

ASTM	IP	Title
D5453		Total Sulfur in Light Hydrocarbons, etc.
D5623		Sulfur Compounds in Light Petroleum Liquids by GC and Sulfur Selective Detection
Chloride		
D5808		Organic Chloride in Aromatic Hydrocarbons and Related Chemicals by Microcoulometry
D5194		Trace Chloride in Liquid Aromatic Hydrocarbons
Nitrogen		
D6069		Trace Nitrogen in Aromatic Hydrocarbon by Oxidative Combustion and Reduced Pressure Chemiluminescence Detection

ASTM	IP	Title
Acidity		
D847		Acidity of Benzene, Toluene, Xylenes, Solvent Naphthas, and Similar Industrial Aromatic Hydrocarbons
D1093		Acidity of distillation Residues or Hydrocarbon Liquids
D2896		Total Base Number of Petroleum Products by Potentiometric Perchloric Acid Titration
Water		
D1364		Water in Volatile Solvents (Fischer Reagent Titration Method)
E203		Water Using Karl Fischer Reagent
E1064		Water in Organic Liquids by Coulometric Karl Fischer Titration

Bibliography

Durrans, T. H., *Solvents*, Chapman & Hall, London, 1957.

Marsden, L., *Solvents Guide*, 2nd ed., Cleaver-Hume, London, 1963.

Mellan, I., *Handbook of Solvents*, Vol. 1, Reinhold, New York, 1957.

Mellan, I., *Industrial Solvents*, 2nd ed., Reinhold, New York, 1950.

Mellan, I., *Industrial Solvents Handbook*, Noyes Data Corp., Newark, NJ, 1970.

Modern Petroleum Technology, Institute of Petroleum, London, 1962.

Petroleum Products Handbook, V. B. Buthrie, Ed., McGraw-Hill, New York, 1960.

Reynolds, W. W., *Physical Chemistry of Petroleum Solvents*, Reinhold, New York, 1963.

14

White Mineral Oils

C. Monroe Copeland[1]

WHITE MINERAL OILS ARE VISCOUS LIQUIDS that are derived from petroleum.[2] They are complex mixtures of saturated hydrocarbons including straight chain, branched chain, and ring structures, some of which have all three configurations. The relative concentration of straight or branched chain hydrocarbons (paraffins) and the saturated ring structures (naphthenes) determines the oils' identity as being paraffinic or naphthenic. Many different standard-setting groups in the food, medical, and cosmetic industries, as well as other individual trade associations representing specific applications, define the chemical and physical properties of white mineral oils. Uses and applications are identified by national regulatory agencies along with specific material specifications. *As defined by USP-31, white mineral oil is a "purified" mixture of liquid hydrocarbons obtained from petroleum. It may contain a suitable stabilizer.*

MANUFACTURE

Two methods, separately or in combination, are used to produce finished white mineral oils. The older method involves treating the oil with fuming sulfuric acid to sulfonate the aromatics and polynuclear aromatics, which are removed after neutralization. The more common method today of two-stage hydrotreating completely saturates all double-bond–containing compounds. These treatments may be repeated under different levels of selective severity until the desired purity is attained. White mineral oils were first made in Russia from a light-colored crude oil using the sulfonation process and used for treatment of constipation. It was the grandfather of our current white mineral oils. Today, the dual-stage hydrotreatment provides very severe treatment of the petroleum fractions to remove all but trace reactive materials to give a pure, colorless, odorless, and nonreactive petroleum product known as white mineral oil [USP, NF, 172.878. 178.3620(a) White Mineral Oil]. Additionally, typical manufacturing practices with proper handling and packaging yield white oil products that are allergen free, without genetically modified organisms (GMOs) and Restriction of Hazardous Substances (RoHS) (EU Directive 2002/95/EC) compliant for global acceptance in many applications. More viscous United States Pharmacopeia (USP) products with low volatility also have application for use in the manufacture of products in contact with food.

PURITY GUARDIANSHIP

It is a combination of these four properties, unique to these severely refined petroleum fractions, that has allowed expansion of the applications of white mineral oils from a simple medicinal use to their widespread use in cosmetics, pharmaceuticals, plastics, agricultural and animal sprays, food processing and protection, animal feed, environmental, refrigeration and electrical equipment, chemical reagent and reaction media, and precision instrument lubricants. With this expansion of applications, the scope of purity requirements and special properties for these uses has also become more diverse. In the United States, the UPS and National Formulary (NF) have defined the properties of white mineral oils that allow them to be used in various industries. In the United Kingdom, the British Pharmacopoeia (BP) became responsible for establishing, maintaining, and strengthening the rigid standards for white mineral oil purity. The requirements of the two pharmacopoeias are substantially the same. The prime objective of specifications for pharmaceutical oils is to ensure that the product is as inert as possible and free of toxic materials. Differences in the key specifications for white mineral oils among the two pharmacopoeias and the NF are shown in Table 1. Further, the U.S. Food and Drug Administration (FDA) provides guidelines for specific uses of the white mineral oil based on the properties defined by USP/NF. From these basic white mineral oil requirements, many other agencies, such as the U.S. Department of Agriculture, NSF International, kosher organizations, CTFA, and others, have added specifics to ensure proper application of the white mineral oils in their particular areas.

The USP specification includes an ultraviolet light absorbance test as a requirement for mineral oil that is essentially the same as ASTM Evaluation of White Mineral Oils by Ultraviolet Absorption (D2269). The procedure measures the absorbance of a dimethyl sulfoxide extract.

Because, during manufacture, natural inhibitors in petroleum oils are removed, the resultant white mineral oils are not stable to oxidation. Both the USP and the NF permit the addition of a suitable stabilizer. The BP permits the addition of DL-α-tocopherol (vitamin E) or butylated hydroxytoluene (BHT) in a proportion not greater than 10 ppm.

Besides pharmaceutical uses, the non reactivity of a white mineral oil makes it suitable as (1) a carrier for alkali dispersions for use in organic chemical reactions; (2) in textile and precision instrument lubricants where nongumming is desired; (3) a tobacco de-suckering oil; and (4) in many human, plant, and animal preparations where nontoxicity and freedom from irritation are desired. Each of these uses may require unique qualities.

ASSESSMENT OF QUALITY

Many test methods are available to determine whether the quality for white mineral oils exists as prescribed by the

[1] *Lubricants Compliance Manager, CITGO Petroleum Corporation, Houston, TX 77077.*

[2] Few changes have occurred in the white oil industry and manufacturing processes since the last edition with the exception of significant USP-NF monograph testing and property descriptions and requirements.

TABLE 1—Pharmaceutical Grades of White Mineral Oils

	Mineral Oil, USP 31, 2008	Light Mineral Oil, NF 26, 2008	Liquid Paraffin, BP 1980
Description	Mineral oil is a "purified" mixture of liquid hydrocarbons obtained from petroleum. It may contain a suitable stabilizer.	Light mineral oil is a "purified" mixture of liquid hydrocarbons obtained from petroleum. It may contain a suitable stabilizer.	Transparent, colorless, oily liquid free from fluorescence in daylight, almost odorless and tasteless.
Identification	Infrared absorption <197F>	Infrared absorption <197F>	
	Viscosity test <911>	Viscosity test <911>	
Specific gravity, 25°C	0.845 to 0.905	0.818 to 0.880	64 min at 37.8°C
Kinematic viscosity, mm^2/s (cSt) at 40 ± 0.1°C	34.5 to 150.0	3.0 to 34.4	...
Solid paraffin	Black line (0.5 mm) clearly visible through a 2 5-mm layer of oil	Black line (0.5 mm) clearly visible through a 2 5-mm layer of oil	...
DELETE Test: Neutrality			
Acidity test	No pink color with the addition of phenolphthalein and not more than 1.0 mL of 0.01N NaOH required to change the color to pink.	No pink color with the addition of phenolphthalein and not more than 1.0 mL of 0.01N NaOH required to change the color to pink.	
Limit of sulfur compounds[A]	Negative to sodium plumbite test	Negative to sodium plumbite test	...
Readily carbonizable substances	Maximum color of acid layer stipulated by test	Maximum color of acid layer stipulated by test	...
Limit of polynuclear compounds; absorbance 260 to 350 nm[B]	Absorbance at any wavelength in range of 260 to 350 nm is not greater than one-third of the absorbance at 275 nm of the standard solution.	Absorbance at any wavelength in range of 260 to 350 nm is not greater than one-third of the absorbance at 275 nm of the standard solution.	...
Absorbance; l-cm 240 to 280 nm, 2.0 % solution	...	...	0.10 max

[A] Test reinstated to USP-NF requirements by USP31-NF26.
[B] Absorbance of dimethyl sulfoxide extract.

pharmacopoeias, the NF and consumers. The following paragraphs outline the significance of these tests.

Readily Carbonizable Substances: ASTM D565/USP 31-NF 26 <271>

The acid test, ASTM Test for Carbonizable Substances in White Mineral Oil (Liquid Petrolatum) (D565), is a measure of the purity of a white mineral oil and its freedom from aromatic, unsaturated, and other material that will react with sulfuric acid.

In the test, equal portions of oil and 94.7 ± 0.2 % sulfuric acid are shaken in a prescribed manner at 100°C (212°F) for 10 min, and the color of the acid layer is compared with that of a standard color solution. The more fully refined the oil, the lighter the color of the acid layer.

Most white mineral oil manufacturers, for their own information, protection, and process control, have gone beyond this pharmacopoeia requirement and developed rating systems to measure the degree of color of the acid layer and thus ensure the highest levels of the oil purity.

Aniline Point: ASTM D611

The ASTM Test for Aniline Point and Mixed Aniline Point of Petroleum Products and Hydrocarbon Solvents (D611/IP 2) provides a measure of the aromaticity and naphthenicity of an oil. In the test, equal volumes of aniline and oil are mixed and heated until a miscible mixture is formed. On cooling at a prescribed rate, the temperature at which the mixture becomes cloudy is recorded and identified as the aniline point. While not typically applicable to pharmaceutical grade white mineral oils, it is employed in technical oil specifications as a measure of degree of refinement and type of base oil stock. For any particular oil fraction, a higher degree of refinement is reflected by an increase in aniline point. Aniline point also increases with the average molecular weight of the oil as well as with increasing proportions of paraffinic hydrocarbons to naphthenic hydrocarbons. Aniline point specifications also identify light technical grade white mineral oils that may be used for agricultural sprays where the presence of aromatic hydrocarbons might cause foliage damage.

Cloud Point: ASTM D2500

ASTM Text for Cloud Point of Petroleum Oils (D2500/IP 219) provides the temperature at which wax appears in an oil. This information is significant for oils to be used at low temperatures where precipitation of wax might affect the performance of the oil. Examples include refrigeration and hydraulic oils.

The cloud point test is considered to be more meaningful than the solid paraffins test called for by the pharmacopoeias, which determines whether or not wax precipitates at 0°C (32°F).

Color and Transparency: ASTM D156

By definition, a pharmaceutical grade white mineral oil must be colorless and transparent. On the Saybolt scale, water has a value of +25, while typical white mineral oils have color ratings of +30. In other words, pharmaceutical grade white mineral oils have less color than water.

Freedom from color and transparency are the unique properties of white mineral oils that make them useful in their many cosmetic, pharmaceutical, agricultural, food, and industrial applications. Among the latter are textiles, plastics, precision instruments, leather, household specialty products, and printing ink.

White mineral oil color is determined by means of ASTM Test for Saybolt Color of Petroleum Products (Saybolt Chromometer Method) (D156). In this instrument, the height of a column of the oil is decreased by levels corresponding to color numbers until the color of the sample is clearly lighter than that of the standard. The color number immediately above this level is recorded as the Saybolt color of the oil.

Dielectric Breakdown Voltage: ASTM D877

Dielectric breakdown voltage is the measure of the ability of an oil to withstand electrical stress. White mineral oils used in the electrical or electronics industries as a coolant, insulating fluid, lubricant, or cleaner must have a high-dielectric strength. Dielectric breakdown voltage is determined by ASTM Test for Dielectric Breakdown Voltage of Insulating Liquids Using Disk Electrodes (D877).

Distillation Range: ASTM D2887, D86, D1160, D477

The distillation range for white mineral oils provides information on volatility, evaporation rates, and residue remaining after evaporation. Such data are important for agricultural and household sprays, agricultural product processing, and printing inks. The baking and plastic industries often include initial boiling point temperature or the minimum allowable temperature at which the first several per cent ages of the oil come overhead during distillation, or both, as part of their specifications for white mineral oils.

The distillation range on low-viscosity grades of mineral oils is determined by ASTM Test for Distillation of Petroleum Products (D86/IP 123). The distillation range for higher boiling grades is measured by ASTM Test for Distillation of Petroleum Products at a Reduced Pressure (D1160). More modern techniques utilizing gas chromatography to determine the distillation range is performed using ASTM Test for Boiling Range Distribution of Petroleum Fractions by Gas Chromatography (D2887).

Flash and Fire Point: ASTM D92 or D93

Flash and fire points of white mineral oils are measures of their fire hazard. High flash and fire points are particularly important in many industrial applications where the oils are used at high temperatures.

Flash and fire points are determined by ASTM Test for Flash Point by Pensky-Martens Closed Tester (D93/IP 34) or by ASTM Test for Flash and Fire Points by Cleveland Open Cup (D92/IP 36). The Cleveland open cup-method is also used to determine fire point.

Both methods measure flash point at the minimum temperature to which the oil must be heated to provide a sufficient amount of vapor to ignite. Since the Pensky-Martens tester is a closed system, flash point values are lower with it than those determined by the Cleveland method.

The fire point is the temperature at which the oil ignites and burns for 5 s in the open flash tester.

Fluorescence

The BP states that the oil shall be free from fluorescence by daylight. Fluorescence is a measure of purity and indicates the presence of aromatic hydrocarbons.

Examination of a sample bottle of oil under ultraviolet light is an even more sensitive test, which many manufacturers employ as a rapid means of detecting incompleteness of refining or the presence of petroleum contamination. While it is possible to obtain fluorescent oils that are free of polynuclear aromatic compounds, oils that do not fluoresce are virtually certain to be free of such compounds.

Neutrality Test Name Changed to Acidity Test

In the USP, the test for neutrality was changed in the latest version of USP 31-NF 26 to be identified as the Acidity Test using phenolphthalein.

This is an extremely sensitive test in that the presence of as little as 0.01 mg potassium hydroxide/g is sufficient to turn the phenolphthalein pink. Ten milliliters of filtrated aqueous layer has 0.1 mL of phenolphthalein added and the solution should remain clear. No more than 1.0 mL of 0.01N NaOH is required to change the color to pink to pass the test. The BP uses 10 mL of alcohol, using 5 g of oil. The test is largely a relic of earlier times when it was thought that detectable amounts of chemicals used in processing could remain in the finished product.

Odor and Taste

The BP requires that white mineral oil be almost odorless and almost tasteless. Both tests are highly subjective, but, with practice, many inspectors can detect the slightest trace of an off odor or taste.

Peroxide Value

The peroxide value is an accelerated test to measure the oxidation stability of an oil. Although stability is not an expressed requirement of the pharmacopoeias, sufficient stability is needed to ensure adequate shelf life without the development of rancidity, acidity, or color degradation.

Many consumers of white mineral oils in the pharmaceutical, cosmetic, textile, and coatings industries prescribe oxidation stability tests in their purchase specifications. When white mineral oils are used as heating fluid baths or for other uses at elevated temperatures where medicinal purity is not prescribed, more stabilizer may be used to increase stability.

Pour Point: ASTM D97

Pour point provides a means of determining the type petroleum feedstock from which the white mineral oil was manufactured or its previous processing history. It also reflects the presence of wax or paraffinic hydrocarbons. In any application where the white mineral oil is handled, stored, or used at low temperatures, the pour point is important and, in many industrial applications, critical.

In the ASTM Test for Pour Point of Petroleum Oils (D97/IP 15), the oil is heated to a specified temperature, which is dependent on the anticipated pour-point range, cooled at a specified rate, and examined at 3°C (5°F) intervals for flow. The lowest temperature at which no movement of the oil is detected is recorded. The 3°C (5°F) temperature value immediately preceding the recorded temperature is defined as the pour point.

Refractive Index: ASTM D1218

The refractive index is the ratio of the velocity of light in air to its velocity in the substance under examination. It is used, together with density and viscosity measurements, in calculating the paraffin-naphthene ratio in white mineral oils. Because refractive index is indicative of aromaticity and unsaturation on a given stock, manufacturers also use it as a means of process control.

Refractive index is measured using ASTM Test for Refractive Index and Refractive Dispersion of Hydrocarbon Liquids (D1218).

Smoke Point

Smoke point is of particular interest to industries, such as the baking industry, whose processes expose the oil to extremely high temperatures.

The smoke-point test is conducted in a black box suitably vented and illuminated to permit detection of white vapors. The oil is carefully heated under specified conditions until the first consistent appearance of vapors is detected. The temperature of the oil at that time is recorded as the smoke point.

Gravity, Density, and API Gravity: ASTM D4052, D1298, D287/USP31-NF26 <841>

Density, specific gravity, or American Petroleum Institute (API) gravity may be determined by any one of several different tests such as ASTM Test Method for Density, Relative Density (Specific Gravity), or API Gravity of Crude Petroleum and Liquid Petroleum Products by Hydrometer Method (D1298/IP 160), or a more common method, Density and Relative Density of Liquids by Digital Density Meter ASTM D4052/IP 365. The terms are defined as follows:

Specific gravity–The ratio of the mass of a given volume of liquid at 15°C (60°F) to the mass of an equal volume of pure water at the same temperature.

Density–The mass (weight in vacuum) of liquid per unit volume at 15°C as stated for the specified hydrometer.

API gravity–This is a special function of specific gravity that was arbitrarily established to permit gravity calculations in whole numbers. It is related to specific gravity by the following formula:

$$\text{API gravity, deg} = \frac{141.5}{\text{Sp gr at } 15/15^\circ\text{C}(60/60^\circ\text{F})} - 131.5 \quad (1)$$

Density, specific gravity, and API gravity values permit conversion of volumes at the measured temperature to volumes at the standard petroleum temperatures of 15°C (60°F). Calculation to weight is possible where compositions are formulated on a weight basis. At a given a viscosity, density, specific gravity, or API gravity provides a means of determining whether a white mineral oil is derived from a paraffinic or naphthenic stock.

Limit of Sulfur Compounds: ASTM D4952

Reinstated by USP 31-NF 26, the total sulfur content of white mineral oils should be verified by the simple sodium plumbite test. Severe hydrotreating processes generally ensure sulfur levels well below 1 ppm as opposed to former acid treating processes. The FDA typically requires this test on all food-grade white mineral oil.

Unsulfonated Residue: ASTM D483

ASTM Test for Unsulfonated Residue of Petroleum Plant Spray Oils (D483) provides a crude measure of the reactable hydrocarbons present in an oil. It is often included in specifications for oils intended for use as foliage sprays where aromatic and unsaturated hydrocarbons would cause foliage damage. For high-quality white mineral oils, the Readily Carbonizable Substance Test is used because it is a much more sensitive measure for such impurities.

Ultraviolet Absorption Test: ASTM D2269

The possible presence of carcinogenic polynuclear aromatic hydrocarbons in white mineral oils has always generated serious concern. Because of this fear, much effort has been expended to develop tests to detect the smallest traces of these carcinogens.

ASTM Test Method D2269 measures the absorbance over the wavelength range of 260 to 350 nm in a 10-mm cell of a dimethyl sulfoxide extract of the oil. The polynuclear aromatic hydrocarbons present in the white mineral oils are concentrated in the extract during testing. The FDA has recognized that there is little likelihood of carcinogenic hydrocarbons being present if the absorbance of the extract in the wavelength range is no greater than 0.1. This level of absorbance is roughly equivalent to 5 ppm of polynuclear hydrocarbons with a test sensitivity of ±0.3 ppm in the extract.

Viscosity: ASTM D445/IP 71 USP31-NF26 <911>

Viscosity is one of the most important properties to be considered in the evaluation of a white mineral oil. From a functional point of view, requirements for viscosity vary widely based on the application for which the oil is intended. White mineral oils are produced in the ISO viscosity range of 4 to 100 cSt. White mineral oil for internal use generally should have high viscosity to minimize possibilities of leakage, aspiration, or both. Other white mineral oil applications use various viscosities depending on exact need and requirement of use. Some applications do require higher viscosities based on migration claims.

ASTM Test Method for Kinematic Viscosity of Transparent and Opaque Liquids (and the Calculation of Dynamic Viscosity) (D445/IP 71) is used for the determination of viscosity of white mineral oils.

Wax Precipitation Point

The wax precipitation point test is unique to the refrigerator and freezer industries where oil is added to the Freon refrigerant to provide lubrication during the expansion and compression cycles. Just prior to expansion, the refrigerant bearing the oil passes through a length of capillary copper tubing. It is critical to the process that no wax precipitates and clogs the capillary tubing.

Solid Paraffin Test USP 31

The solid paraffin test indicates the wax content of liquid paraffin oil. The wax content can indicate the operational temperature range, clarity, molecular size, and other properties of the white mineral oil, thus helping determine its suitability for application in various medicinal or other manufacturing processes. To meet USP requirements, a white mineral oil must pass the solid paraffin test (black line clearly visible through mineral oil in a glass bottle).

Applicable ASTM/IP Standard

ASTM	IP	Title
D86	123	Distillation of Petroleum Products
D92	36	Flash and Fire Points by Cleveland Open Cup Tester
D93	34	Flash Point by Pensky-Martens Closed Cup Tester
D97	15	Pour Point of Petroleum Oils
D156		Saybolt Color of Petroleum Products
D287		API Gravity of Crude Petroleum and Petroleum Products (Hydrometer Method)
D445	71	Kinematic Viscosity of Transparent and Opaque Liquids
D447		Distillation of Plant Spray Oils
D483		Unsulfonated Residue of Petroleum Plant Spray Oils
D565		Carbonizable Substances in White Mineral Oil (Liquid Petroleum) USP <271>
D611	2	Aniline Point and Mixed Aniline Point of Petroleum Products and Hydrocarbon Solvents
D877		Dielectric Breakdown Voltage of Insulating Liquids Using Disk Electrodes
D1160		Distillation of Petroleum Products at Reduced Pressure
D1218		Refractive Index and Refractive Dispersion of Hydrocarbon Liquids
D1298	160	ASTM Test Method for Density, Relative Density (Specific Gravity), or API Gravity of Crude Petroleum and Liquid Petroleum Products by Hydrometer Method
D2269		Evaluation of White Mineral Oils by Ultraviolet Absorption
D2500	219	Cloud Point of Petroleum Oils
D2887		Boiling Range Distribution of Petroleum Fractions by Gas Chromatography
D4052	365	Density and Relative Density of Liquids by Digital Density Meter
D4952		Limit of Sulfur Compounds—Standard Test Method for Qualitative Analysis for Active Sulfur Species in Fuel and Solvents (Plumbite—Doctor's Test)
		• Fluorescence (BP)
		• Acidity (USP 31)
		• Odor and Taste (BP)
		• Peroxide Value (industry)
		• Smoke Point (bakery industry—modified flash/fire point)
		• Wax Precipitation Point (refrigeration industry)
		• Solid Paraffin (USP 31)

Bibliography

API Mineral Oil Review, Health and Environmental Sciences Departmental report, American Petroleum Institute, Washington, DC, January 1992.

British Pharmacopoeia, 1980.

Dunstan, A. E., Nash, A. W., et al., *The Science of Petroleum,* Vol. IV, Oxford University Press, London, 1938.

Franks, A. J., "New Technology of White Mineral Oils," *Am. Perfumer,* Vol. 76, No. 2, 1961.

Meyer, E., *White Mineral Oil, Petrolatum and Related Products,* Chemical Publishing Co., New York, 1968.

"Mineral Oil," 21 CFR 1783620, 2005.

"Odorless Light Petroleum Hydrocarbons, " 21 CFR 172.884, 2005.

"Petroleum Wax," 21 CFR 172.886, 2005.

The National Formulary, Vol. 26, 2000.

United States Pharmacopoeia, Vol. 31, 2000.

"White Mineral Oil," 21 CFR 172.878, 2005.

15

Lubricant Base Oils[1]

Jennifer D. Hall[2]

THIS CHAPTER WAS GENERATED TO GIVE AN overview of base oils. Base oils are components of many finished products, including lubricating oils, automotive and diesel engine oils, lubricating greases, white mineral oils, rubber processing oils, industrial oils, and refrigeration and heat transfer fluids, of which most are discussed in detail in other chapters. Lubricant base oils are commonly referred to as base oils, base stocks, mineral oil base fluids, mineral oils, lubricant base fluids, and hydrocarbon base oils throughout the industry. Base oils are manufactured from various refining processes, including both the refining of crude oil and the re-refining of used oils. Oil properties vary based on the feedstock, refining process, and manufacturing parameters. This chapter represents the first step in describing properties of lubricant base fluids affecting product performance. Test methods have been identified to characterize the composition and performance of base oils in addition to verifying their consistency.

BASE OIL COMPOSITION OVERVIEW

Since formulated lubricants typically contain over 80 % base oil by volume, the physical and chemical properties of base oil play a key role in defining the ultimate performance profile of a finished lubricant and the additive technology that must be employed in its formulation. As a result, the development of lubricants for transportation and industrial applications requires a thorough understanding of the chemical composition of base oil and how this influences other properties such as viscosity, viscosity index, volatility, and long-term thermal and oxidative stability.

Hydrocarbon base fluids commonly used in the formulation of engine oils, industrial lubricants, greases, and other products are composed of a broad spectrum of molecular species, including paraffinic, cycloparaffinic (naphthenic), and aromatic molecules. Separation methods and analytical techniques have made it possible to accurately characterize and quantify the hydrocarbon types that exist in base oils while relating these compositional parameters to crude oil source and refining conditions. For example, ASTM Test Method D2549-02 (2007), Standard Test Method for Separation of Representative Aromatics and Nonaromatics Fractions of High-Boiling Oils by Elution Chromatography, exploits column chromatography to separate a base oil into saturate (paraffinic and naphthenic) and aromatic fractions. Mass spectrometry techniques D2786-91 (2006), Standard Test Method for Hydrocarbon Types Analysis of Gas-Oil Saturates Fractions by High Ionizing Voltage Mass Spectrometry, and D3239-91 (2006), Standard Test Method for Aromatic Types Analysis of Gas-Oil Aromatic Fractions by High Ionizing Voltage Mass Spectrometry, further characterize each column-separated fraction into more specific molecular types. ASTM D2007-03 (2008), Standard Test Method for Characteristic Groups in Rubber Extender and Processing Oils and Other Petroleum-Derived Oils by the Clay-Gel Absorption Chromatographic Method, provides similar information as D2549-02 (2007) on aromatics and saturates but also determines the "polar aromatic/heterocyclic" fraction of base oils. As a result of these methods, the physical and chemical properties of base fluids can be better understood, and in some cases can be predictive of the performance of a formulated lubricant directly from base oil composition using statistical methods.

BASE OIL REFINING

Base oil manufacture is a complex process that separates molecules based on molecular weight (size, boiling range) and composition. The major objectives of refining are to fractionate crude oil into viscosity grades appropriate for lubricant oils, to reduce the level of undesirable aromatic compounds, and to lower the level of sulfur and nitrogen-containing molecules that can negatively influence lubricant stability performance. The heaviest or nonboiling fraction, separated by propane de-asphalting, also finds application in certain lubricants. Sequeria [1] reviews the processing stages for base oil in great detail. In addition to removing undesirable molecular species, refining generally increases the viscosity index (VI) of an oil. VI reflects the relationship between base oil viscosity and temperature and is calculated from kinematic viscosities recorded at 40 and 100°C (as determined by ASTM D445-06, Standard Test Method for Kinematic Viscosity of Transparent and Opaque Liquids), using an empirically derived procedure defined in ASTM D2270-04, Standard Practice for Calculating Viscosity Index from Kinematic Viscosity at 40 and 100°C. All hydrocarbon-based mineral oils decrease in viscosity with increasing temperature, and the magnitude of viscosity loss with temperature is related to the VI. Higher VI oils exhibit a lower degree of viscosity decrease with increasing temperature than do lower VI oils. For example, comparing two oils that have an identical viscosity at 40°C, the higher VI oil will have a higher viscosity at 100°C and a lower viscosity at 0°C compared to the lower VI product. Typical "paraffinic" base oils,

[1] In preparation of this chapter, the contents of the seventh edition were drawn upon. The author acknowledges the author of the seventh edition, Arthur J. Stipanovic; Director, Analytical and Technical Services, Faculty of Chemistry, State University of New York, College of Environmental Science and Forestry. The current edition will review and update the topics as addressed by the previous author, introduce new technology that has been developed, and include up-to-date references.

[2] *Quality and Specialty Products Development Manager; Holly Refining and Marketing, Tulsa, OK 74107.*

produced from conventional crude oils, have a VI in excess of 90 and are "dewaxed" to remove linear paraffins that can crystallize at low temperatures, causing fluidity problems.

Annual U.S. production of hydrocarbon-based mineral oils is about 20 % designated as "naphthenic" and 80 % "paraffinic." The widely used term "naphthenic" base oil represents those stocks produced from predominately naphthenic crude as defined by the U.S. Bureau of Mines Classification of Crude Oils. In this system, the American Petroleum Institute (API) gravity is measured on two crude fractions that fall within typical base oil boiling ranges. They are considered naphthenic if they have API gravities <33 for a boiling range of 250 to 275°C and <20 for a boiling range 275 to 300°C at 1 atm. Typical paraffinic oils for the same boiling ranges provide API gravities of >40 and >30, respectively. API gravity or density [defined by any of the following ASTM methods: D287-92 (2006); D6822-02 (2008); D1298-99 (2005); D4052-96 (2002); D5002-99 (2005); D1481-02 (2007); D1480-07; D1217-93 (2007)] is sensitive to the relative ratio of paraffins, cycloparaffins, and aromatics in a crude or base oil. For the same boiling range material a higher paraffinic character increases API gravity. A significant and commercially important feature that distinguishes most naphthenic crudes is a very low natural level of linear paraffins (wax), which enables the production of base oils with low wax pour points (having a viscosity influenced pour point), which is desirable in applications such as refrigeration oils. Note that many modern refrigeration oils are now made with paraffinic base oils containing a pour point depressant additive. Due to their low VI, naphthenics have become unsuitable for modern engine oils but many find application in process oils.

Although traditional synthetic fluids such as polyalphaolefins (PAOs) and polyolesters have grown in their importance in many lubricants, including engine oils, this chapter will focus on mineral oil base fluids. For most lubricant product lines such as engine oils and industrial process oils, petroleum base stocks account for greater than 90 % of the commercial product volume, although the penetration of non–petroleum-based synthetics is increasing. In recent years, the term "synthetic" has also been applied to the high VI Group III base oils.

ASTM BASE OIL TESTS

From the perspective of lubricant applications, a broad portfolio of base oil tests have been defined by ASTM that help to optimize the performance of a fully formulated product that contains base oil plus an additive package designed for a specific application. These methods may generally be divided into tests of base oil composition (percentage aromatics, percentage saturates, etc.), chemical properties (oxidation, solvency, etc.), physical properties (viscosity, pour point), and "bench tests" that mimic the important performance features of the lubricant in actual service. It should be emphasized that many tests developed for fully formulated or "finished" lubricants are routinely applied to base oils as well, in order to "benchmark" the performance of different base oils. In these cases, the base oils being evaluated might contain one or more additives so that their performance in the ASTM test is more characteristic of a formulated product. For these benchmarking studies, the normal repeatability and reproducibility specified in the ASTM method might not hold, however, so some care must be taken when applying "finished" lubricant tests to unadditized or marginally additized base oils. A summary of test methods strictly appropriate for base oils is given in ASTM D6074-08, Standard Guide for Characterizing Hydrocarbon Lubricant Base Oils.

BASE OIL CHARACTERIZATION METHODS

Hydrocarbon-Type Compositional Analysis

The chemical composition of lubricant base oil is strongly dependent on a number of factors. These include crude oil source, molecular weight range (generally higher-molecular-weight crudes are richer in multi-ring aromatics), the refining process, the degree of refining, and the effectiveness of the finishing process.

Determining the chemical compositional profile of base oil is usually initiated with a chromatographic separation procedure to isolate the aromatic and saturate fractions of the sample. ASTM Method D2549-02 (2007) involves an open column procedure using an activated bauxite and silica gel column packing with polar and nonpolar organic solvents to elute the aromatic and saturate fractions, respectively. In this procedure, a sample of base oil is applied to the top of the column, which is then flushed sequentially with solvents of increasing polarity. Saturated molecules, including both paraffins and naphthenic molecules, bind to the column less strongly than aromatic compounds, and, as a result, they elute from the column with *n*-pentane (nonpolar solvent), while the aromatics require "stronger" polar solvents (chloroform/ethyl alcohol) to be eluted.

Base oils contain a preponderance of saturated compounds. Paraffinic base oils exhibit a higher percentage of paraffinic carbons (typically >65 %) than cycloparaffinic carbons (typically <30 %) and have an aromatic carbon content below 5 %. Base oils processed from hydrotreating naphthenic crudes typically exhibit equal cycloparaffin and paraffin contents and a lower level of aromatics (<20 %). This is easily calculated using ASTM D2140-03, Standard Test Method for Carbon Type Composition of Insulating Oils of Petroleum Origin, or D3238-95 (2005), Standard Test Method for Calculation of Carbon Distribution and Structural Group Analysis of Petroleum Oils by the n-d-M Method.

The column chromatography method, ASTM D2007-03 (2008), is based on a clay-silica gel adsorption procedure that resolves a base oil sample into aromatic, saturate, and polar fractions. In this procedure, the following solvents are used to elute the hydrocarbon fractions: *n*-pentane (elutes saturates), toluene (elutes aromatics), and toluene/acetone (elutes polars). ASTM D2007-03 (2008) was chosen as the preferred method for base oil characterization in the "Base Oil Interchangeability Guidelines" for engine oils established in the API Publication 1509, Engine Oil Licensing and Certification System. Under these guidelines, base oils are grouped into five categories according to composition, viscosity index, and sulfur content as shown in Table 1. Group IV is reserved for PAOs, and Group V includes other base stocks not included in Groups I–IV. However, this clay-silica gel chromatography method is not useful for measuring the aromatic content in Group II and III base oils because of the precision of the method.

Spectrometric Identification of Base Oils Components

Although nuclear magnetic resonance (NMR) spectroscopy, ultraviolet (UV) spectroscopy [ASTM D2008-91 (2006, Standard Test Method for Ultraviolet Absorbance and

TABLE 1—API Lubricant Base Stock Categories

API Group	% Saturates, ASTM D2007-03 (2008)	Viscosity Index, ASTM D2270-04		% Sulfur[A]
I	<90	80–119	And/or	>0.03
II	≥90	80–119	And	≤0.03
III	≥90	≥120	And	≤0.03
IV	All polyalphaolefins (PAOs)			
V	All base stocks not in Groups I-IV (naphthenics, re-refined, non-PAO synthetics)			

[A] Sulfur content by any of the following: ASTM D1552, D2622, D4294, D4927, D3120.

Initiation:	I-I ⇨ I• (Free Radical)
	R-H + I• ⇨ I-H + R•
Propagation:	R• + O_2 ⇨ RO_2•
	RO_2• + R-H ⇨ ROOH + R•
Termination:	R• + R′O_2 ⇨ ROOR′
	2 RO_2• ⇨ R-O_4-R
	R-O_4-R ⇨ O_2 + Non-Radical Products

Fig. 1—Hydrocarbon oxidation by free radicals.

Absorptivity of Petroleum Products], and infrared spectroscopy can be successfully employed to characterize the compositional profile of lubricant base oils, certain mass spectrometry (GCMS) techniques have evolved into standard ASTM methods suitable for base oil analysis. Once base oil saturate and aromatic fractions are isolated by the column chromatography procedure [ASTM D2549-02 (2007)], each fraction can be subjected to further analysis and subcategorization by GCMS using ASTM Methods D2786-91 (2006) and D3239-91 (2006). ASTM D2786-91 (2006) is useful when analyzing nonolefinic compounds having average carbon numbers 16 through 32 and less than 5 % monoaromatic hydrocarbons. D3239-91 (2006) covers a broader carbon range, but samples must still be nonolefinic, be low in sulfur (<1 %), and not contain more than 5 % monoaromatic hydrocarbons. The aromatic types can be subdivided by ASTM D3239-91 (2006) such that the volume percent of materials like chrysenes, naphthalenes, dibenzoanthracenes, dibenzothiophenes, among others, can be determined.

Although the color of a lubricant base oil does not directly reflect its bulk composition, since the presence of a very small amount of a highly colored species can determine its overall color, it may generally be concluded that oils rich in aromatic and olefinic structures are more highly colored (yellow, brown colors) than paraffinic structures. ASTM Method D1500-07, Standard Test Method for ASTM Color of Petroleum Products (ASTM Color Scale), provides a visual index by which the color of undyed base oils can be estimated and compared.

Base Oil Oxidation

Hydrocarbon oxidation chemistry is of great importance in the ultimate application of base oils and lubricants since, at the high temperatures typical of most lubricating environments, oxidation can lead to undesirable oil thickening, the formation of insoluble sludge and varnish deposits, and the creation of potentially corrosive organoacid compounds. In summary, hydrocarbon oxidation proceeds through a number of steps involving free radicals as shown in Fig. 1, where I is an initiator species, R-H represents a base oil hydrocarbon molecule, and various RO_x species are oxidation products.

A number of ASTM tests are used to characterize the oxidation characteristics of finished lubricant products, and many can also be applied to determine the relative stability of a base oil by including a standard "dose" of antioxidant or additive package, or both, in each oil. The long-term oxidation resistance of base oils employed in industrial lubricants (steam turbine oils, hydraulic fluids, etc.) can be evaluated using ASTM D943-04a, Standard Test Method for Oxidation Characteristics of Inhibited Mineral Oils, which monitors the length of time (typically in months) required to reach a specific neutralization number (D3339-08, Standard Test Method for Acid Number of Petroleum Products by Semi-Micro Color Indicator Titration, or D664-09, Standard Test Method for Acid Number of Petroleum Products by Potentiometric Titration) in the presence of oxygen, water, copper, and iron metals. Method D2272-02, Standard Test Method for Oxidation Stability of Steam Turbine by Rotating Pressure Vessel, is also commonly employed to evaluate base oils to which a turbine oil antioxidant package has been added. This method, also called the Rotary Bomb Oxidation Test (RBOT), utilizes oxygen at high pressure to accelerate base oil oxidation in the presence of a copper coil catalyst and water at 150°C. The duration of this test is, typically, minutes compared to the much longer period of the D943-04a test.

Base oils employed in automotive lubricants can be benchmarked in the presence of an appropriate additive package using the Thin Film Oxygen Uptake Test (TFOUT) ASTM D4742-08, Standard Test Method for Oxidation Stability of Gasoline Automotive Engine Oils by Thin-Film Oxygen Uptake. This test, conducted at 160°C, utilizes a bomb pressurized with oxygen and a metal catalyst package, a fuel catalyst, and water to simulate the conditions of a gasoline combustion engine. In addition, an instrumental method based on high-pressure differential scanning calorimetry (DSC), originally developed for use on greases, is commonly applied to base oils that contain a stabilizer package that delays oxidation onset. The original test method, ASTM D5483-05, Standard Test Method for Oxidation Induction Time of Lubricating Greases by Pressure Differential Scanning Calorimetry, subjects the base oil to 3.5 MPa of oxygen pressure at temperatures between 155 and 210°C. Test duration is usually also on the order of minutes. ASTM Guide D4871-06, Standard Guide for Universal Oxidation/Thermal Stability Test Apparatus, highlights additional protocols for defining the thermal performance of lubricants. An additional test method that focuses on both oxidation and the development of metal corrosion capabilities in base oils and lubricants is D4636-99 (2004), Standard Test Method for Corrosiveness and Oxidation Stability of Hydraulic Oils, Aircraft Turbine Lubricants and Other Highly Refined Oils.

Although hydrocarbon oxidation can have a profound effect on the properties of base oil and lubricants, this process can, fortunately, be controlled by the use of antioxidant additives. Very generally, antioxidants function either by scavenging radical species (see Fig. 1; I•, R•, or RO_2• species) or by decomposing peroxides (ROOR′ molecules) to unreactive products. Radical scavengers include hindered phenols such as *t*-butylphenol, quinones, and certain amines. Zinc dialkyldithiophosphate, a common antiwear compound, also serves in a peroxide-decomposing capacity as well.

Impact of Sulfur Compounds on Base Oil Chemistry

Throughout the evolution of base oil refining and processing technology, sulfur content has been employed as an indicator of product quality and a predictor of lubricant performance. Although current manufacturing trends suggest that very low sulfur content is most desirable for base oil, slightly higher sulfur levels directionally improve base oil oxidation stability as well as friction and wear performance. In this context, the term "sulfur" relates to soluble, organosulfur compounds that occur naturally in crude oil in contrast to elemental, yellow sulfur. Current domestic base oil sulfur concentrations typically range from <0.005 weight percent to 0.1 weight percent for light base stocks and up to 0.05–1.0 weight percent for heavier viscosity grades. In other regions of the world, it is not uncommon to observe sulfur levels in excess of 1.0 weight percent. From the perspective of base oil chemistry, how the organosulfur content impacts lubricant performance is very complex and depends, to a large extent, on the ultimate application and the nature of the additives used in the formulation.

It is generally observed that desulfurized base oils exhibit relatively poor oxidation inhibition compared to the original base oil and that certain sulfur compounds, specifically organosulfides, can significantly reduce the rate of oxidation at high temperatures. Sulfur-containing aromatic molecules, however, such as dibenzothiophene, do not appear to act as antioxidants and may actually increase the rate of oxidation [2]. The role of sulfur compounds as naturally occurring base oil stabilizers has some very interesting implications. It is well known that untreated, sulfur-free synthetic base oils, such a PAOs, oxidize more readily in tests such as the RBOT (ASTM D2272-02) than conventional mineral oils that contain some sulfur compounds. However, if appropriate antioxidants are added to both types of base fluids, the PAOs typically respond with better long-term oxidation stability. Severely hydrocracked, low-sulfur Groups II and III base oils exhibit similar behavior in their response to additive technology.

A number of ASTM methods are commonly employed to determine the sulfur content of base oils for API group classification. Other tests are designed to analyze lubricants that may be formulated with sulfur-containing additives. Standard Test Methods D2622-08, Sulfur in Petroleum Products by Wavelength Dispersive X-Ray Fluorescence Spectrometry, and D4294-08a, Sulfur in Petroleum Products by Energy Dispersive X-Ray Fluorescence Spectroscopy, are generally used for base oils ranging in sulfur content from 0.005 to 5 mass percent. Additized lubricant formulations that may contain sulfur plus certain other elements can be characterized using ASTM D4927-05, Standard Test Methods for Elemental Analysis of Lubricant and Additive Components–Barium, Calcium, Phosphorus, Sulfur and Zinc by Wavelength Dispersive X-Ray Fluorescence Spectroscopy.

Nitrogen Compounds in Base Oil

Nitrogen-containing organic molecules found in base oils can accelerate oxidation, promote varnish and sludge formation, and ultimately reduce the useful lifetime of lubricants. Multiple ASTM techniques can determine the total level of nitrogen (both organic and inorganic) in base oils and expected concentrations will help determine which is best suited for the specific oil analyzed. D4629-08, Standard Test Method for Trace Nitrogen in Liquid Petroleum Hydrocarbons by Syringe/Inlet Oxidative Combustion and Chemiluminescence Detection, is effective for concentrations of nitrogen, 0.3 to 100 mg/kg. A modified Kjeldahl technique, ASTM D3228-08, Standard Test Method for Total Nitrogen in Lubricating Oils and Fuel Oils by Modified Kjeldahl Method, can detect even higher nitrogen concentrations, 0.03 to 0.1 mass percent. For nitrogen concentrations above 0.1 mass percent, either D5291-02 (2007), Standard Test Method for Instrumental Determination of Carbon, Hydrogen, and Nitrogen in Petroleum Products and Lubricants, or D5762-08, Standard Test Method for Nitrogen in Petroleum and Petroleum Products by Boat-Inlet Chemiluminescence (which covers nitrogen concentrations of 40 to 10,000 ppm) can be used.

Olefins in Base Oil

Although crude oils are generally relatively low in olefin content, base oil processing techniques can introduce olefins, especially at high temperatures, due to "cracking" reactions. In the presence of heat or UV light, olefins can polymerize to form higher-molecular-weight products that can color the base oil or actually cause sediment. In general, olefins can be removed during the process of hydrofinishing or by clay treatment. Depending on the boiling range of the base oil, olefin contents may be determined by their reactivity with bromine according to the following methods based on electrometric titration: ASTM D1492-08, Standard Test Method for Bromine Index of Aromatic Hydrocarbons by Coulometric Titration, D2710-99 (2004), Standard Test Method for Bromine Index of Petroleum Hydrocarbons by Electrometric Titration, and D5776-07, Standard Test Method for Bromine Index of Aromatic Hydrocarbons by Electrometric Titration. It should be noted that, strictly speaking, these olefin methods apply to oils with a boiling point under 288°C. Most lubricant base oils distill at higher temperatures under vacuum and boil above 288°C. Sulfur compounds may also interfere in these methods, so these techniques are probably best viewed as "indicative" rather than "absolute" for base oils.

Evaporative Loss/Volatility Characteristics of Base Oils

An important base oil characteristic incorporated into the specifications for passenger car and diesel engine oils is a limit on evaporative loss at elevated temperature for reasons of emissions reduction and catalytic converter longevity. ASTM Test Method D2887-08, Standard Test Method for Boiling Range Distribution of Petroleum Fractions by Gas Chromatography (GC), describes a "simulated distillation" procedure for determining the boiling range of petroleum products (up to a final boiling point of 538°C [1,000°F], and

D6352-04, Standard Test Method for Boiling Range Distribution of Petroleum Distillates in Boiling Range from 174 to 700°C by Gas Chromatography, can be used for products with final boiling points up to 700°C (1,292°F). These can be used to determine the fraction of base oil that will evaporate at a specific temperature. Related to ASTM D2887-08 and D6352-04 is ASTM D6417-03, Standard Test Method for Estimation of Engine Oil Volatility by Capillary Gas Chromatography, which may be applied to both base fluids and finished lubricants. Since not all finished lubricant additives may elute, the calculation in this method makes assumptions and may bias results high.

Volatility can also be determined using an international method, DIN 51.5681, the Noack Evaporative Test, which does not require a gas chromatograph. A version of this procedure was adopted as an ASTM technique, D5800-08, Standard Test Method for Evaporation Loss of Lubricating Oils by the Noack Method. For Procedure A of this test protocol, a 65-g sample of base oil is placed in a cup under vacuum (20-mg Hg pressure), and it is exposed to high temperature (250°C). Test results are reported as weight percent loss after 60 min. Although this method enjoys reasonably good repeatability (5.8 %), it suffers from poor lab-to-lab reproducibility (18.3 %). To a lesser extent, the volatility of base oils can be expressed in terms of flash point, which can be determined by ASTM Methods D92-05a, Standard Test Method for Flash and Fire Points by Cleveland Open-Cup Tester, and D93-08, Standard Test Method for Flash by Pensky-Martens Closed-Cup Tester. In both cases, semiautomated and automated apparatuses are available. Flash point is generally defined as the temperature at which a vapor will ignite when an ignition source, flame, or an electric igniter is brought to within a specified distance from the surface of the fluid. Flash point, however, is a poor determinant of volatility as small amounts of volatile contamination are difficult to detect.

Base Oil Pour Point and Rheology

The ability of base oil to sustain a lubricating film in mechanical equipment at both high and low temperatures, under a broad spectrum of shear rate conditions, is a defining element in its ultimate performance. The rheology, or flow properties, of a lubricant strongly reflects base oil composition as well as any polymeric additives that are included in the formulation to reduce viscosity loss at elevated temperatures. Probably the most fundamental rheological property determined for base oils is kinematic viscosity, determined by capillary viscometry at 40 and 100°C using ASTM Method D445-06. From viscosity values determined at 40 and 100°C, the VI can be computed using D2270-04, which reflects the relative magnitude of viscosity loss upon heating. Comparing hydrocarbons of similar molecular weight, paraffins provide the highest positive contribution to base oil VI, while aromatics and naphthenes, particularly multi-ring structures, strongly decrease VI. Base oils of high VI are generally preferred for finished engine oil lubricants because they provide higher viscosity at high temperatures and lower viscosities at low temperatures provided they are properly treated to inhibit wax crystallization at low temperatures. Since many finished lubricants are required to operate at temperatures in excess of 100°C, methods capable of measuring the viscosity of a base oil or lubricant at 150°C, under conditions of high shear rate ($>10^6$ s^{-1}), such as encountered in an automobile engine, have been developed. These techniques include capillary-based method D5481-04, Standard Test Method for Measuring Apparent Viscosity at High-Temperature and High-Shear Rate by Multicell Capillary Viscometer, and procedures based on a tapered-bearing simulator (TBS), D4683-04, Standard Test Method for Measuring Viscosity at High Shear Rate and High Temperature by TBS, and D4741-06, Standard Test Method for Measuring Viscosity at High Temperature and High Shear Rate by Tapered-Plug Viscometer. ASTM D6616-07, Standard Test Method for Measuring Viscosity at High Shear Rate by TBS Viscometer at 100°C, utilizes the TBS to determine high shear viscosity of engine oils at 100°C. Generally, these high-temperature, high-shear-rate methods are applied to finished lubricants to estimate the extent of "shear thinning" experienced by polymeric VI enhancing additives, but they can be applied routinely to base oils as well, which normally exhibit "Newtonian" rheology where viscosity does not change with increasing shear rate.

The rheology of base oils at low temperatures (below 0°C, for example) is also important in defining the applicability of many lubricants. For most crude oil–derived, hydrocarbon-based lubricants, low-temperature flow properties are defined by VI and the extent of wax crystallization that occurs at a specific temperature. Obvious manifestations of wax crystallization in base oil are turbidity (cloudiness) and failure to flow under modest levels of stress. The onset temperature of cloudiness is the cloud point as determined by ASTM D2500-05: Standard Test Method for Cloud Point of Petroleum Products, while the onset temperature for inhibited fluid flow is called the pour point. Historically, pour point was determined visually as the temperature at which the surface of a tube of base oil or lubricant did not appear to flow when the tube was tilted at a specific angle. This visual method is detailed in ASTM D97-08, Standard Test Method for Pour Point of Petroleum Products. Automated methods have been introduced based on mechanical tilting, rotation, or nitrogen pressure pulsing of the base oil surface. These tests are outlined in the following ASTM methods, respectively: ASTM D5950-02 (2007), Standard Test Method for Pour Point of Petroleum Products (Automatic Tilt Method), D5985-02 (2008), Standard Test Method for Pour Point of Petroleum Products (Rotational Method), and D5949-08, Standard Test Method for Pour Point of Petroleum Products (Automatic Pressure Pulsing Method). In most cases, polymeric pour point depressant (PPD) additives inhibit the onset of wax crystallization, and, under these circumstances, base oils rich in paraffins can exhibit excellent rheological properties at low temperatures. However, even in the presence of a pour point depressant, low temperature extremes can cause base oil to "solidify." This is commonly referred to as a vis-pour as opposed to a wax pour.

In engine oils, poor cold temperature fluidity can reduce pumpability to the extent that oil starvation causes catastrophic engine failure. To protect against this occurrence, the SAE J300 Engine Oil Viscosity Classification specifications include two ASTM methods: D5293-08, Standard Test Method for Apparent Viscosity of Engine Oils Between −5 and −35°C Using the Cold-Cranking Simulator, and D4684-08, Standard Test Method for Determination of Yield Stress and Apparent Viscosity of Engine Oils at Low Temperature, which measure the high shear viscosity and pumpability, respectively, of engine oils at low temperatures to ensure that a motor will start and have sufficient oil fluidity for good lubrication. These tests are generally applied to finished lubricants, but D5293-08 is also commonly used as a base oil specification for viscosity grades utilized for engine oils. In this test, the very high shear rate of the "cold cranking simulator" destroys wax network formation, and a viscosity can be determined at temperatures from −5 to −35°C, well below the pour point of the base oil that is determined at very low shear rate.

A low-temperature viscosity limit is also included in the performance specifications for many gear oils, transmission fluids, power steering fluids, and other lubricants potentially used at low temperatures. As defined in ASTM D2983-04a, Standard Test Method for Low Temperature Viscosity of Lubricants Measured by Brookfield Viscometer, viscosity is determined using a very low shear viscometer operating at +5 to −40°C, depending on viscosity grade. Testing and comparison of base oils (versus formulated lubricants) is normally accomplished by adding a standard amount of a polymeric PPD to the base oil before the Brookfield procedure to prevent wax crystallization. In the absence of a PPD, base oil Brookfield viscosities often exceed the maximum measurable value at low temperature.

Aniline Point and Base Oil Solvency

In any lubricant formulation, base oils are required to dissolve polar additive compounds and to ultimately disperse polar oxidation products that are formed during use. For this reason, the aniline point of base oil can be a critical parameter in defining its compatibility with additives and the by-products of use. Aniline point is determined by ASTM D611-07, Standard Test Methods for Aniline Point and Mixed Aniline Point of Petroleum Products and Hydrocarbon Solvents, and represents the temperature at which aniline (C_6H_5-NH_2), a polar aromatic compound, becomes miscible with a hydrocarbon base oil. At low temperatures, base oil and aniline are not miscible. However, as the temperature is raised, they become a single phase at the aniline point. In general, base oils of low aromatic/high saturate content have high aniline points (>230°F, 110°C; API Group II, III base oils). Conventional solvent-refined base oils, such as Group I oils, exhibit moderate aniline points (200 to 215°F, 93 to 102°C), and naphthenic base oils have low aniline points (<150°F, 65°C). As a result, aniline point can be viewed as a sensitive index of the "solvent power" of base oil toward polymeric VI improvers, organometallic friction modifiers and antiwear agents, detergents, antioxidants, and other molecules. Additives soluble in low-to-medium aniline point base oils might experience limited solubility in high aniline point products. In such cases, improved solubility can be attained by adding a small fraction of a very polar synthetic base stock such as polyolester or an alkylbenzene to the high aniline point base oil, thereby increasing its "solvent power."

Rubber Compatibility

Since there is a chapter specifically dedicated to rubber process oils, this chapter will limit discussion on the subject. In many service applications, lubricant products come in contact with rubber seals, gaskets, o-rings, and other "elastomer" components. Since most rubber compounds include a base oil or synthetic plasticizer, or both, to modify the compound, the ability of a base oil to remove and solubilize the plasticizer is an important consideration. In other cases, certain base oil molecular fractions can actually migrate into the rubber matrix, causing it to swell. Although most lubricant base oils are relatively inert in their ability to deleteriously interact with a variety of elastomeric materials commonly in use, high aniline point products, such as Group IV PAO, can cause elastomer cracking after long periods of exposure at high temperatures due to a loss in plasticity. In many cases, a seal swell agent can be successfully added to PAO to maintain good seal characteristics. At the other extreme, low aniline point naphthenic base stocks can cause seals to swell excessively, also creating operational problems. Most typically, the effect of base oil is evaluated as part of a finished lubricant containing an additive package using ASTM test protocol D4289-03 (2008), Standard Test Method for Elastomer Compatibility of Lubricating Greases and Fluids, which examines elastomer volume and hardness changes upon exposure to a lubricant for 70 h at 100 or 150°C.

Biological Properties of Base Oils

Since lubricants can come in direct contact with humans in some applications, the biological toxicity of base oils is a concern, especially for those products containing high concentrations of polycyclic aromatic hydrocarbons (PAHs). All oils contain some level of PAHs. Some of these compounds have been found to be mutagenetic and carcinogenic, and the International Agency for Research on Cancer (IARC) has classified 18 PAHs as cancer causing. From a molecular structure perspective, these 18 molecules each have in common a 4 to 6 fused aromatic ring system and a so-called three-sided concave "bay region" [3].

Probably the most well-established method to assess the potential mutagenicity/carcinogenicity of base oils is the IP 346 method that uses dimethyl sulfoxide (DMSO) extraction to isolate a subset of PAHs known as polycyclic aromatics (PCAs) up to 15 mass percent in unused additive-free lubricant base oils. By correlation to mouse skin painting studies, it was determined that base oils exhibiting DMSO extractable levels of PCAs exceeding 3 weight percent should be considered carcinogenic. ASTM Method E1687-04, Standard Test Method for Determining the Carcinogenic Potential of Virgin Base Oils in Metalworking Fluids, represents a standard procedure for determining the dermal carcinogenic potential of virgin base oils to be used in metalworking fluids as defined by an Ames *Salmonella* mutagenicity assay.

The ability of a lubricant to degrade naturally in the environment is an increasingly important property that is directly related to the composition and molecular weight of the base oil employed in its formulation. Currently, several standard experimental protocols are utilized to evaluate the biodegradability of a lubricant: ASTM D5864-05, Standard Test Method for Determining Aerobic Aquatic Biodegradation of Lubricants or Their Components, is suitable for fully formulated lubricants and base oils, while D6006-97a, Standard Guide for Assessing Biodegradability of Hydraulic Fluids, focuses on a specific product. A related European method, CEC-L-33-T-82, is used internationally. The CEC test [4] shows that hydraulic fluids based on hydrocarbon mineral oils only degrade to an extent of 40 % after 25 days, while synthetic, ester-based fluids are degraded almost entirely (>90 %). Synthetic hydrocarbons and polyethylene glycols were found to display intermediate behavior with maximum degradation levels of 70 and 55 %, respectively.

In addition to composition, the size of a base oil molecule can influence its biodegradability. Evidence suggests that a limiting molecular size exists above which an organism cannot adsorb and biodegrade hydrocarbon molecules.

Miscellaneous Base Oil Properties

In cases where base oil might be formulated into a product that is applied at very high temperature where volatilization or pyrolysis, or both, might occur, the residual carbon (coke) remaining after thermal exposure is an important parameter since it could result in solid deposit formation. Two ASTM methods give comparable results for carbon residue: D189-06, Standard Test Method for Conradson Carbon Residue of Petroleum Products, and D4530-07, Standard Test Method for Determination of Carbon Residue (Micro Method). The second of which is a "micro" technique, which as indicated by the name requires less sample than D189-06. A third method provides carbon residue results that can be "correlated" to the Conradson numbers: D524-04, Standard Test Method for Ramsbottom Carbon Residue of Petroleum Products. For all carbon residue measurements, it should be noted that inorganic ash present in the sample will contribute to the quantity of residue determined. For samples where high ash levels are suspected, it is recommended that D482-07, Standard Test Method for Ash From Petroleum Products, be applied along with carbon residue determination. The resulting ash value (mass percent) can then be used to adjust the apparent carbon residue values to account for the presence of noncarbonaceous inorganic compounds.

In many lubricant applications, mechanical agitation in the presence of air can cause air entrainment and foaming. In order to characterize the potential of a base oil or lubricant to create foam, ASTM Test Method D892-06, Standard Test Method for Foaming Characteristics of Lubricating Oils, is used and applied over the temperature range of 24 to 93.5°C, while at 150°C, D6082-06, Standard Test Method for High Temperature Foaming Characteristics of Lubricating Oils, can be employed. In most cases, lubricant foaming is commonly the result of additives rather than the base oil itself, but exceptions may exist.

International Standards

Many of the ASTM experimental test procedures outlined in this chapter have analogs in the IP standard test methods (formerly Institute of Petroleum now Energy Institute) published in the United Kingdom. Several examples are given at the end of the chapter, where the ASTM and EI organizations share a single test method.

In many other cases, ASTM and EI have developed methods to determine the same lubricant property (sulfur content, low temperature viscosity, base oil aromatics, etc.) through the application of somewhat different procedures. Each organization maintains an Internet website with extensive capabilities of searching for specific methods. The addresses for each site are Energy Institute, www.energyinstpubs.org.uk, and ASTM International, www.astm.org.

SUMMARY

This chapter reviewed a broad spectrum of ASTM test methods that can be applied to characterize the chemical, physical, and operational "in-service" properties of hydrocarbon base oils. In addition, it was emphasized that several other standard methods can be used to speciate the molecular types typically found in base oil, enabling the correlation of many performance properties to fundamental molecular structure. As a result, within the context of the extensive portfolio of ASTM techniques, it has become possible to actually predict base oil performance from measurable compositional parameters using statistical modeling. It should also be emphasized that, in actual commercial practice, many tests originally designed for formulated lubricants can also be applied to "benchmark" base oils by including a common dose of additive package or single additive component. Care should be exercised with this approach, however, since the original test method repeatability and reproducibility may not hold for such base oil formulations.

Joint ASTM/IP Lubricant Test Methods

ASTM	IP	Title
D92-05a	36/84(89)	Flash and Fire Points by Cleveland Open-Cup Tester
D93-08	34/99	Flash Point by Pensky-Martens Closed-Cup Tester
D97-08	15/95	Pour Point of Petroleum Products
D189-06	13/94	Conradson Carbon Residue
D445-06	71/1/97	Kinematic Viscosity and Calculation of Dynamic Viscosity
D482-07	4/96	Ash from Petroleum Products
D524-04	14/94	Ramsbottom Carbon Residue
D611-07	2/98	Aniline Point and Mixed Aniline Point Determination
D664-09	177/96	Acid Number Determination
D892-06	146/200	Lubricant Foaming
D1298-99 (2005)	160/99	Density, Specific Gravity, and API Gravity
D1500-07	196/97	ASTM Color
D2270-04	226/91(95)	Calculating Viscosity Index from Kinematic Viscosity at 40°C and 100°C
D2500-05	219/82	Cloud Point of Petroleum Oils
D2710-99 (2004)	299/92(98)	Bromine Index by Electrometric Titration
D2887-08	406	Simulated Distillation (Volatility)
D4052-96 (2002)	365/84(86)	Density and Relative Density by Digital Densitometer
D4629-08	379/88	Trace Nitrogen

References

[1] Sequeira, A., Jr., *Lubricant Base Oil and Wax Processing*, Marcel Dekker, Inc., New York, 1994.

[2] Dennsion, G. H., and Conduit, P. C., "Oxidation of Lubricating Oils–Mechanism of Sulfur Inhibition," *Indus. Engin. Chem.*, Vol. 37, No. 11, 1945, pp. 1102–1108.

[3] Nessel, C. S., Coker, D. T., King, A. G., and Mumford, D. L., "Carcinogenic Assessment of Lubricant Base Oils by Proton NMR," *Preprints of the Division of Petrol. Chem. of the Am. Chemical Soc.*, Vol. 42, No. 1, 1997, pp. 251–254.

[4] Voltz, M., "Biodegradability of Lubricant base Stocks and Fully Formulated Products," ACS Petroleum Division Symposium Proceedings, San Diego, CA, 1994.

Bibliography

Harpp, D. N., Robertson, J., Laycock, K., and Butler, D., "Organosulfur Antioxidants in Hydrocarbon Oils," *Sulfur Reports*, Vol. 4, No. 6, 1985, pp. 195–227.

Korcek, S., and Jensen, R. K., "Relation Between Base Oil Composition and Oxidation Stability at Increased Temperatures," *ASLE Transactions*, Vol. 19, No. 2, 1975, pp. 83–94.

Speight, J. G., *The Chemistry and Technology of Petroleum*, 2nd ed., Marcel Dekker, Inc., New York, 1991.

Stipanovic, A. J., Smith, M. P., Firmstone, G. P., and Patel, J. A., "Compositional Analysis of Re-Refined and Non-Conventional Lubricant Base Oils: Correlations to Sequence VE and HIE Gasoline Engine Tests," SAE Publication 941978, 1994.

Yoshida, T., Igarashi, J., Stipanovic, A. J., Thiel, C. Y., and Firmstone, G. P., "The Impact of Basic Nitrogen on the Oxidative and Thermal Stability of Base Oils in Automotive and Industrial Applications," SAE Publication 981405, 1998.

16

Lubricating Oils[1]

Dave Wills[2]

THE MAJOR FUNCTION OF LUBRICATING OILS IS the reduction of friction and wear by the separation of surfaces, metallic or plastic, which are moving with respect to each other. The oils also act as carriers for many special chemicals such as corrosion inhibitors, antiwear agents, load-carrying friction modifiers, and foam suppressors. Performance requirements can also include cooling, the dispersion and neutralization of combustion products from fuels, and the transport of contaminants to a filter. The high-quality and improved properties of present-day lubricants have enabled engineers to design machines with higher power-to-weight ratios that generally have higher stresses, loads, and operating temperatures than before. Thus, it has been possible to develop automobile engines, turbines, gear sets, etc. that are capable of higher speeds and higher specific power output per unit weight of machinery. In a very different field, lubricants with increased resistance to the effects of radiation have been developed for nuclear power stations.

From a historical standpoint, solvent-refined paraffinic oils have been the most widely used base stock for lubricating oil applications. In recent years, alternative refining processes such as catalytic isomerization and deep hydrogenation have been developed to yield higher purity base oils that are better suited to withstand severe operating conditions. Research in the field of additives has, in turn, produced lubricant formulations that can operate under the higher piston-ring belt temperatures of super-charged automotive diesel engines and provide the dispersion required to prevent the formation of low-temperature sludge in gasoline engines for stop-start, short-distance motoring.

Despite the increasing temperatures, loads, and other requirements imposed on lubricants, mineral oils are likely to continue to be employed in the foreseeable future for the majority of automotive, industrial, and marine applications. However, in the aviation field, synthetic lubricants are used extensively. There are also a growing number of critical automotive, industrial, and marine applications where the use of synthetic lubricants can be justified on the basis of total performance or fire resistance.

ASSESSMENT OF QUALITY

New uses and formulation technology for lubricating oils necessitate a constant review of the methods for assessing the quality of both new and used lubricating oils. The traditional physical and chemical tests are still applied, but these are being supplemented, and in some cases, replaced by instrumental techniques based on physicochemical methods, which include infrared absorption, ultraviolet absorption, emission spectroscopy, X-ray absorption, and fluorescence methods. It is convenient to consider these tests in five categories: physical tests, chemical tests, physicochemical tests, laboratory bench tests, and engine and rigs tests. The first three determine the characteristics and compositions of lubricants, and the last two are laboratory tests that simulate service conditions.

Physical Tests

Physical tests are comparatively simple laboratory procedures that define the nature of the product by measuring physical properties. Examples are viscosity, flash point, specific gravity, color, and appearance.

Chemical Tests

Chemical tests define the composition of the lubricating oil by determining the presence of such elements as sulfur, chlorine, phosphorus, and metals, which often are related significantly to the additive content of the lubricant.

Physicochemical Tests

Tests in this classification are either (1) those that determine the presence of elements using instrumented physical procedures, (2) those that give information on the molecular structure of the components of the lubricant, or (3) those that give pH, acidity, and alkalinity values.

Physicochemical tests are used generally to characterize products for specific applications, provide quality control at blending plants, and check the suitability of used lubricants for further service.

Laboratory Bench Tests

Laboratory bench tests subject the lubricating oil to individual environmental conditions, which are designed normally to exceed the appropriate service requirements. These include such glassware tests as thermal stability, oxidation stability, and corrosion resistance, which are used frequently for screening formulations during the development of a new product and are also used to determine whether a used lubricant still provides adequate protection.

Also falling in this category are test machines for measuring wear and load-carrying properties. Examples are the Four-Ball, Timken, FZG, and Falex equipment.

Engine and Rig Tests

Mechanical tests are used to assess the effects on various properties of lubricating oils that will be produced by the

[1] I would like to recognize the following individuals who previously contributed significantly to the improvement and upgrade of the seventh edition of this chapter: Dr. Shirley Schwartz, retired from GM, Warren, MI; Paul Michael, Milwaukee School of Engineering, Milwaukee, WI; Don Clason, retired from Lubrizol Corporation, Wickliffe, OH; and Barbara Parry, Mohawk Lubricants, North Vancouver, BC, Canada.

[2] *Retired from Sauer-Danfoss, Ames, IA.*

environment in which the lubricant will function. The equipment is set up in a prescribed manner on laboratory stands, and tests are carried out under carefully controlled conditions. Such tests generally are designed to correlate with actual service, but, for new products, laboratory mechanical testing usually is followed by field evaluation.

SIGNIFICANCE OF TESTS

Physical tests, chemical tests, laboratory bench tests, and engine tests are extremely valuable as tools for attempting to predict how a specific lubricant formula will perform in full-size machinery under many different operating conditions. They must always be used and interpreted, however, with the full realization that they are not infallible or foolproof. The ultimate decision as to the success or failure of a lubricating fluid can be made only on the basis of its behavior in the end-use item such as production engines, pumps, gear drives, hydraulic systems, etc. It is usually on the basis of this ultimate performance that many equipment manufacturers will "recommend" a brand name product for use in their equipment.

COMPOSITION AND MANUFACTURE

Petroleum-base lubricating oils are present in the boiling residue above 370°C (698°F) from the atmospheric distillation of selected crude oils of both paraffinic and naphthenic types. This residue is further distilled, under conditions of high vacuum, into a series of fractions to provide high- to low-viscosity lubricating oil stocks. The number of fractions depends on the type of crude oil and the requirements of the refiner, but four to five is a typical number. These base oils are further refined, either by solvent extraction or by catalytic processes to produce oils suitable for incorporation into finished lubricants. The individual refined stock oils from one or more crude sources are blended in various proportions to provide lubricating oils suitable for a wide range of applications. The blending process can either be mechanical or air agitation using a batch or continuous in-line method.

Only for the less severe uses is it possible to employ nonadditive mineral oils. In the majority of cases to meet specific application requirements, chemical additives are used to enhance the properties of base oils. Additives are used to improve such characteristics as oxidation resistance, change in viscosity with temperature, low-temperature flow properties, corrosion and radiation resistance, and load-carrying capacity. Lubricants frequently contain a number of additives to achieve a balance of properties suitable for the intended application. These must be compatible with the base oils, other additives present, and lubricants blended by others who manufacture products intended for similar uses. Thus, the proper selection of the components for a lubricating oil formulation requires knowledge of the most suitable crude sources for the base oils, the type of refining required, the types of additives necessary, and the possible interactions of these components on the properties of the finished lubricating oil.

Control of product quality at the blending plant is usually based on a supplier's own internal standards. The number of tests applied varies with the complexity of the product and the nature of the application. More important tests, such as viscosity, flash point, and infrared analysis, are usually performed on every batch. Other tests may be on a statistical basis, dependent on data developed at the individual blending plant. Physical property measurements such as viscosity and flash point are useful for verifying the base oil composition of a finished lubricant. Additive levels are usually evaluated though spectrochemical methods such as infrared analysis, which can be presented in the graphical form. Comparison of infrared "fingerprints" with a known standard can be used as a check on the composition.

GENERAL PROPERTIES

Before describing the quality criteria for some of the more important types of lubricating oils, it will be fruitful to discuss general properties common to most lubricating oils and the methods used to determine these properties. In this discussion, pertinent American Society for Testing and Materials and Institute of Petroleum (ASTM/IP) test methods are listed after the text.

Viscosity

The viscosity of a lubricating oil is a measure of its flow characteristics. To meet the requirements of an application, viscosity is generally the most important controlling property for manufacture and selection. While the viscosity of a mineral oil changes with temperature, it usually does not change with shear rate unless non-Newtonian additives are used to modify viscosity or temperature characteristics, an aspect that is discussed more fully in the section on automotive engine oils.

There are several methods for measuring the viscosity of lubricating oils. The most widely utilized method is ASTM D445, Standard Test Method for Kinematic Viscosity of Transparent and Opaque Liquids. In this test, the time is measured for a fixed volume of liquid to flow under gravity through the capillary of a calibrated viscometer at a closely controlled temperature. The kinematic viscosity is the product of the measured flow time and the calibration constant of the viscometer. In 1975, a cooperative effort between ASTM, ASLE, BSI, and DIN resulted in ISO 3448, Viscosity Classification for Industrial Liquid Lubricants. This classification is commonly referred to as ISO viscosity grades. The ISO system classifies lubricants solely on kinematic viscosity measured at 40°C. The choice of 40°C as the reference temperature is a compromise between maximum operating and ambient temperatures for many industrial lubrication applications.

The ISO classification system is made up of 18 ISO viscosity grades, usually written as ISO VG. The grades start with ISO VG 2 and go up to ISO VG 1500. Each grade is named by the whole number, which is the rounded, midpoint viscosity of its associated range of viscosity. Each range is ±10 % of the midpoint viscosity (Table 1).

The ISO viscosity grades are not continuous. Each ISO VG is approximately 50 % more viscous than the next lower grade. Since the ISO grades are only ±10 % around the midpoint viscosity, gaps exist between grades. A fluid with a viscosity that does not fall into an ISO VG range cannot be formally classified. This is a simple viscosity classification system that addresses only the fluid's kinematic viscosity at one temperature.

Because the main objective of lubrication is to provide a film between load-bearing surfaces, the selection of the correct viscosity for the oil is aimed at a balance between a viscosity high enough to prevent the lubricated surfaces from contacting and low enough to minimize energy losses through excessive heat generation caused by having too viscous a lubricant.

TABLE 1—ISO 3448 Viscosity Grades: Kinematic Viscosity, 40°C (mm^2/s)

ISO Viscosity Grade	Midpoint	Minimum	Maximum
2	2.20	1.98	2.42
3	3.20	2.88	3.52
5	4.60	4.14	5.06
7	6.80	6.12	7.48
10	10.0	9.00	11.0
15	15.0	13.5	16.5
22	22.0	19.8	24.2
32	32.0	28.8	35.2
46	46.0	41.4	50.6
68	68.0	61.2	74.8
100	100	90.0	110
150	150	135	165
220	220	198	242
320	320	288	353
460	460	414	506
680	680	612	748
1000	1000	900	1100
1500	1500	1350	1650

The "classic" hydrodynamic theory for moderately loaded bearings predicts complete separation between metallic surfaces with a comparatively thick layer of fluid oil, while highly loaded gears are considered to be in a state of boundary lubrication in which opposing surface irregularities cause metal-to-metal contact to occur. However, at very high pressures (several thousand psi) the viscosity of mineral oils increases considerably with an increase in pressure. The extent of the viscosity change depends on the crude source of the oil and on the molecular weight of the constituent components. Modern elastohydrodynamic theory for lubricated surfaces takes into account that because of the high pressure generated, the viscosity of the oil increases considerably and elastic deformation of the surfaces occurs. Under these conditions, it has been shown that the lubricant film approaching "boundary" conditions is thicker than previously was supposed.

Viscosity increase in a used oil, that is, an oil in service, usually indicates that the oil has deteriorated by oxidation or contamination, while a decrease usually indicates dilution by a lower viscosity oil or fuel, or by shear of the viscosity index (VI) improver. Viscosity blending charts may be used to estimate the amount of dilution. The extent of the viscosity change permissible before corrective action is required differs in various applications.

The ASTM Viscosity-Temperature Charts for Liquid Petroleum Products (D341) are useful for estimating the viscosity of an oil at the various temperatures that are likely to be encountered in service.

Viscosity Index

The viscosity of petroleum-base oils decreases with a rise in temperature, but this rate of change depends on the composition of the oil. The VI is an empirical, unitless number indicating the effect of temperature change on the kinematic viscosity of an oil. It compares the rate of change of viscosity of the sample with the rates of change of two types of oil having the highest and lowest VIs at the time (1929) when the VI scale was first introduced. A standard paraffinic oil was given a VI of 100, and a standard naphthenic oil, a VI of 0. Equations were generated connecting the viscosity and temperature for these two types of oil, and, from these equations, tables were prepared showing the relationship between viscosities at 40°C (104°F) and 100°C (212°F) for oils with a VI between 0 and 100. With these tables and viscosity data at low and high temperatures, the VI of an oil can be calculated. A high VI denotes a low rate of change of viscosity with temperature.

The use of additives and modern refining techniques allows oils to be produced with high VIs regardless of the type of crude oil from which they originated. At the same time, viscosity improvement additives can produce oils with VIs greater than 100. Initially, this problem was solved by simply extrapolating the original tables, but, as VIs rose ever higher, this produced anomalies.

In 1964, ASTM adopted an extension to the tables based on an equation developed for the purpose. Values derived from this equation are designated as VI_E, to distinguish them from the original VI. This method has since been adopted by IP jointly with ASTM, calculating VI from kinematic viscosity at 40 and 100°C (D2270/IP 226). This replaces the former method ASTM D567/IP 73. (VIs below 100 have not been affected by this revision.)

The VI of an oil is of importance in applications where an appreciable change in the temperature of the lubricating oil could affect the start-up or operating characteristics of the equipment. The automatic transmission for passenger vehicles is an example of equipment where high VI oils using VI improvers are employed to minimize differences between a viscosity low enough to permit a sufficiently rapid gear shift when starting under cold conditions and a viscosity adequate at the higher temperatures encountered in normal running.

Cloud and Pour Points

Petroleum oils contain components with a wide range of molecular structure and thus do not have a sharp freezing point. They become semiplastic solids when cooled to sufficiently low temperatures.

The cloud point of a lubricating oil is the temperature at which paraffinic wax and other readily solidifiable components begin to crystallize out and separate from the oil under prescribed test conditions per ASTM Test for Cloud Point of Petroleum Oils (D2500/IP 219). Cloud point is of importance when narrow clearances might be restricted by accumulation of solid material (for example, suction line strainers, small-size oil-feed lines, or filters).

The pour point is the lowest temperature at which the oil will flow under specified test conditions per ASTM Test for Pour Point of Petroleum Oils (D97/IP 15) and is roughly equivalent to the tendency of an oil to cease to flow from a gravity-fed system or container. Since the size and shape of the containers, the head of the oil, and the physical structure of the solidified oil all influence the tendency of the oil to flow, the pour point of the oil is a guide to, and not an exact measure of, the temperature at which flow ceases under the service conditions of a specific system.

The pour point of wax-containing oils can be reduced by the use of low-flow improvers or pour-point depressants, which inhibit the growth of wax crystals. It is a recognized property of oils of this type that previous thermal history may affect the measured pour point. ASTM Method D97/IP 15 includes a section that permits some measurement of this thermal effect on wax crystals.

The importance of the pour point, to the user of lubricants, is limited to applications where low temperatures are likely to influence oil flow. Obvious examples are refrigerator lubricants and automotive engine oils in cold climates.

Flash and Fire Points

The flash-point test gives an indication of the presence of volatile components in an oil, and it is the temperature to which the oil must be heated under specified test conditions to give off sufficient vapor to form a mixture that will ignite in the presence of an open flame.

The fire point is the temperature to which the product must be heated under somewhat similar test conditions to cause the vapor/air mixture to burn continuously on ignition. The ASTM Test for Flash and Fire Points by Cleveland Open Cup (D92/IP 36) can be used to determine both flash and fire points of lubricating oils, and it is the most generally used method for this purpose in the United States. In the United Kingdom, the ASTM Flash Point by Pensky Martens Closed Tester (D93/IP 34) and Open Flash Points (IP 35) are used widely.

Not only for the hazard of fire, but also as an indication of the volatility of the oil, the flash and fire points are significant in cases where high-temperature operations are encountered. In the case of used oils, the flash point is employed to indicate the extent of contamination by fuels or a more volatile oil. The flash point also can be used to assist in the identification of different types of base blends.

Relative Density (Specific Gravity) and API Gravity

Relative density and American Petroleum Institute (API) gravity are alternative, but related, means of expressing the weight of a measured volume of a product. Relative density, ASTM test for density, relative density (specific gravity), or API Gravity of Crude Petroleum and Liquid Petroleum Products by Hydrometer Method (D1298/IP 160), also known as specific gravity, is used widely outside the United States. In the United States, API gravity is used throughout the petroleum industry. The API gravity, ASTM Test for API Gravity of Crude Petroleum and Petroleum Products (Hydrometer Method) (D287/IP 192), is based on a hydrometer scale, which may be readily converted to the relative density basis by use of tables or formulas.

Both types of gravity measurements are used as manufacturing control tests. In conjunction with other tests, gravimetric measurements are also used for characterizing unknown oils, since they correlate approximately with hydrocarbon composition and, therefore, with the nature of the crude source of the oil.

Color

The color of a sample of lubricating oil is measured in a standardized glass container by comparing the color of the transmitted light with that transmitted by a series of numbered glass standards, Test for ASTM Color of Petroleum Products (ASTM Color Scale) (D1500). The test is used for manufacturing control purposes and is important since the color is readily observed by the customer. The color of a lubricating oil is not always a reliable guide to product quality and should not be used indiscriminately by the consumer in writing specifications for purchases. Where the color range of a grade is known, a variation outside the established range indicates possible contamination with another product.

AUTOMOTIVE ENGINE OILS

Properties

The crankcase oil of automotive gasoline and diesel engines is used to lubricate and cool the pistons, cylinders, bearings, and valve train mechanisms. Thus, the duty performed by an automotive engine oil is highly complex, and the oil needs to be formulated appropriately. As examples, the oil must contain sufficient oxidation inhibitors, because even the best mineral oils react with oxygen at high temperatures to form sludge and varnish. Good detergent-dispersants are needed to suspend sludge and varnish forming material until they are removed by draining the oil. Water and combustion acids form during engine operation, and the corrosive wear and rusting that they cause must be counteracted with corrosion inhibitors, which are usually designed to impart alkalinity to the oil. Many modern engines have high valve train loadings, which require special antiwear additives in the lubricating oil. The compatibility of the various additives is also an important consideration.

Crankcase oil also has an impact on the control of exhaust emissions in today's ecology-conscious world. The operation of the positive crankcase ventilation valve, which prevents blowby gas venting to the atmosphere, is influenced by the quality of the oil used. Crankcase oils also can influence combustion chamber deposits, spark plug life, valve operation, engine wear, catalyst efficiency, oil consumption, and other factors influencing exhaust emissions.

To meet a particular ambient temperature condition, the viscosity of the engine oil is a main controlling property for manufacture and selection. Engine oils generally are recommended by vehicle manufacturers according to the SAE viscosity classification (J300). This classification sets the limits for the viscosity at 100°C (212°F) for all grades of oil, and at −5 to −35°C for W grades, and at 150°C under high shear conditions, for multigrade conditions.

Multigrade engine oils, for year-round service, are sufficiently fluid at a low temperature to permit easy starting of the engine in winter conditions and still have an adequate viscosity at operating temperatures to provide for lubrication and to control oil consumption. The VI of multigrade oils is typically in the range 120 to 160, while single-grade oils are usually between 85 and 105.

Multigrade oils containing VI improver do not behave as Newtonian oils. The improved viscosity and temperature characteristics of multigrade oils enable, for example, an SAE 10W-30 oil to be formulated to have characteristics of SAE 10W oils at low temperature and SAE 30 grade oils at higher temperatures. The result is that the viscosity of multigrade oils generally is higher at −17.8°C (0°F) than is predicted by extrapolation from 100°C (212°F) and 40°C (KMT) viscosity values. The extent of the deviation varies with the type and amount of the VI improver used. To better correlate with actual engine conditions, the viscosity of the SAE W grades is based on a measured viscosity using the Cold Cranking Simulator (ASTM D2602) and the Mini-Rotary Viscometer (ASTM D3829) tests.

A number of factors are at work in an engine to change the viscosity of the oil. Multigrade oils containing polymeric VI improver are subject to mechanical shearing, resulting in a decrease in viscosity. The extent of this is dependent on the type of VI improver used, since the base oil is relatively shear stable. The viscosity of automotive engine oils in service may also be decreased by fuel dilution and water, but increased by oxidation and combustion products. It should be noted that solution-forming contaminants such as gasoline cause oil's viscosity to decrease. Emulsion-forming contaminants such as water or alcohol (from alcohol fuel) cause the engine oil's viscosity to increase. Detergent or dispersant oils can keep these contaminants in suspension, but undesirable deposits (sludge, varnish, rust, etc.) can be formed on critical parts within the engine if oil drain periods are extended indiscriminately. Recommendations for oil-change intervals are usually made by the vehicle manufacturer, or, in the case of commercial vehicle fleets with known patterns of operation, the user may establish his own optimum change periods, which are dependent on the quality level of the lubricating oil employed. In addition, computerized systems are being developed (oil-change indicators) that calculate the rate of oil degradation during service and inform the driver when an oil change is needed.

Engine Test Specifications and Procedures

Engine test methods used in the development of new formulations and for purchase specifications have originated from the following sources:

1. API/ASTM/SAE engine service classifications
2. Institute of Petroleum tests (IP)
3. Coordinating European Council (CEC) tests
4. Heavy-duty diesel engine manufacturers' tests
5. Military specifications (U.S. Army and Navy, British Ministry of Defense, NATO, etc.)

Originally, the API defined the type of service for which an engine oil was designed, and the "MS" sequence tests were used to describe each type ("MS" denoted a service condition that was most severe for gasoline engines). In 1972, a joint API/ASTM/SAE system of nomenclature was adopted as a guide for the selection of engine oils for different service conditions.

Gasoline engine service conditions are designated by the letters SA, SB, SC, SD, SE, SF, SG, SH, SJ, SL, and SM. Diesel engine oils are designated by CA, CB, CC, CD, CD-2, CE, CF, CF-2, CF-4, CG-4, CH-4, CL-4, and CJ-4. Performance criteria have been established for each designation using procedures that include published engine test Sequences IIA, IIB, IIC, IID, IIIA, IIIB, IIIC, HID, HIE, IV, V, VB, VC, VD, VE, L-38, and IH2 for gasoline engines. For diesel engines, tests are the L-4, L-38, L-l, LTD, IIA, IIB, IIC, IID, IH, IH2, ID, IG, IG2, T-6, T-7, NTC-400, 6V53T, 6V92TA, IK, and IM. The designation and performance criteria are fully described in the SAE Standard Report J183, ASTM Research Report on Engine Oil Performance Classification (D2-1002), and ASTM Performance Specification for Automotive Engine Oils (D4485).

The engine tests used to evaluate engine oil performance are included in Multicylinder Test Sequences for Evaluating Automotive Engine Oils, ASTM STP 315; Single Cylinder Engine Tests for Evaluating the Performance of Crankcase Lubricants, ASTM STP 509, ASTM D5119, ASTM D5302; ASTM Research Report (RR) D2:l 194, RR D02 1273; and other ASTM Committee Section D.02.130.02 reports. Many of the performance designations (oil categories) are obsolete in that they include performance tests where engine parts or test fuel, or reference oils are no longer generally available and the tests are no longer being monitored by the test developer or ASTM. Letter designations, where these situations are applicable, are SB, SC, SD, SE, SF, CA, and CB.

The energy-conserving potential of engine oils has also become important to the vehicle manufacturers. While not a part of the service condition designations, ASTM test methods (Five Car Test and Sequence VI) and performance categories have been established to designate energy-conserving engine oil performance levels. Most vehicle manufacturers recommend energy-conserving engine oils.

An important U.S. Army specification is the MIL-PRF-2104, which is used as a performance reference for commercial automotive engine oils. Vehicle manufacturers generally specify the most current performance criteria and specification engine oils during their warranty periods for cars and trucks.

Engine and bench tests for the evaluation of lubricants are also developed in Europe, Japan, and other countries. A discussion of these tests and their related applications can be found in SAE Information Report J 2227.

Examination of Used Oils

An examination of used oil during and after service is beneficial in determining the useful life of a lubricant in a particular application and in providing indicators of equipment condition.

Fuel dilution can contribute to viscosity reduction and may contribute to performance problems. Diesel fuel dilution, resulting from low-temperature or short-distance stop/start operation, can be estimated from measurements as determined by ASTM Test for Diesel Fuel in Used Lubricating Oils by Gas Chromatography (D3524). Gasoline dilution can be measured by a distillation procedure, ASTM Test for Gasoline Dilution in Used Gasoline Engine Oils by Distillation (D322/IP 23), or determined by ASTM Test for Fuel Dilution in Gasoline Engine Oils by Gas Chromatography (D3525).

Low-temperature service conditions may also result in water vapor from combustion products condensing in the crankcase. Rapid cooling of equipment at shutdown and equipment operating in high humidity environments may also experience internal water-related corrosion conditions. The water content of a used oil can be measured by distillation techniques per ASTM Test for Water in Petroleum Products and Bituminous Materials by Distillation (D95/IP 74).

Coolant contamination reduces the ability of oil to lubricate and can promote varnish and deposit formation. The presence of coolant in used oil can be detected by ASTM Test for Trace Ethylene Glycol in Used Engine Oil (D4291), a quantitative chromatograph method, or qualitatively by ASTM Test for Glycol-Based Antifreeze in Used Lubricating Oils (D2982), a colorometric test.

The extent and nature of the contamination of used automotive engine oil by oxidation and combustion products can be ascertained by determining the amounts of materials present in the lubricating oil, which are insoluble in *n*-pentane and toluene, ASTM Test for Insolubles in Used Lubricating Oils (D893). Both are expressed as percentage by weight. Pentane insolubles include some oil-soluble materials. Toluene insolubles include material from external contamination such as dirt, fuel carbon, and material from degradation of fuel, oil and additives, and engine wear and corrosion materials.

Where highly detergent or dispersant oils are under test, coagulated pentane insolubles and coagulated toluene insolubles may be determined by using methods similar to those just described, but employing a coagulant to precipitate the very finely divided materials, which may otherwise be kept in suspension by the detergent or dispersant additives.

Size discrimination of insoluble matter may be used to distinguish between finely dispersed, relatively harmless matter, and the larger potentially harmful particles in oil by ASTM Test for Pentane Insolubles by Membrane Filtration (D4055).

The metallic constituents (barium, boron, calcium, magnesium, tin, silicon, zinc, aluminum, sodium potassium, etc.) of new and used lubricating oils can be determined by a comprehensive system of chemical analysis.

For new lubricating oils, ASTM Test for Sulfated Ash from Lubricating Oils and Additives (D74/IP 163) can be employed to check the concentration of metallic additives. These standard chemical procedures are time consuming to carry out, and, where the volume of samples justifies the purchase of such equipment, emission spectrographs, X-ray fluorescence spectrometers, atomic absorption, and other instruments that are much more rapid are available (see applicable ASTM/IP standards).

ASTM Test for Determination of Additive Elements, Wear Metals and Contaminants in Used Lubricating Oils and Selected Elements in Base Oils by Inductively Coupled Plasma Emission Spectrometry (D5185) may be used to monitor wear metals in used oil and, when monitoring equipment over time, wear trends.

Reserve alkalinity protects engine components from corrosion by neutralizing the acids formed during combustion. The amount of reserve alkalinity in the used oil can be determined by using ASTM Test for Base Number of Petroleum Products by Potentiometric Perchloric Acid Titration (D2896/IP 276) or ASTM Test for Base Number Determination by Potentiometric Titration (D4739). These are titration methods where, because of the nature of the used oil, an electrometric, instead of a color, end point is used.

MARINE DIESEL ENGINE OILS

From the lubrication viewpoint, marine diesel engines are of two principal types. These are trunk-piston engines, in which the crankcase oil also lubricates the bearings and cylinders, and crosshead engines, in which the cylinders are separately lubricated. Marine diesel engines can also be classified as low-, medium-, and high-speed types with speeds of 0 to 50, 250 to 1,000, and over 1,000 rpm, respectively. The lower speed engines are less sensitive to fuel quality and can operate satisfactorily on residual fuels with high-sulfur contents, while the higher speed engines are generally more similar in design to automotive engines and used distillate fuels. Medium-speed engines vary in their fuel requirements according to the design.

The oil used in trunk-piston engines must have a viscosity suitable for lubricating the bearings and the cylinders. The trend in marine diesel engine design is toward smaller and lighter engines of higher specific output so that proportionately less space in a ship is required for the propulsion unit. Turbo-charging is used more frequently as a means of increasing the amount of power obtained from a given size of engine. This increases the heat input and oils with comparatively high levels of detergents, and antioxidants are required to maintain satisfactory engine cleanliness. The oxidation stability of trunk-piston engine oils can be measured by the single-cylinder CRC L-38 and the Petter W-l tests. Detergency level is determined by one of the single-cylinder Caterpillar procedures.

Alkaline additives are required in the formulation of marine engine oils to neutralize potentially corrosive acids formed as a result of blow-by gases entering the crankcase. These additives must be capable of maintaining the alkalinity of the oil throughout the life of the charge. This is particularly important when high-sulfur (above 1 %) residual fuels are used. Samples of oil drawn from the crankcase can be tested to assess the reserve of alkalinity remaining by determining the total base number of the oil. Other tests performed on the used engine oil, which serve as a guide to the suitability of the oil for further service, are the viscosity, flash point, pentane and benzene insolubles, and sulfated ash.

In the crosshead-type engines, the crankcase oil lubricates the bearings and may also be used to cool the pistons. As well as having the appropriate viscosity, the oil also must have satisfactory oxidation resistance and good antifoam and anticorrosion properties. Small amounts of acidic contaminants entering the crankcase oil from the cylinders can be neutralized by using crankcase oils with a low level of alkalinity.

The cylinders of crosshead diesel engines are lubricated separately on an oil-loss basis. The oil is injected into the cylinders through feed points around the cylinder and distributed by the scraping action of the piston rings. Excess oil collects in a scavenge space and runs off to the exterior of the engine. The oil used for this purpose is exposed to particularly high temperatures. As crosshead engines usually operate on residual fuels that frequently contain relatively high levels of sulfur, it is important that the cylinder oil has a sufficiently high level of alkalinity to neutralize acidic combustion products formed and thus to minimize the occurrence of corrosive wear. Some low-specific output engines are prone to exhaust-port blocking with carbonaceous combustion products and may also require special lubricating oils to reduce this tendency. Crosshead cylinder oils may be true solutions, in which the additives are dissolved in the oil, or they may contain finely divided additives suspended in the oil.

The deposit-forming tendencies due to thermal instability can be assessed by the Panel Coking Test when used in conjunction with other tests, and the available alkalinity of these oils can be measured.

The high-speed engines for fishing vessels, pleasure craft, or auxiliaries in large vessels require lubricating oils of a type similar to the higher detergency level automotive diesel engine lubricants. In some cases, automotive grades are recommended for use in these marine engines. However, since the sulfur contents of the fuels normally used in marine applications are higher than those of the fuels used for automotive purposes, it is important to ensure that the reserve of alkalinity is sufficient to neutralize the additional amounts of corrosive acids that may be formed.

INDUSTRIAL AND RAILWAY ENGINE OILS

A wide range of sizes and designs are employed for industrial engines, but the types of lubricants used and their related test procedures are similar to those described under the sections on automotive and marine engines. Crankcase systems for land-based engines may be larger in capacity than for marine purposes, and the oil charge, in these cases, would be expected to extend for longer periods of operation.

Engines with large cylinder bores (above 15.2 cm^3 or 6-in. diameter) usually have higher cylinder wall temperatures and require oils with high viscosities. Intermediate-size engines generally use SAE 30 and 40 grade oils, while for the large engine SAE 50 oils may be recommended.

Because of space limitations, locomotive engines have higher specific ratings than their industrial counterparts. The resultant higher bearing and cylinder temperatures, coupled with operating practices that include long idling periods followed by rapidly increased speed and load, make the lubricant performance criteria for locomotives relatively severe.

The importance of keeping railway locomotives in service for as long as possible before overhaul has led to the use of the regular checking of oil samples for the presence of wear metals. For this purpose, direct reading spectrographs requiring a minimum of operator time have been developed. A sudden increase in the amount of a particular metallic element present in the oil may indicate an incipient-bearing failure.

New railway diesel oil formulations must pass specific tests of the railway diesel engine manufacturers. Ultimate acceptance of an oil is based on extended satisfactory service performance.

GAS-TURBINE LUBRICANTS

Gas-turbine engines, originally developed for aviation use, are now being employed increasingly for industrial, marine, and automotive applications. The electricity generation industry, for example, uses a considerable number of gas turbines for standby and periods of peak operation. Although more and more installations have begun to depend on this type of engine for full-time service, the requirement to conserve fuel resources may reverse this trend.

From the lubrication viewpoint, the most important features of gas turbines are the large volumes of high-temperature combustion gases which flow through the engine and the comparatively high unit loads on the gearing as a result of the need to reduce weight to a minimum, particularly with turbines in aircraft. These features dictate the use of a lubricant of high thermal and oxidative stability combined with good load-carrying properties. The latter property is particularly important where the engine oil also has to lubricate reduction gearboxes, such as those used for turbo-propeller or helicopter transmissions.

For aviation purposes, British and U.S. military specifications have had a strong influence in establishing quality criteria and performance levels for gas-turbine lubricants. The continuous development of engines for military and commercial use has resulted in increasingly high bearing temperatures, dictating the adoption of synthetic fluids rather than mineral oils.

Synthetic aircraft turbine lubricants were first used in the United States in 1952. The U.S. Air Force composed the first military performance specification, MIL-PRF-7808. Various changes in performance of qualified fluids have resulted in several revisions of this specification as indicated by a letter suffix, that is, MIL-PRF-7808 A, B, C, D, and F.

Pratt and Whitney engine tested most MIL-PRF-7808 type fluids in 1957 and subsequently formed their own qualified products list for commercial applications. General Electric Company (GE) and the Allison Division of General Motors Corporation also formed qualified products lists. The U.S. Navy and Pratt and Whitney initiated aircraft-turbine lubricant improvement programs in 1961. Their programs are generally referred to as Type "1-1/2" and Type II, respectively. The military specification covering Type "1-1/2" turbine lubricants is MIL-PRF-23699 (WEP), published by the U.S. Navy in early 1963. The Pratt and Whitney specification covering Type II aircraft-turbine lubricants is PWA 521-B Type II, published in mid-1963. The most significant difference between these two specifications is that MIL-PRF-23699 (WEP) requires qualified fluids to be shear stable, whereas the Pratt and Whitney specification does not. Otherwise, both specifications require almost identical performance.

The Type I synthetic lubricants were produced to meet the requirements of engines in the 1950s, but, with further advances in engine design, together with the demand for increased periods between engine overhauls, the more thermally and oxidatively stable Type "1-1/2" and II lubricants were developed. Some of the physical and performance characteristics of the Type I, "1-1/2", and II synthetic lubricants are shown in Table 2.

These specifications also include requirements for resistance to oxidation and corrosion. Other important properties of aviation gas-turbine lubricants are low volatility, foam resistance, seal compatibility, and hydrolytic stability.

TABLE 2—Physical and Performance Characteristics of Type I, "1-1/2," and II Synthetic Lubricants

General Commercial Designation	Military Designation	Viscosity cSt at 99°C (210°F)	Limiting Low-Temperature Viscosity	Load (Ryder Gear Test) lb/in. (kg/cm)—Carrying	Use Temp Bulk Oil
Type I	MIL-PRF-7808-F	3.0 (min)	13,000 cSt max at −53°C (−65°F)	1,900 to 2,200 (339 to 393)	to 149°C (300°F)
Type "1-1/2"	MIL-PRF-23699 (WEP) (USN)	5.0 (min)	13,000 cSt max at −40°C (−40°F)	2,400 to 2,700 (428 to 482)	to 204°C (400°F)
Type II	None[A]	5.0 (min)	13,000 cSt max at −40°C (−40°F)	1,900 to 2,400 (339 to 428)	to 204°C (400°F)

[A] Pratt and Whitney, PWA 521-B.

Because of the safety aspect and the need to prolong periods between engine overhauls, the in-service condition of aviation lubricants is usually monitored by drawing regular samples for analysis. Metals analysis by emission spectroscopy can check whether or not any particular metal shows a sudden change in concentration in the used oil. This can indicate increased wear of a particular engine component and a possible need for replacement.

Industrial and marine gas-turbine installations vary in their severity of operating conditions relative to each other and aircraft service. They may run satisfactorily on high-quality, steam-turbine mineral oils or may need synthetic oils according to the particular service requirements. Automotive gas-turbine engines are being developed with the objective of using mineral oils, but, in most cases, high bearing temperatures, particularly from heat soak along the drive shaft immediately after stopping the engine, have necessitated the use of either synthetic oils or mineral oil synthetic oil blends. In certain installations aboard ship or in industry, the very real danger of a fire from leakage of the lubricant may require the use of a fire-resistant fluid.

GAS ENGINE OILS

In areas where natural or liquefied petroleum gas (LPG) is available at a reasonable price level, these gases are finding increasing use as fuels for industrial engines. Gas engines range from large, relatively low output, low-temperature engines to small, high-speed, supercharged engines. Lubricant requirements vary with the engine design and operating conditions from uninhibited mineral oil, through mildly alkaline oxidation-inhibited detergent oils, to ashless highly detergent oils. Combustion in the large, low-output engines using natural gas or LPG fuels is relatively clean, and, since crankcases for these engines usually contain large quantities of oil, the operating conditions for the lubricant tend to be comparatively mild. For example, under certain conditions, an oil change life of several years may be achieved.

The principal difference between the requirements of gas and other internal combustion engine oils is the necessity to withstand the degradation that can occur to the oil from accumulation of oxides of nitrogen that are formed during combustion. The condition of lubricating oils in large gas engines can be followed by measuring oil viscosity increase and determining changes in the neutralization number. Analytical techniques such as infrared spectroscopy and membrane filtration also can be used to check for nitration of the oil and buildup of suspended carbonaceous material.

The smaller gas engines generally operate at higher crankcase oil temperatures than in the larger types, and lubricant degradation in these engines can be tracked by viscosity, neutralization number, and insolubles determinations.

In many areas, the comparatively high fuel cost for gas engines dictates that they operate at the maximum efficiency. In the case of two-cycle gas engines, relatively small amounts of port plugging (about 5 %) can increase fuel costs to such an extent that engine overhaul becomes necessary. In these engines, port plugging can be caused by carbonaceous lubricant deposits and solid impurities in the combustion air. The lubricant should minimize port plugging from either of these causes and thus help to increase operation time between overhauls.

GEAR OILS

The range of uses for gear oils is extremely wide and includes industrial, automotive, marine, and aviation applications. Gears can vary from large, open types used in quarrying to very small instrument gears used for the control of aircraft. However, the primary requirement is that the lubricant provide satisfactory low wear and control or minimize other forms of damage such as pitting, scuffing or scoring, or rusting by maintaining a lubricant film between the moving surfaces.

Although gears are of many types, including spur, helical, worm, bevel, hypoid, etc., they all function with some combination of rolling and sliding motion. The contact between the mating surfaces may be either along a line (as in the case of spur gear) or at a point (as for nonparallel, nonintersecting helical gear shafts). Although deformation of the metal will broaden the dimensions of the line or point contacts to areas of contact under service conditions, these areas are small in relation to the load on them. Therefore, the unit loadings of gear-tooth surfaces are relatively high compared with ordinary bearing surfaces. Gear teeth often transmit peak loads of 250,000 psi, $3{,}556 \times 10^3$ kg/cm^2, whereas sleeve-bearing loads are usually in the range of 10 to 500 psi (142 to 7,112 kg/cm^2) and seldom exceed 3,000 psi (426.74 kg/cm^2).

Where the gear loadings are comparatively light, straight mineral oils may be used as the lubricant, but, with increased unit loading, it is necessary to incorporate antiwear additives. For very highly loaded conditions in which shock is also experienced, special extreme-pressure compounds are included in gear oil formulations.

It is necessary for a satisfactory balance to be maintained between the properties desired in the lubricant and the components used. For example, overactive chemical additives may promote undesirable wear by chemical attack. This type of wear is known as corrosive wear and results from the progressive removal of chemical compounds formed at the elevated temperatures on the tooth surface. Other types of gear wear are caused by fatigue, abrasion, and welding. In the case of metal fatigue, contact between the surfaces is not necessary for its occurrence. If the gear surfaces are subjected to stresses that are above the fatigue endurance limit of the metal, subsurface cracks can develop that may lead eventually to surface failure. Abrasion of the gear surface is caused either by the harder surface of the two cutting into the other or a hard contaminant or wear particle acting as the abrasive medium. Welding occurs when the severity of the load is high enough to cause complete breakdown of the lubricant film and metal-to-metal contact occurs that results in transfer of metal between the surfaces. The correct choice of lubricant and satisfactory standards of cleanliness will minimize these wear effects.

Various methods are used to apply gear lubricants. Application may be by drip feed, splash, or spray. For large open gears, the lubricant must possess a sufficiently high viscosity and good adhesive properties to remain on the metal surface. This is particularly the case for applications such as strip coal mining and in the cement industry, where gears may have to operate under wet and dirty conditions.

Large industrial gear sets using spur or helical gears operating under moderate loads are usually lubricated by circulating systems. The heavier viscosity turbine oils are generally suitable for this application. In enclosed systems, the temperatures reached may be high enough to necessitate the use of oils of good oxidation resistance. The oils should also possess satisfactory antifoam properties as well as good antirust and demulsification characteristics. Where higher loadings are encountered in industrial gears, lead, sulfur, and phosphorus compounds commonly are used to improve the load-carrying capacity of the lubricant. While lead compounds may still be in use for some high-load industrial applications, their use is diminishing because of environmental concerns.

For highly loaded spiral bevel, worm, or hypoid gears, where sliding contact predominates over rolling contact between gear teeth, lubricating oils with special extreme pressure additives are used. Sulfur, chlorine, lead, and phosphorus compounds are used widely for this purpose.

The amount of active elements in new and used gear oils can be determined by "wet" analytical methods described previously, but there are also a number of instrumental techniques that enable the results to be obtained much more rapidly. Among these tests are polarographic, flame photometric, X-ray fluorescence, atomic absorption, and plasma emission spectrometric methods.

The analytical techniques described for measuring the elements associated with the load-carrying compounds present in gear oils do not identify the specific additives used; nevertheless, they are useful for controlling the quality of the finished products at blending plants. The same techniques can be used to determine similar additives and contaminants that may be present in used oils.

Since active sulfur is necessary for some extreme-pressure applications, ASTM Detection of Copper Corrosion from Petroleum Products by the Copper Strip Tarnish Test (D130/IP 154) can indicate whether the formulation has a satisfactory balance of sulfur activity and copper corrosion protection. The Copper Strip Test is used widely for the quality control of gear oils at blending plants.

Mechanical tests are used for assessing the extreme pressure, friction, and antiwear properties of gear oils in the laboratory. Examples of these are the Institute of Automotive Engineers (IAE) and Ryder gear rigs, the Timken Lubricant Tester, the SAE, David Brown and Caterpillar Disc (or Roller), Four-Ball, Falex, and Almen machines. The selection of the appropriate test machine depends on the application being considered. In spite of the variety of equipment available, the correlation of laboratory tests with practice is not precise, and, hence, the final evaluation of the lubricant needs to be made under controlled field conditions.

A number of industry and military specifications exist for automotive extreme pressure gear lubricants. The most common of these are API GL-5 and MIL-L-2105D. It is estimated that nearly 80 % of all automotive gear oils produced are of API GL-5 quality. The six tests that are used to define GL-5 are explained in Table 2. A complete description of each test may be found in ASTM STP 512A, Laboratory Performance Tests for Automotive Gear Lubricants Intended for API GL-5 Service. The British Ministry of Defence Specification CS 3000B is based on the performance of a blended oil in full-scale axle tests as well as a comprehensive series of laboratory bench tests including oxidation, foam, and storage stabilities. The axle tests specified are the High Torque Test (IP 232) and a modified High Speed Shock Test (IP 234).

In the case of a heavy-duty truck, farm tractor, and highway construction equipment transmission and wet-brake applications, lubricant recommendations fall into several different types. Gear oils continue to be recommended for many heavy-duty manual transmissions. In large, heavy-duty, off-road construction equipment, some manufacturers recommend gear oils for combined axle-wet brake systems serviced by a common sump.

However, many major motor manufacturers today require the use of specialized fluids quite different from common automotive gear oils. These oils require special frictional characteristics to avoid the occurrence of severe vibration or "chatter" when braking is applied, while providing lubrication properties appropriate for the transmission and other hydraulic applications on the equipment. Equipment manufacturer's recommendations should be consulted to ensure choice of the proper lubricant.

AUTOMATIC TRANSMISSION FLUIDS

The fluids used to lubricate automatic transmissions for passenger cars should facilitate the satisfactory operation of such components as the torque converter, planetary or differential gearing, wet clutches, servo-mechanisms, and control valves. The viscosity characteristics of the oil are extremely important. Since a minimum change of viscosity with temperature is desirable, VI improvers are incorporated. Low viscosity improves the efficiency of the torque converter, but the lower limit is dictated by the viscosity required to protect the gearing. It is usual to employ fluids in the range of SAE 5W to 20W for automatic transmissions.

The shearing forces exerted by the automatic transmission's components, such as the pumps and clutches, tend to reduce the viscosity of the polymeric viscosity improvers incorporated in these fluids. It is important, therefore, that automatic transmission fluids have adequate shear stability. To determine shear stability, full-scale road, dynamometer, or laboratory bench tests may be used with the viscosity compared before and after shearing. Because of the comparatively high temperatures reached in service, oils with very good oxidation and thermal stability are also necessary.

The two most important specifications for automatic transmission fluids in the United States have been the Ford MERCON® and the GM DEXRON®. The DEXRON® III, and MERCON® V specifications supersede the GM DEXRON® IIE and Ford MERCON® specifications, respectively. The new specifications place greater emphasis on oxidation stability, frictional durability, and antiwear and thermal stability characteristics of the fluid (Table 3).

Because of the complex range of functions performed, automatic transmission fluids are highly complex products. Additives are used to enhance their performance. Load-carrying additives are used to protect the gearing and thrust washers. Smooth clutch engagements dictate the use of additives for foam control. Specific additives may be used to control the frictional properties of the lubricant since, with most automatic transmissions, static and dynamic friction properties are important for satisfactory operation of the clutches and brake bands. During selection of additives, a prime consideration is fluid inter-compatibility.

STEAM-TURBINE OILS

The primary purpose of lubricating oils for steam-turbine circulating systems is to provide satisfactory lubrication and cooling of the bearings of turbines and generators. Depending on the application, steam-turbine oils also are often used

TABLE 3—History of ATF Specifications

Specification	Number	Year	Company
Type A		1949	GM
Type A, Suffix A		1957	GM
Type F		959	Ford
DEXRON®		1967	GM
DEXRON®-II[A]	6137M	1973	GM
ATF+3	MS7176	1980	Chrysler
ATF+4	MS9602	1995	Chrysler
ATF+4	Licensing system	2005	Chrysler
MERCON®		1987	Ford
DEXRON®-IIE	6137-M	1990	GM
DEXRON®-IIE-Rev.	6137-M	1992	GM
DEXRON®-III	6297-M	1993	GM
MERCON® V		1996	Ford
MERCON C		2005	Ford
MERCON LV		2008	Ford
DEXRON®-IIIG	6417-M	2000	GM
DEXRON®-IIIH	6417-M	2003	GM
DEXRON VI	6418-M	2005	GM

[A]DEXRON®-II was originally released using "C" qualification numbers. In 1975, after suppliers rolled over approvals due to GM mandating a fix for the cooler corrosion problem, the qualification numbers were preceded by a "D." This led to referring to the fluids as DEXRON®-II "C" or "D."

in auxiliary equipment such as pumps, hydraulic governor and control systems, gears in geared turbine systems, and compressors. The first important property of a steam-turbine oil is that it has the proper viscosity at operating temperature to provide effective lubricating films and an adequate load-carrying ability to protect against wear in heavily loaded mechanisms and under boundary lubrication conditions. Viscosity ranges of the steam-turbine oils for which the turbines are designed are generally specified by the turbine manufacturer. It is also a common practice to design the turbine to have the lubricating oil also function as a governor-hydraulic fluid. The viscosity of the oil is important for both of these functions, but the loading of the gear is the major factor in the choice of lubricant. A sufficiently thick film of oil must be maintained between the load-bearing surfaces; the higher the load on the gears, the higher the viscosity required. However, as the circulated oil also acts as a coolant for the bearings, it is necessary to have as low a viscosity as possible consistent with the lubrication requirements.

Since steam-turbine oils generally are required to function at elevated temperatures (for example, 71.1°C [160°F]), it is most important that the oxidation stability of the oil is satisfactory, otherwise the service life of the oil will be unduly short. The oxidation stability of steam-turbine oils is frequently assessed by ASTM tests Oxidation Characteristics of Inhibited Mineral Oils (D943) and Determination of the Sludging Tendencies of Inhibited Mineral Oils (D4310). Oxidation inhibitors are added to the base oil to improve this characteristic. Lack of oxidation stability results in the development of acidic products, which can lead to corrosion (particularly of bearing metals) and also affect the ability of the oil to separate from water. Oxidation also causes an increase in viscosity and the formation of sludges, which restricts oil-ways, impairs circulation of the oil, and interferes with the function of governors and oil relays. Correctly formulated turbine oils have excellent resistance to oxidation and will function satisfactorily for years without changing the system charge.

Turbine-oil systems usually contain some free water as a result of steam leaking through glands and then condensing. Marine systems may also have salt water present due to leakage from coolers. Because of this, rust inhibitors are almost always incorporated in the formulation. Rust preventing properties may be measured by ASTM Test for Rust-Preventing Characteristics of Inhibited Mineral Oil in the Presence of Water (D665/IP 135).

The presence of water in turbine systems tends to lead to the formation of emulsions and sludges containing water, oil, oil oxidation products, rust particles, and other solid contaminants that can seriously impair lubrication. The lubricating oil, therefore, should have the ability to separate from water readily and resist emulsification. Approximate guides to the water-separating characteristics can be gained by ASTM Test for Water Solubility of Petroleum Oils and Synthetic Fluids (D1401/IP 79).

Although systems should be designed to avoid entrainment of air in the oil, it is not always possible to prevent. The formation of a stable foam increases the surface area of the oil that is exposed to small bubbles of air, thus assisting oxidation. The foam can also cause loss of oil from the system by overflow. Defoamants are usually incorporated in turbine oils to decrease their foaming tendency. Foaming tendency can be measured by ASTM Test for Foaming Characteristics of Lubricating Oils (D892/IP 146). Air release is also an important property if a soft or spongy governor system is to be avoided. A careful choice of type and amount of defoamant will provide the correct balance of foam protection and air release properties.

Marine turbine gearing design has advanced to the stage where increased loading has permitted a decreased size of the gear train. This saving in space is very desirable, particularly in naval vessels, but turbine oils for such applications may also require a moderate level of extreme-pressure properties. Load-carrying properties for turbine oils may be measured by ASTM Test for Load-Carrying Capacity of Petroleum Oil and Synthetic Fluid Gear Lubricants (D1947), which uses a Ryder gear rig, or by IP 166, which utilizes an IAE gear rig.

HYDRAULIC OILS

The operation of many types of industrial machines can be controlled conveniently by means of hydraulic systems that consist essentially of an oil reservoir, a pump, control valves, piping, an actuator, and sometimes an accumulator. The wide range of hydraulic applications encountered necessitates the use of a variety of pump designs including gear, vane, axial, and radial piston types, which, in turn, utilize various metallurgical combinations. The hydraulic fluid is required to transmit pressure and energy; minimize friction and wear in pumps, valves, and cylinders; minimize leakage between moving components; and protect the metal surfaces against corrosion.

To obtain optimum efficiency of machine operation and control, the viscosity of the oil should be low enough to minimize frictional and pressure losses in piping. However, it also is necessary to have a sufficiently high viscosity to provide satisfactory wear protection and minimize leakage of the fluid. High VI fluids help to maintain a satisfactory viscosity over a wide temperature range. The antiwear properties of high-quality hydraulic oils usually are improved by the incorporation of suitable additives in the formulation.

Since the clearances in pumps and valves tend to be critical, it is important to provide adequate filtration equipment (full flow or bypass or both) to maintain the system in as clean a condition as possible and thus minimize wear. The oil should have good oxidation stability to avoid the formation of insoluble gums or sludges; it should have good water separation properties, and, because air may be entrained in the system, the oil should have good air-release properties and resistance to foaming. Similarly, good rust protection properties will assist in keeping the oil in a satisfactory condition. Over the years, ASTM has established a variety of test methods that are useful in the design and evaluation of specific fluid properties required in hydraulic applications. Some of the most commonly referenced methods are listed in Table 4.

TABLE 4—Common ASTM Hydraulic Fluid Tests

ASTM Test Method	
D665	Test Method of Rust-Preventing Characteristics of Inhibited Mineral Oil in the Presence of Water
D892	Test Method for Foaming Characteristics of Lubricating Oils
D943	Test Method for Oxidation Characteristics of Inhibited Mineral Oils
D1401	Test Method for Water Separability of Petroleum Oils and Synthetic Fluids
D2070	Test Method for Thermal Stability of Hydraulic Oils
D2619	Test Method for Hydrolytic Stability of Hydraulic Fluids (Beverage Bottle Method)
D6973	Test Method for Indicating the Wear Characteristics of Petroleum Hydraulic Fluids a High Pressure Constant Volume Vane Pump
D3427	Test Method for Air Release Properties of Petroleum Oils
D5621	Test Method for Sonic Shear Stability of Hydraulic Fluids
D6046	Classification of Hydraulic Fluids for Environmental Impact
D6080	Practice for Defining the Viscosity Characteristics of Hydraulic Fluids
D6158	Specification for Mineral Hydraulic Oils

While petroleum oils, properly formulated, are excellent hydraulic fluids, they are flammable, and their use should be avoided in applications where serious fires could result from oil leakage contacting an open flame or other source of ignition. Fire-resistant fluids should be used in such applications. The major fire resistant fluids are listed in Table 5.

Over the years, a number of tests have been used to evaluate the fire-resistant properties of such fluids under a variety of conditions. Most tests involve dripping, spraying, or pouring the liquid into a flame or on a hot surface of molten metal. ASTM recently has developed and published a method in which the fluid is impregnated into ceramic fiber media, and the linear flame propagation rate, used for the comparison of relative flammability, is measured. This ASTM test is D5306, Standard Test Method for Linear Flame Propagation Rate of Lubricating Oils and Hydraulic Fluids.

Fire-resistant fluids, or safety devices on hydraulic systems, are used widely in the coal mining industry. The use of such fluids also is expanding in the metal cutting and forming, lumber, steel, aluminum, and aircraft industries.

Another group of hydraulic fluids drawing much attention is the environmentally acceptable "EA" or environmentally friendly fluids. These fluids are biodegradable and nontoxic to varying degrees, yet provide outstanding performance characteristics. Three different types of base oils have been employed: vegetable oil (rape seed base), polyglycols, and synthetic esters. The primary use for these types of fluids is in hydraulic systems, where leakage of conventional mineral oil could result in environmental damage. Some of its applications are agriculture, forestry, construction, mining, dockside cargo handling, and ski grooming, to name a few.

Currently there are several standards in existence that help define biodegradability and toxicity, but the challenge before ASTM is to develop performance standards. This work is currently ongoing in ASTM D2 Subcommittee N on Hydraulic Fluids.

TABLE 5—Types of Fire-Resistant Hydraulic Fluids

Symbol	Classification	Commercial Descriptions
HFAE	Oil-in-water emulsions $>$ 80 % water containing typically	Soluble oils
HFAS	Chemical solutions in water containing typically $>$ 80 % water	High water based fluids
HFB	Water-in-oil emulsions containing approximately 45 % water	Invert emulsions
HFC	Water-polymer solutions containing approximately 45 % water	Water glycol
HFDR	Synthetic fluids containing no water and consisting Phosphate esters	Phosphate ester
HFDU	Synthetic fluids containing no water and of other compositions	Polyolesters

OTHER LUBRICATING OILS

While many of the properties that have been discussed earlier apply to other types of lubricating oils, there are a number of industrial applications for which special performance requirements are demanded. The following are examples.

Air Compressor Oils

In addition to possessing the correct viscosity for satisfactory bearing and cylinder lubrication, very good oxidation resistance is required to avoid degradation of the lubricant in the presence of heated air. This is particularly important where discharge temperatures are high, since carbon and oxidized oil deposits may autoignite if exposed continuously to temperatures above 148°C (300°F). The fire potential that exists under these conditions makes low volatility and high-autoignition values equally or more important than high flash or fire points. As an added safeguard against compressor fires, a fire-resistant fluid (phosphate ester) should be considered.

In air compressor lubrication, condensed water is frequently present. For this reason, the oil must possess properties that ensure that the oil rather than water wets the metal surfaces. Also, to avoid the accumulation of invert water-in-oil emulsions in the after-coolers, the water should separate out rather than form an emulsion.

Refrigerator Compressor Oils

The efficiency of compression-refrigeration systems can be influenced directly by the properties of the lubricant. This is because the oil used for cylinder lubrication tends to be carried over into the system where it can have a detrimental effect on the efficiency of the evaporator and associated equipment. The lubricating oil, therefore, must effectively minimize friction and wear in the compressor and prevent the formation of undesirable deposits. In addition, the lubricant should have no adverse effects on the operation of the condenser expansion valve or evaporator. Low-temperature viscosity must be balanced against the need to protect the cylinder against wear. Adequate oxidation resistance and thermal stability to resist the high temperatures encountered at the compressor discharge are also necessary. In ammonia systems, it is important that the pour point of the oil is below the evaporator temperatures to prevent congealing of the oil on evaporator heat-transfer surfaces. With Freon systems, a low-Freon floe point of the oil indicates freedom from the likelihood of waxy deposits that could otherwise interfere with the satisfactory operation of the expansion valve and lower the rate of heat transfer.

Steam Cylinder Oils

Steam engines, pumps, forging hammers, and pile drivers are among the equipment using steam cylinders. The performance of this type of machinery is affected directly by the efficiency of the lubrication of the valves, piston rings, cylinder walls, and rods. The selection of the lubricating oil is influenced by the steam temperatures encountered, the moisture content of the steam, the cleanliness of the steam (that is, possible contamination by solids), and the necessity for the oil to separate from the exhaust steam or condensate. Excessive oxidation of the oil in service could cause a build up of deposits in the stem and rod packings and result in shutdown of the equipment for cleaning.

Machine Tool Tableway Lubricating Oils

Satisfactory lubrication of the ways and slides of machine tools is important in maintaining the precision of equipment designed to work to close tolerances. The movement of work-tables, workheads, tool holders, and carriages should be facilitated by the lubricant so that the control is smooth and precise. The characteristics required for these oils include a suitable viscosity to enable ready distribution of the oil to the sliding surface, while ensuring that the necessary oil films are formed at traverse speeds under high-load conditions. Static friction must be minimized, and the oil should prevent the alternate sticking and slipping of moving parts, particularly at very low speeds. Especially when the position of these surfaces is in the vertical plane, good adhesive properties are required to maintain an adequate film on intermittently lubricated surfaces.

SUMMARY

This chapter is a brief overview of those "significant" tests that are used to assess the quality and performance of lubricating oils. General lubricating oil properties and specific lubricating properties of several classes of oils are covered, including automotive engine oils, marine diesel engine oils, industrial and railway engine oils, gas turbine lubricants, gas engine oils, gear oils, automatic transmission fluids, and hydraulic oils. Also, briefly described are air compressor oils, refrigeration compression oils, steam cylinder oils, and machine tool tableway lubricating oils. It is important to recognize that, although tests such as "physical," "chemical," "laboratory bench," and "engine tests" are extremely valuable, the successful application of a lubricating oil can only be determined with certainty through actual field testing.

Applicable ASTM/IP Standards

ASTM	IP	Title
	19	Demulsification Number-Lubricating Oil
	35	Flash Point (Open) and Fire Point by Means of the Pensky-Martens Closed Cup Tester
D92	36	Flash and Fire Points by Cleveland Open Cup
D93	34	Flash Point by Pensky-Martens Closed Cup Tester
D95	74	Water in Petroleum Products and Bituminous Materials by Distillation
D97	15	Pour Point of Petroleum Oils
	110	Barium in Lubricating Oil
	111	Calcium in Lubricating Oil
	114	Oxidation Test for Turbine Oils
	117	Zinc in Lubricating Oil
	120	Lead, Copper and Zinc in Lubricating Oils
D129	61	Sulfur in Petroleum Products (General Bomb Method)
	148	Phosphorus in Lubricating Oil, Additives, and Concentrates

ASTM	IP	Title
D130	154	Detection of Copper Corrosion from Petroleum Products by the Copper Strip Tarnish Test
	166	Load-Carrying Capacity Test for Oils—IAE Gear Machine
	175	Engine Cleanliness—Petter WI Spark-Ignition Test
	176	Oil Oxidation and Bearing Corrosion—Petter WI Spark-Ignition
D322	23	Gasoline Diluent in Used Gasoline Engine Oils by Distillation
D341		Viscosity-Temperature Charts for Liquid Petroleum Products
D445	71	Kinematic Viscosity of Transparent and Opaque Liquids (and the Calculation of Dynamic Viscosity)
D665	135	Rust-Preventing Characteristics of Inhibited Mineral Oil in the Presence of Water
D808		Chlorine in New and Used Petroleum Products (Bomb Method)
D811		Chemical Analysis for Metals in New and Used Lubricating Oils
D874	163	Sulfated Ash from Lubricating Oils and Additives
D892	146	Foaming Characteristics of Lubricating Oils
D893		Insolubles in Used Lubricating Oils
D943	157	Oxidation Characteristics of Inhibited Mineral Oils
D974	139	Neutralization Number by Color-Indicator Titration
D1091		Phosphorus in Lubricating Oils and Additives
D1317	118	Chlorine in New and Used Lubricants
D1401	79	Water Solubility of Petroleum Oils and Synthetic Fluids
D1500		ASTM Color of Petroleum Products (ASTM Color Scale)
D1947		Load-Carrying Capacity of Petroleum Oil and Fluid Gear Lubricants (Erdco, Ryder, WAAO Machines)
D2270	226	Calculating Viscosity Index from Kinematic Viscosity at 40 and 100°C
D2271		Preliminary Examination of Hydraulic Fluids (Wear Test)
D2272		Continuity of Steam-Turbine Oil Oxidation Stability by Rotating Bomb

ASTM	IP	Title
D2422		Recommended Practice for Viscosity System for Industrial Fluid Lubricants
D2500	219	Cloud Point of Petroleum Oils
D2602		Apparent Viscosity of Engine Oils at Low Temperature Using the Cold-Cranking Simulator
D2619		Hydrolytic Stability of Hydraulic Fluids (Beverage Bottle Test Method)
D2670		Measuring Wear Properties of Fluid Lubricants (Falex Method)
D2711		Demulsibility Characteristics of Lubricating Oils
D2782		Measurement of Extreme-Pressure Properties of Lubricating Fluids (Timken Method)
D2783		Measurement of Extreme-Pressure Properties of Lubricating Fluids (Four-Ball Method)
D2893		Oxidation Characteristics of Extreme-Pressure Oils
D2896	276	Total Base Number of Petroleum Products by Potentiometric Perchloric Acid Titration
D2982		Detecting Glycol-Base Antifreeze in Used Lubricating Oils
D2983		Low-Temperature Viscosity of Automotive Fluid Lubricants Measured by Brookfield Viscometer
D3119		Mist Spray Flammability of Hydraulic Fluids
D4172		Test Method for Wear Preventive Characteristics of Lubrication Fluid (Four-Ball Method)
D4293		Standard Specification for Phosphate Ester Based Fluids for Turbine Lubrication
D4310		Test Method for Determination of the Sludging and Corrosion Tendencies of Inhibited Mineral Oils
D5182		Test Method for Evaluating the Scuffing Load Capacity of Oils (FZG Visual Method)
D5534		Test Method for Vapor-Phase Rust-Preventing Characteristics of Hydraulic Fluids
D6006		Guide for Assessing Biodegradability of Hydraulic Fluids
D6046		Classification of Hydraulic Fluids for Environmental Impact
D6547		Test Method for Corrosiveness of a Lubricating Fluid to a Bi-Metallic Couple
D6813		Standard Guide for Performance Evaluation of Hydraulic Fluids for Piston Pumps

17

Passenger Car Engine Oil and Performance Testing[1]

Raj Shah[2] and Theodore Selby[3]

IN THE CREATIVE BEGINNING OF THE DEVELOPMENT of the automotive engine, it was clearly recognized that lubrication of the engine components was a most critical and basic aspect to continued engine operation. This understanding has always paced engine development throughout its history. Once the early engine had been brought to a level of application, automotive transportation was an obvious but challenging area of use.

EVOLUTION OF AUTOMOTIVE ENGINE OIL—A BRIEF HISTORY

The engines of more than a century ago were noisy, unreliable, and somewhat dangerous to start. Starting an engine, when using a hand crank, could cause a broken arm for the person doing the cranking when the suddenly firing engine whipped the crank unexpectedly against the extended arm.

Engine lubrication was primitive. Engine oil and other lubricants in a vehicle needed to be changed or refreshed at exceedingly short intervals, in comparison to today's change intervals [1]. A vehicle owner (or someone hired to take care of the lubrication) could be performing some lubrication function on a daily basis [1]. In some cases, the engines consumed so much oil that supplemented oil kept the oil fresh. However, great care had to be exercised since, if the oil level got too low, engine failure would readily occur.

In the early days of automobiles, there were no standards for engine oil; there was only the practical experience that oils made from Pennsylvania crudes were not as susceptible to oxidation and were easier to start. Pennsylvania oils became identified with quality for years until oil additives and refining techniques began to even the field after the lubrication advances during World War II.

Efforts at describing engine oil characteristics based on performance began in 1947. In this early system, when describing the quality of engine oil, so-called regular engine oils did not contain performance-enhancing additives. In contrast, premium oils typically contained oxidation inhibitors and detergents. In 1952 the additives for diesel engine oil began to diverge from those used in gasoline engine oil.

Although starter motors had long replaced hand cranking ones, understanding the role of the engine oil in affecting startability–particularly at lower ambient temperatures–still had to go through considerable development in the early 1950s. Without such transformation, the utility of the automobile was restricted by weather, starter motor size, and battery power. The educative process involved weaning automotive and engine oil manufacturers away from presumed levels of engine oil viscosity to actual measurement of this property at engine starting temperatures. The same experience was ultimately found necessary in several other engine-oil/engine relationships such as oxidation, wear, oil volatility, detergency, etc.

Specifically, as engines became more sophisticated, valve-train wear problems emerged in the 1950s that were related to the introduction of overhead valve engines. With growth in the size and power output of V-8 engines, sludge formation and oil starvation were also problems in the late 1950s. Oil thickening and oil starvation were also associated with extended high-temperature operation in the early 1970s.

These problems highlighted the growing need for better engine oils. It is not surprising that improvement and standardization of engine oils subsequently have followed a winding pathway as engineers of the automotive, petroleum, and additive industries continue to learn what basestocks and additives need to be formulated into engine oils to enhance their performance. Standardized laboratory and dynamometer test methods were gradually and painstakingly developed and improved over a period of many decades. In substance, as automobiles became more commonplace but complex, and engine designs became increasingly refined, major efforts were–and continue to be–directed toward making these engines more durable and maintenance free with the engine oil playing a major supportive role.

In North America, early attempts to provide information on refined mineral oil for the engine focused on oil properties. Even as early as 1911, properties of interest included the temperature at which the oil would burn or flash in the combustion chamber and the amount of ash residue that might form on the piston, chamber, and piston rings. Viscosity was certainly of strong interest as it was clearly a factor in engine durability. Other properties became of interest as laboratory skills and instrumental techniques were developed at later dates. World War II and the needs of the

[1] In preparation of this chapter, the contents of the sixth edition were drawn upon. The current edition will review and update the topics as addressed by the previous author, introduce new technology that has been developed and include up-to-date references. This chapter is adapted and is updated from a previous chapter ("Automotive Engine Oil and Performance Testing") by Shirley E. Schwartz and Brent Calcut in the sixth edition of the *Manual on Significance of Tests for Petroleum Products*. Credit and acknowledgment go to Shirley E. Schwart and Brent Calcut, General Motor Ltd., also for this work.

[2] *Director, Koehler Instrument Company, Bohemia, NY 11716*
[3] *President, Savant Inc., Midland, MI 48642.*

military groups involved provided a great incentive for the development of acceptable engine oil properties.

In 1969, creation of a system of standardized performance tests for different types of service was started. The effort was technically spearheaded by a "tripartite" association of the American Society for Testing and Materials (ASTM), Society of Automotive Engineers (SAE), and American Petroleum Institute (API).

In the 1980s, another group was organized by the automotive industries of North America and Japan to expedite the development of new engine oils and tests to meet the needs of increasingly fuel-efficient, smaller, but powerful engines. This group was called the International Lubricant Standardization and Approval Committee (ILSAC) and, in balanced voting cooperation with engine oil producers, this became the primary group responsible for setting specifications of automotive engine oils for engine manufacturers in North America and Japan.

As a further consequence of this fully conjoint responsibility for engine oil specification, the previous API "donut" symbol on the engine oil container identifying acceptable engine oils to the consumer was augmented by a new "starburst" symbol indicating that the contained engine oil also met the standards desired by ILSAC. Such oils were given designations of GF-1, GF-2, etc., where numbers advanced with engine oil specifications and, as they progressed, made earlier numbers obsolete.

Thus, the current ILSAC system of test methods to qualify engine oil for use in a vehicle is descended from the recognition and inception of the need for standardization in 1969. For those engineers involved from the automotive, petroleum, and additive industries, the struggle to develop appropriate test methods for engine oils has been a long, expensive, and technically sophisticated effort. However, judging from the performance and durability of modern engines, the struggle has, to a great extent, been successful.

FUNCTIONS OF ENGINE OIL

Engine oil has many functions in a vehicle and is often compared to the many functions of blood in the body. First, if the oil's viscosity is sufficient, it provides a fluid film on which a moving surface, pressed against a stationary opposing surface, can be moved easily and without abrasion and wear–a process called hydrodynamic lubrication. The oil also cools the operating engine by ferrying heat away from hotter areas to the relatively cool oil sump. Engine oil also transports any particles of wear or other contaminants that enter the oil to the engine oil filter, thus preventing such particles from further abrasion of lubricated surfaces.

The one adverse effect of the engine oil in the operating engine occurs not to the engine but to the owner's pocketbook. Engine oil is the major source of power loss because of the viscous drag that must be overcome in generating the power needed to propel the vehicle. This is manifested in fuel efficiency–the higher the viscosity, the greater is the power loss. Fuel efficiency is thus also affected by the operating temperature of the engine. Short trip driving (to be discussed), in which the engine does not have the opportunity to warm up, is less fuel efficient. Fuel efficiency is lower generally in colder climates or colder times of the year. Thus, choice of the operating viscosity of the engine oil must be balanced between having acceptable hydrodynamic lubrication to engine wear reduction, good antiwear additive technology (to be discussed shortly), and good fuel efficiency. Engine oil formulation is, above all, a balance of desirable and undesirable but temporarily tolerable properties and thus the creation of better base stocks and additives to push the balance in a more favorable direction, such as the development of additives or base stocks that favorably modify the viscosity-temperature relationship.

In addition to the physical benefits of the oil, there are a number of chemical benefits that are imparted by the proper choice of base stock and additives for the formulated engine oil. For example, if a vehicle is driven under conditions producing high operating temperatures such as going up steep mountains in a hot climate, fluid films can become very thin, since viscosity (and, thus, hydrodynamic lubrication) is reduced severely at higher operating temperatures. On heavy load–bearing surfaces such as the cam–lifter interface, chemical agents in engine oil lay down a protective film (typically a phosphate) on mating iron surfaces, which greatly reduces wear.

Another example of the oil's chemistry is that oxidation resistance of the oil is incorporated by proper oil formulation with the choice of additives and base stocks. Similarly, oil additives neutralize acids produced by the slight but continuous passage of combusted fuel bypassing the piston rings into the oil. These additives also reduce the oxidation effects at hotter spots in the operating engine.

EFFECTS OF DIFFERENT TYPES OF SERVICE

As the gasoline automotive engine became increasingly reliable and simple to use, it has become virtually indispensable for both the transportation of people and their possessions. Examples of the effects of various driving styles are discussed and illustrated in the following sections of this chapter.

Freeway Driving

When the vehicle is used on the freeway networks of roads, the engine oil performs its many functions with very little degradation and very little wear on the engine. Engine operating temperatures are hot enough that combustion is complete, and fuel and combustion products (water, carbon dioxide, acids) do not accumulate to any significant extent in the engine oil. However, the engine oil is not so hot that it suffers rapid or abnormal oxidation and thermal decay. Thus, the engine is operating in a mode that promotes long engine-oil life.

High Temperature, High Load

Under high-temperature, high-load conditions, such as pulling a trailer up a steep mountain road, the engine oil degrades primarily as a consequence of high-temperature oxidation. The antioxidant in the engine oil is designed to sacrificially react with oxygen from the air before the oxygen has a chance to significantly damage the bulk of the engine oil. That is, by reducing the harmful effects of oxygen, the antioxidant prolongs the life of the engine oil until, of course, the antioxidant is expended.

Engine oils' base stocks (especially in the case of mineral oil) consist of hydrocarbon molecules with a range of molecular weights and chemical structures that have varying degrees of vulnerability to oxidation. When oxidized during high-temperature operation, they form unwanted oxidation-generated compounds such as organic acids, aldehydes, ketones, and alcohols. As a consequence, under such operation, engine oil formulations that use base stocks more highly resistant to oxidation are more desirable.

Organic acids and nitrates (that are created on exposure of the oil to heat and air) can form deposits in an engine and can corrode susceptible metals. The rate of high-temperature, high-load engine oil degradation can range from just a little faster than is observed in freeway driving to several times faster, depending on how hard the engine works and how hot the engine oil becomes. In addition to these chemical effects, as previously mentioned, the engine oil's viscosity is lower at high temperature. Thus, fluid films are thinner, hydrodynamic lubrication less effective, and dependence on the protective antiwear chemistry of the oil is greater.

City Driving

In city driving, such as commuter times greater than an hour or taxi service, somewhat different mechanisms of oil degradation occur. Oil temperatures are typically cooler than they might be if a vehicle had been pulling a trailer up a mountain. Thus, the chemical reactions that occur in city driving are, to some extent, slower than those reactions that occur in high-temperature, high-load service.

For example, in city driving, the engine oil is fully warm. But, periodic engine idling generates partially combusted fuel products that can react with the engine oil and cause the oil to degrade. As a consequence, in city driving, the oil's detergent becomes inactivated at a faster rate than occurs during easy freeway driving, as illustrated later in this chapter.

Short-Trip Driving

Under conditions in which a vehicle is driven only for very short trips (5 or 10 min per trip, during which the engine oil is never completely warm), water, fuel, and combustion products, such as organic acids, condense in the engine oil. These condensed contaminants, once the engine oil becomes sufficiently degraded, contribute to rust formation and corrosion of various metals in the engine. When gasoline fuel (which is completely soluble in engine oil) condenses in the oil during short trips in winter, it can cause the oil's viscosity to drop to less than half of the viscosity of the fresh oil at the same temperature, and thus reduce wear protection from the viscosity-dependent hydrodynamic lubrication.

Comparison of Different Types of Service

Figure 1 provides an example of differences in severity when comparing oil analysis results from a cross-country road test (circles and triangles) to results from a short-trip winter driving test (squares). The circles are the pooled oil analyses from two vehicles, primarily freeway driving, but including some city and some mountain driving. The triangles represent a vehicle on the same test trip that pulled a 1,000-kg trailer. The squares represent data from a short-trip test in winter. The oxidation induction time of the engine oil (obtained using ASTM D5483, Oxidation Induction Time of Lubricating Greases by Pressure Differential Scanning Calorimetry) is a good measure of severity of service shown in this figure.

It can be seen that engine oil life was the longest under conditions that were primarily freeway driving. When pulling a trailer weighing 1,000 kg, the rate of degradation of the oil was approximately 10 % faster than was the case for the two vehicles that were experiencing the same road conditions but were not pulling a trailer. A heavier trailer would have caused a faster decrease in oil life. Short-trip winter driving caused the engine oil to degrade at a much faster rate (for a given distance traveled) than freeway driving, as shown in the left-most curve of Fig. 1 [2]. Part of the loss of oxidative stability during short-trip driving is related to the fact that fuel and combustion products condense in engine

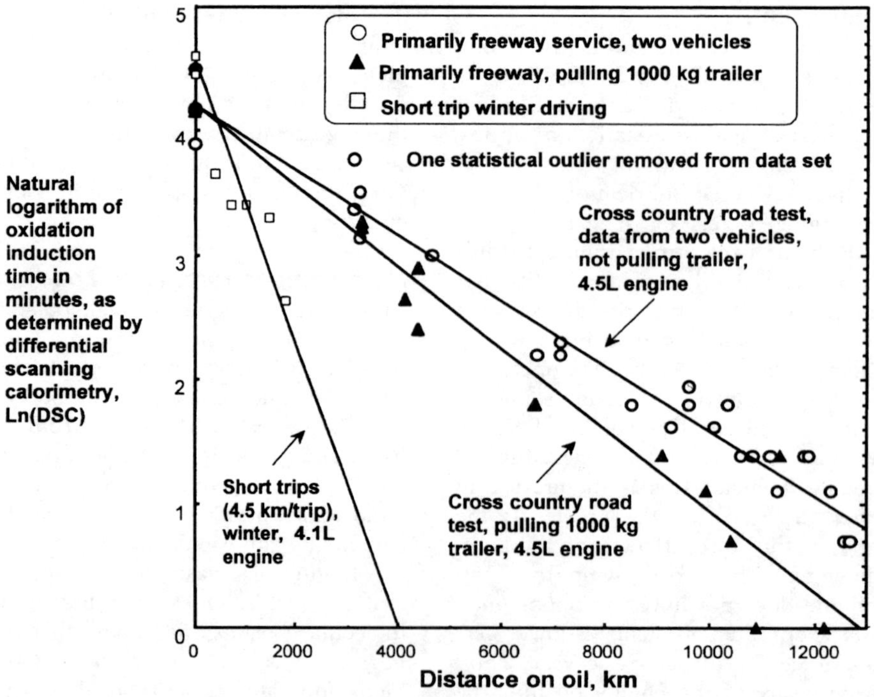

Fig. 1—Influence of trip length, service conditions, and condensation of contaminants in engine oil on loss of oxidative stability of engine oil.

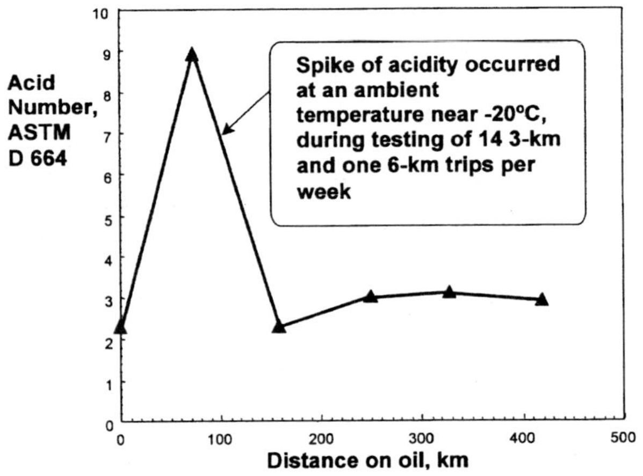

Fig. 2—Condensation and later evaporation of volatile acidic fuel combustion products during short trips and extremely cold weather.

oil when the oil is cold. Since fuel is easily oxidized, the presence of fuel in the engine oil reduces the overall oxidative stability of the oil. If trips had been shorter than those shown in Fig. 1, or the temperatures were colder, the slope of the short-trip driving curve would be even steeper. If the outside temperature had been warmer, the slope of the short-trip curve would be less steep.

During extremely cold weather and under short-trip driving conditions, acids from partial combustion of the fuel condense into engine oil. The presence of these fuel-derived acids can easily be documented, as indicated by the spike in Fig. 2 [3]. Volatile fuel-derived acids were not observed in the engine oil at higher ambient temperatures, since fuel and fuel reaction products boil out of the engine oil once the oil warms. Therefore, the extent of degradation of engine oil during short-trip driving is somewhat reversible if the engine is later operated at sufficiently high temperatures.

The rate of oil degradation during typical city driving falls between the values for freeway and short-trip driving and depends on the severity of the driving pattern.

Interesting insights can sometimes be gained when modifying one or more fundamental characteristics of a system as complicated as an internal combustion engine. A number of tests were conducted with an alternative fuel (methanol, also known as methyl alcohol) to identify fuel effects (in contrast to engine or service effects). In some cases, it became possible to distinguish service effects from fuel effects for those tests in which alcohol fuel behaved differently from gasoline.

For example, Fig. 3 indicates that in freeway service, the rate of oil degradation when using gasoline was the same as for the case in which the fuel was M85 (85 % methanol, 15 % gasoline). Thus, the rate of oil degradation can be considered an engine effect in both cases [3].

In city service, the engine oil degraded faster with gasoline than with M85 (the driving cycles were identical, since the test vehicles were all operated on a chassis dynamometer). This difference for city service is very likely caused by the fact that the partial combustion products of methanol are more volatile than those of gasoline and therefore do not remain in the engine oil during city service. Thus, fuel and its reaction products played a significant role in the degradation of the engine oil of the gasoline-fueled vehicle in city service [3].

Alcohol fuels such as M85 or E85 (85 % ethanol, 15 % gasoline) generate more water per kilometer of service than is generated by gasoline [4]. During very short trips in a long and harsh winter, water and fuel condense in the engine oil. Contaminant concentration (water plus fuel plus products of incomplete combustion) as high as 50 % was observed in engine oil when driving short trips using methanol fuel (median trip length 3 km, three trips per day, outside temperature below freezing). In contrast, as shown by the squares in Fig. 4, slightly more than 30 % volatile contaminants accumulated in the engine oil of the gasoline fueled vehicle.

Alcohol fuel and water from the combustion process, in the presence of the engine oil's dispersant, form an emulsion in engine oil. Emulsion formation causes the oil to thicken (that is, increase in viscosity), as shown in Fig. 5. The engine oil's dispersant and the shearing action of flow through tight clearances in the engine greatly increases the

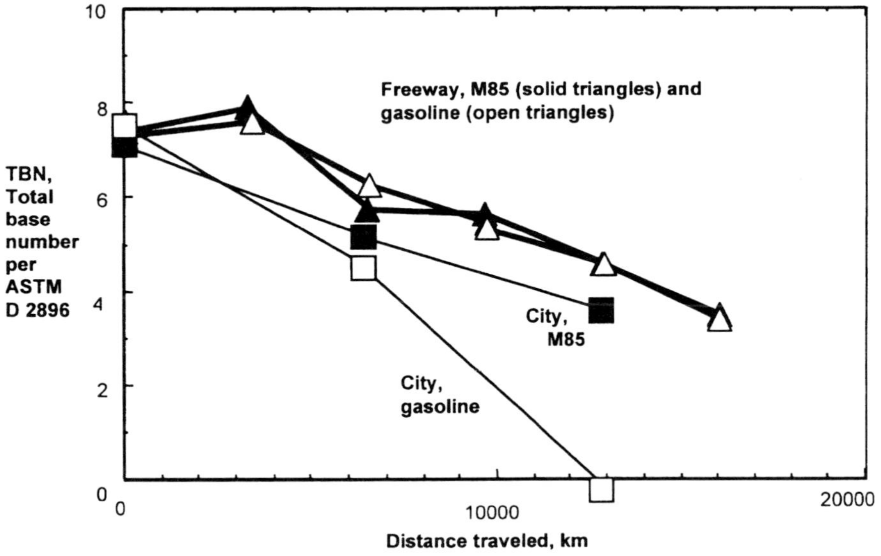

Fig. 3—Effects of fuel type and service on the rate of loss of engine oil alkalinity.

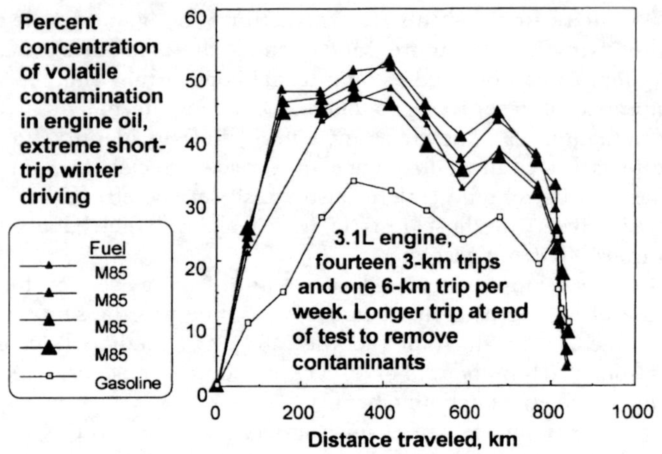

Fig. 4—Entry of contaminants (fuel, water, reaction products) into engine oil during hort-trip, winter driving.

stability of the emulsion that is formed. Gasoline forms a true solution in engine oil, which, in contrast, causes a reduction in the viscosity of the oil/gasoline mixture, compared to the viscosity of the fresh oil, as seen in the curves designated with squares. At the end of the test (when weather had become warmer), a longer trip was taken, volatile fuel-related contaminants evaporated as the oil warmed to a normal operating temperature, and the engine oil in each of the vehicles returned to a near-normal viscosity.

These examples highlight the fact that engine oil can suffer various degradation mechanisms. Other automotive lubricants such as automatic transmission fluid or hypoid gear oil typically are not exposed to the reactive chemicals that reach the engine oil as a consequence of partial combustion of the fuel during city or short-trip operation. In addition, some automotive lubricants may have to endure high temperatures, but it is unlikely that the heat will be as harsh as small amounts of engine oil experience in the flame front during the combustion process. Surprisingly, the bulk of a good engine oil will withstand these constant small-volume exposures to a flame front and the acquisition of contamination from fuel and combustion products and still protect the engine.

COMPOSITION AND FUNCTION OF ENGINE-OIL ADDITIVES

The physical characteristics of the engine oil, such as viscosity and density, are provided largely by the base stock as modified by additives such as viscosity modifiers. Also, regarding oxidation, there are significant differences between base stocks falling into various categories such as mineral, highly paraffinic mineral, synthetic, and combinations of these. However, beneficial chemical properties of engine oil are primarily a function of the characteristics of the additives in the oil. These additives typically include antioxidants, detergents, dispersants, viscosity modifiers, pour-point depressants, and defoamers.

Antiwear/Antioxidant Additives

Antiwear agents in the engine oil are the most important additives in the oil since, when hydrodynamic lubrication is not sufficient to prevent wear, these additives come into action to prevent serious wear from occurring under boundary lubrication conditions by laying down a protective antiwear film on a metal surface. In addition, some antiwear additives also have antioxidant properties and slow the rate of oil oxidation by acting as a sacrificial agent that is more easily oxidized than the oil itself, thus promoting long oil life as they provide wear protection. An important antiwear/antioxidant agent is zinc diorganodithiophosphate (ZDDP). Additional antioxidant additives are also often used in engine oil formulations.

Detergents

The detergent in the oil's additive package is an oil-soluble alkaline agent that is designed to neutralize acids and inactivate by-products that form in hot spots in the engine or in the heat of combustion. Thus, detergents reduce the extent of deposit formation and help keep engine surfaces clean.

Dispersants

Dispersants keep contaminant particles from growing in size and prevent oil-oxidation products from clumping, which reduces the possibility of clogging critical oil flow passages. Dispersants are particularly important for engine oils used in diesel engines, in which significant amounts of soot can be generated. Dispersants ensure that soot particles do not agglomerate to the extent that engine failure can occur.

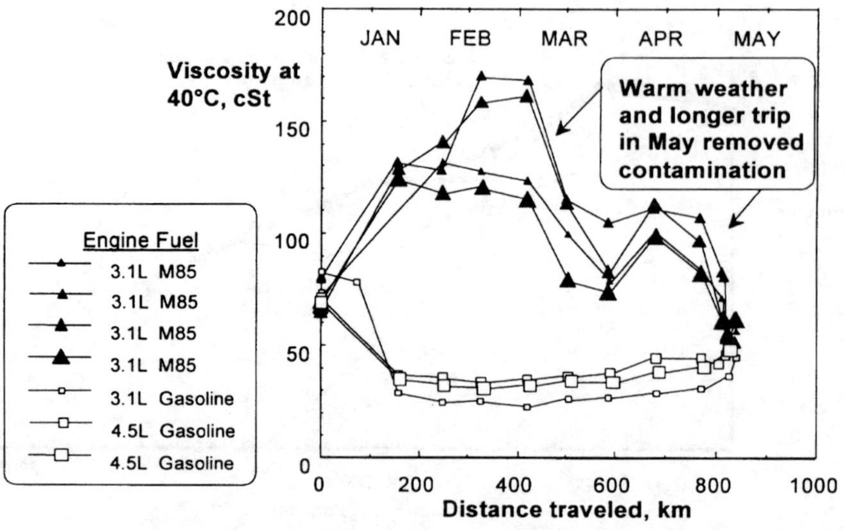

Fig. 5—Viscosity of engine oil during short-trip test.

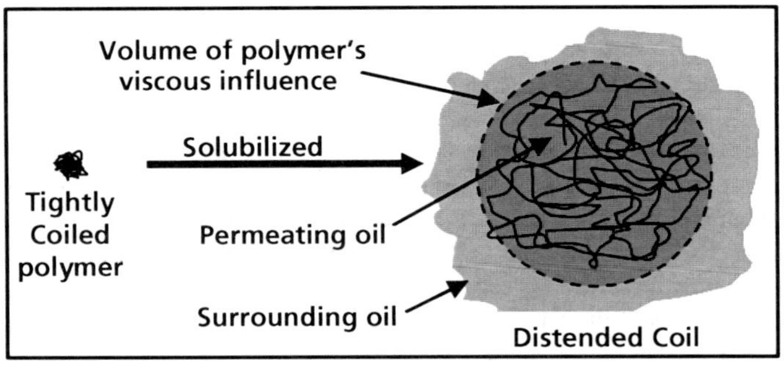

Fig. 6—Oil-solubilized viscosity index improver.

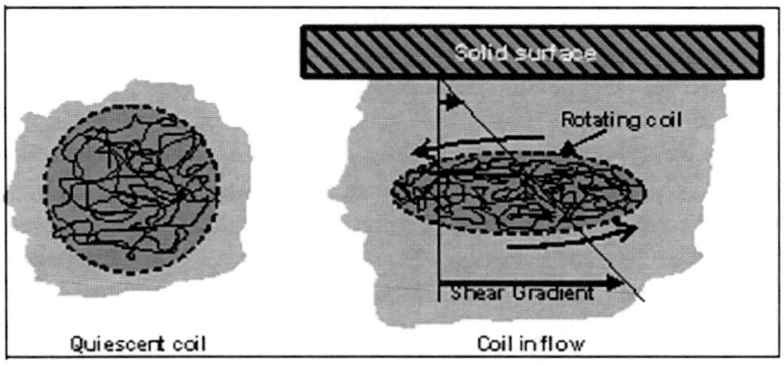

Fig. 8—Polymer orientation in flow and temporary loss of some of its viscous contribution.

Viscosity Index Improvers

Considering the critical role of engine oil viscosity and its dependence on temperature, World War II saw the development of Viscosity Index Improvers (VIIs) of various types. These VIIs are mostly long-chain, very oil-soluble polymers that simply thicken by unfolding in the engine oil, as shown in Fig. 6. Their highly extended state obstructs the flow of the relatively very small oil molecules.

Most VIIs have high solubility in mineral oil and their viscosity-increasing effect is proportionately the same at all temperatures, as shown in the lower portion of Fig. 7. For this reason, these VIIs have been called thickeners [5]. However, other VIIs are more soluble in oil at higher temperatures than at lower temperatures, as shown in the upper portion of Fig. 7. That is, when the oil is hot, this form of VII distends and markedly increases the viscosity of the oil, while at lower temperatures, the VII is less soluble in the oil and the polymer coil contracts. This form of VII has been termed a Viscosity Temperature Improver [5] because of its helpful effects on the viscosity-temperature relationship.

However, a less positive aspect of all VIIs is their susceptibility to high shear rate in which their viscous contribution to the oil is partially lost as a consequence of distortion or orientation of the polymer coil under tractive viscous forces of the flowing oil surrounding the coil, as shown in Fig. 8. This viscosity loss is regained when the viscous forces subside and the VII coil regains its previous dimensions. This form of viscosity loss is called Temporary Viscosity Loss.

Of greater concern, depending on the size and configuration of the VII, is when the tractive forces of the oil on the coil become too great, causing the polymer coil to rupture and lose a portion of its viscosity-modifying effect. This viscosity loss is not recoverable and is called permanent viscosity loss.

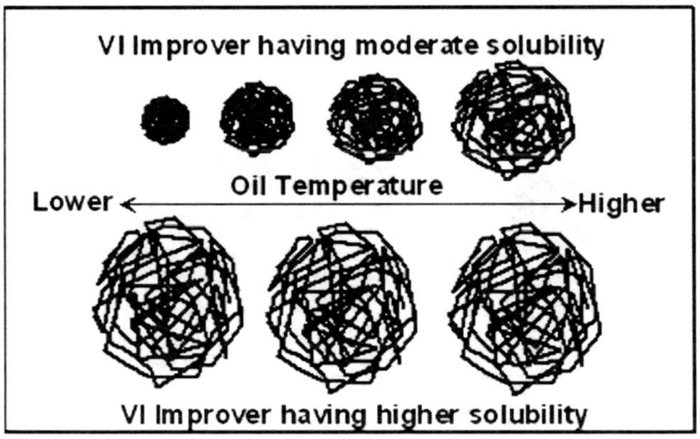

Fig. 7—Solubility response of different Viscosity Index improvers.

The combination of temporary and permanent viscosity losses can be 30 % or more of an oil's viscosity.

Pour-Point Depressants

Pour-point depressants are irregularly shaped molecules that become integrated in the initially forming gelation structure. Their irregular shape prevents other more regularly shaped oil molecules from adding to the growth of gelation structures, which would ultimately form a gelation mass in the crankcase. Thus, with only small packets of gelated oil, the oil is still able to flow to the engine areas needing lubrication, pick up heat from the operating engine, and carry this heat back to the crankcase, preventing the catastrophic failure of the engine caused by oil gelation. More simply, pour-point depressants allow the oil to cool without solidifying (forming smaller waxy agglomerates that can flow, for example, like marbles in a sack, rather than allowing the oil to solidify completely).

Defoamers

Defoamers prevent foam by reducing the surface tension of the oil. They consist of chemical compounds that are marginally soluble in engine oil, and, as a consequence, some of the defoamant material migrates to the oil surface. When air bubbles rise to a surface containing a defoamer, the surface tension of the oil is lowered, allowing these foam bubbles to rapidly break at the surface and not grow in height to allow foam-containing oil to destroy hydrodynamic lubrication and markedly increase engine wear.

BENCH TEST METHODS FOR ENGINE OIL

There are many reasons for developing bench tests to simulate results obtained in engines. Some of these are as follows:

1. Generally, they have a much lower cost to develop and relevantly maintain than engine tests.
2. There usually is considerably greater precision.
3. When necessary, they are usually much simpler to modify with changing conditions or engines.
4. There is greater reliability.
5. They involve relative ease of inserting or modifying test parameters.
6. There is continuity of test parameters and test equipment through engine model changes.

However, the engine is almost always necessary to establish the data and parameters or to provide a basis with which to generate correlation for the bench test.

To prolong the life of an engine, it is desirable to create test methods for engine oils that ensure that these oils will have suitable properties and perform the functions desired throughout their expected working life. These test methods fall into several categories: physical properties, chemical properties, and performance.

Tests for measurement of physical properties tend to be less complicated than other types of tests and are typically less expensive to perform. They include tests for such properties as engine oil viscosity, pour point, and volatility. These values indicate whether the engine oil is in an appropriate range for use in a vehicle. Measurements of chemical properties include a determination of the chemical composition of the oil and its additives, the extent to which the chemical properties have changed during use, and the measurement of various contaminants that may have entered the engine oil. Tests for performance in engines on test stands or in the field tend to be highly complicated and are often expensive to develop and conduct with many intrusive parameters to be controlled while measuring those desired. Performance tests typically mimic extreme examples of various driving conditions that an engine might experience, such as severe taxi service during city driving or pulling a trailer up a mountain.

AREAS OF PHYSICAL TESTS FOR ENGINE OIL AND THEIR SIGNIFICANCE

High-Temperature, High Shear Rate Viscosity

All engines require adequate high shear rate viscosity to operate. For many years, engine oil viscosity was categorized by low shear rate capillary viscometers at temperatures reflecting engine operation. These earlier oils were Newtonian in behavior–that is, did not change in viscosity with shear rate. However, the advent of VIIs and multigrade engine oils for easier cold starting changed the viscous nature of the engine oil. These VII-containing oils were non-Newtonian and, as discussed earlier, were shown to lose viscosity with increasing shear rate. Since hydrodynamic lubrication is a high shear rate process, concern was raised about what the actual viscosity of the lubricant really was at higher shear rates and operating temperatures. After much work and discussion, it became evident that protection of the engine required the measurement of oil viscosity at high shear rates and engine operating temperatures.

In the 1970s this work began, and by 1980 a commercial high shear rate rotational viscometer was made available [6] and later standardized as ASTM D4683, Measuring Viscosity at High Shear Rate and High Temperature by the Tapered Bearing Simulator. The slightly tapered rotational geometry shown in Fig. 9 permitted operation at constant shear rate, produced dynamic (centipoise) viscosity values, and the slight matching tapers of the rotor and stator permitted operation as an absolute viscometer.

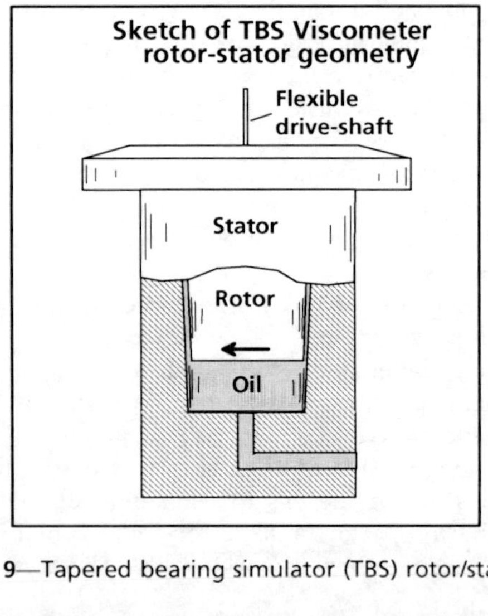

Fig. 9—Tapered bearing simulator (TBS) rotor/stator.

The development and importance of high-temperature, high-rate (HTHR) viscometry were confirmed by direct correlation with engine bearing oil film clearances in which low shear capillary viscometry was considerably poorer [7]. This HTHR correlation allowed establishment of minimum high shear rate limits for engine oils under severe operating temperatures, which became part of the SAE Engine Oil Viscosity Classification, J-300.

Low Shear Rate Viscosity

The low shear capillary viscometer gives values in *kinematic* viscosity, which is defined as dynamic (true) viscosity multiplied by density. It has been used universally to measure the viscosity of many kinds of fluids including engine oils. Very early in the specification of engine oils, this method was standardized as ASTM D445, Kinematic Viscosity of Transparent and Opaque Liquids, where the test measures the rate of flow of a fluid through a calibrated capillary at a specified temperature. Figure 10 shows a photograph of a viscosity bath in conformance to ASTM D445 containing a capillary viscometer.

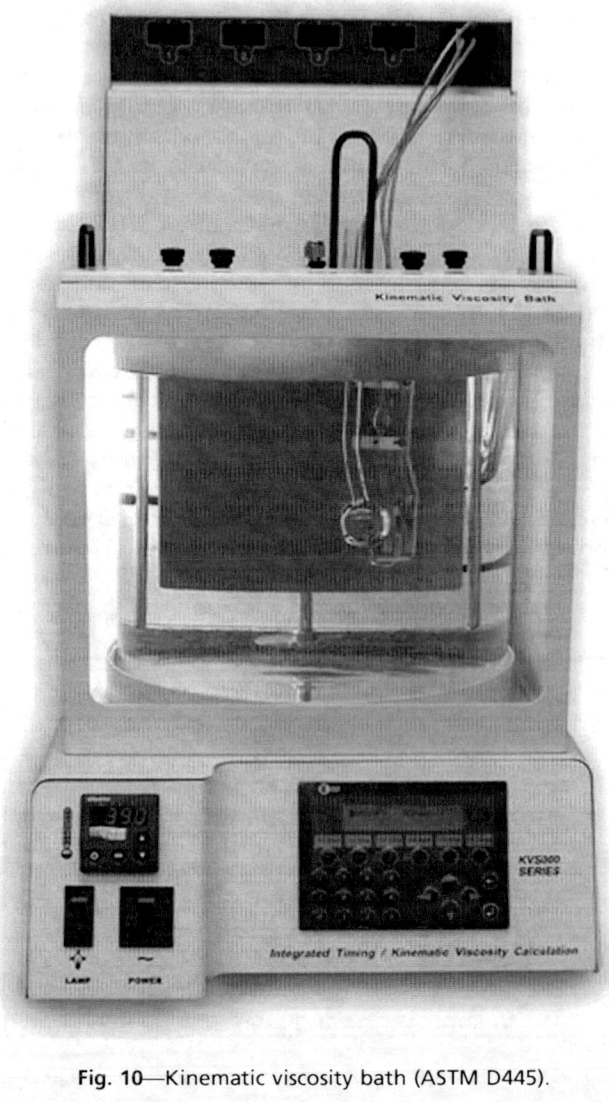

Fig. 10—Kinematic viscosity bath (ASTM D445).

Even though kinematic viscosity is no longer considered as highly meaningful to hydrodynamic lubrication in the engine, it is still one of the parameters measured in SAE J-300. Moreover, as shown previously in work on the effects of driving patterns on engine oil, kinematic viscometry can provide interesting and significant information.

As previously noted in several sections, kinematic viscosity measurements have provided useful information about the extent and nature of engine-oil degradation. In some cases, kinematic viscosity results are surprising, as described previously in the discussions related to Figs. 4 and 5 and the effects of exceedingly harsh short-trip winter service.

The extreme thickening due to emulsion formation (a "milkshake" effect) had not been anticipated. In that particular study, equations were developed to mathematically describe the effects on this low shear rate form of viscosity due to the simultaneous presence of both solution-forming and emulsion-forming contaminants. Despite these opposing kinematic viscosity effects, it was still possible to predict oil viscosity.

The equations, and the thought process that went into their development, are described in Schwartz (1992) [4]. Once the weather warmed, much of the fuel and water evaporate from the oil and the oil's kinematic viscosity returned to more normal values. The viscosity values at the end of the test were somewhat lower than at the start of the test, perhaps indicating some residual heavy ends of gasoline in the oil or perhaps shear of a VII.

Because an accurate measurement tool of kinematic viscosity, ASTM D445, was available, investigators were able to identify and quantify unusual interactions between fuel and oil. A literature search indicated that the equations developed to describe these unusual viscosity effects had not been envisioned previously. It would be interesting to measure these effects in low shear rate dynamic viscometry.

Low-Temperature Viscometry and Gelation

If the oil is too viscous, or becomes too viscous during service, it may not flow adequately into highly constrained spaces (for example, journal-to-bearing clearances). If so, using this example, bearing failure can result. One obvious condition in which an otherwise acceptable engine oil at operating temperature can become very viscous is at low ambient temperatures.

Four methods that address the issue of low-temperature viscosity in order of development and publication include:

1. ASTM D5293 (originally D2602), Apparent Viscosity of Engine Oils Between −5 and −35 °C using the Cold-Cranking Simulator
2. ASTM D3829, Predicting the Borderline Pumping Temperature of Engine Oils
3. ASTM D5133, Low Temperature, Low Shear Rate Viscosity/Temperature Dependence of Lubricating Oils Using a Temperature Scanning Technique
4. ASTM D4684, Determination of Yield Stress and Apparent Viscosity of Engine Oils at Low Temperature

The latter two are engine oil pumpability tests with the former commonly known as the Scanning Brookfield Technique and the latter as the MRV TP-1.

ASTM D5293 was developed to predict the low-temperature startability of automotive engines, whereas ASTM D5133 and D4684 were developed to predict the low-temperature, low-shear rate pumpability of engine oils. Since low-temperature startability problems have been virtually eliminated today, engine oil pumpability has become the most important low-temperature engine oil measurement.

In the winter of 1979–1980, Sioux Falls, South Dakota, experienced an epidemic of engine failures because the engine oil would not flow. Failure to flow was caused by the formation of a gelated mass of the engine oil in the crankcases. In a following study [8], it was found that a particular set of cooling conditions was responsible for the response of the engine oil and that it would be difficult to anticipate the varied weather conditions that could be presented by nature. It was also shown that the pumpability test that had been previously developed and applied (ASTM D3829) failed to catch this massive problem.

The Scanning Brookfield Technique

Immediately after the massive epidemic, efforts were made to develop a more sensitive low-temperature test. A new approach to low-temperature viscometry was developed and presented in 1981 [9] in which the engine oil is first heated to 95°C, cooled quickly to −5°C and its viscosity is measured continuously at a cooling rate of 1°C/h using a sensitive rotational viscometer. Analysis of the viscosity-temperature curve produced determines the tendency of the oil to form gelation as well as the severity of gelation [10].

Since the test scans the entire temperature range of interest, only one test is required. Moreover, scanning the temperature range of interest tends to force the engine oil to show gelation tendencies (if any exist) [10]. The test was standardized as ASTM D5133 and is also known as the Scanning Brookfield Technique (SBT). Figure 11 shows the response of the SBT when applied to two engine oils. One of these is Newtonian, with its behavior showing the typical exponential increase of viscosity with decreasing temperature. The second starts out its cooling curve in a Newtonian manner but, at a temperature of about −13°C, the viscosity begins to increase more rapidly than if it continued the exponential relationship. Obviously, some additional presence is causing this increase, and that presence is the beginning of the formation of viscosity-increasing gelated structures in the oil.

Analyzing the data collected in Fig. 11 using the first derivative of a linear (and empirical) viscosity-temperature equation gives the information in Fig. 12 showing that the Newtonian oil plots, as expected, as a horizontal line, while the non-Newtonian oil shows a peak of 22 at −14.3°C. This peak is a measure of the maximum amount of relative gelation that this oil will form and the peak value is called the gelation index.

For comparison to the oils causing the Sioux Falls epidemic, the lowest gelation index measured on a field-failing oil was 16 and ILSAC specifications today call for a safe maximum gelation index of 12.

Applying the SBT to a set of reference oils exposed in cold-room engine pumpability tests, correlation was found to be very good, as shown in Fig. 13, using the relationship between the development of gelation and the occurrence of engine air-binding reported from field correlation [8].

Later work with the SBT with sooted diesel engine oils showed that the technique was equally applicable [11].

The Mini-Rotary Viscometer

In 1985, another pumpability test was developed using the original instrument for ASTM D3829 in a considerably different manner than the earlier test.

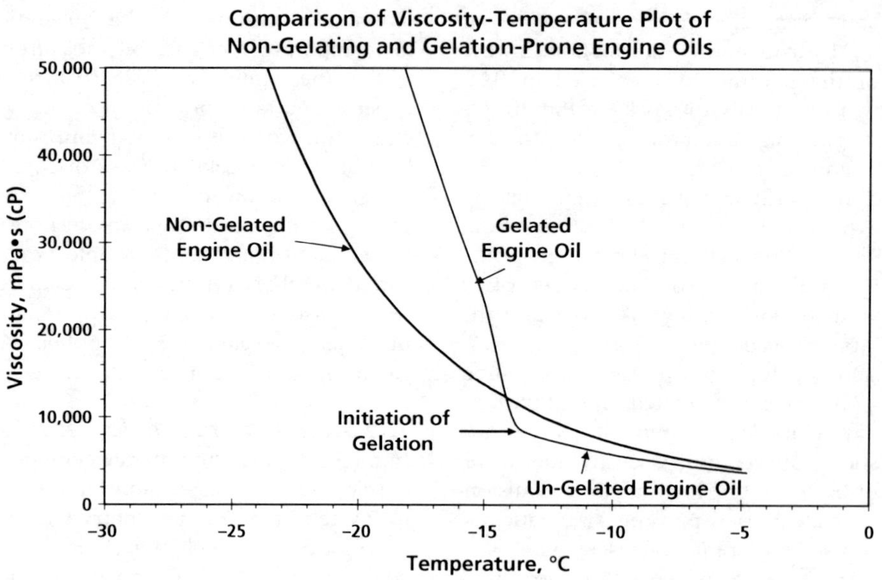

Fig. 11—Scanning viscosity curves of gelating and nongelating engine oils.

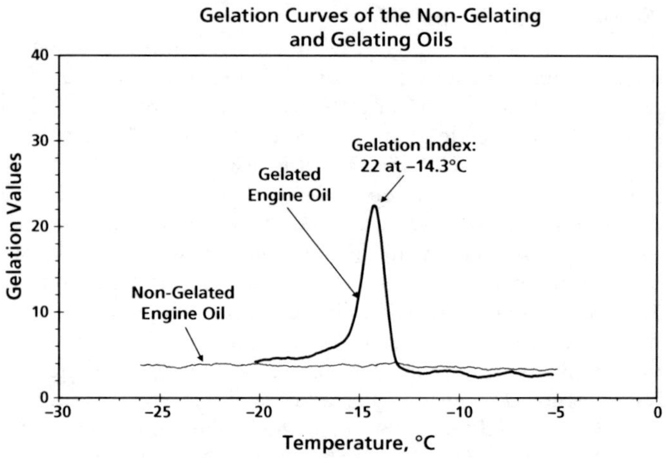

Fig. 12—Gelation curves from analysis of gelating and nongelating engine oils in Fig. 10.

After extensive work [12], a temperature profile, TP-1, was found that permitted the instrument to detect the presence of gelation by the development of an inability to turn the rotor in the stator under moderate loads. This viscometric response is known as yield stress since it requires a certain amount of applied stress to force the structure in the fluid to yield.

This test became ASTM D4684 and lists the following procedure: "An engine oil is held at 80°C then cooled at a pre-determined rate to a final test temperature. A low torque is applied to the rotor shaft to measure the yield stress. A higher torque is then applied to determine the apparent viscosity of the sample." A photograph of the mini-rotary viscometer is show in Fig. 14.

The two low-temperature pumpability tests are complementary in the viscometric data they produce. However, the Mini-Rotary Viscometer (MRV) TP-1 protocol tests for the presence of gelation at a given temperature under quiescent

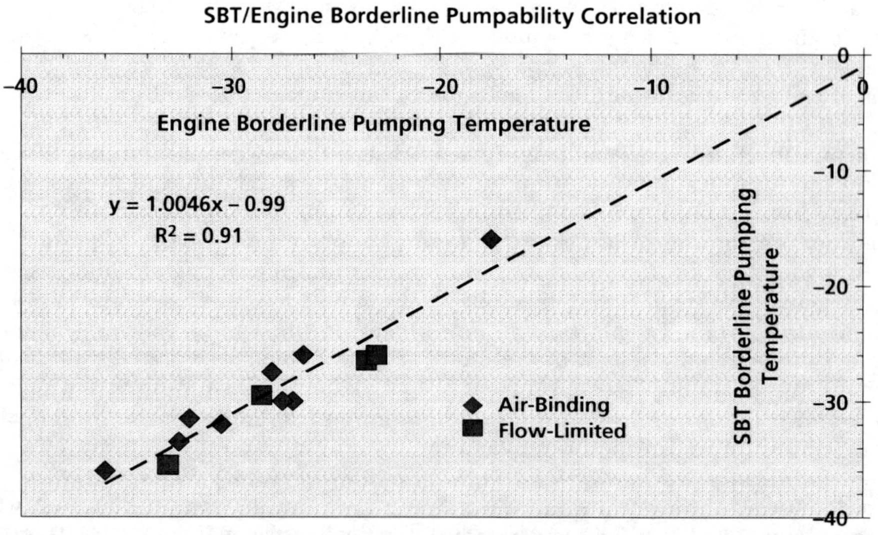

Fig. 13—Correlation of Scanning Brookfield Technique (SBT) data and engine pumpability studies.

Fig. 14—Mini-Rotary Viscometer (ASTM D4684).

conditions while the SBT tests for the tendency of the engine oil to form gelation over a temperature range under gentle dynamic motion. Thus, in regard to the occurrence of gelation in an oil, the two instruments give complementary views that include the frequency that the SBT will find gelation not apparent to the MRV/TP-1 and the latter will find gelation that takes time to develop at a given temperature that registers only mildly in the SBT.

Cold-Cranking Simulator

The cold-cranking simulator was developed to simulate the response of engine oil in resisting the engine cranking motor during starting. It was originally accepted by the ASTM as D2602, but this was withdrawn in favor of a more advanced method in 2002: ASTM D5293.

In the cold-cranking simulator test, temperature is maintained using refrigerated coolant. A rotor is set in a relatively close-fitting stator. The rotor is turned by a small motor with limited torque output, and rotor speed is measured as a function of the fluid viscosity on the basis of a previously determined calibration using oils of known viscosity.

AREAS OF CHEMICAL TESTS FOR ENGINE OIL AND THEIR SIGNIFICANCE

Chemical tests describe the composition of engine oil and help investigators interpret changes that have affected the oil during service. Some of the chemical questions that might be of interest to an investigator include the following:

- Can a chemical test be substituted for an engine test?
- Are all the proper additives in the oil?
- Are the additives present at an appropriate concentration?
- To what extent have the additives or the engine oil degraded?
- Are the additives so badly degraded that they are no longer effective?
- Have fuel and fuel reaction products entered the engine oil?
- Have contaminants such as wear debris or corrosion residues entered the oil?

Chemical tests for engine oils include measurement of the concentration of the various oil additives and the chemical elements (such as sulfur and phosphorus) that are in the additives to ensure that an appropriate oil was used. The extent of oil degradation can be documented using such tests as remaining antioxidant protection, remaining alkalinity, acidity of the oil, and the presence of contaminants in the oil. The extent of oil contamination can be assessed by measuring acidity, the amount of fuel and water (formed during combustion of the fuel) in the oil, and the extent to which engine materials or corrosion products have entered the oil. In a number of cases, more than one test is available for a given type of measurement.

Base and Acid Number

One example is a measurement of the remaining alkalinity in the engine oil, using ASTM D2896, Base Number of Petroleum Products by Potentiometric Perchloric Acid Titration, or by ASTM D4739, Base Number Determination by Potentiometric Titration. ASTM D2896 employs a very strong acidic reagent (perchloric acid). The test has been used in studies assessing the end of useful engine oil life, since the test is sensitive to low alkalinity values that would not easily be detected using ASTM D4739 [13]. On the other hand, some investigators consider that ASTM D2896 is too sensitive with regard to marginally alkaline agents.

ASTM D664, Acid Number of Petroleum Products by Potentiometric Titration, can be used to determine the extent to which engine oil has acidified during use. Acidification of the engine oil occurs primarily as a consequence of exposure of the engine oil to hot spots in the engine. A photograph of an automatic potentiometric titration instrument capable of performing ASTM D664, D2896, and D4739 is shown in Fig. 15.

Oxidation Induction Time

Remaining antioxidant protection, ASTM D5483, Oxidation Induction Time of Lubricating Greases by Pressure Differential Scanning Calorimetry, has been correlated to the severity of service that an engine oil has experienced. Thus, this oxidation test has become a useful tool in assessing the rate of engine oil degradation, as was shown in Fig. 1 [2,14].

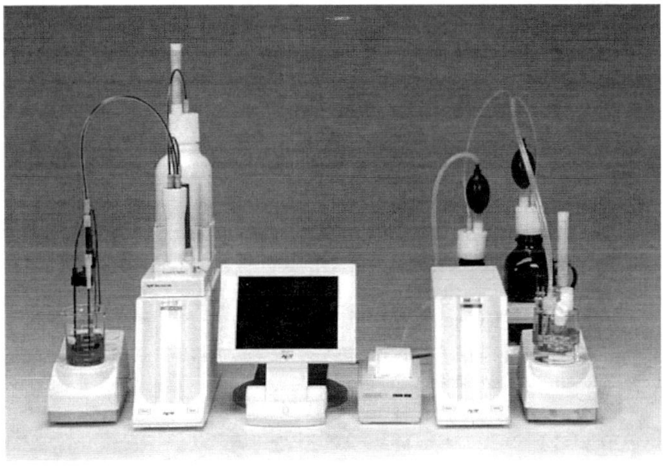

Fig. 15—Automatic potentiometric titration apparatus (ASTM D664, D2896, and D4739).

A Chemical Test That Replaced an Engine Dynamometer Test: Ball Rust Test

Bench tests can be considerably less expensive than engine tests. A bench test that has recently been developed is known as the Ball Rust Test, ASTM D6557, Evaluation of Rust Preventive Characteristics of Automotive Engine Oils. This chemical test measures the extent to which an engine oil should be able to protect an engine against rust during extreme short trips in which fuel and water enter the engine oil. Previously, an engine dynamometer test, ASTM D5844, Evaluation of Automotive Engine Oils for Inhibition of Rusting (Sequence IID) had been used to measure rust inhibition.

Developers of the Ball Rust Test were given the challenge of identifying appropriate reagents and test conditions such that it would be possible to substitute a bench test for the Sequence IID engine test. In addition, it was essential for the developers to prove that the bench test correlated to the engine test and provided adequate rust protection to the engine. Proving correlation between an engine test and a bench test is not always easy. If one includes enough variables, one can sometimes obtain an accidental correlation. If an accidental correlation has been used as the basis for justifying a bench test, results of the test may not provide useful predictions regarding oil properties. Thus, it is important to identify critical parameters that increase the probability that a bench test is a predictor for engine performance. For example:

- Material properties used in the bench test (composition, geometry, hardness of the test specimens, etc.) should be the same as in the real-world application.
- Chemical properties and reactivity of the reagents used in the bench test should be similar to those in the engine test.
- Reactive chemicals that are generated during engine operation should be incorporated into the bench test.

In the Ball Rust Test, a ball (an actual engine component) is exposed to the types of chemicals that are generated during extreme short-trip cold-start driving. The extent of rusting of the ball is assessed using a computer-imaging technique. The development of this bench test was a major effort that included identifying the corrosive chemicals that were formed during the engine test, determining the test conditions that would permit the bench test to be representative of the engine test, and determining evaluation techniques that standardized the interpretation of test results. An extensive testing program was conducted to ensure that this bench test correlated to engine results. Advantages of this test include: the test results are independent of engine design, the test can probably be used for many years without the necessity of an upgrade, and the test method evaluates the inherent properties of the engine oil.

When investigating whether a true correlation occurs between a test method and a real-world phenomenon, it must be confirmed that mechanisms of degradation have not changed. In addition, it may be necessary to develop a separate test for each different aspect of a degradation mechanism (e.g., physical properties, chemical properties, service characteristics).

Developers of the Ball Rust Test used an actual engine component, created a fluid mixture that had similar chemical characteristics to those found in real engines in short-trip service, and developed test conditions that created the same corrosive environment as occurred in extreme short-trip winter driving. Thus, the test developers incorporated the essential physical, chemical, and thermal properties in their test. The consequences of this development process are that the bench test that was developed correlates to field experience (legitimately, not accidentally). In addition, the Ball Rust Test requires considerably less effort and expense than the Sequence IID test.

Standard Performance Tests for Engine Oils

It is essential to determine whether an engine oil will protect the engine under various severe driving conditions. Dynamometer tests, under a variety of operating conditions, are used to determine whether the protective properties in an engine oil are acceptable.

The Sequence HIE test, ASTM D5533, Evaluation of Automotive Engine Oils in the HIE Spark-Ignition Engine, is an engine test that measures the extent to which an engine oil has degraded under extreme high-temperature conditions and the extent to which the engine has been damaged. This test requires leaded fuel. A concern that troubled users of the test in recent years was the fact that leaded fuel was no longer appropriate for most North American vehicles. That is, the chemical environment in the engine had changed. Thus, it became desirable to modify the standard high-temperature test and convert it to a test that had more relevance for current usage. However, a number of fundamental issues had to be resolved. Unleaded fuel is significantly less harsh than leaded fuel. Therefore, the chemical environment of the new test would have to be fundamentally different from that of the previous test. Since chlorine and bromine react readily with iron, the new fuel would not contain some of the chemicals that contributed to interactions between the fuel and the engine.

An issue with regard to test development was whether the decrease in severity, by not using leaded fuel, could be compensated for by such tactics as increasing the test temperature or increasing the test duration. An increase in test temperature runs the risk of changing the fundamental mechanisms of oil degradation and engine damage to the extent that the test method no longer represents the mechanisms that occur in actual operating vehicles. An increase in the duration of the test means that the cost of the test will increase. Neither of these alternatives is desirable. Related concerns had to be resolved when modifying the Sequence V test (representing city or taxi driving) for use with unleaded fuel. These examples illustrate the difficulties that must be addressed when any fundamental aspect of a standard test procedure needs to be modified. As occurred in the transition from the Sequence IID test to the Ball Rust Test, it was essential to answer the question: What are the fundamental degradation mechanisms in this test? Issues such as temperature, chemical reactions, severity as a consequence of loads or pressures, and material composition all had to be resolved during the test development process. The Sequence IIIE high-temperature engine test has now been upgraded to Sequence IIIF, ASTM D6984, and again to Sequence IIIG, ASTM D7320, to evaluate oils under current operating conditions.

The Sequence VE test, ASTM D5302, Evaluation of Automotive Engine Oils for Inhibition of Deposit Formation and Wear in a Spark-Ignition Internal Combustion Engine Fueled with Gasoline and Operated under Low-Temperature, Light-Duty Conditions, is an engine test that represents an extreme example of city driving. This deposit test has since been upgraded to the Sequence VG test.

Fuel economy is measured in the Sequence VIB test, ASTM D6837, Standard Test Method for Automotive Engine Oils on the Fuel Economy of Passenger Cars and Light-Duty Trucks in the Sequence VIA Spark Ignition Engine. High-temperature bearing wear is measured using ASTM D5119, Evaluation of Engine Oils in the CRC L-38 Spark-Ignition Engine. The CRC L-38 test is also known as the Sequence VIII test, ASTM D6709.

A sampling of automotive engine oil tests has been described in the previous sections. A comprehensive list of standard ASTM tests for automotive engine oil is provided in ASTM D4485, Standard Specification for Performance of Engine Oils.

CONCERNS RELATED TO TEST METHOD DEVELOPMENT

Various insights regarding relationships among fluid composition and test performance are available in the technical literature. For example, the relationship between remaining antioxidant protection of an engine oil during use has been correlated to the number of times the engine oil has been exposed to the explosion of the gasoline in an engine cylinder [2,3]. In addition, chemical wisdom suggests that the rate of chemical reactions increases with an increase in oil temperature. Thus, an investigator can surmise that oil temperature and engine speed (which has a major influence on oil temperature) will be important parameters in the selection of test conditions.

One concern, however, is to identify whether there is a point at which a given fundamental mechanism of oil degradation suddenly shifts to an alternative mechanism. If this shift occurs in a test method but does not occur under real-world operation, the validity of the test method is questionable. If test conditions are such that a "shift point" is exceeded, false conclusions may be drawn regarding the results of a given test. It is therefore essential, in test method development, that a bench or engine test mimics the conditions of interest and does not measure some other unrealistic condition. Therefore, it is necessary to create test methods such that (1) the failure mechanisms of the test are identical to those in field service and (2) inappropriate test conditions are not used.

An example of a shift point is provided in Schwartz (1992) [4] in which it was found that, during short-trip service, if the engine oil temperature never exceeded 50°C, the water content of the engine oil increased. If the engine oil temperature rose above 50°C, the water content of the engine oil began to decrease. Thus, 50°C was a "shift point" separating extreme short-trip driving conditions (in which oil contamination effects were a major cause of oil degradation and engine damage) from city or highway driving conditions (in which the rate of oil degradation was largely a function of oil temperature and engine speed).

Alternatively, an automobile manufacturer may find it desirable to create a test method to identify how robust an engine design is, or a lubricant developer may want to learn the limits of stability of a newly developed formulation. Thus, a test method would be required that could identify the conditions at which a shift point occurred. If a test is structured such that severity is increased until failure occurs, the person conducting the test can then learn whether materials, components, or lubricants need to be strengthened. Therefore, by determining the conditions at which a shift point or a failure occurs, that knowledge can be used to develop components that are more durable. For example, at the high end of the temperature spectrum, there may be a temperature at which an oil additive self-destructs or an engine component suffers an unacceptable decrease in hardness. Testing to failure can determine whether component properties and engine designs are sufficiently robust.

The tests described above are examples of material property tests and standard performance requirements for engine oil. The descriptions provide an indication of the extreme complexity and sophistication required for the characterization of engine oil. Test methods have been refined over many years, so that today, an engine oil that has passed the current qualification tests provides protection to the engine for a much longer time than was the case before these tests were available. In addition, the component developer has an arsenal of test methods to probe the durability of vehicle hardware.

FUTURE ISSUES

As can be seen by looking at the history of vehicle evolution, vehicles change. The car of today is fundamentally the same but very different from the car of 80 years ago. In addition to the many changes in the vehicles, lubricants for the vehicles are also far different from those 80 years ago. Thus, the issues of today are different from the issues of the past.

A considerable amount of wisdom is available in chemical and automotive literature relative to test methods. However, whenever anything in a system changes (such as an alternative power source or a mandate to use or discontinue using a given chemical component), new knowledge needs to be developed. Desirable directions for generating this new wisdom relative to oil and engine durability include:

- Understand various fundamental mechanisms of wear and oil degradation (thermal decay, chemical reaction and catalysis, surface effects that occur in narrow spaces), especially under any test conditions that may not have been explored extensively in the past.
- Identify and investigate test conditions under which a shift point occurs, in which there is an observable change in the mechanisms of engine damage or oil degradation.
- Identify the extent to which simple bench tests can be substituted for complicated engine tests.

A study of interactions among physical effects, chemical changes, properties of materials, service effects, etc. could involve investigating all possible combinations of engine and operating conditions to look for synergistic benefits or disadvantages. Such a study would be prohibitively expensive, but even a partial understanding of these interactions would promote improved explanations regarding the conditions that generate problems or failures. Multidimensional maps, indicating the various technical domains in which a given failure mode occurred, could be a big asset with regard to understanding failure mechanisms. It is estimated that failure mechanisms would often overlap more than one technical domain (physics, chemistry, materials science, etc.).

Thus, desirable directions for future investigations include increasing our understanding of interactions and shift points, in which degradation modes can change from one fundamental mechanism to another as a consequence of changes in test conditions. One example is the 50°C oil temperature shift point that marked the transition from the domain of low-temperature corrosion to fully warm operation, described previously. Temperature shift points such as this can be used as guidelines for test-method developers. Temperatures at which material failure occurs (on the high-temperature end of the spectrum) should also be included in a temperature map, as should the temperatures at which various oil additives spontaneously self-destruct. Documentation of shift points and failure modes should include changes that occur in all pertinent technical disciplines, otherwise an investigator or test method developer may miss some important concern.

The trend toward substitution of a bench test for an engine test is a desirable direction in which to aim, but the investigator must be aware that such substitutions may not remain valid whenever engine materials, designs, operating temperatures, or geometries have changed. However, it is much easier and much less demanding in time, cost, and test precision to adapt a bench test to changing engine parameters and conditions.

FUTURE DIRECTIONS RELATED TO STANDARD TESTS FOR ENGINE OILS

Development of appropriate test methods for automotive engine oils has provided essential information to ensure acceptable engine oil performance in vehicles. Changes in government regulations, pollution concerns, and the increasing need for fuel efficiency continue to put pressure on those who create standard tests for engine oils. The supply of oil in the world is finite. Thus, at some point, alternative vehicles, alternative fuels, and alternative lubricants will need to become available. As a consequence, an entire new arsenal of test methods will need to be developed.

ASTM Test Methods

ASTM	Title
D445	Kinematic Viscosity of Transparent and Opaque Liquids (the Calculation of Dynamic Viscosity)
D664	Acid Number of Petroleum Products by Potentiometric Titration
D1298	Relative Density (Specific Gravity), or API Gravity of Crude Petroleum and Liquid Petroleum Products by Hydrometer Method
D2602	Apparent Viscosity of Engine Oils Between −5 and −35 °C Using the Cold-Cranking Simulator
D4485	Standard Specification for Performance of Engine Oils
D4683	Measuring Viscosity at High Shear Rate and High Temperature by Tapered Bearing Simulator
D4684	Determination of Yield Stress and Apparent Viscosity of Engine Oils at Low Temperature
D4739	Base Number Determination by Potentiometric Titration
D5119	Evaluation of Engine Oils in the CRC L-38 Spark-Ignition Engine
D5133	Low Temperature, Low Shear Rate Viscosity/Temperature Dependence of Lubricating Oils Using a Temperature Scanning Technique
D5293	Apparent Viscosity of Engine Oils Between −5 and −35 °C using the Cold-Cranking Simulator
D5302	Evaluation of Automotive Engine Oils for Inhibition of Deposit Formation and Wear in a Spark-Ignition Internal Combustion Engine Fueled with Gasoline and Operated Under Low-Temperature, Light-Duty Conditions
D5483	Oxidation Induction Time of Lubricating Greases by Pressure Differential Scanning Calorimetry
D5533	Evaluation of Automotive Engine Oils in the Sequence HIE, Spark-Ignition Engine
D5844	Evaluation of Automotive Engine Oils for Inhibition of Rusting (Sequence IID)
D6202	Standard Test Method for Automotive Engine Oils on the Fuel Economy of Passenger Cars and Light-Duty Trucks in the Sequence VIA Spark-Ignition Engine
D6557	Evaluation of Rust Preventive Characteristics of Automotive Engine Oils
D6837	Measurement of Effects of Automotive Engine Oils on Fuel Economy of Passenger Cars and Light-Duty Trucks in Sequence VIB Spark-Ignition Engine
D6891	Evaluation of Engine Oils in the Sequence IVA Spark-Ignition Engine
D6984	Evaluation of Engine Oils in the Sequence IIIF Spark-Ignition Engine
D7320	Evaluation of Engine Oils in the Sequence IIIG Spark-Ignition Engine

References

[1] Ludema, K. C., "Tribology at the Transition of Millennia," *Lubrication Engineering*, Vol. 56, No. 10, 2000, pp. 54–57.

[2] Schwartz, S. E., "A Model for the Loss of Oxidative Stability of Engine Oil During Long-Trip Service, Part I: Theoretical Considerations," *STLE Tribology Transaction*, Vol. 35, No. 2, 1992, pp. 235–244.

[3] Schwartz, S. E., and Mettrick, C. J., "Mechanisms of Engine Wear and Engine Oil Degradation in Vehicles Using M85 or Gasoline," SAE International Paper 942027, 1994.

[4] Schwartz, S. E., "A Comparison of Engine Oil Viscosity, Emulsion Formation, and Chemical Changes for M85 and Gasoline-Fueled Vehicles in Short-Trip Service," SAE International Paper 922297, 1992.

[5] Selby, T. W., "The Non-Newtonian Characteristics of Lubricating Oils," *ASLE Transactions*, Vol. 1, No. 1, pp. 68–81, Pergamon Press, New York, 1958.

[6] Piasecki, D., Selby. T. W., and Smith, M. F., "Development and Performance of a Tapered Bearing Simulator for High Temperature, High Shear Viscometry," SAE 811225.

[7] Spearot, J. A., Ed., ASTM Symposium on High-Temperature, High-Shear Oil Viscosity Measurement and Relationship to Engine Operation, ASTM STP 1068, ASTM International, West Conshohocken, PA, 1989.

[8] Stambaugh, R. L., and O'Mara, J. H., "Low Temperature Flow Properties of Engine Oils," *SAE International Automotive Congress and Exposition*, Detroit, MI, SAE 820509, Warrendale, PA, Feb. 1982.

[9] Selby, T. W., Discussion of Ref. 8, *SAE International Automotive Congress and Exposition*, Detroit, MI, SAE 820509, Warrendale, PA, Feb. 1982.

[10] Selby, T. W., and Miller, G. C., "Thermal History of the Engine Oil and Its Effects on Low-Temperature Pumpability and Gelation Formation," *SAE Powertrain and Fluid Systems Conference*, SAE 2008-01-2481, Chicago, 2008.

[11] Selby, T. W., and McGeehan, J. A., "Low-Temperature Rheology of Soot-Laden Heavy-Duty Engine Oils Using the Scanning Brookfield Technique," *SAE International Automotive Congress and Exposition*, Detroit, MI, SAE 2006-01-3362, Warrendale, PA, April 3–6, 2006.

[12] Henderson, K. O., et al., "New Mini-Rotary Viscometer Temperature Profiles that Predict Engine Oil Pumpability," *SAE International Automotive Congress and Exposition*, Detroit, MI, SAE 850443, Warrendale, PA, Feb. 25-Mar. 1, 3–6, 1985.

[13] Schwartz, S. E., and Smolenski, D. J., "Development of an Automatic Oil-Change Indicator System," SAE International Paper 870403, 1987.

[14] Schwartz, S. E., "A Model for the Loss of Oxidative Stability of Engine Oil During Long-Trip Service, Part II: Vehicle Measurements," *STLE Tribology Transactions*, Vol. 35, No. 2, 1992, pp. 307–315.

18

Petroleum Oils for Rubber[1]

John M. Long[2] and John H. Bachmann[3]

HISTORICAL BACKGROUND

Petroleum products covering a molecular weight range that includes light solvents, lubricating oils, waxes, and residual materials have been used with rubber for at least 150 years. Today, these same products are still used, and, since the advent of synthetic rubber, their use has reached greater proportions than ever before. This chapter, however, will be limited to petroleum oils.

Why are such products usable when one normally thinks of oils as the enemy of rubber? The explanation is that the degrading effect of oils on vulcanized rubber is put to use as a softener for unvulcanized rubber. Rubber, both natural and synthetic, is somewhat elastic in the raw state. It is broken down by mechanical or chemical means to a more plastic state prior to adding the necessary compounding ingredients. Oils are added to the rubber compound that make the rubber compound more workable and pliable. Materials that cause this softening effect are called *softeners*. They also are called *processing aids* or *plasticizers*.

Natural Rubber

In the early days of the rubber industry, natural rubber was the only rubber polymer used. It is characteristic of natural rubber to break down fairly easily when milled, and, as a consequence, little softener is required. While petroleum oils are usable for this purpose, it was found that naval stores derivatives, such as pine tar, not only aided breakdown but also facilitated the development of tack in the mixture. Tack is the ability to adhere, and it is a highly desirable property for certain rubber applications. For this reason, pine tar became a preferred softener in all natural rubber compounds, and it is used to some extent today. In current practice, however, natural rubber is used frequently in blends with synthetic rubber. Because petroleum oils are less expensive and work well with this combination, they now are used generally for plasticization.

Synthetic Rubber

The preparation of a synthetic rubber equivalent to natural rubber has long been a research laboratory goal. This has been particularly true in countries vulnerable to having their rubber supply shut off in time of war. With this motivation, both Germany and the United States devoted much effort to the problem prior to World War II. This culminated in the development of synthetic rubber in both countries, with some of the German technology licensed in the United States.

Synthetic rubber (styrene-butadiene rubber [SBR]) reached large-scale production during World War II. Although it was a reasonable substitute, it was not a duplicate for natural rubber. One place where the lack of similarity between natural and synthetic rubber showed up was in factory processing. Synthetic rubber proved to be much more difficult to process. This problem offered a real technical challenge to the rubber compounder when synthetic rubber was first commercialized. While many different schemes to overcome this deficiency were tried, the use of softener, plus better quality control in the polymer plants, proved to be a reasonably satisfactory solution. Improvements in polymerization, changing from batch process to a continuous process toward the end of World War II, improved the consistency and efficiency of SBR production. In the late 1940s, it was found that polymerization at 5°C produced a superior polymer compared to the original polymerization at 50°C.

Early efforts focused on the hunt for the best softener to use with the SBR that was chosen for commercialization in this country. Based on performance, cost, and availability, petroleum oils proved to be the best of several alternatives.

During the investigations on the use of oils, it was found that more oil was required for synthetic rubber than for natural rubber. Thus, oils and their properties became more important to the compounder both from the processing standpoint and the effect on physical properties of the final rubber product. From the standpoint of the supplier, the increased demand for rubber oils made them an important specialty product.

Oil-Extended Rubber

At the close of World War II, the synthetic rubber plants in production were adequate to supply the needs of the United States. An ingenious technical innovation was found during the early 1950s that production of very high-molecular-weight polymer could be combined with oil prior to coagulation, resulting in a polymer with superior properties that could still be processed in existing equipment. High-molecular-weight polymer could not be processed without the addition of oil, and the addition of oil after coagulation or in a mixer would have also been difficult if not impossible. The amount of oil added may represent 35 % or more of the extended polymer; the most common level is 27.27 % (37.5 PHR). The

[1] In preparation of this chapter, the contents of previous editions were drawn upon. The present authors acknowledge the author of the sixth edition, J. S. Sweely from Sun Oil Company, and the authors of the seventh edition, J. M. Long from Lion Copolymer LLC, formerly DSM Copolymer Inc., and A. D. Recchuite from Sunoco, Inc. The current edition will review and update the topics as addressed by the previous authors, introduce new technology that has been developed, and include up-to-date references.

[2] *Technical Service Manager, Lion Copolymer LLC, North Canton, OH 44720.*

[3] *Product Specialist, Rubber Process Oils, Holly Refining and Marketing–Tulsa LLC, Tulsa, OK 74121.*

advent of this oil-extended polymer also increased the output of the polymer plants. It averted any potential shortage that had been foreseen during the Korean War era.

Today, both extended and nonextended synthetic rubbers are produced. In the case of SBR, production of extended exceeds nonextended.

COMPOSITION OF RUBBER OILS

Petroleum oils for use in rubber are generally in the lubricating oil viscosity range, although residual materials are sometimes employed. The importance of composition of these oils lies in the effect composition has on their compatibility with rubber.

Rubber-plasticizer systems may be thought of as solutions in which plasticizing oils are not used in sufficient quantity to form a liquid external phase. As with all such systems, solubility is dictated by the chemical nature of the constituents. In the case of rubber-oil mixtures, insolubility is manifested by "sweatout" or exudation of oil to the surface of a rubber article usually most pronounced after vulcanization. To avoid the sweatout problem and benefit the rubber compound in a variety of ways, good compatibility is desirable to the rubber compounder. Benefits include lowered volatility of oil from polymer, faster incorporation of compounding ingredients during mixing, better dispersion of reinforcing pigments (which correspondingly aids physical properties), reduction of oil take-up time and heat generation, and numerous other effects.

As the interest in processing oils and their usage grew, composition assumed more importance. Two basic methods for obtaining compositional information about rubber oils, came into use: molecular- and carbon-type analyses.

Molecular-Type Analysis

Molecular-type analysis separates an oil into different molecular species. These are asphaltenes, polar compounds, and saturates, as shown in Table 1. Aromatics are calculated by difference. The molecular composition is an important property that determines oil-polymer compatibility. A molecular-type analysis is the so-called clay-gel analysis. In this method, group separation is achieved by adsorption in a percolation column with selected grades of clay and silica gel as the adsorption media. This procedure is now standardized as an ASTM Test for Characteristic Groups in Rubber Extender and Processing Oils and Other Petroleum-Derived Oils by the Clay-Gel Absorption Chromatographic Method (D2007).

TABLE 1—Classification of Oil Types by ASTM D2226

Type	Asphaltenes, max, %	Polar Compounds, max, %	Saturated Hydrocarbons, %
101	0.75	25	20 max
102	0.5	12	20.1 to 35
103	0.3	6	35.1 to 65
104[A]	0.1	1	65 min

[A] Type 104 oils are further classified into two subtypes: 104A and 104B for SBR polymers only. Type 104B oils have a viscosity-gravity constant of 0.820 max; Type 104A have a viscosity-gravity constant greater than 0.820.

Carbon-Type Analysis

Although informative, molecular-type analytical methods are cumbersome and time consuming, there has been great interest in the correlation of chemical composition with physical properties. Among the most definitive of these efforts was the refractive index-density-molecular weight (n-d-M) method reported by Van Nes and Van Westen [1]. They demonstrated that the n-d-M method is a reliable procedure for determining carbon-type composition of oil fractions. The method has been standardized as an ASTM Test for Calculation of Carbon Distribution and Structural Group Analysis of Petroleum Oils by the n-d-M Method (D3238). Carbon-type composition gives the breakdown of total carbon atoms between various structures.

Kurtz and associates developed a correlative method for determining carbon-type distribution from the viscosity-gravity constant (VGC) and the refractivity intercept (RI) [2]. Both of these functions are independent of molecular weight, which makes them well suited for characterizing oils of widely different viscosities. These workers showed that it is possible to superimpose lines of constant VGC and RI on a triangular graph of carbon-type composition. Then, if VGC and RI are calculated for an oil whose composition is desired from viscosity, gravity, and refractive index, a point is fixed on the graph that uniquely establishes the percentage of aromatic ring structures (C_A), naphthene ring structures (C_N), and paraffinic structures (C_P) for the oil. This method has been standardized as the ASTM Test for Carbon-Type Composition of Insulating Oils of Petroleum Origin (D2140).

Kurtz and associates were also the first to apply physical property methods for determination of composition of rubber process oils [3]. Since the physical properties required are easily obtainable, they present a convenient way for the rubber technologist whose laboratory is not set up for analysis of petroleum oils to obtain compositional information on process oils. Usually, physical property methods are applied to the analysis of the whole oil; however, if more detailed information is required, the oil can be given a molecular-type separation into an aromatics and a saturates fraction and the carbon-type composition of each determined.

As a summary for this section on composition, analysis based on molecular or carbon-type distribution provides the rubber compounder with sound information. For research work, instrumental methods such as high-resolution mass spectrometry, nuclear magnetic resonance, high-pressure liquid chromatography, supercritical fluid chromatography, and infrared spectroscopy are being used more frequently. In addition, new analytical techniques for polar compounds are much faster than conventional methods.

IMPORTANCE OF COMPOSITION TO COMPOUNDERS

What does the determination of oil composition mean to the rubber compounder? First, both molecular and carbon-type compositions give a measure of aromaticity of oils. The aromatic content of an oil determines its compatibility with many rubbers.

For natural rubber compounds, if petroleum oils are used, they are Type 103 or 104; however, the oil level is generally low. When blended with SBR or polybutadiene, Type 101, 102, or 103 oils are frequently used depending on the application and whether or not discoloration or staining is a consideration.

For SBR, polybutadiene, or Neoprene, Type 101, 102, or 103 oils are used depending on the application. When color stability or U.S. Food and Drug Administration applications require the use of an approved oil, specially treated versions of Type 103 oils are normally specified. In Neoprene, normally small amounts are used, whereas for SBR and polybutadiene the amount used may exceed 25 % of the compound. Paraffinic oils should be avoided in these polymers because of poor solubility and the consequent tendency to bleed or bloom to the surface.

Recent European Union (EU) legislation will put some restrictions on the types of extender oils that may be used in the production of tires or component parts of tires in the future. After January 1, 2010, European Union Directive 2005/69/EC prohibits the use of extender oils in the production of tires or parts of tires, if they contain more than a specified amount of eight listed polycyclic aromatic hydrocarbons (PAHs), which will preclude the use of the DAE oils that have been used in the past. TDAE, MES, RAE, and heavy naphthenic oils are being used by many tire companies as suitable replacements. In Europe, MES or TDAE oils have been used to replace DAE oils. MES and TDAE oils are not readily available in the United States, where heavy naphthenic oils or RAE oils have been used to replace DAE oils.

When substituting highly aromatic oil with MES or heavy naphthenic oils in SBR-containing compounds, modifications are necessary to maintain desirable properties. These modifications can be in the form of changing the styrene level of the polymer or modifications of the compound ingredients to obtain the desired properties. The major impact of these changes is on properties related to the glass transition temperature (Tg) of the compound and the related compound properties.

EPR and butyl predominately use Type 104 oil, although Type 103 oils are also frequently used, particularly in the case of the halogenated butyl polymers.

EPDM compounds are probably best served by Type 104 oils, but Type 103 oils are probably the more widely used.

Petroleum oils have limited use in the more polar types of polymers such as butadiene-acrylonitrile rubbers and polyvinyl chloride.

Asphaltenes present in an oil are determined by pentane precipitation. They indicate the presence of resinous oxidation products or residual material that may be undesirable to the rubber compounder since such products often lack batch-to-batch uniformity and have undesirable effects on the compound, particularly for fatigue resistance and reduced tensile strength.

All petroleum oils contain oxygen, nitrogen, and sulfur compounds to some degree. If these nonhydrocarbons or polar compounds are reactive, as would be the case with acidic or basic compounds, they could impart an adverse effect on the vulcanization characteristics of the rubber with which they are used. They may also cause discoloration and staining of light-colored rubber goods under exposure to sunlight or ultraviolet light. For these reasons, oils containing high levels of polar compounds usually are checked for any adverse effects before acceptance for general usage.

CLASSIFICATION OF RUBBER OILS

To aid the user of rubber process oils, a classification system has been developed based on some of the properties just discussed. This classification, which is given in ASTM Classification for Various Types of Petroleum Oils for Rubber Compounding Use (D2226), is shown in Table 1.

Table 1 does not list aromatics; however, an indirect measure of aromaticity is given in the saturates limits. Saturates are used instead of aromatics for classification purposes because aromatics must be desorbed from adsorbent in the clay-gel method, whereas saturates are determined directly after removal of solvent. Aromatics can be determined by the difference as described in ASTM D2007.

Observe also in Table 1 that VGC is used to differentiate between Type 104A and 104B oils. In this case, VGC is used as an estimate of aromaticity. In addition to its use for determining carbon-type composition in the viscosity-gravity constant/refractivity intercept method, VGC by itself has long been used to indicate aromaticity of rubber oils. Increasing VGC is generally characteristic of increasing aromatic content. Kurtz and associates used VGC to group oils by type as shown in Table 2 [3].

VGC is a function independent of molecular weight and quite useful for comparing different oils. The viscosity-gravity constant has been standardized as ASTM Calculation of Viscosity-Gravity Constant (VGC) of Petroleum Oils (D2501).

One other measure of aromaticity that has received a considerable degree of acceptance in the rubber industry is ASTM Test for Aniline Point and Mixed Aniline Point of Petroleum Products and Hydrocarbon Solvents (D611). Unlike VGC, aniline point generally decreases with increasing aromatic content. Unfortunately, aniline point is not molecular-weight independent. This means that oils with the same aromatic content can have different aniline points because their viscosities are different. Other methods such as D2007 may be used if a more precise measure of aromaticity is required.

PHYSICAL PROPERTIES OF RUBBER OILS

In general, rubber compounders use tests to control uniform quality of rubber oils and predict how the oils will affect the rubber compound or oil-extended rubber. Standard physical property tests used by the petroleum industry are intended to satisfy these requirements.

The following paragraphs describe the significance of physical property tests that normally are used for rubber oils. Details on individual tests can be found in the appropriate *Annual Book of ASTM Standards*.

TABLE 2—Original Classification of Rubber Oils by VGC Calculated by ASTM D2501

Oil Type	VGC Range
Paraffinic	0.790 to 0.819
Relatively paraffinic	0.820 to 0.849
Naphthenic	0.850 to 0.899
Relatively aromatic	0.900 to 0.949
Aromatic	0.95 to 0.99
Very aromatic	1.00 to 1.05
Extremely aromatic	>1.05

Note: Original classification by Kurtz and associates [3]. Different VGC ranges and different descriptions for oil types have been used for this method of classifying oils.

Specific Gravity

Specific gravity provides a means to convert volumes to weights and vice versa. Since rubber compounding is done on a weight basis, a considerable volume effect in rubber can be obtained with oils of different gravities. Knowledge of specific gravity permits the calculation of the magnitude of the difference.

For oils of similar viscosity, specific gravity can be used as a rough measure of aromaticity with higher specific gravity corresponding to higher aromatic content.

As described in an earlier paragraph, specific gravity is also needed to calculate VGC and refractivity intercept in the determination of carbon type composition.

ASTM Test for Density, Relative Density (Specific Gravity), or API Gravity of Crude Petroleum and Liquid Petroleum Products by Hydrometer Method (D1298) may be used to determine specific gravity or, alternately, ASTM Test for Density and Relative Density of Liquids by Digital Density Meter (D4052) may be used.

Viscosity

Viscosity gives a measure of the flow properties of the oil, which, in turn, determine the ease of handling at various temperatures. Viscosities of oils at low temperature have been shown to correlate with low-temperature properties of rubber containing these oils. Because viscosity is a measure of molecular weight, it can be used to estimate the compatibility of oils in polymers. Polymers with a critical tolerance for oil will not hold large volumes of oil of high molecular weight. Viscosity for rubber oils is measured by ASTM Test for Kinematic Viscosity of Transparent and Opaque Liquids (the Calculation of Dynamic Viscosity) (D445).

Color

Oil color is an indicator of rubber oil suitability for use in light color rubber compounds. It also influences polymer color in oil-extended polymers. Initial oil color cannot be used as a measure of color stability of an oil itself or of a rubber compound containing it. Instead, other methods such as ASTM Test for Rubber Deterioration-Heat and Ultraviolet Discoloration of Light Colored Surfaces (D1148) and for Staining of Surfaces (Contact, Migration, and Diffusion) (D925) may be used to help quantify the effect of oil discoloration on polymers. The degree of oil discoloration has been related to oil composition. Dimeler and associates reported that oil discoloration is primarily due to heterocyclic compounds, principally nitrogen and sulfur heterocyclics in the oil [4]. Color is determined by ASTM Color of Petroleum Products (ASTM Color Scale) (D1500).

Ultraviolet Absorptivity

For oils of a similar type, ultraviolet (UV) absorptivity is an indicator of the resistance of an oil to discoloration under exposure to artificial or natural light. This is based on the assumption that the more reactive highly condensed aromatic ring structures satisfactorily index color instability. The absorptivity for these structures is larger than for the less reactive monocyclic and dicyclic aromatic ring structures, and thus lower ultraviolet absorptivity of an oil should relate to better color stability. Oils with low absorptivities at 260 nm have been found to impart good color stability to light-colored rubber compounds. ASTM Test for Ultraviolet Absorbance and Absorptivity of Petroleum Products (D2008) is used to measure UV absorptivity.

Refractive Index

ASTM Test for Refractive Index and Refractive Dispersion of Hydrocarbon Liquids (D1218) and ASTM Test for Refractive Index of Viscous Materials (D1747) are convenient tests for establishing batch-to-batch continuity in rubber oil shipments. A refractive index is also needed to calculate the refractivity intercept in the determination of carbon-type composition. The refractive index of an oil generally increases as the aromatic content of the oil increases. As such, refractive index may also be an indicator of oil composition.

Aniline Point

ASTM Test for Aniline Point and Mixed Aniline Point of Petroleum Products and Hydrocarbon Solvents (D611) gives a rough measure of aromaticity of an oil, which is useful in the prediction of swelling characteristics of rubbers exposed to oil and oil-polymer compatibility. This test should be used with caution since aniline point is molecular-weight dependent. This means that oils with the same aromatic content can have different aniline points because their viscosities are different. The aniline point is the temperature at which an oil is miscible in aniline. Since aniline is an aromatic material, more highly aromatic oils are miscible at lower temperatures. Therefore, aniline points generally decrease as the aromatic content of oils increases.

Flash Point

Rubber compounders often use flash point as a measure of oil volatility. Volatility is important because rubber products are exposed to elevated temperatures during mixing operations and, oftentimes, in service. Although flash point has a certain utility for this purpose, it gives no indication of the amount of low-boiling material present. Therefore, a distillation range should be used or ASTM D972 for evaporative loss (discussed below) where volatility is a critical factor. Alternately, the rubber compound containing the oil could be tested for volatile losses under conditions simulating actual processing or service conditions.

Flash point is also used to determine fire hazard aspects, particularly when use of oils with low flash points are being considered. Flash point is determined by ASTM Test for Flash and Fire Points by Cleveland Open Cup Tester (D92/IP 36) or by ASTM Test for Flash Point by Pensky-Martens Closed Cup Tester (D93/IP 34).

Evaporation Loss

ASTM Test for Evaporation Loss of Lubricating Grease and Oils (D972) gives a measure of oil volatility under controlled conditions and is used frequently for specification purposes. However, because volatility of oil from a rubber compound may be influenced by its compatibility with the rubber, a volatility test of the compound often is made under laboratory test conditions pertinent to the intended service.

Pour and Cloud Point

Pour and cloud point tests are useful in establishing low-temperature characteristics of an oil. Pour point is mainly of concern in connection with handling of oils at low temperatures. It is the lowest temperature at which flow of the oil specimen is observed. Cloud point is useful in showing whether any wax is present in an oil. It is the temperature at which a cloud of wax crystals is first observed at the bottom of the test jar. ASTM Test for Pour Point of Petroleum Oils (D97) or ASTM Test for Pour Point of Petroleum Products

(Automatic Tilt Method) (D5950) normally is used to determine pour point of rubber oils.

ASTM Test for Cloud Point of Petroleum Products (D2500) may be used to measure cloud point.

SPECIALTY APPLICATIONS OF PROCESS OILS

While rubber affords a substantial outlet for processing oils, it is by no means the only outlet for materials of this type. Other outlets where mineral oils can be used include caulks and sealants, textile specialties, resin extenders, polyvinyl chloride, and other polymer plasticizers and adhesives.

In these product areas, a wide variety of specialized oils are used. Requirements range from high-priced oils that satisfy U.S. Food and Drug Administration regulations to the cheapest oil available where product cost is the primary consideration. Process oil producers are sufficiently versatile to meet any reasonable requirements. ASTM test methods and specifications simplify the task.

ASTM/IP Standards

ASTM	IP	Title
D92	36	Flash and Fire Points by Cleveland Open Cup
D93	34	Flash-Point by Pensky-Martens Closed-Cup Tester
D97	15	Pour Point of Petroleum Oils
D445	71	Kinematic Viscosity of Transparent and Opaque Liquids (the Calculation of Dynamic Viscosity)
D611		Aniline Point and Mixed Aniline Point of Petroleum Products and Hydrocarbon Solvents
D925		Staining of Surfaces (Contact, Migration, and Diffusion)
D972		Evaporation Loss of Lubricating Greases and Oils
D1148		Rubber Deterioration-Heat and Ultraviolet Light Discoloration of Light-Colored Surfaces
D1218		Refractive Index and Refractive Dispersion of Hydrocarbon Liquids
D1298	160	Density, Relative Density (Specific Gravity), or API Gravity of Crude Petroleum and Liquid Petroleum Products by Hydrometer Method
D1500	196	ASTM Color of Petroleum Products (ASTM Color Scale)
D1747		Refractive Index of Viscous Materials
D2007		Characteristic Groups in Rubber Extender and Processing Oils and Other Petroleum-Derived Oils by the Clay-Gel
D2008		Ultraviolet Absorbance and Absorptivity of Petroleum Products
D2140		Carbon-Type Composition of Insulating Oils of Petroleum Origin
D2226		Classification for Various Types of Petroleum Oils for Rubber Compounding Use
D2500		Cloud Point of Petroleum Products
D2501		Viscosity-Gravity Constant (VGC) of Petroleum Oils
D3238		Calculation of Carbon Distribution and Structural Group Analysis of Petroleum Oils by the n-d-M Method
D4052	365	Density and Relative Density of Liquids by Digital Density Meter
D5950		Pour Point of Petroleum Products (Automatic Tilt Method)

References

[1] Van Nes, K., and van Westen, H. A., *Aspects of the Constitution of Mineral Oils,* Elsevier, New York, 1951.

[2] Kurtz, S. S., King, R. W., Stout, W. J., Partikian, D. G., and Skrabek, E. A., "Relationship Between Carbon-Type Composition, Viscosity-Gravity Constant, and Refractivity Intercept of Viscous Fractions of Petroleum," *Anal. Chem.*, Vol 28, 1956, pp 1928–1936.

[3] Kurtz, S. S., King, R. W., and Sweely, J. S., "Hydrocarbon Composition and Viscosity-Gravity Constant of Rubber Processing Oils", *Ind. Eng. Chem.*, Vol 48, 1956, pp 2232–2234.

[4] Dimeler, G. R., Boyer, J. S., and Mills, I. W. "The Effect of Process Oil Components on the Pigment Requirements of Light Colored Rubber Stocks," presented at a meeting of the Rubber Division, American Chemical Society, Los Angeles, California, April 29–May 5, 1967.

19

Lubricating Greases[1]

Raj Shah[2]

THE ESSENTIAL FUNCTION OF ANY LUBRICANT IS to prolong the life and increase the efficiency of mechanical devices by reducing friction and wear. Secondary functions include the dissipation of heat, protection from corrosion, power transmission, and removal of contaminants. Generally, fluid lubricants are difficult to retain at the point of application and must be replenished frequently. If, however, a fluid lubricant is thickened, its retention is improved and the lubrication intervals can be extended. A lubricating grease is simply a lubricating fluid that has been gelled with a thickening agent so that the lubricant can be retained more readily in the required area.

Lubricating greases have a number of advantages over lubricating fluids. Some of these are:

- Dripping and spattering are significantly reduced
- Less frequent applications may be required
- Greases can be easier to handle
- Less expensive seals can be used
- Greases can form a seal in many cases and keep out contaminants
- They can adhere better to surfaces
- Some greases may reduce noise and vibration
- Some grease remains even when relubrication is neglected
- Greases are cost-effective

Grease was previously defined as a gelled lubricating fluid. Although this simplistic definition conveys the general concept of a grease, a more extensive discussion is required to provide a fuller understanding of just what constitutes a lubricating grease. A *lubricating grease is a semifluid to solid product of a dispersion of a thickener in a liquid lubricant.* Wherever the term *grease* without the qualifier, *lubricating,* is used in this publication, it shall mean lubricating grease and exclude commercial grease. Tallow, lard, and other animal fats are items of commerce called *grease.* Even if a commercial grease is used to lubricate, it does not always satisfy the requirements for a lubricating grease, because it is not a liquid lubricant thickened with a dispersed solid. Additives, either liquid or solid, are usually included to improve grease properties or performance.

By definition, grease is a lubricant. It is also essentially a two-phase system–a solid-phase, finely divided thickener, which is uniformly dispersed in a liquid-phase lubricant. The liquid is immobilized by the thickener dispersion, which must remain relatively stable with respect to time and exhibit predictable characteristics during use. At normal operating temperatures, thickeners are insoluble or, at most, only slightly soluble in the liquid lubricant. There must be some affinity between the solid thickener and the liquid lubricant in order to form a stable, gel-like structure; otherwise, the dispersion will separate into two phases. The thickener can be constituted of fibers (such as various metallic soaps) plates, or spheres (such as certain nonsoap thickeners). The essential requirements are that the particles should be extremely small, uniformly dispersed, and capable of forming a relatively stable, gel-like structure with the liquid lubricant. The sizes of grease thickener particles vary a great deal, from about a tenth of a micron for aluminum soaps to about 100 μm long by 1 μm wide for fibrous sodium soaps.

Greases are different from lubricating pastes, which can appear grease-like. In these products, the primary lubricating constituent is the solid, and the liquid component, usually petroleum oil, is used to provide the requisite consistency. In some instances, a lubricating grease is used as the liquid component of a paste. A number of lubricating solids can be used (e.g., molybdenum disulfide, graphite, fluorocarbon polymers, etc). Pastes are mostly solids, generally about 70 to 95 % solids but sometimes merely wetted solids. In comparison, the solid thickener concentrations of greases range from about 3 to 30 %. Also, for pastes, affinity between the solid and liquid phases is not essential; neither is it necessary that a stable, gel-like structure be formed. The manufacturing methods also differ. Pastes are simply solid-liquid mixtures formed using low-shear mechanical methods (e.g., stirring). Grease manufacturing requires chemical processing.

The basic steps required to make a grease include mostly in situ formation and dispersion of the thickener system in the fluid, mixing thoroughly, cooling the mixture to permit proper crystallization of the thickener, blending in the desired additives at the proper time and temperature, and then final finishing and processing. The additives enhance the performance of the grease. Homogenization or deaeration, or both, may be included in the finishing and processing steps when needed.

COMPOSITION

The lubricating fluids that can be thickened to form greases vary widely in composition and properties. By far, the largest volume of greases in use today are those made with

[1] In preparation of this chapter, the contents of the sixth edition were drawn upon. The current edition will review and update the topics as addressed by the previous author, introduce new technology that has been developed, and include up-to-date references. This chapter is completely adapted and is merely updated from Chapter 9 ("Lubricating Greases") by T. M. Verdura in the sixth edition of the *Manual on Significance of Tests for Petroleum Products.* Credit and acknowledgment go to T. M. Verdura, General Motors R&D (retired); Glen Brunette, Citgo Petroleum Products; and Wayne Goldenberg and Krystin Wolstromer, Koehler Instrument Company, also for this work.

[2] *Director, Sales, Marketing, and Technical Services, Koehler Instrument Company, Bohemia, NY 11716.*

petroleum oils thickened with soaps. Many types of petroleum oils are used (e.g., naphthenic, paraffinic [solvent refined, hydrotreated, or hydrocracked], etc.). In addition to petroleum oils, other lubricating fluids, such as esters, diesters, silicones, polyesters, and synthetic hydrocarbons, can be used. Of the synthetic fluids used in grease manufacturing, the most common type is poly(alpha)olefin (PAO). Because it is more expensive, its usage is limited compared to petroleum oil, but has increased because PAO greases often provide cost-effective performance when exceptional service over a wide-temperature range is required. Table 1 lists some of the raw materials used to manufacture the most common lubricating greases.

Many products have been used as thickeners for grease. Soaps were the first thickeners and still have the widest application. Other thickeners that have been employed include polymers, clays, silica gel, pigments, etc.

Soap is generally formed by the neutralization of an acid and base to form a salt and water. When the acid is a fatty acid, its salt is called a soap, and the reaction is called saponification. Soaps can also be formed by alkaline hydrolysis of fatty esters. When the acidic component has a narrow range of molecular weight, a simple soap is made, such as lithium stearate. If the metal is polyvalent, reacting the metallic base with two dissimilar acids of widely different molecular weights will form a complex soap, such as calcium stearate acetate. Mixed-base greases consist of a mixture of two different thickener systems, such as lithium-calcium or aluminum complex-clay.

The natural fatty materials used for soap formation can be of animal or vegetable origin. Pure or synthesized fatty acids can be used also, but they are more expensive. The type of fatty materials used affects the properties of the soap and the grease. Improved or special properties or both can be obtained by the use of mixtures or blends of fatty materials.

Soaps are present in greases in the form of fibers. The structure and size of these fibers (i.e., thickness and length) depend on the metallic moiety and the conditions under which they are formed. In general, fibers can vary from about 1 to 100 μm in length with about a 10:1 length-to-diameter ratio. Large, coarse fibers do not absorb fluids as well as fine, closely knit fibers. Thus, higher percentages of coarse fibered soaps are required to make greases having the same consistency as those made with fine fibered soaps. Nonsoap thickeners are generally smaller, even colloidal, and have either a spheroidal or platy structure.

The most common basic components used to neutralize the fatty acids are hydroxides of lithium, calcium, and sodium. For aluminum, special compounds (e.g., isopropoxides) must be used as the base. The resulting greases are named according to the basic component and the acidic component used to form the soap (e.g., calcium stearate grease, lithium 12-hydroxys-tearate grease, etc).

GREASE SELECTION

When selecting a grease, one should first determine the type and viscosity of the lubricating fluid required, because it is the oil that must satisfy the lubricating requirements of the system. Otherwise, greases are generally determined by the characteristics of the thickener system (except for low-temperature properties, which are influenced significantly by the viscosity and type of the fluid component). Additional discussion of the properties of the various soap-type greases will be provided in subsequent paragraphs.

GENERAL CHARACTERISTICS

Describing the general characteristics of a grease based on the thickener type can be difficult, because the base oil components and the performance additives affect the characteristics of the grease as much or more than the thickener type. Therefore, assigning characteristics to a grease based on thickener type has limited value; however, some characteristics inherent to the thickener type are discussed below. Table 2 lists the general properties of the most common grease types. This table can be used as a guide for the preliminary selection of a grease for a particular application, but the reader should be aware that some of the inherent characteristics of a thickener type can be modified by the inclusion of performance additives.

Aluminum Soap Greases

Aluminum soap greases are the easiest to make. Although greases can be produced with aluminum soaps made in situ, the use of preformed soap (usually aluminum tristearate) is preferred. It is dissolved in hot oil in a mixing grease kettle,

TABLE 1—Raw Materials for Lubricating Greases

	Soaps		
Liquid Lubricants	**Acids**	**Bases**	**Nonsoaps**
Petroleum oil	Hydroxy fatty acids	Hydroxides, oxides, or carbonates of: aluminum calcium lithium sodium	Polyureas
Synthetic hydrocarbons	Fatty acids		Organo-clay
Poly(alpha)olefins	Natural fats		Fumed silica
Alkylated aromatics			PTFE
Esters			Calcium sulfonate
Polyglycols			Others
Silicones			
Fluorinated fluids			
Vegetable oil			

TABLE 2—General Characteristics of Lubricating Greases

Thickener Property	Aluminum	Sodium	Calcium Hydrated	Calcium Anhydrous	Lithium	Aluminum Complex	Calcium Complex	Lithium Complex	Polyurea	Organoclay	Calcium Sulfonate	PFPE-PTFE
Dropping point, °C	100–110	160–205	95	140	180–205	260+	260+	260+	245	260+	316+	140
°F	210–230	325–400	200	285	355–400	500+	500+	500+	475	500+	601	284
Maximum usable Temperature, °C	80	120	80	110	135	175	175	175	175	175	218	300
°F	175	250	175	230	275	350	350	350	350	350	425	572
Water resistance	Excellent	Poor to fair	Good to excellent	Excellent	Good	Excellent	Fair to excellent	Good to excellent	Good to excellent	Fair to excellent	Excellent	Excellent
Shear stability	Poor	Fair	Fair to good	Good to excellent	Good to excellent	Good to excellent	Fair to good	Good to excellent	Poor to good	Fair to good	Excellent	Good
Oxidation stability	Excellent	Poor to good	Poor to excellent	Fair to excellent	Fair to excellent	Fair to excellent	Poor to good	Fair to excellent	Excellent	Fair to good	Good to excellent	Excellent
Rust protection	Good to excellent	Good to excellent	Poor to excellent	Poor to excellent	Poor to excellent	Good to excellent	Fair to excellent	Fair to excellent	Poor to excellent	Good to excellent	Excellent	Good to excellent
Oil retention	Good	Fair to good	Poor to good	Good	Good to excellent	Good to excellent	Good to excellent	Good to excellent	Good to excellent	Good to excellent	Good	Good to excellent
Pumpability	Poor	Poor to fair	Good to excellent	Fair to excellent	Fair to excellent	Fair to good	Poor to fair	Good to excellent	Good to excellent	Good to excellent	Poor to fair	Excellent
Appearance	Smooth and clear	Smooth to fibrous	Smooth and buttery	Smooth and buttery	Smooth and buttery	Smooth and buttery	Smooth and buttery	Smooth and buttery	Smooth and buttery	Smooth and buttery	Brownish red liquid	White solid
Common usage	Thread lubricant	Rolling contact bearings	General uses for economy	Multiservice military	Multiservice industrial and automotive	Multiservice industrial and automotive (wet and high temp.)	Industrial and automotive	Multiservice industrial and automotive	Multiservice industrial and automotive (long life)	Multiservice industrial and automotive (long life)	Food Industry	Nonstick coating for cookware

and the hot mixture is poured into pans to gel and cool. The cooling rate affects the final consistency. The final product is a smooth, transparent grease having poor shear stability but excellent oxidation and water resistance.

Aluminum soap greases are used as thread and way lubricants. The poor shear stability of aluminum-soap greases is used as an advantage in electrically conducting conveyers of electroplating systems; the grease shears down during use, allowing good electrical contact in the track rollers while still providing good lubrication.

Calcium Soap Greases

The earliest known greases were made with calcium soaps. Greases thickened with hydrated calcium soaps (also known as conventional calcium or lime soaps) are water resistant and inexpensive. The greatest shortcoming of calcium soap grease is its low maximum operating temperature, limited by the loss of water from the system and destabilization of the thickener. Typically, they have a dropping point of about 95°C (200°F). Anhydrous calcium soaps (usually calcium 12-hydroxystearate) are somewhat more temperature resistant, having a dropping point of about 140°C (284°F).

Anhydrous calcium soaps are not widely used in industry; they find their greatest usage (when made with low-viscosity base oil) in military applications where operation over a wide range of climatic conditions is essential because of logistical considerations.

Sodium Soap Greases

Sodium (soda) soap greases have higher dropping points (about 185°C [365°F]) than calcium greases. However, they are not water resistant and emulsify in the presence of water and have poor mechanical stability. In earlier times, these greases were the most temperature resistant. They are still used in moderately high temperature applications. Sodium soap greases are not normally compatible with other greases.

Lithium Soap Greases

Lithium greases were the first so-called multipurpose greases; they provide both the water resistance of calcium soap greases and the high-temperature properties of sodium soap greases. Lithium 12-hydroxystearate greases have dropping points of about 190°C (375°F). They also have good mechanical stability; that is, they do not soften much when worked. Lithium greases are generally the most cost-effective and have the widest application. Lithium grease production exceeds that of all other types combined and is the most common type of grease used today.

Complex Soap Greases

Complex soap greases are noted for their high dropping points (260°C [500°F] and above), although most are not recommended for use up to the dropping point temperature due to limitations of the oils and additives they contain. Fully formulated complex soap greases have excellent oxidation characteristics, good to excellent resistance to water, provide good protection against rust, and usually have good mechanical stability.

Aluminum complex greases have excellent water wash-off and spray-off characteristics as well as high-temperature resistance. Consequently, they are widely used in steel mill applications and in automotive components subject to these conditions.

Calcium complex greases have inherent EP (extreme pressure) properties and provide good friction and wear performance. Low temperature properties are not as good as other complex soap greases. Some calcium complex greases tend to harden in long-term storage or under pressure in lubrication systems. Used within their limitations, calcium complex greases can be cost-effective.

Lithium complex grease performance is generally similar to that of lithium 12-hydroxysterate grease, except the dropping points are about 70°C higher. Lithium complex greases provide a balance of good low-temperature performance and excellent high-temperature life performance in tapered roller bearings.

Polyurea

Polyurea greases are similar to the complex soap greases with respect to high temperature performance; dropping points are about 245°C (475°F). Although used in all types of bearings, they have proved especially useful for the lubrication of sealed-for-life ball bearings used in electric motors and automotive wheel bearings where superior oxidation resistance and high-temperature life are essential. Some polyurea greases also provide outstanding fretting wear protection.

Organo-Clay Greases

Clay thickened greases are often referred to as nonmelting greases, because they tend to decompose before reaching their dropping point (about 290°C [550°F]). They usually have poor mechanical stability properties, vis-à-vis soap greases, and are more difficult to inhibit with extreme pressure additives. Although water resistant, they can be susceptible to severe degradation from other contaminants, such as brine. Low-temperature performance could be considered satisfactory, but many clay greases are formulated with high-viscosity base oils for high-temperature applications. Such greases will have poor low-temperature properties. Clay greases are not normally compatible with other greases.

CLASSIFICATION AND SPECIFICATIONS

The road to develop a grease classification system and grease specifications for industry-wide applications was a long and difficult one. However, in 1989 after about 15 years of development, ASTM D4950, Standard Classification and Specification for Automotive Service Grease, was published. Development of this standard was a cooperative effort by the American Society for Testing and Materials, (ASTM), National Lubricating Grease Institute (NLGI), and Society of Automotive Engineers (SAE). It is the first and only cooperatively developed grease specification accepted by the American industry. Development of this standard was made possible only because several grease performance tests were developed specifically for automotive applications. After determining which grease properties are needed for automotive service applications and evaluating grease known to be either satisfactory or unsatisfactory, it was concluded that meaningful specifications could not be written solely with existing test methods. The development of new test methods was essential. These performance tests were developed as industry standards: D3527, Test Method for Life Performance of Automotive Wheel Bearing Grease; D4170, Test Method for Fretting Wear Protection by Lubricating Greases; D4289, Test Method for Compatibility of Lubricating Grease with Elastomers; D4290, Test Method for Determining the Leakage Tendencies of Automotive Wheel Bearing Grease Under Accelerated Conditions; and D4693, Test Method for Low-Temperature Torque of

TABLE 3—Guide to Requirements for ASTM D4950 Grease Categories

ASTM Test	Description	LA	LB	GA	GB	GC
D217	Penetration		✓	✓	✓	✓
D566[A]	Dropping Point	✓	✓	✓	✓	✓
D1264	Water Washout				✓	✓
D1742	Oil Separation		✓		✓	✓
D1743	Rust Protection		✓		✓	✓
D2266	Four Ball Wear	✓	✓		✓	✓
D2596	Four Ball EP		✓			✓
D3527	High Temperature Life				✓	✓
D4170	Fretting Wear		✓			
D4289	Elastomer Compatibility	✓	✓		✓	✓
D4290	Leakage				✓	✓
D4693	Low Temperature Torque		✓	✓	✓	✓

[A] ASTM D2265 may be substituted.

Grease-Lubricated Wheel Bearings. These standards will be discussed later in this chapter.

D4950 classifies automotive service greases into two chassis grease and three wheel bearing grease categories based on the performance needs of several service conditions. A guide to the requirements of these categories is given in Table 3.

The NLGI established a licensing procedure for greases qualifying for ASTM D4950, categories GC and LB. In conjunction with licensing, three certification symbols can be used only with the highest-performance categories of chassis and wheel bearing greases (Fig. 1). NLGI specifically prohibits the use of the symbols with lesser-performance grease categories.

Industry response to the licensing system has grown and now it is an established symbol. Containers of grease bearing the certification symbols are commonly available in the aftermarket, and their availability is widespread. At the present time, most U.S. automakers recommend the use of NLGI Service Greases GC, LB, and GC-LB for scheduled maintenance of chassis and wheel bearings of passenger cars and light-duty trucks.

TEST METHODS AND THEIR SIGNIFICANCE

Greases are used for particular lubrication applications because of their intrinsic properties. Users and producers alike need a common means to describe the properties required for grease performance for particular applications. Test methods are devised to describe the requisite properties. When the usefulness of a test becomes known throughout the industry, it is developed, through cooperative effort, into an industry standard–an ASTM standard.

ASTM standard tests have been devised to evaluate the properties of formulating materials and finished products. The standard tests used to determine properties of petroleum oils are commonly applied to grease-base stocks as well. Among these are kinematic viscosity (ASTM D445), flash and fire points (ASTM D92), pour point (ASTM D97), and aniline point (ASTM D611). The significance of these tests is discussed elsewhere in this manual.

There are no standard tests for the evaluation of grease thickeners, per se, because most thickeners (soaps) are not used independently of the grease and they are most often formed in situ. Consequently, there is little need for standard tests to evaluate neat thickeners. Work was conducted in ASTM D02.G.7 Research Committee a few years ago using the Penn State Micro-Oxidation Test to study oxidation characteristics of grease, but it only limited insight into the evaluation of the thickener. Also, pressure differential scanning calorimetry (PDSC) techniques (such as ASTM D5483) can also be used for this purpose if the thickener material is isolated.

ASTM International, International Standard Organization (ISO), and Energy Institute (EI) have developed and standardized a number of tests to describe the properties and performance characteristics of lubricating greases. Because these tests are conducted in laboratories under well-defined conditions, they are used primarily as screening tests. Some of the grease tests do indicate how a grease might perform in service; however, direct correlation between laboratory and field performance is often unattainable because the tests cannot duplicate the exact range of conditions that a grease will experience in service.

Consistency

Consistency can be defined as the degree to which a plastic material, such as a lubricating grease, resists deformation under an applied force. It is a measure of the firmness or rigidity of the thickener structure of the grease. The standard method for measuring grease consistency is the penetration test. Consistency is reported in terms of ASTM cone penetration, NLGI consistency number, or apparent viscosity. Cone penetrations and NLGI number are discussed in the following paragraphs. Apparent viscosity is included in the section on "Shear Stability."

Cone Penetration

ASTM D217 Test Method for Cone Penetration of Lubricating Grease is the universal standard for the determination of

Fig. 1—The three approved versions of the NLGI certification mark.

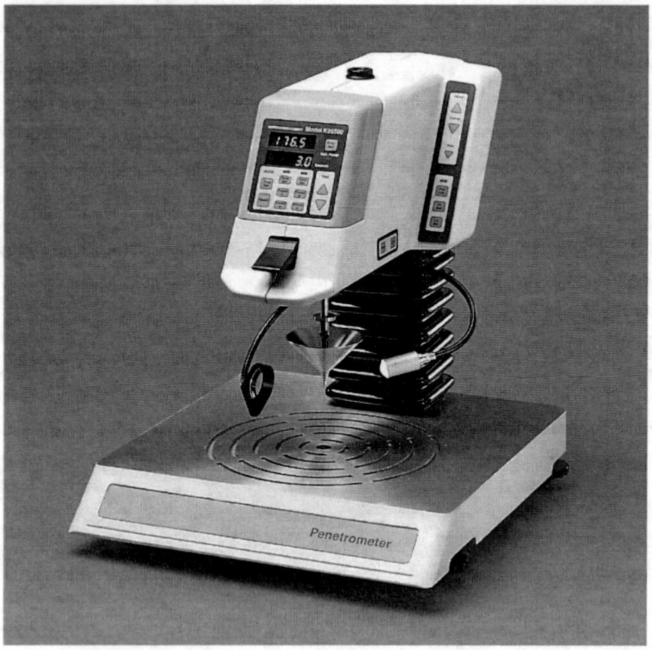

Fig. 2—Automatic penetrometer (ASTM D217).

Fig. 3—One-quarter and one-half scale grease workers and cones (ASTM D1403).

the penetration of a normal grease sample. (In this context, "normal" greases are those that are neither too soft nor too hard to be measured by this method.) In this test, a double-tapered cone of prescribed construction sinks under its own weight into a sample of grease at 25°C (77°F) for 5 s. Fig. 2 shows a photograph of an automatic penetrometer used in laboratories today. The depth of penetration, measured in tenths of a millimeter, is the penetration value. (Although not strictly correct, it is common practice to omit the units when reporting or specifying penetration values.) Firm greases have low-penetration values, whereas soft greases have high-penetration values.

Full-scale penetration tests require about 450 g (1 lb) of sample. The penetration number for smaller samples of grease can be determined by using ASTM D1403, Test Methods for Cone Penetration of Lubrication Grease Using One-Quarter and One-Half Scale Cone Equipment. Figure 3 shows one-quarter and one-half scale grease workers and cones used for ASTM D1403. The one-quarter and one-half scale tests require about 5- and 50-g samples, respectively. As might be expected, the reduced-scale tests are not as precise as the full-scale test, meaning that there is more variability between repeat tests on the same sample and especially between different laboratories. Reduced-scale penetration values are not normally used; instead, equations are given to convert the reduced-scale penetration values to equivalent full-scale results.

The following paragraphs describe the several types of penetration measurements, and Fig. 4 shows the typical cone and needles used in various ASTM penetration test methods.

Unworked Penetration

This value is obtained when a penetration measurement is made on a grease transferred from the original container to the standard grease worker cup, with only a minimum amount of disturbance. This result is not always reliable, because the amount of disturbance cannot be controlled or repeated exactly. This measurement may be significant to indicate consistency variances in transferring grease from container to equipment and for comparison with the worked penetration result.

Worked Penetration

Worked penetration is the standard penetration measurement for a grease. It is measured after a grease has been worked for 60 double strokes in the standard grease worker. This method is more reliable than unworked penetration, because the disturbance of the grease is standardized by the prescribed working process. A significant difference between unworked and worked penetration can indicate poor shear stability and indicate a need for further evaluation by prolonged working or roll testing (see the section on "Shear Stability").

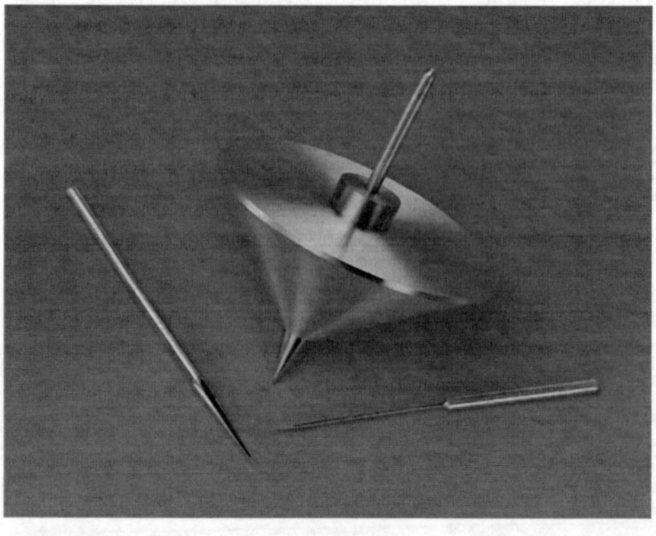

Fig. 4—Cone/needles used in ASTM penetration test methods.

Prolonged Worked Penetration

This value is obtained after a sample has been worked for a prolonged period in the grease worker (e.g., 10,000, 50,000, or 100,000 double strokes). After prolonged working, the sample and worker are brought back to the penetration test temperature of 25°C (77°F) in 1.5 h. It is then worked for a further 60 double strokes, and the penetration is measured. (The difference between the original 60-stroke worked penetration value and the prolonged-worked penetration value provides an indication of the shear stability of the grease. The number of worker strokes is usually specified by the user.) Over time, the industry trend has been to prolong the number of strokes to which the grease is subjected.

Block Penetration

If a grease is firm enough to hold its shape, it does not need to be transferred to a grease container for testing. Instead, the penetration is determined on three faces of a freshly cut 50-mm (2-in.) cube of grease.

Undisturbed Penetration

Undisturbed penetration is measured on a grease sample as received in a container without any disturbance. This measurement was formerly a requirement in D217, but because of dubious values, it now is included merely as an information item. Such measurements can be used to assess the degree to which a grease develops false body or sets with prolonged storage.

NLGI Consistency Numbers

On the basis of worked penetration, the NLGI has standardized a numerical scale as a means of classifying greases in accordance with their worked consistency. This scale is shown in Table 4 in order of increasing hardness. The majority of greases used in automotive and industrial applications fall in the range of NLGI No. 1 to NLGI No. 3.

Consistency Stability

The consistency of a grease may change with its history. Some greases may harden with age; others may change with wide fluctuations in temperature. Evaluation of these changes needs to be done on an individual basis. That is, the test grease needs to be subjected to controlled aging or temperature fluctuations, with penetration measurements taken periodically.

TABLE 4—NLGI Consistency Classification

NLGI Consistency No.	ASTM Worked Penetration at 25°C
000	445 to 475
00	400 to 430
0	355 to 385
1	310 to 340
2	265 to 295
3	220 to 250
4	175 to 205
5	130 to 160
6	85 to 115

The consistency of greases may also change in service, due to changes in the size and dispersion of thickener particles resulting from mechanical shearing. The ability of a grease to resist changes in consistency during mechanical working is referred to as consistency stability, shear stability, work stability, or mechanical stability. Two test methods have been standardized to evaluate the stability of a grease resulting from two degrees of relatively low-shear working.

Prolonged Worked Penetration and Low Temperature Penetration

ASTM D217, described previously, is used before and after prolonged working in a grease worker to determine the change in grease consistency. Because shear rates are low, evaluation of shear stability by prolonged working is time-consuming (working 100,000 strokes takes nearly 28 h). Mechanical grease workers with cut-off timers help make this a more manageable task. Work has also been done recently by the IP/ISO groups to study low temperature (below ambient) penetration measurements (ISO 13737).

Roll Stability

Results can be obtained more quickly with the roll stability test, which operates at somewhat higher shear conditions. Roll stability is determined by ASTM D1831, Test Method for Roll Stability of Lubricating Grease. Figure 5 shows a roll stability tester conforming to ASTM D1831. It is used in conjunction with the reduced-scale penetration test (D1403).

After a worked penetration has been measured on a grease sample, 50 g of unworked grease is transferred into a horizontally mounted cylinder containing a 5-kg (11-lb) steel roller. The cylinder is rotated at 165 rpm for 2 h at room temperature. The inner roller moves over the grease, working it during the test. After the test, the worked penetration of the grease is once again measured by ASTM D1403, and the difference in worked penetration values gives an indication of shear stability. The roll stability test can also be conducted at elevated temperatures (often above 100°C) and for prolonged

Fig. 5—Roll stability tester for ASTM D1831 (roll stability of lubricating grease).

periods (e.g., 96 h), to help evaluate the shear stability of greases at higher temperatures.

The roll stability test is widely used in specifications. Test results are significant insofar as they can show a directional change in consistency that could occur during service. No accurate correlations between roll-test results and actual service performance have been established. In both shear stability tests (ASTM D217 and D1831), the change in consistency is reported as either the absolute change in penetration values or the percentage change as outlined in ASTM D1831. If absolute values are to be reported, the quarter-scale or half-scale values are first converted to full-scale results.

Flow Properties

The consistency of a grease is a critical parameter that helps define the ability of a grease to perform under given operating conditions. Consistency, as measured by penetration, is affected by temperature. However, the penetration test is not suitable for determining the minor, yet sometimes significant, changes in consistency as the grease approaches temperatures at which phase changes in the base fluid and thickener occur. Penetration is basically a flow measurement. In addition, there are other flow-measurement tests that can be utilized to evaluate this property under other conditions.

Apparent Viscosity

Grease is a non-Newtonian fluid. Flow is not initiated until a stress (or force) is applied. Increases in shear stress (or pressure) produce disproportionate increases in flow. The term *apparent viscosity* is used to describe the observed viscosity of greases; it is measured in poises using ASTM D1092, Test Method for Apparent Viscosity of Lubricating Greases. Since apparent viscosity varies with both temperature and shear rate, the specific temperature and shear rate must be reported along with the measured viscosity.

In this test, a sample of grease is forced through a capillary tube by a floating piston actuated by a hydraulic system using a two-speed gear pump. From the predetermined flow rate and the force developed in the system, the apparent viscosity is calculated using the classic Poiseuille equation. A series of eight capillaries and two pump speeds provide 16 shear rates for the determination of apparent viscosities. The results are expressed in a log-log graph of apparent viscosity as a function of shear rate at a constant temperature, or apparent viscosity at a constant shear rate as a function of temperature.

This apparatus also has been used to measure the pressure drops of greases under steady-flow conditions at constant temperature. Such information can be used to estimate the pressure drop or required pipe diameters in distribution systems. Also, apparent viscosity data are useful for evaluating the ease of handling or pumping in dispensing systems at specified temperatures; it is often used to evaluate pumpability at low temperatures. (The NLGI can provide a group of charts that relate pressure drop, apparent viscosity, shear rate, and pipe-flow data.) Apparent viscosity is also used to provide an indication of the directional value of starting and running torques of grease-lubricated mechanisms. Specifications may include limiting values of apparent viscosity for greases to be used at low temperature.

Over the past 10 years, an ASTM round robin was conducted to rewrite the U.S. Steel Grease Mobility Test Method under the aegis of ASTM International. Work is currently stalled on this project and it has been put on hold for a while. Figure 6 shows a photograph of a Grease Mobility Tester capable of conforming to the U.S. Steel Grease Mobility Test Method. This test is easier to conduct than ASTM D1092 and uses an apparatus similar to the ASTM D1092 apparent viscosity test method. The Low Temperature Grease Mobility Test Method requires only one capillary tube (versus eight for D1092), maintained at various temperatures. The amount of grease collected in a specific time (g/min) is measured and used as an indication of the mobility of the grease.

Fig. 6—Grease mobility tester for U.S. Steel Grease Mobility Test.

Low-Temperature Torque (Ball Bearings)

Greases designed for low-temperature applications must not stiffen so much that excessive resistance to rotation is produced. However, greases inherently harden and become more viscous as the temperature is lowered. Sometimes, the grease can become so rigid in the bearing that excessive torque is required for rotation. In extreme cases, the grease can solidify to the point of bearing lock-up. Two standard test methods are available to measure bearing torque at low temperatures. Both operate on the same principle–the restraining force or torque is measured while grease-lubricated test bearings are run at low speed. The tests differ in the size and types of bearings, the intended applications, and types of greases (i.e., in viscosity characteristics).

ASTM D1478, Test Method for Low-Temperature Torque for Ball Bearing Greases, measures the starting and running torques of lubricating greases packed in small ball bearings at temperatures as low as −54°C (−65°F). In this procedure, fully-packed bearings are installed on a spindle that can be rotated at 1 rpm. The assembly is inserted in a cold box. The outer race is connected by a cord assembly to a spring scale (a load cell may be used in addition to a spring scale), which measures the restraining force. When the motor is started, the initial peak restraining force is recorded. After running for 10 min, the restraining force is recorded again. The force values are multiplied by the length of the lever arm, and the products are reported as the starting and running torques in gram-centimeters (g-cm). (Because the centimeter-gram-second [cgs]) metric unit is nearly universally used in grease specifications, g-cm is the standard unit of torque measurement for this test. Recently, ASTM has moved for all values in its standards to be stated in SI units. An exception, however, was made for torque values given in cgs-metric units because of the aforementioned universal use in grease specifications.)

Because this method was developed using greases with extremely low-torque characteristics at −70°C (−158°F) or lower in some cases, it may not be applicable to other greases, speeds, or temperatures. If a machine has significantly more power available than is actually required, torque is not an important consideration. On the other hand, it can be very important in low-powered equipment. This test is significant because it provides a means of comparing the low-temperature torque effects of widely different greases. It is useful in the selection of greases for low-powered mechanisms, such as instrument bearings used in aerospace applications. The suitability of this method for other applications requiring different loads, speeds, and temperatures should be determined on an individual basis. Usually, test conditions are substantially different from those found in service, so test results may not correlate with actual service performance.

Low-Temperature Torque (Tapered Roller Bearings)

For applications using larger bearings or greater loads, the torque requirements exceed the limits of D1478. In such cases, D4693, Test Method for Low-Temperature Torque of Grease-Lubricated Wheel Bearings, is better suited than D1478.

ASTM D4693 was developed specifically to evaluate the low-temperature performance of wheel bearing greases. If the torque requirements are too high for the prevailing temperature, the bearings may be distressed. It is used to accurately predict the performance of greases in automotive wheel bearings operating at sub-zero temperatures.

This test method will differentiate among greases having distinctly different low-temperature characteristics. It is used for specification purposes, and it is one of the automotive grease performance tests required by D4950. This method is not known to correlate with other types of service. It should be noted that greases having characteristics that permit evaluations by either D1478 or D4693 will not give the same values in the two test methods (even when converted to the same torque units) because the apparatus and test bearings are different. Figure 7 portrays a test rig designed to run both D1478 and D4693 (different test bearings) at low temperatures.

ASTM D4693 determines the extent to which a grease retards the rotation of a specially manufactured, tapered roller bearing assembly. The test unit is a model device that closely duplicates an automotive wheel bearing assembly. Additionally, it employs a spring-loading mechanism to improve test repeatability; it is not intended to simulate any service-load condition. Although the test assembly and torque-measuring transducer are necessarily inside the cold-test chamber, the drive mechanism can be either inside or outside.

In this test, a sample of test grease is stirred and worked, and a specified amount is packed into the two test bearings. The test assembly is heated to mitigate the effects of grease history and then cooled at a specified rate to the desired temperature, for example −40°C (−40°F). A drive mechanism rotates the spindle at 1 rpm, and the torque required to prevent rotation of the hub is measured by a strain gauge load cell at precisely 60 s after start of rotation. The results are recorded on a strip chart. (Although it may be recorded, starting torque was found to be less repeatable, and it is not required to be reported in this test.)

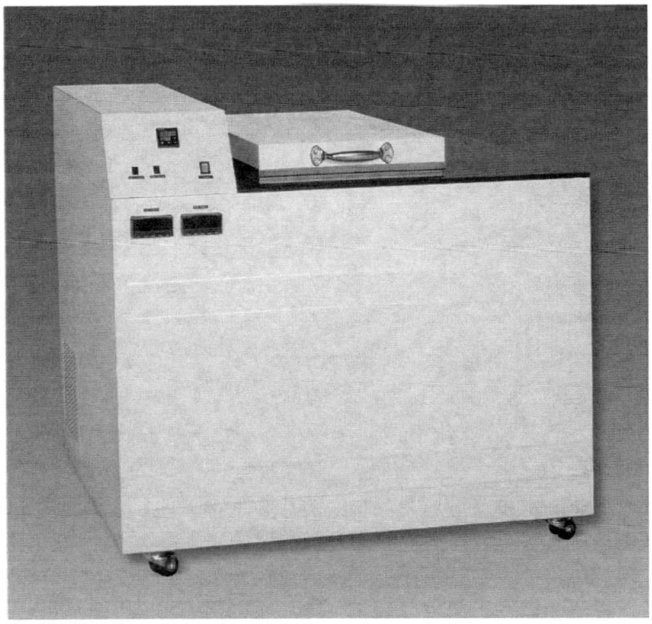

Fig. 7—Test rig for ASTM D1478 to D4693 (low temperature torque tests).

Heat Resistance

Heat affects greases in several ways. As the temperature increases, most greases soften and flow more readily; oxidation rate increases (see description of ASTM D942); oil evaporation increases (see descriptions of ASTM D972 and D2265); and the thickener may melt or lose its ability to retain oil (see descriptions of ASTM D566 and D2595). Sometimes, these phenomena are all involved simultaneously (see descriptions of ASTM D3336, D3337, and D3527).

ASTM D3527, Life Performance of Automotive Wheel Bearing Grease, is one of the automotive grease performance tests required by the D4950 specification. In this test, the test grease is distributed in the bearings of a modified automobile front wheel hub-spindle-bearings assembly. While the bearings are thrust-loaded to 111 N, the hub is rotated at 1,000 rpm and the spindle temperature is maintained at 160°C for 20 h on, 4 h off operating cycle. The test is terminated when changes to the grease cause the drive motor torque to exceed a calculated cut-off value. Grease life is expressed as the accumulated on-cycle hours. This test method is intended to differentiate among wheel bearing greases having significantly different high-temperature characteristics. A high-temperature wheel bearing grease tester conforming to ASTM D3527 is shown in Fig. 8. This tester is also capable of performing ASTM D4290, Leakage Tendencies of Automotive Wheel Bearing Grease Under Accelerated Conditions. ASTM D4290 is discussed in detail later in this chapter.

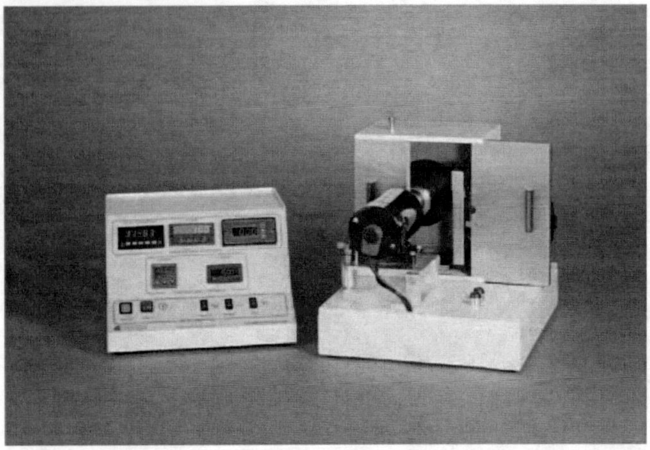

Fig. 8—High-temperature wheel bearing grease tester (ASTM D3527 and D4290).

Dropping Point

The dropping point of a grease is the temperature at which it passes from a semisolid to a liquid. Although a thickener can have a definite melting point, the resulting grease does not. Rather, the thickener loses its ability to function as a grease thickener as the temperature is increased. As the temperature is raised, the thickener may soften to the extent that it loses its ability to hold on to the oil or maintain its structure. As the temperature is increased, further free oil may be released from the thickener matrix, or the thickener structure itself may collapse (allowing the grease to flow under its own weight). The temperature at which either phenomenon takes place, releasing material from the bottom of a standard cup under standard conditions, is called the dropping point.

Two similar procedures are used to determine the dropping point of grease. In both methods, a prescribed layer of grease is coated on the inner surface of a small cup whose sides slope toward a hole in the bottom. With ASTM D566 (IP132), Test Method for Dropping Point of Lubricating Grease, the sample is heated at a prescribed rate until a liquid drop falls from the cup. In ASTM D2265, Test Method for Dropping Point of Lubricating Grease Over Wide Temperature Range, the sample is introduced into a preheated environment so that the heating rate is controlled more uniformly. Figure 9 shows the High temperature Dropping Point Apparatus used to run ASTM D2265. In both tests, the difference in temperature between the grease in the cup and the environment is taken into account when calculating the dropping point of the grease. Some greases containing non-soap thickeners may not separate oil or melt.

Cooperative testing indicates that dropping points by ASTM Methods D566 and D2265 generally agree up to about 260°C (500°F). In cases where results differ, there is no known cause.

The dropping point is useful (1) in establishing bench marks for quality control, (2) as an aid in identifying the type of thickener used in a grease, and (3) as an indication of the maximum temperature to which a grease can be exposed without complete liquefaction or excessive oil separation.

Greases normally do not perform satisfactorily at temperatures near the dropping point because other factors are involved. High temperature performance can depend on the application method and frequency, whether a softened grease is retained at the point of application by proper seals, and whether the high temperature is continuous or intermittent. High temperature stability and evaporation properties of the grease also can affect performance. Maximum usable temperatures have been determined from experience.

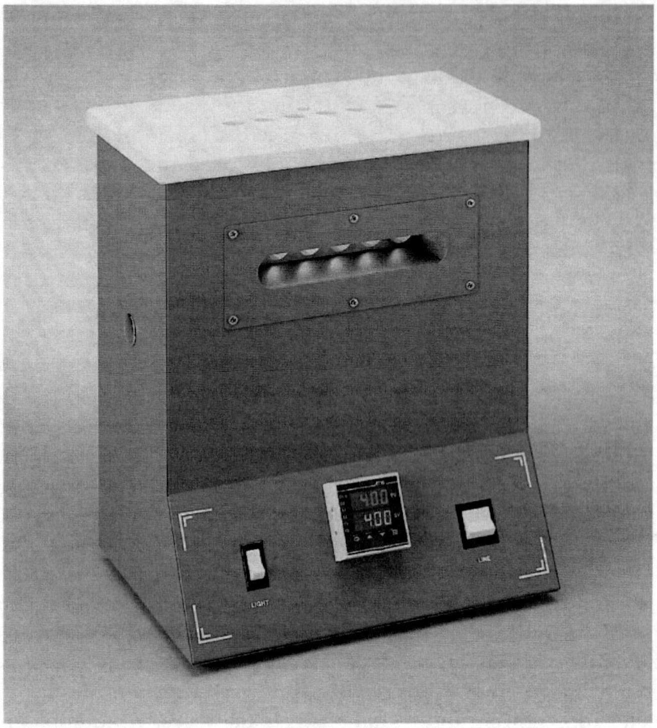

Fig. 9—High-temperature dropping point apparatus (ASTM D2265).

Dropping point is most useful as a quality-control tool. Unless correlation has been established, dropping point has no direct bearing on service performance. Performance at high temperatures is better evaluated with one of the performance-type tests or by actual experience.

Evaporation Loss

Exposure of a grease to high temperatures can cause evaporation of some of the liquid lubricant, causing the remaining grease to become drier and stiffer or leading to other undesirable changes in the grease structure. Greases containing low-viscosity oils for improved low-temperature performance may be susceptible to evaporation losses at higher temperatures. Evaporation can cause additional problems when vapors may be hazardous or combustible or interfere with operations. In most applications, even at high temperatures, evaporation is not a serious problem because of effective sealing. However, when it is necessary to evaluate evaporation loss, two ASTM test methods are available.

Test Method for Evaporation Loss of Lubricating Greases and Oils

ASTM D972 determines mass evaporative losses from greases or oils in the temperature range of 100 to 150°C (210 to 300°F). A weighed sample of lubricant is placed in an evaporation cell in an oil bath at the desired test temperature. Heated air at a specified flow rate is passed over the sample surface for 22 h, after which the loss in sample mass is determined. Figure 10 shows a photograph of an evaporation loss test cell conforming to ASTM D972.

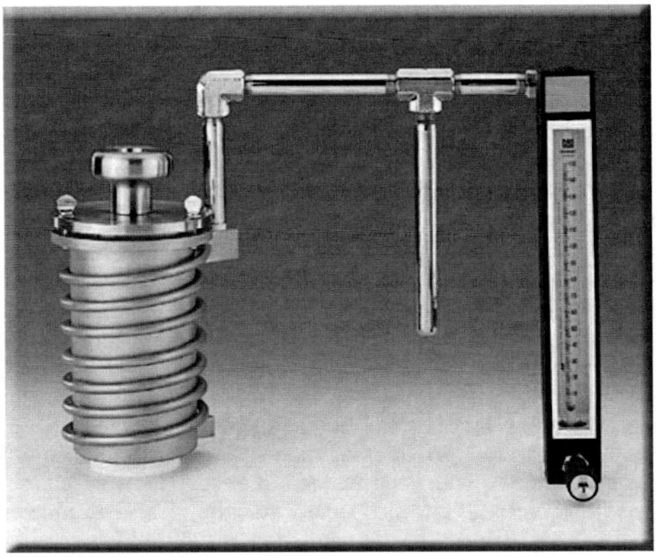

Fig. 10—Evaporation loss test cell (ASTM D972).

Test Method for Evaporation Loss of Lubricating Greases Over Wide Temperature Range

This method augments ASTM D972, which is limited to 150°C (300°F), and was developed because of higher service temperatures. D2595 can be used to determine the loss of volatile materials from a grease over a temperature range of 93 to 316°C (200 to 600°F). This test uses the same sample cup as ASTM D972, but the rest of the apparatus is markedly different. It uses an aluminum block heater, instead of an oil bath, to achieve much higher temperatures. A photograph of this apparatus is shown in Fig. 11. The other test conditions (i.e., air flow rate and test duration) remain the same. Laboratories equipped with D2595 do not need D972; the results will be similar, but not necessarily identical.

Within their respective temperature constraints, both tests can be used to compare evaporation losses of greases intended for similar service. All other factors being equal, greases having the lowest evaporative losses will probably perform longer in high-temperature service. However, the results of these evaporation tests may not be representative of volatilization, which can occur in service.

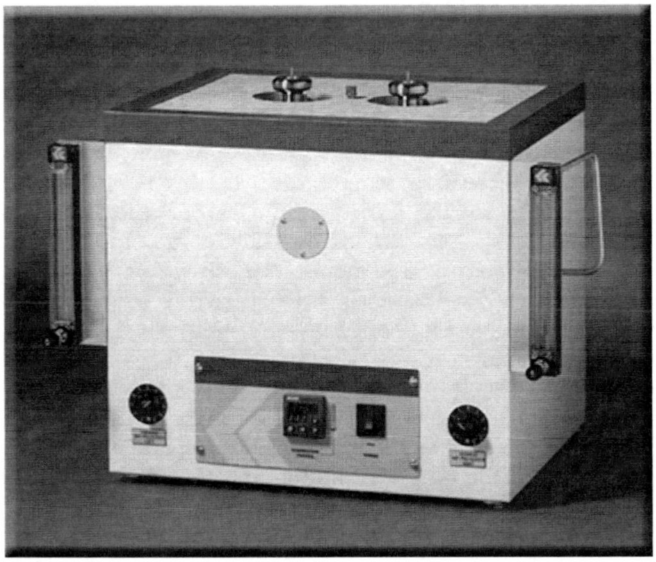

Fig. 11—High-temperature evaporation loss tester (ASTM D2595).

Oil Separation (Static Bleed Tests)

Nearly all greases will release some oil during storage, but they differ markedly in the amounts that are liberated. If a grease releases too much oil during storage, the grease may harden to the extent that subsequent lubrication performance will be affected. Therefore, excessive liberation of free oil during storage is to be avoided. Oil can be released from a grease at varying rates, depending on the gel structure, the nature and viscosity of the lubricating fluid, and the applied pressure and temperature.

ASTM D1742, Test Method for Oil Separation from Lubricating Greases During Storage, is used to determine the tendency of lubricating greases to separate oil when stored at 25°C (77°F) under an applied air pressure of 1.72 kN/m^2 (0.25 psi). It gives an indication of the oil retention characteristics of lubricating greases stored in original containers. The apparatus conforming to ASTM D1742 is shown in Fig. 12. This test is not suitable for use with greases softer than NLGI No. 1 consistency because of a tendency for the grease to seep through the screen.

The test is useful because the results correlate directly with oil separation, which occurs in 16-kg (35-lb) containers of grease stored at room temperature. Storage in other containers gives similar results. This test should not be used to predict the oil separation of grease under dynamic service conditions. ASTM D6184, Standard Test Method for Oil Separation from Lubricating Grease (Fig. 13, conical sieve method) is virtually the same as Federal Test Method (FTM) 791C Method 321.3, Oil Separation from Lubricating Grease (Static Technique), also known as the cone bleed test.

In this method, a 10-g sample of grease is placed in a 60-mesh wire cone, which is then suspended in a covered beaker and placed in an oven at 100°C (212°F) for 30 h (note that some specifications may require different temperatures or durations). The percentage of separated oil is

Fig. 12—Oil separation apparatus (ASTM D1742).

Fig. 13—Oil separation apparatus (ASTM D6184).

calculated from the weight of oil that bleeds from the grease collected in the beaker. Some specifications require the cone and residual grease to be weighed also, and the additional weight loss is reported as evaporation loss.

Despite the lack of a precision statement, Test Method 321.3 has been used for many years in military and industrial grease specifications to determine fluid separation tendencies. The ASTM D6184 standard reflects the procedures and precision determinations used in a cooperative study in ASTM Subcommittee D02.G Lubricating Grease and is intended to supersede the FTM method.

Oil Separation (Centrifuge Test)

ASTM D4425 describes a procedure for determining the tendency of lubricating grease to separate oil when subjected to high centrifugal forces. The results can be related to grease performance in shaft couplings, universal joints, and rolling element thrust bearings subjected to large or prolonged centrifugal forces. Results correlate well with actual service performance.

In this test, pairs of centrifuge tubes are charged with test grease and placed in a high-speed centrifuge. The grease samples are subjected to a centrifugal force equivalent to a relative centrifugal acceleration at a G-value of 36,000 at 50°C. (The units for the G-value are awkward and not used, but they have acceleration dimensions of length/time^2.) The normal test duration is 24 h, but it can be extended to 48 or 96 h. At specified time intervals, the centrifuge is stopped, and the amount of separated oil is measured and the volume percent calculated. The resistance-to-separation index, called the *K36* value, is reported as the volume percent of separated oil/total test hours (both actual values are reported; the value is not to be reduced by division).

Leakage from Wheel Bearings

There are two tests to evaluate leakage of grease from wheel bearings at high temperatures. The older test, D1263, Test Method for Leakage Tendencies of Automotive Wheel Bearing Greases, utilizes a modified automotive front hub assembly (1940s vintage design and bearings). The two bearings are packed with a specified amount of test grease, and an additional 85 g is distributed in the hub. The assembly is run at 660 rpm for 6 h at 104°C (220°F [spindle temperature]). After the test, the amount of grease that leaked into the hubcap and collector is weighed. The bearings are washed and examined for varnish, gum, and lacquer-like material.

ASTM D1263 provides a means to differentiate among grease products with significantly different leakage characteristics. In addition, skilled operators can observe important changes in other grease characteristics that may have occurred during the test. However, these observations are somewhat subjective and cannot be used for quantitative ratings. The test does not distinguish well between wheel bearing greases having similar or borderline leakage.

Accelerated Leakage from Wheel Bearings

ASTM D4290, Test Method for Determining Leakage Tendencies of Automotive Wheel Bearing Grease Under Accelerated Conditions, uses the same principle of operation as does D1263. It is, however, a more recent test that uses a model front wheel-hub-spindle assembly employing tapered roller bearings. The test apparatus is identical to that used in D3527; however, the test conditions are markedly different. In D4290, the test temperature, 160°C (320°F), is significantly higher, and the duration, 20 h, is much longer than that of D1263.

D4290 was developed to evaluate wheel bearing grease for use in vehicles equipped with disk brakes, which involve much higher operating temperatures. It is used in grease specifications for such applications, and it is one of the automotive grease performance tests required by D4950. This test is not known to correlate with any other type of service.

OXIDATION STABILITY

Bomb Oxidation Test

The Standard Test Method for Oxidation Stability of Lubricating Greases by the Oxygen Bomb Method, D942 (IP 142), determines the resistance of lubricating greases to oxidation when stored statically in an oxygen atmosphere in a sealed system at an elevated temperature. In this test, five glass dishes are filled with 4 g of grease each, for a total of 20 g. These dishes are then racked and sealed in a bomb, which is then pressurized to 758 kPa/cm^2 (110 psi) with oxygen being kept in a bath at 99°C (210°F) to accelerate oxidation. The amount of oxygen absorbed by the grease is recorded in terms of pressure drop over a period of 100 h and, in some cases, 500 or 1,000 h. The pressure drop is a net result of the consumption of oxygen by oxidation of the grease and the gain in pressure due to any gases or volatile by-products released from the grease. A photograph of an oxidation bath with oxidation bombs is shown in Fig. 14.

Care must be exercised in the interpretation of data derived from the oxidation bomb test. Additives incorporated into the grease can produce misleading results,

Fig. 14—Oxidation bath with oxidation bombs (ASTM D974).

because they also can react with oxygen. As an example, sodium nitrite is sometimes added to grease to serve as a rust inhibitor. In the oxidation bomb test, this material consumes oxygen to form sodium nitrate. In this instance, the drop in pressure is not indicative of the amount of oxidation of the grease alone. Also, greases containing excess carbonate can release carbon dioxide whose vapor pressure will tend to offset the pressure decrease due to oxygen consumption by reaction.

The bomb oxidation test was originally designed to predict shelf storage life of greases in prepacked bearings. However, experience has shown little correlation with the stability of grease films in bearings or on other parts. It predicts neither the stability of greases stored in containers for long periods nor those used under dynamic conditions. Its primary usefulness is for quality control to indicate batch-to-batch uniformity. It can be used to estimate relative oxidation resistance of greases of the same type, but it should not be used to compare greases of different types.

Although widely used for specification purposes, it is important to note that D942 has been severely criticized for its potential for misleading results and for having no relation to oxidation in service. There are no standard dynamic oxidation tests. For dynamic tests that are influenced to a greater or lesser extent by the oxidation resistance of the grease, see descriptions of ASTM D3336, D3337, and D3527.

PDSC Oxidation Test

A number of grease test methods have recently been developed by the ASTM D02.G, Grease Committee, but grease test methods are not the exclusive property of D02.G Lubricating Grease. ASTM D02.9 Oxidation of Lubricants, Oxidation Committee, has responsibility for oxidation tests, including those that evaluate greases. One of these is ASTM D5483 Standard Test Method for Oxidation Induction Time of Lubricating Greases by Pressure Differential Scanning Calorimetry. In this test, a few milligrams of grease are placed in a sample pan in a chamber, which is pressurized to 3.5 MPa (500 psi) with oxygen and regulated at that pressure until an exothermic reaction occurs. From a plot of heat as a function of test time, the oxidation induction time (determined as from the *extrapolated onset time*) is determined. This method of evaluating oxidation stability has several advantages over D942: (1) It is considerably faster, generally less than an hour, versus 100 h or more, and (2) unlike D942, this method is not subject to false values from greases containing materials that give off carbon dioxide or other gases during heating. The disadvantages are that the apparatus is expensive and that no correlation has been determined between test and service performance (similar to ASTM D942). Its primary value is for product development and quality control, although it is used in a few grease specifications.

Another test that may be considered is the Penn State Thin Film Micro-oxidation Test that has been shown in literature to provide a good indication of the oxidation stability of a grease sample. The apparatus for this pending test method is shown in Fig. 15.

Greases in Ball Bearings at Elevated Temperatures

ASTM Test Method for Performance Characteristics of Lubricating Greases in Ball Bearings at Elevated Temperatures (D3336) is used to evaluate the performance characteristics of lubricating greases in ball bearings operating under light loads at high speeds and elevated temperatures for extended periods. Correlation with actual field service cannot be assumed. This method has been criticized for having two test spindles qualified for use that purportedly do not give the same results.

In this test, the lubricating grease is evaluated in a 20 mm, SAE No. 204, heat resistant, steel ball bearing rotated at 10,000 rpm under light loads of 22 to 67 N (5 to 15 lb-ft) at a specified elevated temperature up to 370°C (700°F). The test is run on a specified, test temperature–dependent, operating cycle until lubrication failure or completion of a specified time. (Unless automatic controls are used, a 72-h weekend shutdown is required.) With some very high performance greases, tests can last up to several thousand hours. Multiple tests need to be run because the results follow Weibull, rather than normal, distributions.

Greases in Small Bearings

The computer and aircraft industries have used miniature bearings (<10-mm bore) for many years. As the trend toward miniaturization increased in other industries, a

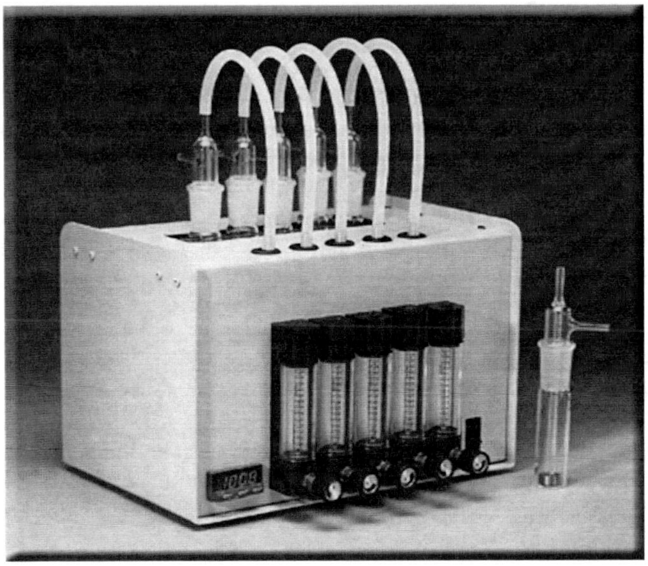

Fig. 15—Micro-oxidation bath.

suitable test was needed to evaluate lubricating greases in small bearings. ASTM D3337, Test Method for Evaluation of Greases in Small Bearings, was developed to serve this purpose. However, it is not currently under the jurisdiction of D02.G Lubricating Grease. Although this test is not the equivalent of long-term service-performance tests, it can be used to predict relative grease life at high temperatures in a reasonable test period. Also, this test can measure running torque at both 1 rpm and 12,000 rpm if this property is significant for the intended application. It will not differentiate among greases of closely related characteristics.

ASTM D3337 determines grease life and torque in a small (6.35-mm bore) R-4 ball bearing. In this test, the bearing is run at 12,000 rpm with a 2.2 N ($^1/_2$ lb-ft) radial load and a 22 N (5 lb-ft) axial load. While a test temperature of 250°C (or 500°F) may typically be specified, the equipment is capable of testing up to 315°C (600°F) if high-temperature bearings are used.

Wheel Bearing Grease Life

With the advent of automotive disk brakes in the 1960s, then-current test methods proved inadequate for evaluating greases for this high-temperature application. A specific, correlating test method was needed, and after several years of development, one was standardized: ASTM D3527, Standard Test Method for Life Performance of Automotive Wheel Bearing Greases. It evaluates grease life in tapered roller wheel bearings in a model front wheel assembly run at 1,000 rpm, under a thrust load of 111 N, at 160°C, using a cycle of 20 h on and 4 h off. The test apparatus is the same as that of D4290, but the operating conditions and measured parameters are quite different.

Motor torque is monitored, and the test is terminated at a calculated, preset torque value.

Grease life is indicated by the number of "on" hours (or number of cycles) to failure. This is a severe test; the results are influenced by a combination of grease properties, such as oxidation stability, shear stability, and volatility. It has been difficult to get good repeatability and reproducibility with this test; however, this test method is used for specification purposes and is required by ASTM D4950.

Extreme Pressure and Wear

A lubricant functions by completely separating bearing surfaces. If the separation is always complete, parts would never wear. However, the integrity of the lubricant film cannot be maintained under all conditions, and contact occurs to varying degrees. Such contact depends on operating conditions (such as load and speed), lubricant properties (such as fluid viscosity and grease consistency), and chemistry (such as the presence of wear inhibitors and extreme pressure additives). Several test methods are available for evaluating the antiwear and load-carrying properties of greases.

Extreme Pressure Timken Method

The ASTM D2509, Test Method for Measurement of Extreme Pressure Properties of Lubricating Grease (Timken Method), can be used to determine the load-carrying capacity of a grease at high loads. Nonstandard techniques have been devised to measure wear at lighter loads, but they are not discussed here.

In the Timken test, a tapered roller bearing cup is rotated against a stationary, hardened steel block. During the test, the parts are continually lubricated with fresh grease by means of a feed mechanism. Using a lever system with a tenfold mechanical advantage, fixed weights apply a force to the block in line contact with the rotating cup. Loads are applied stepwise until lubrication failure occurs, as evidenced by inspection for scoring or welding. The *OK value* is the maximum load the lubricant film will withstand without rupturing and causing scoring in the contact zone after a 10-min run. Figure 16 shows a photograph of the Timken Tester conforming to ASTM D2509.

This test is a fairly rapid method, which can be used to differentiate between greases having low, medium, or high levels of extreme pressure properties. It is widely used for specification purposes; however, the results may not correlate with service performance.

Extreme Pressure Four-Ball Test

ASTM D2596, Test Method for Measurement of Extreme Pressure Properties of Lubricating Grease (Four-Ball Method), is another test used to determine the load-carrying properties of lubricating greases. With this procedures, two evaluations are made: (1) the Load-Wear Index (formerly called Mean-Hertz Load), and (2) the Weld Point. The apparatus for ASTM D2596 is shown in Fig. 17. The Four-Ball Tester also conforms to ASTM D2266 Test Method for Wear Preventive Characteristics of Lubricating Grease (Four-Ball Method), discussed in more detail later in the chapter.

This test was developed to evaluate the extreme pressure and anti-weld properties of a lubricant. The tester is operated with one steel ball rotated under load against three like balls

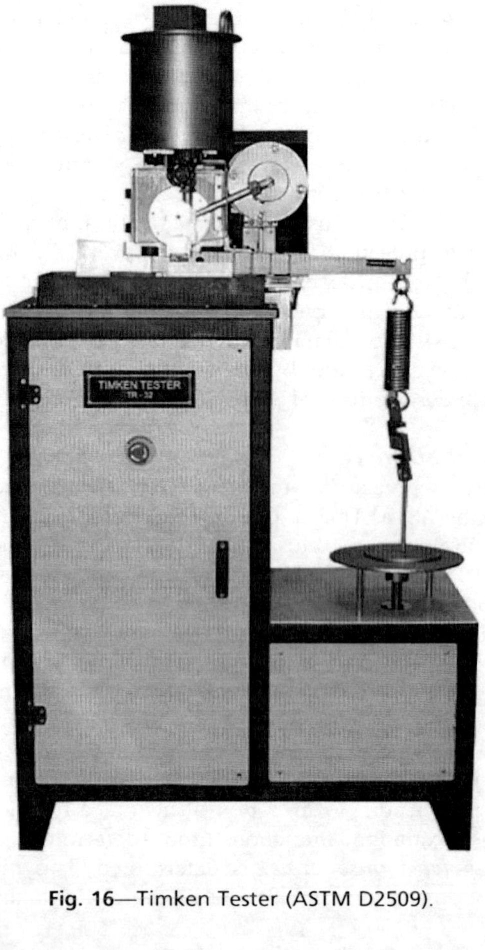

Fig. 16—Timken Tester (ASTM D2509).

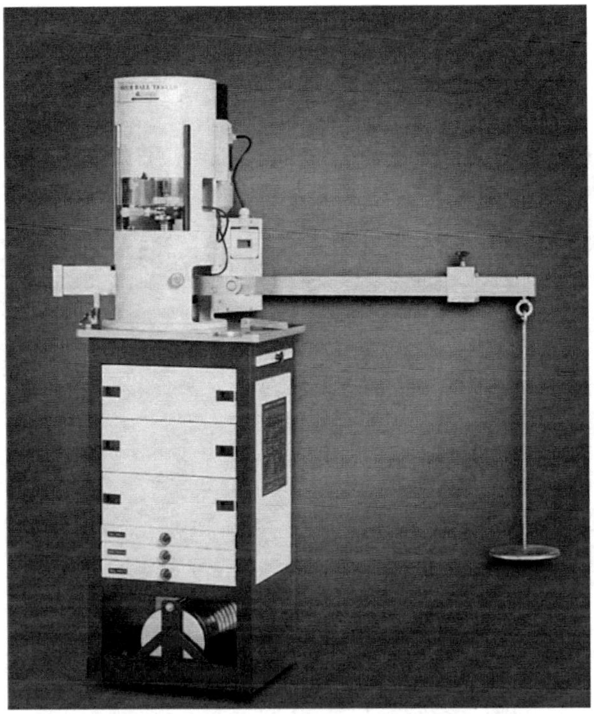

Fig. 17—Four-Ball Tester (ASTM D2266, D2596).

held stationary in the form of a cradle. The grease under test covers the contact area of the four balls. Loads up to 800 kgf (7,845 N, 1,760 lb-ft) can be applied to the balls to achieve unit pressures up to 6.9×10^6 kPa (1,000,000 psi).

The procedure involves the running of a series of 10-s tests over a range of increasing loads until welding occurs. During a test, scars are formed in the surfaces of the three stationary balls. The size of the scar depends on the load, speed, test duration, and lubricant. The scars are measured with a microscope having a calibrated grid. From the scar measurements, the Load-Wear Index is calculated. The lowest load at which the rotating ball seizes and then welds to the stationary balls is called the Weld Point; it indicates that the load-carrying capacity of the grease has been exceeded.

The significance of this test is that it is a relatively quick method, which can be used to differentiate among greases having low, medium, or high levels of extreme pressure properties. It is widely used for specification purposes, but the results may not correlate with service performance. Because of their poor lubricity, some lubricating greases containing a silicone or a halogenated silicone fluid component are not suitable for evaluation by this method.

Wear Preventive Characteristics of Grease

ASTM D2266, Test Method for Wear Preventive Characteristics of Lubricating Grease (Four-Ball Method), is used to determine the wear preventive characteristics of greases in sliding steel-on-steel applications. This test does not distinguish between EP and non-EP greases.

As in ASTM D2596, a four-ball configuration is used, but there are few other similarities (the apparatus and operating conditions are significantly different). Wear prevention qualities are evaluated from the dimensions of the wear scars that occur on the stationary balls during the test.

The test is significant because it can be used to determine the relative wear-preventing properties under the test conditions. It is useful for grease development. D2266 is widely used in grease specifications, but its actual usefulness is suspect because of the following limitations:

1. If test conditions are changed, the relative ratings may change.
2. Wear characteristics are not predicted for metal combinations other than AISI (American Iron and Steel Institute) E52100 steel unless nonstandard balls of other materials are used.
3. No differentiation can be made between EP and non-EP greases.
4. No correlation can be inferred between the results of the test and field service unless such correlation has been established.

Fretting Wear

Fretting wear is a form of abrasive wear caused by vibratory or oscillatory motion of small amplitude. It is characterized by the removal of finely divided particles from the rubbing surfaces. Air can cause immediate local oxidation of the wear particles produced by fretting wear, and moisture can hydrate the oxidation product. In the case of ferrous metals, the oxidized wear debris is abrasive iron oxide (Fe_2O_3) having the appearance of rust, which gives rise to the nearly synonymous terms *fretting corrosion* and *friction oxidation*.

Fretting is a serious problem in industry. If severe enough, it can cause destructive vibrations, premature failures, and parts seizure. No grease can give total protection if fretting conditions exist, but greases vary significantly in their ability to protect against fretting wear. ASTM D4170, Test Method for Fretting Wear Protection by Lubricating Grease, evaluates grease performance in a proprietary test machine (Fafnir Friction Oxidation Tester), which oscillates two grease-lubricated, ball thrust bearings under specified conditions of load, speed, and angle. Fretting wear is determined by measuring the mass loss of the bearing races. (The balls and retainers are not included.)

A related, but somewhat different, phenomenon often accompanies fretting wear. *False brinelling* is localized fretting wear that occurs when the rolling elements of a bearing vibrate or oscillate with small amplitude while pressed against the bearing race. The mechanism proceeds in stages. First the asperities weld, are torn apart, and the wear debris that is subsequently formed is oxidized. Then, because of the small-amplitude motion, the oxidized detritus cannot readily escape, because it is abrasive and it accelerates the wear. As a result, wear depressions are formed in the bearing races. These depressions appear similar to the Brinell depressions obtained with static overloading.

This test correlates with the fretting performance of greases in wheel bearings of passenger cars shipped long distances. ASTM D4170 also has been used to predict grease performance in automobile drivelines. It is used for specification purposes and is one of the performance tests required by ASTM D4950.

Oscillating Motion

There is another standard wear test involving oscillatory motion, namely, ASTM D3704, Test Method for Wear Preventive Properties of Lubricating Greases Using the Block on Ring Test Machine in Oscillating Motion. Ring and block parts, similar to those of the Timken tester (ASTM D2509), are operated under varying conditions of load, speed, oscillation angle, time, temperature, and specimen surface finish and hardness to simulate service conditions. This test can distinguish among greases of low, medium, and high wear preventive properties using oscillating motion and can be used for grease development. The user should determine whether test results correlate with service performance or results from other bench test machines.

Oscillating Wear (SRV) Tests

A high-load, high-frequency, low-amplitude, high-speed oscillating test, using the SRV *(Schwingung, Reibung, Verschleiss)* apparatus, simulates high-speed vibrational or start-stop motions that occur in many mechanisms. Two procedures, using a ball-on-disk configuration, have been developed: ASTM D5706 (EP Test) to measure the ability of a grease to carry loads under extreme pressure and ASTM D5707 (Wear) to measure wear-protection qualities and coefficient of friction. (This apparatus can use a variety of materials and configurations to suit different applications.) Both procedures have been correlated with grease performance in automotive drive-line mechanisms and are used in grease specifications for these applications. These procedures can be used to evaluate lubricants and materials for other applications of similar motion.

CORROSION

Copper Corrosion

Lubricated parts that contain copper alloys, such as copper or brass electrical components or bronze gears and bearings, may be susceptible to the corrosive effects of formulated greases. For example, such corrosion can cause high resistance in electrical contacts or premature bearing failure from chemical attack.

D4048, Test Method for Detection of Copper Corrosion from Lubricating Grease by the Copper Strip Tarnish Test, is the grease analog of the D130 method used to evaluate oils. Figure 18 shows the Copper Strip Corrosion Test Bath used to run D4048. In this test, a prepared copper strip is totally immersed in test grease and heated at specified conditions, usually 100°C (210°F) for 24 h.

At the end of the test period, the strip is removed, washed, and compared with the ASTM Copper Strip Corrosion Standards (Adjunct No. AdJD0130). Although this method is used in specifications, correlation between test results and actual service performance must be established by the user. This test does not determine the ability of a grease to inhibit copper corrosion caused by factors other than the grease itself; neither does it determine the stability of grease in the presence of copper. The sole determination is the chemical staining of copper by lubricating grease.

Fig. 18—Copper Strip Corrosion Test Bath (ASTM D4048).

Rust Prevention

Greases must not be corrosive to metals they contact and should not become corrosive with aging or oxidation. A method for assessing rust prevention by greases is ASTM D1743, Test Method for Corrosion Preventive Properties of Lubricating Greases. An apparatus to run ASTM D1743 is shown in Fig. 19.

In this method, a tapered roller bearing is packed with grease, and following a short run-in period, they are exposed to distilled water and stored at 100 % relative humidity at 52°C (125°C) for 48 h. The bearing is then cleaned and examined for corrosion. The test requires three bearings,

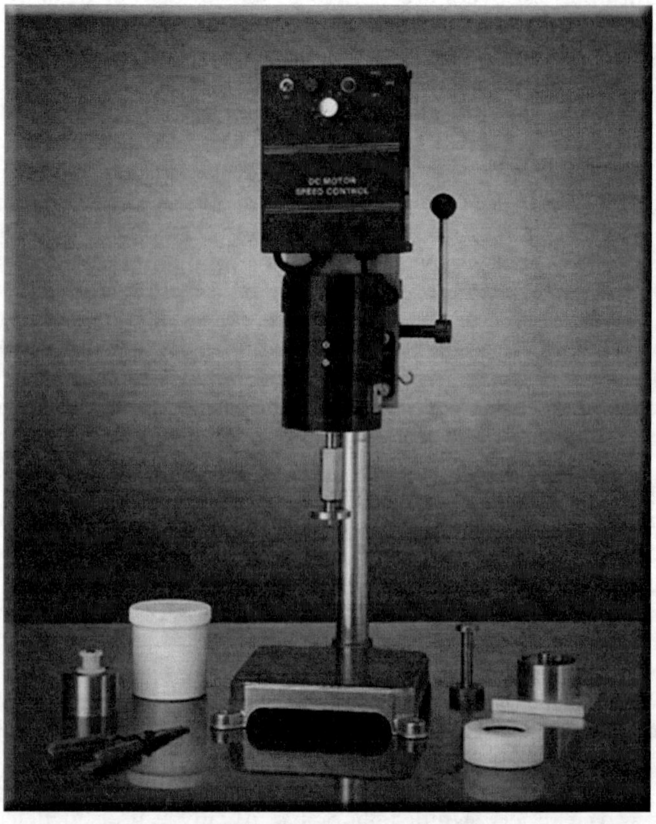

Fig. 19—Corrosion preventive properties apparatus (ASTM D1743).

and the result is determined by a pass or fail on at least two of the bearings. (An earlier version of this test used a 1, 2, or 3 rating scale, rather than a pass/fail rating.)

The significance of this test is that it indicates those greases capable of preventing rust in static or storage conditions. The test is widely used in grease specifications. The correlation with service conditions, particularly under static conditions, is considered to be quite good.

Accelerated Rust Tests

Two standard test methods, which were developed in the past few years, are used to evaluate the corrosion protection properties of greases under severe conditions. These are:

1. A significantly modified version of ASTM D1743 using synthetic sea water known as ASTM D5969, Test Method for Corrosion Preventive Properties of Lubricating Greases in the Presence of Dilute Synthetic Sea Water Environments.
2. Two procedures of the IP 220/DIN 51802 dynamic rust test, commonly known as the EMCOR test. One procedure uses distilled water and another uses several concentrations of salt water used in D6138, Test Method for Determination of Corrosion Prevention Properties of Lubricating Greases Under Dynamic Wet Conditions (EMCOR Test). Figure 20 shows the EMCOR Test rig used extensively for dynamic rust testing.

The EMCOR Grease Test method determines the corrosion-preventive properties of greases. It tests lubricated ball bearings partially immersed in water (distilled, sodium chloride solution, or synthetic sea water) at a speed of 83 ± 5 rpm at a cycle of running and stopping for a week. The bearing rings are then studied to find the level of corrosion.

Fig. 20—EMCOR Grease Testing Machine (ASTM D6138).

Effect of Water

Contamination by water can affect greases and grease performance in several ways. Corrosion or rust protection, previously discussed, is one. Other effects include changes in consistency, texture, or adhesiveness. An emulsion can be formed, which would probably be an inferior lubricant, or the grease could be washed away. Attempts to standardize means of evaluating these effects have had mixed success. However, two standard tests do exist: the water washout test and the water spray-off test.

Water Washout

The ability of a grease to resist washout under conditions where water may splash or impinge directly on a bearing is an important property in the maintenance of a satisfactory lubricating film. ASTM D1264, Test for Water Washout Characteristics of Lubricating Greases, evaluates the resistance of a lubricating grease in a bearing to washout by water. A Water Washout Tester is shown in Fig. 21.

This test method uses a standardized bearing, available from ASTM International. It is a 204 K Conrad-type, ball bearing equipped with shields, but without seals. The bearing is packed with 4 g of test grease, then rotated at 600 rpm while a jet of water, at either 38°C (100°F) or 79°C (175°F), impinges on the bearing housing for 1 h. The bearing is then dried, and the mass percent grease loss is determined.

This test method is of questionable value. It serves only as a relative measure of the resistance of a grease to water washout. It should not be considered the equivalent of a service evaluation, unless such correlation has been established. Test results are affected by grease texture and consistency. Test precision is relatively poor, especially with soft greases. Although widely used, this test can give misleading results. Even comparative results between similar greases may not predict the relative performance of the two greases in actual service.

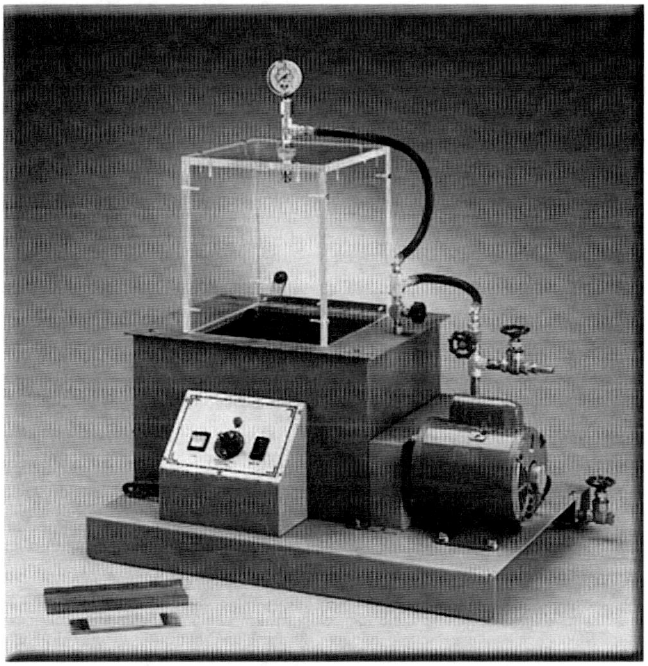

Fig. 21—Water washout tester (ASTM D1264).

Water Spray-Off

ASTM D4049, Test Method for Resistance of Lubricating Grease to Water Spray, is used to evaluate the ability of a grease to adhere to a metal panel when subjected to direct water spray. Test results correlate directly with operations involving direct water impingement, such as steel mill roll neck bearing service and certain automotive body hardware applications.

In this test, a 0.79-mm (1/32-in.) film of test grease is uniformly coated onto a stainless steel panel. Then, water at 38°C (100°F) is sprayed directly on the panel for 5 min. The spray is controlled by a specified spray nozzle, pump, and plumbing. After the spraying period, the panel is dried, weighed, and the percentage of grease spray-off is determined. Figure 22 shows a photograph of a water spray apparatus.

ASTM D7342, Shear Stability of Lubricating Grease in Presence of Water (Water Stability Test), is used because water can affect the shear stability of some greases. This test includes having a grease sample and a small amount of water subjected to low shear at 20 to 35°C for a certain amount of time or number of strokes in a standard grease worker or roll stability apparatus. The difference between the cone penetration results before and after measures the shear stability of grease in presence of water.

Fig. 22—Water spray apparatus (ASTM D4049).

MISCELLANEOUS

Contamination/Compatibility

ASTM D1404, Test Method for Estimation of Deleterious Particles in Lubricating Grease, defines a deleterious particle as one that will scratch a polished plastic surface. The test is applicable to all greases, even those containing fillers. In fact, it can be used to test fillers, such as graphite, if they are dispersed into a grease (or petrolatum) that is known to be free of deleterious particles. It can also be used to test other semi-solid or viscous-liquid substances. Figure 23 shows a photograph of the Deleterious Particles Determination Apparatus.

In this method, the test material is placed between two clean, highly polished acrylic plastic plates held rigidly and parallel to each other in metal holders. The assembly is pressed together by squeezing the grease into a thin layer between the plastic plates. Any solid particles in the grease larger than the distance of separation of the plates, and harder than the plastic, will become imbedded in the opposing plastic surfaces. The apparatus is so constructed that one of the plates can be rotated about 30° with respect to the other, while the whole assembly is under pressure. This will cause the imbedded particles to form characteristic arc-shaped scratches in one or both plates. The relative number of such solid particles is estimated by counting the total number of arc-shaped scratches on the two plates.

The test has significance because it is a rapid means for estimating the number of deleterious particles in a lubricating grease. However, a particle that is abrasive to acrylic plastic may not be abrasive to steel or other bearing materials. Therefore, the results of this test do not indicate the quality of performance in field service.

Elastomer Compatibility

Nearly all mechanisms lubricated with grease have elastomeric seals to retain lubricant and exclude contaminants. In order for these seals to function properly, the grease must be compatible with the rubber-like elastomer seal. ASTM D4289, Test Method for Compatibility of Lubricating Grease with Elastomers, is a simple total immersion test designed to evaluate the compatibility of grease with elastomers cut from standard sheets. It can also be used as a guide to evaluate compatibility of greases with rubber products not in standard sheet form.

Fig. 23—Deleterious particles determination apparatus (ASTM D1404).

Elastomer specimens are cut from standard ASTM sheets (D3182), immersed in test grease for 70 h at either 100 or 150°C. Compatibility is evaluated by determining the changes in volume and Durometer A hardness (ASTM D2240). (Volume is determined by the water displacement method, ASTM D471.)

Unlike other standard compatibility tests, which are designed to evaluate elastomers in standard fluids, the emphasis of ASTM D4289 is the evaluation of grease. The changes in volume and hardness determined in this test do not duplicate the changes that occur in rubber seals in actual service conditions. However, they can be correlated in many instances. For example, the volume-change values correlated very well (r^2 = 0.99) with those that occurred in a vehicle test. Because of wide variations in grease and elastomer formulations and service conditions, correlations between this test and particular applications should be determined on an individual basis.

This method provides for optional testing with two Reference Elastomers to evaluate relative compatibility. The results can be used to judge a service characteristic of lubricating greases; in this respect, the test method is useful for specification purposes. ASTM D4950 requires testing with Reference Elastomer CR (polychloroprene) and Reference Elastomer NBR-L (acrylonitrile-butadiene).

Grease Compatibility

Mixing of two different grease types often occurs when a mechanism is service lubricated with a type of grease different from that already in the bearing. If the two greases are incompatible, it is likely that lubrication will be inadequate or the lubrication life will be greatly shortened. The compatibility of the greases can be very important for equipment manufacturers who must stand behind their products, and the users of those products who depend on their reliability. The consequences of mixing incompatible greases can range from changes in consistency that are entirely acceptable, to changes that are utterly catastrophic, depending on the particular application. The problem of incompatible grease mixtures has long been known. Foreknowledge of the chemistry of the grease is not often reliable in predicting compatibility. Compatibility needs to be judged on a case-by-case basis.

There have been several nonstandard means worthy of consideration. However, the repeatability of the nonstandard test methods must be taken into consideration when making determinations about compatibility.

There are two approaches to evaluating the compatibility of grease mixtures. One is to determine whether such mixtures meet the same specification requirements as the constituent components. The other approach is specification independent; it describes the evaluation of compatibility on a relative basis using specific test methods. This latter approach was taken in developing ASTM D6185, Standard Practice for Evaluating Compatibility of Binary Mixtures of Lubricating Greases. This standard describes a protocol for evaluating the compatibility of three binary mixtures (concentrations of 10:90, 50:50, and 90:10 mass ratios) of lubricating greases by comparing their properties or performance relative to those of the neat greases comprising the mixture.

Although most incompatible mixtures could be detected by testing 50:50 mixtures, for a better evaluation, the testing of other concentrations was found necessary. For some grease mixtures, the incompatibility occurs at other concentrations. Incompatibility is indicated if any mixture tests worse than the poorer of the neat greases; the repeatability of the test methods must be considered when making such determinations. Two constituent greases are blended in three specific ratios. A 50:50 mixture simulates a ratio that might be experienced when one grease (Grease A) is installed in a bearing containing a previously installed different grease (Grease B), and no attempt is made to flush out Grease B with Grease A. The 10:90 and 90:10 ratios are intended to simulate ratios that might occur when attempts are made to flush out Grease B with Grease A. The latter two dilute ratios, prescribed by this standard, are considered representative of the careless flushing practice found in the real world.

These three mixtures and the two neat greases are then tested in the Primary Testing Protocol, which uses three independent, standard test methods:

1. Dropping point (ASTM D566 or D2265)
2. Shear stability (either by prolonged working, 100,000-stroke worked penetration, ASTM D217, or by the roll test, ASTM D1831)
3. Storage stability at elevated-temperature by change in 60-stroke penetration (D217) after prolonged storage (such as 1 to 6 months)

ASTM D6185 applies only to those lubricating greases having characteristics suitable for evaluation by the suggested test methods. If the scope of a specific test method limits testing to those greases within a specified range of properties, greases outside that range cannot be tested for compatibility by that method. An exception to this would be when the tested property of the neat, constituent greases is within the specified range, but the tested property of a mixture is outside the range because of incompatibility.

Chemical Analysis

ASTM D128, Test Methods for Analysis of Lubricating Grease, is the current standard for grease analysis. This procedure gives flow diagrams and details for the analysis of conventional greases (i.e., those made of soap-thickened petroleum oils). The constituents that can be determined are soap, unsaponifiable matter (petroleum oil, etc.), water, free alkalinity, free fatty acid, fat, glycerin, and insolubles. A supplementary test method is provided for application to greases that cannot be analyzed by conventional methods because of the presence of nonpetroleum oils or nonsoap thickeners.

These test procedures can be used to identify and estimate the amount of some of the constituents of lubricating greases. The methods are applicable to many, but not all, greases. Composition should not be considered as having a direct bearing on service performance, unless such correlation has been established.

ASTM D128 references several useful, but nonstandard, methods that can be used for grease analysis. Infrared (IR) spectroscopic analysis is commonly used for research purposes, quality control, and specification purposes. However, an IR method has not been standardized because it frequently has to be adapted to the specific grease being analyzed.

Although D128 is the general analytical method for greases, there are other test methods for specific constituents. These include the following: ASTM D95, Test Method for Water in Petroleum Products and Bituminous Materials by Distillation; ASTM D129, Test Method for Sulfur in Petroleum Products (General Bomb Method); ASTM D808, Test Method for Chlorine in New and Used Petroleum Products (Bomb Method), ASTM D1317; Test Method for Chlorine in

New and Used Lubricants (Sodium Alcoholate Method); and ASTM D3340, Test Method for Lithium and Sodium in Lubricating Greases by Flame Photometer.

ASDM D7303, Standard Test Method for Determination of Metals in Lubricating Greases by Inductively Coupled Plasma Atomic Emission Spectrometry (ICP-AES), determines metals in unused lubricating greases. A grease sample is weighed and subjected to sample dissolution. The diluted acid solutions are examined by using ICP-AES. The emission intensities of elements in the test sample are measured with the calibration standards to calculate the concentration of metallic elements in the test sample.

This test is significant because there are over a dozen metallic elements blended into lubricating greases, and the content of metal can indicate the amount of thickeners in the grease as well as potentially help trouble shoot field problems.

DISCONTINUED STANDARDS

Several standards have been discontinued in the past two decades. Unless otherwise noted, the reason for discontinuance of the following test methods was lack of interest (i.e., the standards were no longer being used or supported by the industry):

D1262	Test Method for Lead in New and Used Greases (Discontinued 1991)
D1402	Test Method for Effect of Copper on Oxidation Stability of Lubricating Greases by the Oxygen Bomb Method (Discontinued 1985)
D1741	Test Method for Functional Life of Ball Bearing Greases (Discontinued 1991)
D3428	Test Method for Torque Stability, Wear, and Brine Sensitivity Evaluation of Ball Joint Greases (Discontinued 1990)

INTERNATIONAL STANDARDS

In 1987, the ISO published ISO6743-9, an international classification system for grease. This standard classifies lubricating greases according to the operating conditions of the use–unlike most other ISO product standards, which are classified according to specific end use. The multifunctional nature of greases allows a specific grease to be used in many applications. This makes it impractical to classify grease by end use, and a description of properties is a reasoned alternative. Consequently, users are advised to use ISO6743-9 to define the requisite grease properties, but they are cautioned not to rely solely on the standard for grease selection for a particular application. Rather, users are advised to consult with the grease supplier, as well. In this classification system, each grease will have one symbol only. This symbol should correspond to the most severe conditions of temperature, water contamination, and load in which the grease can be used. Grease products are designated in a uniform manner, with each character having its own significance. The line call-out code will have the following format:

$$ISO - L - XS_1S_2S_3S_4N \quad (1)$$

where:

ISO = identification of the standards organization,
L = designator for class (lubricants),
X = designator for family (grease),
S_1 = symbol for lower operating temperature,
S_2 = symbol for upper operating temperature,
S_3 = symbol for water contamination,
S_4 = symbol for EP
N = NLGI consistency number.

Symbols (letters) are used to designate the four operating conditions. The lower operating temperature is defined by symbols representing requirements at 0, −10, −20, −30, −40, and <−40°C; the upper operating temperature is defined by symbols representing seven temperatures from 60 to >180°C Nine symbols are used to define the effects of water, which include both water contamination and antirust requirements. Only two symbols are used for the EP or load-carrying requirement.

Limits have been tentatively established for the operating conditions. However, full implementation of ISO 6743-9 depends on the development of ISO standard test methods. National standards must be converted to ISO format and approved by the ISO community. The national standards proposed to ISO are as follows:

Lower operating temperature	NFT 60-171 (France), a low-temperature penetration test.
Upper operating temperature	ASTM D566/D2265 and D3336
Water effect	ASTM D1264 for Water Contamination and DIN 51802 (Germany)/ IP 220 (United Kingdom) for Rust Protection
EP (load-carrying capacity)	ASTM D2596

ASTM/Energy Institute Standards for Evaluating Lubricating Grease Properties and Performance

ASTM	IP	Title
D217	50	Test Methods for Cone Penetration of Lubricating Grease
D566	132	Test Method for Dropping Point of Lubricating Grease
D942	142	Test Method for Oxidation Stability of Lubricating Greases by the Oxygen Bomb Method
D972	183	Test Method for Evaporation Loss of Lubricating Greases and Oils
D1092		Test Method for Apparent Viscosity of Lubricating Greases
D1263		Test Method for Leakage Tendencies of Automotive Wheel Bearing Greases
D1264	215	Test Method for Water Washout Characteristics of Lubricating Greases
D1403	310	Test Methods for Cone Penetration of Lubricating Grease Using One-Quarter and One-Half Scale Cone Equipment
D1404		Test Method for Estimation of Deleterious Particles in Lubricating Grease

ASTM	IP	Title
D1478		Test Method for Low-Temperature Torque of Ball Bearing Greases
D1742		Test Method for Oil Separation from Lubricating Grease During Storage
D1743		Test Method for Corrosion Preventive Properties of Lubricating Greases
D1831		Test Method for Roll Stability of Lubricating Grease
D2265		Test Method for Dropping Point of Lubricating Grease Over Wide Temperature Range
D2266	239	Test Method for Wear Preventive Characteristics of Lubricating Grease (Four-Ball Method)
D2509	326	Test Method for Measurement of Extreme Pressure Properties of Lubricating Grease (Timken Method)
D2595		Test Method for Evaporation Loss of Lubricating Greases Over Wide-Temperature Range
D2596	239	Test Method for Measurement of Extreme-Pressure Properties of Lubricating Greases (Four-Ball Method)
D3336		Test Method for Performance Characteristics of Lubricating Greases in Ball Bearings at Elevated Temperatures
D3337		Test Method for Evaluation of Greases in Small Bearings
D3527		Test Method for Life Performance of Automotive Wheel Bearing Grease
D3704		Test Method for Wear Preventive Properties of Lubricating Grease Using the (Falex) Block on Ring Test Machine in Oscillating Motion
D4048		Test Method for Detection of Copper Corrosion from Lubricating Grease by the Copper Strip Tarnish Test
D4049		Test Method for the Resistance of Lubricating Grease to Water Spray
D4170		Test Method for Fretting Wear Protection by Lubricating Greases

ASTM	IP	Title
D4289		Test Method for Compatibility of Lubricating Grease with Elastomers
D4290		Test Method for Determining the Leakage Tendencies of Automotive Wheel Bearing Grease Under Accelerated Conditions
D4425		Test Method for Oil Separation from Lubricating Grease by Centrifuging (Koppers Method)
D4693		Test Method for Low-Temperature Torque of Grease-Lubricated Wheel Bearings
D4950		Classification and Specification for Automotive Service Greases
D5438		Practice for Collection of Floor Dust for Chemical Analysis
D5706		Test Method for Determining Extreme Pressure Properties of Lubricating Greases Using a High-Frequency, Linear-Oscillation (SRC) Test Machine.
D5707		Test Method for Measuring Friction and Wear Properties of Lubricating Grease Using a High-Frequency, Linear-Oscillation (SRV) Test Machine.
D5969		Test Method for Corrosion Preventive Properties of Lubricating Greases in the Presence of Dilute Synthetic Sea Water Environments
D6138		Test Method for Determination of Corrosion Prevention Properties of Lubricating Greases Under Dynamic Wet Conditions (EMCOR Test)
D6184	98	Standard Test Method for Oil Separation from Lubricating Grease (Conical Sieve Method)
D6185		Standard Practice for Evaluating Compatibility of Binary Mixtures of Lubricating Greases Grease Ready Reference—Principal European Standards
D7303		Standard Test Method for Determination of Metals in Lubricating Greases by Inductively Coupled Plasma Atomic Emission Spectrometry
D7342		Standard Test Method for Shear Stability of Lubricating Grease in Presence of Water (Water Stability Test)

Grease Ready Reference: Principal European Standards

Standards Designation								
	Characteristic/Property	ISO	DIN	IP	NF-T 60	ASTM-D	FTM 791 b	GOST
1	Determination of Dropping Point	2176	51801	132	102	566	1421	6793
2	Determination of Cone Penetration 1/1-1/2-1/4 Cone	2137	51804/1-2	50	132	217	311	5346
3	Determination of Flow Pressure		51805					

Standards Designation								
	Characteristic/Property	**ISO**	**DIN**	**IP**	**NF-T 60**	**ASTM-D**	**FTM 791 b**	**GOST**
4	Corrosion Preventing Properties		51802	220	135	1743	40001.2	5757
5	Corrosive Effects on Copper	2160	51811	112		1261	5309.4	
6	Determination of Ash of Greases (incl. Sulfate)		51803	5	M 07-037	128		1461 6474
7	Neutralization Number		51809/1-2	37/137/139	133/112	128/974		
8	Content of Solid Foreign Matters		51813	134		(1404)	3005.3	6370
9	Effect of Water							
9.1	Water Washout Test	51807/2	215		1264			
9.2	Static Test		51807/1					
10	Oxidation Stability		51808	142		942	3453	5734
11	Oil Separation		51817	121		1742	321/2	7142 1631
12	Content of Base Oil and Soap		51814			128		
13	Content of Li/Na/Ca by Atomic Absorption Spectroscopy		51815		M 07-38			
14	Determination of Li/Na by Flame Photometer			199		3340		
15	Determination of Solids (Graphite or MoS_2)		51831				3720/22	
16	Determination of Particle Size of Solid Lub.		51832					9270
17	Density			59				
18	Evaporation Loss			183		972		9566
19	Content of Water	3733		74	113	95		1044 2077
20	Pumpability Properties							
20.1	Shell-DeLimon Rheometer		51816/1					
20.2	Decompression Characteristics		51816/2					
20.3	Apparent Viscosity				139	1092		
21	Roll Stability					1831		
22	Extreme Pressure Properties							
22.1	Shell-Four-Ball Test		51350-4/5	239		2266	6503.2	
22.2	Timken Test			326		2509		
23	Mechanical Dynamic Testing						331.2/ 333.1	
23.1	SKF-R2F		51806					
23.2	Roller Bearing Performance			168				
23.3	Churning			266				
23.4	FAG-FE 8 (EP Greases)		(51821/2)					
23.5	FAG-FE 9		(51821/1)					
23.6	Wheel Bearing Leakage					1263		
24	Low Temperature Torque			186		1478		
25	Elastomer Compatibility							

Standards Designation								
	Characteristic/Property	**ISO**	**DIN**	**IP**	**NF-T 60**	**ASTM-D**	**FTM 791 b**	**GOST**
25.1	Hardness Change (Shore A)	R868	53505					
25.2	Volume Change	1817	53521				3603.3	
26	Tests on Base Oil							
26.1	Viscosity	3104	51562	71			445	
26.2	Pour Point	3016	51597	15			97	
26.3	Flash Point	2592	51376	36			92	
26.4	Aniline Point	2977	57175/ 51787	2			611	
27	Infrared-Analysis		51820 E					

Bibliography

Annual Book of ASTM Standards, Volumes 05.01, 05.02, 05.03, ASTM International, West Conshohocken, PA, 2002.

Boner, C. J., *Lubricating Greases: Their Manufacture and Use,* Reinhold Publishing Corporation, 1954.

Dyroff, G. V., Ed., "Lubricating Greases (Chapter 9)," *Manual on Significance of Tests for Petroleum Products,* 5th ed., ASTM Manual Series, MNL 1, ASTM International, West Conshohocken, PA, 1989.

Ehrlich, M., Ed., *Lubricating Grease Guide,* 4th ed., National Lubricating Grease Institute, Kansas City, MO, 1996.

Federal Test Method Standard No. 791C, Lubricants, Liquid Fuels, and Related Products: Methods of Testing (September 30, 1986), Global Engineering Documents, Irvine, CA.

Grease Ready Reference, *Principal European Standards;* The Lubrizol Corporation, Wickliffe, OH, 2002.

Labude, K. M., *Index of the NLGI Spokesman:* April 1950-September 1987, National Lubricating Grease Institute, Kansas City, MO, 1987.

Verdura, T. M., "Lubricating Greases," *Manual on Significance of Tests for Petroleum Products,* 6th ed., G. V. Dyroff, Ed., ASTM Manual Series, MNL 1, ASTM International, West Conshohocken, PA, 1993.

20

Petroleum Waxes Including Petrolatums[1]

Alan R. Case[2]

WAXES CAN BE DEFINED AS HYDROCARBONS OR hydrocarbon derivatives that are solid at room temperature, but are low-viscosity liquids at moderate temperatures. Waxes are thermoplastic materials but, due to their low molecular weight, they are normally not considered to be plastics or polymers. Common wax properties are water repellency, smooth texture, low toxicity, low odor, combustibility, solubility in most organic solvents, insolubility in water, and low reactivity. By far the most important waxes in terms of volume produced and economic impact are petroleum waxes. Petroleum waxes are derived from crude oil. Other types of wax include hydrogenated triglycerides made from tallow or plant oils; animal waxes, such as beeswax; plant waxes such as carnauba and candelilla; mineral waxes such as montan; and synthetic waxes such as Fischer-Tropsch and polyethylene. Although this chapter deals only with petroleum waxes, much of the information and many of the test procedures can be applied to other types of wax as well.

In the early years of petroleum processing, the waxy materials that were separated were regarded as having no commercial value. Ultimately, however, wax products were recovered that, with the benefit of additional refining, were found to be useful in many applications. A variety of grades of petroleum wax evolved that covered a broad range of physical properties. Now the petroleum wax product lines are in demand for a wide variety of uses. Modern refining methods have made available select grades of wax of controllable and reproducible quality, and some have unique properties for specialized applications.

THE MANUFACTURING OF PETROLEUM WAXES

A discussion of petroleum waxes must start with their source, crude oil. Crude oil is a complex mixture of hydrocarbons. These hydrocarbons range in size from single carbon atom molecules to molecules containing over 100 carbon atoms. Some of the molecules may contain sulfur or nitrogen, but most consist entirely of hydrogen and carbon in various structures. All crude oils contain some waxy hydrocarbons, but in most the percentage is very low. On average, wax content is about two tenths of 1 % of the total crude oil. Some crude, especially valuable to wax manufacturers, contain over 10 % wax.

The wax is separated from the other components during the refining process. First, the crude oil is subjected to atmospheric pressure distillation. This step removes and separates all of the low boiling point components, such as gases, gasoline, naphtha, kerosene, and diesel fuel. The higher boiling point components remaining are then processed by vacuum distillation. The vacuum distillation produces several distillate cuts and a heavy residue. These cuts are the source of lubricating oils and waxes. The distillate cuts are usually solvent extracted to remove impurities using polar solvents such as furfural (furfuraldehyde, C_4H_3OCHO) and *N*-methyl-2-pyrollidone (C_5H_9NO) solvent. The residue and some intermediates are solvent extracted using a combination of propane, phenol, and cresylic acid. The solvent extraction step removes aromatic and heterocyclic compounds.

To produce high-grade lube oils, these cuts must also be "dewaxed." If the wax were not removed, the oils would thicken too much at low temperatures and not flow properly. There are methods of dewaxing that do not result in the production of finished waxes. These catalytic dewaxing methods either crack the wax into fuels or isomerizes the wax into oil molecules. Since no wax is yielded by those methods, we will not address them further.

One of the original methods used for removing wax from oil, cold settling pressure filtration, consisted of a chilled pressing step, where pressure filters were used to squeeze oil from the wax crystals that formed upon cooling of the oil/wax blend. In a second variation, centrifuges were used to separate the wax crystals from the oil. The most common modern method of dewaxing is called solvent dewaxing. Over the years many different variations of the process have been developed, but all are based on three steps. First is the crystallization of wax from a solvent/oil mixture by chilling (to approximately −28°C), second is filtration to remove the wax crystals, and third is solvent recovery from the wax cake. The resulting oil fraction is now low enough in wax content to meet the standard pour-point specification for lube oil.

The removed wax that is still very high in oil content, about 40 %, and has to be "deoiled" for it to be useful in commerce. Originally the wax cake was deoiled using a sweating process. Applicable only to highly paraffinic streams, the pressed wax was formed on to screens in pans held in a carefully temperature-controlled room. As the temperature in the room was slowly increased, oil would begin to "sweat" out of the wax and flow to the bottom of the pan, where it could be drawn off. Although the sweating process has disappeared from modern refineries, an interesting modification of the process, called the Sulzer process, is in use at the Sasol Wax refinery in Germany. In the Sulzer process a paraffinic slack wax is continuously deoiled in temperature-controlled

[1] In preparation of this chapter, the contents of the seventh edition were drawn upon. The current edition will review and update the topics as addressed in the previous edition, introduce new technology that has been developed, and include up-to-date references. The author wishes to thank Lindsey Barnes, Citgo Petroleum, and Gernot Meyer, Sasol Wax GMBH, for their assistance with the new edition.

[2] *Technical Director, Sasol Wax North America, Richmond, CA 94804.*

columns containing stainless steel media that hold the wax in a thin film on its surface. The process is said to be cost-effective, and has the advantage of using no solvents. A short-coming of the process is that it can only be used on highly paraffinic slack waxes. The most common process for deoiling is to repeat the solvent dewaxing process at different conditions. The waxy oil cut is dissolved in a solvent blend that typically contains methylethylketone (MEK) with either toluene or methylisobutylketone (MIBK), although other solvents are sometimes used. The solution is then chilled (to approximately −5°C); at the low temperature, the wax precipitates from the oil and can be removed using large rotary drum filters. In some cases the deoiling step is repeated a second time. By varying the temperature, solvent selection and ratio, and the number of recrystallizations, waxes of varying oil content can be produced. In general, waxes containing from over 5 to 20 % oil are classified as slack waxes. Waxes of about 1 to 5 % oil are classified as scale waxes. Waxes that have oil contents of 1 % or less (typically 0.5 %) and meet U.S. Food and Drug Administration criteria are classified as fully refined waxes.

The nonboiling residual fraction, or "bright stock" cut, is a source of microcrystalline wax and petrolatum. The distillate cuts yield paraffin waxes when deoiled. The highest boiling distillate cuts are frequently referred to as intermediate lube distillates, and the resulting waxes are called intermediate waxes. This term refers to the position and structure of this wax as falling between true paraffins and microcrystalline waxes. These waxes exhibit some of the characteristics of both types.

To manufacture fully refined (food grade) paraffin waxes, microcrystalline waxes, or petrolatums, a final step of hydrogen finishing, or clay treatment, completes the processing. The traditional method was clay treatment using a selective absorbent like bauxite to remove any remaining polar contaminants. The more modern process is hydrogen finishing, a type of mild hydrogenation using a fixed bed catalyst. These treatments remove (in the case of bentonite), or transform (in the case of hydrogenation), low levels of contaminants that could cause odor and color stability problems.

For a more in depth treatment of this subject the reader is referred to texts such as *Lubricant Base Oil and Wax Processing* by Avilino Sequeira.

DEFINITIONS

An exact classification of waxes is difficult because of the wide variety of waxes that have become available through the expansion of refining technology and the apparent overlapping among the types. However, petroleum waxes often are identified as paraffin wax (including scale, slack, and fully refined), microcrystalline wax (crude or finished), or petrolatum (crude or finished). Fully refined paraffin wax is petroleum wax consisting mainly of normal alkanes, with varying small amounts of cycloalkanes and isoalkanes. Slack and scale paraffin wax may also contain small amounts of aromatic or heterocyclic compounds. Molecular weights are usually less than 450, and the viscosity at 100°C normally will be less than 6 mm^2/s. Either needle- or plate-type crystal structures are common.

Paraffin waxes also exhibit pronounced latent heats of crystallization, and some have transition points (temperatures at which crystal structure modification occurs below the apparent solidification point of the wax).

Microcrystalline waxes contain substantial portions of hydrocarbons other than normal alkanes, and the components usually have higher molecular weights (550 to 800) than paraffin wax. Microcrystalline waxes also have smaller crystals and greater affinity for oil than paraffin waxes. Microcrystalline waxes usually melt between 65.6 and 90.5°C and have viscosities between 10 and 20 mm^2/s at 100°C.

A broad definition of petrolatum is a petroleum wax of unctuous structure containing substantial amounts of oil. The definition embraces the unrefined industrial grades as well as the refined pharmaceutical grades. The semisolid nature of petrolatum is due to the gel-like dispersion of a relatively high proportion of oil (usually above 10 %) in the microcrystalline wax base. Petrolatums resemble microcrystalline waxes in composition, but contain higher proportions of oil-like, branched alkane molecules. Melting points range from about 37.8 to 79.4°C, and viscosities range from 10 to 25 mm^2/s at 100°C. Petrolatum colors range from the almost black crude form to the highly refined white pharmaceutical grade. Product quality and purity characteristics for medicinal uses of petrolatum, paraffin, and microcrystalline waxes are contained in the U.S. Pharmacopoeia and in the National Formulary.

APPLICATIONS FOR WAX

Wax is one of the most versatile products of the oil industry. Wax has long served as a source of light and heat in such applications as candles, fire logs, matches, and flares. In fact, candles have become one of the largest applications for petroleum wax. Large amounts of slack wax, either straight or in emulsion form, are used for sizing particleboard, hardboard, and chipboard. As a coating for a variety of paper, film, and foil substrates in the packaging industry, fully refined paraffin wax provides improved appearance, strength, moisture proofing, food oil resistance, sealing strength, and other desirable effects. The corrugated board industry consumes a large amount of refined petroleum wax either in the form of straight wax or as hot melt blends.

In other uses, the high-gloss characteristic of some petroleum waxes makes them suitable ingredients for polishes. The highly refined waxes have excellent electrical properties for application in the insulation of small transformers, coils, capacitors, and certain cables. Waxes are used for the coating of fruit and cheese and the lining of cans and barrels. Blends of waxes are used by dentists for making dentures and by engineers for mass-producing precision castings. Petroleum wax is also an ingredient in certain hot melt adhesives, where the wax serves both as a viscosity reducer and a set point improver. Other applications include crayons, plastics lubricants, emulsions, and rubber antiozonants.

QUALITY CRITERIA

Table 1, "Comparative Table of International Test Standards Used for Petroleum Waxes Including Petrolatum," lists the description of tests and the applicable ASTM, IP, and other numbered standard test methods used to assess the quality of waxes.

Criteria for judging the quality of a wax must be specific for the type of wax and its intended application.

Three general categories of properties are useful for assessing the quality of waxes: physical properties (characterizations used for quality control); chemical properties and composition (used for basic characterizations); and functional

TABLE 1—Comparative Table of International Test Standards Used for Petroleum Waxes Including Petrolatum

ASTM	IP	Description of Test
D938	IP 76	Congealing Point of Petroleum Wax
D3944		Solidification Point of Petroleum Wax
D4419		Measurement of Transition Temperature of Petroleum Wax by DSC
D127	IP 133	Drop Melt Point of Petroleum Wax And Petrolatum
D87	IP 55	Melting Point of Petroleum Wax (Cooling Curve)
D937	IP 179	Cone Penetration of Petrolatum
D2500	IP 219	Cloud Point of Petroleum Products
D1321		Needle Penetration of Petroleum Waxes
D445	IP 71	Kinematic Viscosity of Transparent and Opaque Liquids
D2669		Apparent Viscosity of Petroleum Wax Compounds with Additives
D3236		Apparent Viscosity of Hot Melt Adhesives and Coating Materials
D1500	IP 196	ASTM Color of Petroleum Products
D156		Saybolt Color of Petroleum Products
	IP 17	Determination of Color—Lovibond Tintometer Method
D721	IP 158	Oil Content of Petroleum Waxes
D3235		Solvent Extractables of Petroleum Wax
D1833	IP 185	Odor in Petroleum Wax
D5442		Analysis of Petroleum Waxes by GC
D2008		UV Absorbance and Absorptivity of Petroleum Products
D3344		Total Wax Content of Corrugated Board
D3708		Weight of Wax Applied During Curtain Coating Operation
D2423		Surface Wax on Waxed Paper or Paperboard
D3521		Surface Wax Coating on Corrugated Board
D3522		Applied Coating Wax and Impregnating (Saturation) Wax in Board Facing
D1465		Blocking and Picking Points of Petroleum Wax
D2534		Coefficient of Kinetic Friction of Wax Coatings
D1832		Peroxide Number of Petroleum Wax
D612		Carbonizable Substances in Petroleum Wax
D92	IP 36	Flash and Fire Point by Cleveland Open Cup
D93	IP 34	Flash Point by Pensky-Martens Closed Tester
D1298	IP 160	Density and Relative Density (Specific Gravity) of Crude Petroleum and Liquid Petroleum by Hydrometer Method
D287		Gravity, API of Crude Petroleum and Petroleum Products (Hydrometer Method)
D94	IP 136	Saponification Number of Petroleum Products
D95		Water in Petroleum Products by Distillation
D482		Ash from Petroleum Products
D974		Acid and Base Number by Color Indicator Titration

properties (empirical evaluations under simulated consumer conditions).

Criteria for Judging Physical Properties

MELTING POINT

Melting point is one of the most widely used tests to determine the quality and type of wax. Petroleum waxes usually do not melt at sharply defined temperatures because they are mixtures of hydrocarbons with different melting points. Paraffin waxes, as relatively simple mixtures, usually have narrower melting ranges than do microcrystalline waxes and petrolatums that are more complex in composition and melting behavior. The melting point of a wax has direct and indirect significance in most applications. There are three basic ASTM methods used for determining the melting point of a wax, as well as instrumental methods that can be used. The methods are described in more detail below.

Note. Some waxes (especially narrow-cut, highly paraffinic types) undergo a phase change or transition point in the solid state that is usually about 14 to 22°C below the melting point. Accompanying this change in crystal structure there may be significant changes in properties such as refractive index, density, flexibility, hardness, coefficients of expansion and friction, tensile strength, sealing strength, and gloss. For applications requiring consistency of properties over a specific temperature range, waxes are selected that do not exhibit a transition point in the critical range.

VISCOSITY

The fluidity of molten wax is important in applications involving coating or dipping processes, since it influences the amount of wax that is deposited on the substrate, as well as the performance effectiveness of the coating obtained. Examples of such applications are in paper-converting, hot-dip anticorrosion coatings, and taper candle manufacturing.

Paraffin waxes do not differ widely in viscosity, typically falling in the range of 3 to 6 mm^2/s at 100°C, although some grades have viscosities as high as 10 mm^2/s. Microcrystalline waxes are considerably more viscous and vary over a wide range, about 10 to 20 mm^2/s at 100°C.

The hot melt blends of petroleum waxes with various additive modifiers cover a wide spectrum of viscosities. With additives of relatively low molecular weight, only minor thickening effects occur. Additives that are polymers, such as the polyethylenes or various co-polymers (such as ethylene vinyl acetate), may have very high thickening power. The components in a hot melt formulation will be selected to give the proper viscosity consistent with the expected application conditions in the designated waxing equipment. Optimum viscosities usually exist for the various dipping operations, roll coaters, curtain coaters, or high-shear coaters. Viscosities of hot melts for coatings may range as high as 20,000 mPa/s at 176.7°C, although 100 to 2,000 mPa/s is the more common range.

HARDNESS

Hardness is a measure of resistance to deformation, abrasion, or damage; hence, it is an important criterion for many wax applications. Narrow-cut paraffin waxes generally have greater hardness than broad-cut waxes of the same melting point, and waxes with higher proportions of normal alkanes have greater hardness than those containing more highly branched alkanes. Hardness is indirectly related to blocking tendency and gloss. Hard waxes usually have higher blocking points and better gloss than waxes of the same average molecular weight but wider molecular weight range.

Hardness is determined by measuring the depth of penetration of a needle or cone into a wax specimen. Techniques for measuring hardness (cone penetration for softer waxes and needle penetration for the harder waxes) have been standardized by ASTM International and the Institute of Petroleum (IP) and will be described in later paragraphs.

DENSITY

The density of a wax (or alternatively the specific gravity or American Petroleum Institute [API] gravity) is both useful for characterization and for the handling of wax. Waxes are normally sold by the pound, but they are usually handled in their liquid state. A determination of density is required to convert from weight units to volumetric units. Readers should be aware of the potential for confusion on the subject of wax density. When a hydrometer is used to measure API gravity (ASTM D287), tables are used to convert the gravity figure measured at a specific molten temperature to one at 60°F. The method ignores the fact that wax would be a solid at this temperature. The specific gravity or density of a solid wax is much higher due to the significant amount of contraction that a wax undergoes as it solidifies.

The actual specific gravity or density of solid wax would be measured using another technique, and normally the value would be reported at a temperature of 20°C. For illustration purposes, paraffin waxes in the liquid state at 75°C range in density from about 0.768 to about 0.775 g/cm^3. The same waxes in the solid state at 25°C range in density from 0.870 to 0.910 g/cm^3.

FLASH POINT

The flash point is important for many applications in which wax is heated. Heating above the flash point results in the risk of fire. The flash point also gives an indication of low molecular weight components (<CN 24) and contaminants such as solvents.

Criteria for Judging Chemical Properties

Chemical properties and the composition of wax give a good indication of its degree of refinement. The properties of greatest significance are color, odor, oil content, sulfur content, amount of solvent extractables, boiling point distribution, molecular weight, ultraviolet absorptivity, peroxide content, and amounts of carbonizable substances. In addition, there are specific chemical property criteria for food grade wax.

COLOR

Paraffin waxes are generally of a translucent white color, whereas microcrystalline waxes and petrolatums range from white to almost black. A fully refined wax should be virtually colorless ("water-white") when examined in the molten state. Typical scale wax usually exhibits a slight yellow or "straw" tint. Absence of color is of particular importance in waxes used for pharmaceutical purposes, candles, or the manufacture of food wrappings. The acceptability of the color of microcrystalline waxes and petrolatums depends on the use for which they are intended. In some applications (for example, the manufacture of corrosion preventives or firelogs), color may be of little importance, but in others it may be critical. For pharmaceutical purposes, the color of petrolatum is recognized in the grade specifications, with colors ranging from white to yellow. Colors in the solid state normally are expressed in descriptive terms such as white, off-white, yellow, amber, brown, etc. However, there is no standard test method for measuring this characteristic. Several methods, described later, can be used to measure the color of waxes in the liquid state.

ODOR

Freedom from odor and taste is particularly important in applications where the wax is likely to contact foodstuffs. When poor wax odor is detected, it may be due to inadequate refining, contamination during transport, storage, or waxing operations, or deterioration in use. Odors due to inadequate refining are usually associated with high oil content or, less frequently, with residual solvent. Odor may be acquired during storage by absorption of odorous material from the surrounding environment. Wax also readily absorbs the odors of products such as cheese or soap if stored in proximity. Wax that has been overheated in use tends to oxidize and develop an odor similar to stearic acid.

Subjective evaluations, such as odor, are difficult to standardize. There is, however, a technique that ensures reasonable concordance within a group whose members have agreed about odor level. Difficulties arise when there is a difference of opinion as to what constitutes an acceptable odor.

OIL CONTENT

Because petroleum waxes are mixtures of hydrocarbons, which cover a range of molecular type and molecular weight, some components of a given wax may be of lower melting point and show more solubility or extractability than other fractions of the wax. These components are identified as the crystallization of the oil content method [ASTM Oil Content of Petroleum Waxes (D721/IP158)]. It should be recognized that there is no clear-cut division between oil and wax with regard to particular chemical species. It is necessary to distinguish between paraffin wax, microcrystalline wax, and petrolatums when considering the significance of oil content because of the different degree of affinity that each of these wax types has for oil. Nevertheless, the oil content consists mainly of short chain length normal alkanes and branched alkanes. Fully refined paraffin wax usually has an oil content of less than 1 % (typically about 0.5 %). Wax containing somewhat more than this amount of oil (that is, 1 to about 5 %) is referred to as scale wax. Slack wax may contain from 5 to over 20 % oil. Excess oil tends to exude from paraffin wax, giving it a dull appearance and a greasy feel. High oil content tends to have an adverse effect on sealing strength, tensile strength, hardness, odor, taste, color, and color stability. High oil content also tends to plasticize wax and increase its flexibility.

Microcrystalline waxes have a greater affinity for oil than paraffin waxes because of their smaller crystal structure and branching. The permissible amount depends on the type of wax and its intended use. The oil content of microcrystalline wax is, in general, much greater than that of paraffin wax and could be as high as 10 %. Waxes containing more than 10 % oil would usually be classed as petrolatums, but the demarcation is by no means precise.

SOLVENT EXTRACTABLES

For waxes containing no more than 15 % oil, ASTM Method D721/IP158 is used to define oil content. For waxy streams that contain more than 15 % oil and various "soft" or soluble fractions (for example, petrolatum and slack waxes), the ASTM Test for Solvent Extractables in Petroleum Waxes (D3235) should be used. It must be remembered that the solvent separations are empirical, and the distinction between "oil content" and "extractables" is associated with the different solvent compositions used in the tests. The terminology should not be used interchangeably. It is erroneous to indicate oil content as being determined by ASTM D3235.

BOILING POINT DISTRIBUTION

Boiling point distribution is useful because it provides an estimate of hydrocarbon molecular weight. The distribution influences many of the physical and functional properties of petroleum wax. To a lesser extent, distillation characteristics also are influenced by the distribution of various molecular types, that is, normal paraffins, branched, or cyclic structures. In the case of the paraffin waxes that are predominantly straight chain, the distillation curve reflects the molecular size distribution.

In the most common distillation test, ASTM Test for Distillation of Petroleum Products at Reduced Pressure (D1160), cuts are obtained under reduced pressure such as at 10 mm for paraffin waxes or at 1 mm for higher molecular weight waxes. The cuts are taken at intervals across the full distillable range, and the complete results may be reported. In some cases, and for brevity, the distillation results are reported as the temperature difference observed between the 5 % off and 95 % off cut points. This would represent a "width of cut" statement, expressed in degrees, at the reduced pressure. Waxes having very narrow width of cut will tend to be more crystalline, have higher melting points, higher hardness and tensile strength properties, and less flexibility.

The boiling range distribution (D5307) of a wax can also be determined using modern methods of gas chromatography.

MOLECULAR WEIGHT

The molecular weight of various waxes may differ according to (1) the source of the wax (whether it originated in the lighter or heavier grade lubricating oils) and (2) the processing of the wax (the closeness of the distillation cut or the fractionation by crystallization). Thus, the average molecular weight of a wax may represent an average of a narrow or a wide band of distribution.

As a generalization, for any series of similar waxes, an increase in molecular weight increases viscosity and melting point. However, many of the other physical and functional properties are more related to the hydrocarbon types and distribution than to the average molecular weight: the procedure for determination of number average molecular weight is described in later paragraphs.

CRITERIA FOR FOOD GRADE WAX

Federal regulation 21 CFR 172.886 contains a specification that describes food grade wax as a "mixture of solid hydrocarbon, paraffinic in nature, derived from petroleum and refined to meet specifications prescribed by this section."

Besides normal guidelines, this U.S. specification and its counterpart in the European Union impose stringent controls on the polynuclear aromatic hydrocarbon content of food grade waxes for direct use in foods or indirect contact with foods in packaging materials. The controls are to ensure the essential absence of polynuclear hydrocarbons, especially those that may be carcinogenic. The rigorous procedure for estimating polynuclear aromatic content of petroleum waxes is described in the section of this chapter on tests for chemical properties.

ULTRAVIOLET ABSORPTIVITY

For some process control purposes or for research purposes it may be desirable to monitor the total aromatic content of a petroleum wax. ASTM Test for Ultraviolet Absorbance and Absorptivity of Petroleum Products (D2008) provides this capability. Although this procedure shows good operator precision, the interpretation of results requires some caution. Since the test does not include any selective fractionation of the sample, it does not distinguish any particular aromatic. It is also subject to the errors arising from interferences or differences in strong or weak absorptivity shown by different aromatics. Therefore, the test is good for characterization, but cannot be used for quantitative determination of aromatic content or any other absorptive component.

CARBONIZABLE SUBSTANCE

The Readily Carbonizable Substance test is no longer in wide use, although it is still a requirement for paraffin wax in the U.S. Pharmacopeia. The test has some utility as a check on the degree of removal of trace reactive materials in the finishing of the wax products, but it is subject to errors and misleading interpretation. In addition, it does not correlate with specific impurities of interest, nor is it a good quantitative measure.

PEROXIDE CONTENT

The presence of peroxides or oxidized products such as aldehydes or fatty acids is found in wax as a result of oxidation and deterioration of wax either in use or storage. Suitable antioxidants, such as butylated hydroxytoluene and butylated hydroxyanisole, may be used to retard oxidation. ASTM Test for Peroxide Number of Petroleum Wax (D1832) is used to measure the peroxide content of wax.

Criteria for Judging Functional Properties

Testing methods for evaluating functional properties are often empirical in nature and based on the simulated conditions of use. The results of such tests are intended to correlate with practice and are used to determine the suitability of a wax for a particular application. The properties of waxed paper or board depend not only on the individual characteristics of the wax and paper components, but also on the manner in which they interact when they are combined in the waxing process. For this reason, it is necessary to conduct certain functional tests on the finished paper. According to the method of application, the wax may first be coated on one or both surfaces of the paper, or it may be impregnated into the pores and fibers of the paper.

WAX CONTENT OF SUBSTRATES

Many of the functional properties of coated board or wax paper are dependent on the amounts of wax present either on the surface or internally. Surface wax on each or both sides of a weighed specimen can be determined by scraping each side and weighing the specimen after each operation. Internal wax is the difference between total wax content and total surface wax. Total wax content can be determined by solvent extraction or by finding the difference between average weight per unit area of waxed and unwaxed specimens.

BLOCKING PROPERTIES

Blocking tendency is an important quality criterion for waxes, especially those used in various paper and paperboard coating applications. Blocking occurs at moderately warm temperatures when waxed surfaces stick together or block. The surfaces of waxed paper block and the surface appearance, gloss, and barrier properties are destroyed when the papers are separated. The blocking and picking points indicate the temperature range at which waxed film surfaces become damaged if contact is made.

In general, a low resistance to blocking is associated with the presence of low molecular weight and nonnormal (isoparaffinic) fractions in the wax, such as oil or soft waxes. Likewise, broad cut waxes generally show somewhat lower blocking points than shown by narrow-cut waxes of the same average molecular weight or viscosity. In addition, the blocking point of a given wax may also be dependent on the conditions under which the particular wax coating is prepared. For example, subchanges in the crystalline nature of the coating can occur with variations in paper substrate quality, coating application temperature, chilling conditions to set the coating, and particulars of conditioning or aging of the coating. Standard testing procedures are designed to control these factors.

Testing for blocking tendency is done usually with a moderate load (2 kPa) applied to two adjacent waxed paper surfaces that are exposed to the range of temperature across gradient heating plate. The test serves to simulate normal packaging quality needs.

Gloss waxed coatings not only provide protection for packaged goods, but they also provide an attractive appearance because of their high-gloss characteristics. The amount of gloss obtained is determined primarily by the nature of the wax coating and the smoothness of the substrate. Storage of coated paper at elevated temperature tends to be injurious to gloss. Measurements of gloss properties and gloss retention are described in later paragraphs.

SLIP PROPERTIES

Friction tests provide an indication of the resistance to sliding exhibited by two waxed surfaces in contact with each other. The intended application determines the degree of slip that is desired. Coatings for packages that require stacking should have a high coefficient of friction to prevent slippage in the stacks. Folding box coatings should have a low coefficient of friction to allow the boxes to slide easily from a stack of blanks being fed to the forming and filling equipment.

ADHESION

The thermoplastic properties of wax are used to good effect in the heat-sealing of waxed paper packages. The strength of the seal is a function not only of the physical properties but also of the chemical properties and composition of the wax.

In some applications, such as for bread wraps, candy wraps, and cereal and cookie inner bags, it is desirable to have only a moderately strong seal that can be parted easily when the package is opened without any tearing action of the paper itself. For these uses, predominantly paraffinic wax blends are used, and seal testing shows a cohesive failure of the wax. In other applications, such as for various heat-sealed folding cartons, frozen food cartons, and flexible packaging bags, extremely strong and durable seals may be desired. In these cases, the wax formulation contains a major proportion of tacky adhesive and high-tensile components such as the resins and polymers compounded in hot melts. During seal testing the wax cohesive and adhesive strength may be so great that the failure occurs in the paper substrate rather than in the wax.

MOISTURE BARRIER PROPERTIES

The ability of wax to prevent the transfer of moisture vapor is a primary concern to the food packaging industry. Moisture must be kept out to maintain the freshness of dry foods, and moisture must be kept in to maintain the quality of frozen foods and baked goods. Both require different procedures for measuring barrier properties. For example, for dry food packaging, the transmission rate is measured at elevated temperatures and high relative humidity. For frozen food, the transmission rate is measured at low temperatures and low relative humidity.

To have good barrier properties, a wax must be applied in a smooth, continuous film and be somewhat flexible to prevent cracking and peeling. Testing in the "flat" condition provides an assessment of the fundamental barrier capability of the wax coating, for example, the resistance of the microstructure to passage of water vapor. Testing in the "creased" condition gives an evaluation of the strength of the thin wax film in maintaining its integrity even when folded to a sharp crease through 180° and reflattened for test. A strong, flexible, adherent film will survive the creasing with minimal damage, while inferior coatings will lose integrity, crack, or flake off at the crease.

TEST METHODS

Quality criteria for wax are dependent on the type wax and its intended application. Three categories of tests–physical, chemical, and functional–are used to assess the quality of waxes. The following paragraphs describe the most common tests applied to waxes and petrolatums.

Physical Properties of Petroleum Wax

ASTM Test for Melting Point of Petroleum Wax (Cooling Curve) (D87/IP 55)–In this procedure, a thermometer is placed in a sample of molten paraffin wax in a test tube that is mounted into a larger tube containing only air. This arrangement is then placed in a room temperature water bath. As the wax cools, wax temperature versus time is recorded and plotted. Since the latent heat of crystallization released during solidification of the wax is sufficient to temporarily arrest the rate of cooling, a plateau occurs in the time/temperature curve. The temperature at which the plateau occurs is the melting point. This procedure is not suitable for microcrystalline wax, petrolatums, or waxes containing large amounts of nonnormal hydrocarbons because a well-defined plateau rarely occurs in cooling curves of such waxes. It should be noted that many labs have modified the apparatus to use a thermistor and chart recorder to record the cooling curve.

ASTM Test for Drop Melting Point of Petroleum Wax, Including Petrolatum (D127/IP 133)–This method can be used for most petroleum waxes and wax-based hot-melt blends. A chilled thermometer bulb is coated with the molten wax (which is allowed to solidify), placed in a test tube, and heated at a specified rate in a water bath. The melting point is the temperature at which the first drop of liquid falls from the thermometer.

ASTM Test for Congealing Point of Petroleum Wax, Including Petrolatum (D938/IP 76)–This procedure can be used for almost all types of petroleum waxes and hot-melt blends. A thermometer bulb is dipped in the melted wax and placed in a heated flask, serving as a jacket. The thermometer is held horizontally and slowly rotated on its axis. As long as the wax remains liquid, it will hang from the bulb as a pendant drop. The temperature at which the drop rotates with the thermometer is the congealing point. The congealing point of a microcrystalline wax or petrolatum is invariably lower than its drop melting point.

ASTM Test Method for Solidification Point of Petroleum Wax (D3944)–This test describes a rapid, instrumental method for determining the solidification point of petroleum wax.

ASTM Test for Kinematic Viscosity of Transparent and Opaque Liquids (and the Calculation of Dynamic Viscosity) (D445/IP 71)–Kinematic viscosity is measured by timing the flow of a fixed volume of material through a calibrated capillary at a selected temperature. The unit of kinematic viscosity is m^2/s, and kinematic viscosities of waxes are usually reported in mm^2/s. These SI units can be converted to Saybolt Universal Seconds (SUS) by using ASTM Conversion of Kinematic Viscosity to Saybolt Universal Viscosity or to Saybolt Furol Viscosity (D2161).

ASTM Test for Apparent Viscosity of Petroleum Waxes with Additives (Hot Melts) (D2669)–This method is suitable for blends of wax and additives having apparent viscosities up to 20,000 mPa/s at 176.7°C. Apparent viscosity is the measurement of drag produced on a rotating spindle immersed in the test liquid. A suitable viscometer is equipped to use interchangeable spindles and adjustable rates of rotation. The wax blend is brought to test temperature in a simple arrangement using an 800-mL beaker in a heating mantle. Viscosities over a range of temperatures may be recorded and plotted on semi-log paper to determine the apparent viscosity at any temperature in the particular region of interest.

ASTM Test for Apparent Viscosity of Hot Melt Adhesives and Coating Materials (D3236)–This procedure is similar to ASTM D2669 since it measures the ratio of shear stress to shear rate through the use of a rotating spindle. However, ASTM D3236 is applicable to hot melts of a broader range of viscosity. Through the use of a special thermally controlled sample chamber, hot melts may be tested at temperatures of 176.7°C or more and at viscosities of 20,000 mPa/s or more.

ASTM Test for Needle Penetration of Petroleum Waxes (D1321)–The hardness or consistency of wax is measured with a penetrometer applying a load of 100 g for 5 s to a standard needle having a truncated cone tip. To prepare a wax for testing, the sample is heated to 16.7°C above its congealing point, poured into a small brass cylinder, cooled, and placed in a water bath at the test temperature for 1 h. The sample is then positioned under the penetrometer needle, which, when released, penetrates into the sample. The depth of penetration in tenths of a millimeter is reported as the test value. This method is not applicable to oily materials or petrolatums that have penetrations greater than 250.

ASTM Test for Cone Penetration of Petrolatums (D937/IP 179)–This test is used for soft waxes and petrolatums. It is similar to D1321 except that a much larger sample mold is used and a cone weighing 102.5 g replaces the needle. The method requires that a 150-g load be applied for 5 s at the desired temperature.

ASTM Standard Practice for Density, Relative Density (Specific Gravity), or API Gravity of Crude Petroleum and Liquid Petroleum Products by Hydrometer Method (D1298)–Density is determined by using a hydrometer at a convenient temperature and converting to relative density at 15°C (specific gravity) or API gravity at 60°F by using international standard tables.

ASTM Standard Test Method for Flash and Fire Points by Cleveland Open Cup (D92)–This test method describes the determination of the flash and fire point of petroleum products by a manual Cleveland open-cup apparatus or an automated Cleveland open-cup apparatus.

ASTM Standard Test Methods for Flash Point by Pensky-Martens Closed-Cup Tester (D93)–This method covers the determination of the flash point of petroleum products in the temperature range from 40 to 360°C by a manual Pensky-Martens closed-cup apparatus or an automated Pensky-Martens closed-cup apparatus.

Chemical Properties of Petroleum Wax

ASTM Test for Saybolt Color of Petroleum Products (Saybolt Chromometer Method) (D156)–Saybolt color is determined on nearly colorless waxes by putting the melted sample in a heated vertical tube that is mounted alongside a second tube containing standard color disks. An optical viewer allows simultaneous viewing of both tubes. Light is reflected by a mirror up into the tubes and to the viewer. The level of the sample in the column is decreased until its color is lighter than that of the standard. The color number above this level is reported.

ASTM Color of Petroleum Products (ASTM Color Scale) (D1500/IP 196)–This procedure is used for waxes and petrolatums that are too dark for the Saybolt colorimeter. Using a standard light source, a liquid sample is placed in the test container and compared with colored glass disks ranging in value from 0.5 to 8.0. If an exact match is not found and the sample color falls between two standard colors, the higher of the two colors is reported.

ASTM Test for Odor of Petroleum Wax (D1833/IP 185)–A preselected panel determines the odor of the petroleum wax. Ten grams of wax are shaved and placed in an odor-free glass bottle and capped. After 15 min the sample is evaluated in an odor-free room by removing the cap and sniffing lightly. A rating of 0 (no odor) to 4 (very strong odor) is given by each member of the panel. The reported value is the average of the individual ratings.

ASTM Test for Oil Content of Petroleum Waxes (D721/ IP 158)–The fact that oil is much more soluble than wax in methyl ethyl ketone at low temperatures is utilized in this procedure. A weighed sample of wax is dissolved in warm methyl ethyl ketone in a test tube and chilled to −31°C to precipitate the wax. The solvent-oil solution is separated from the wax by pressure filtration through a sintered glass filter stick. The solvent is evaporated, and the oil residue is weighed. The method is applicable to waxes containing not more than 15 % oil.

ASTM Test for Solvent Extractables in Petroleum Wax (D3235)–This method is very similar to ASTM D721/IP 158 except that the solvent used is a 1:1 mixture of methyl ethyl ketone and toluene. The method may be used on waxes having high levels of extractables, for example, 15 to 50 % extractables.

ASTM Test for Distillation of Petroleum Products at Reduced Pressure (D1160)–In this standard test the sample is distilled at an accurately controlled reduced pressure (usually 10 or 1 mm) in a distillation flask and column that provides approximately one theoretical plate fractionation. The apparatus includes a condenser and a receiver for collection and measurement of the volume of cuts obtained. The data are reported as the series of vapor temperatures observed (at the reduced pressure) for the specified intervals of volumetric percentage distilled.

ASTM Test for Boiling Range Distribution of Petroleum Fractions by Gas Chromatography (GC) (D2887)–A sample of the test wax is dissolved in xylene and introduced into a GC column that is programmed to separate the hydrocarbons in boiling point order by raising the temperature of the column at a reproducible, calibrated rate. When wax samples are used, a flame ionization detector is used to measure the amount of eluted fraction. Data obtained in this procedure are reported in terms of percentage recovered at certain fixed temperature intervals. Full description of the carbon number distribution cannot be obtained with this method because the relatively short column used does not provide sufficient resolution to distinguish individual peaks. In addition, the method is not applicable to higher chain-length paraffin waxes or to microcrystalline waxes.

ASTM Test for the Analysis of Petroleum Waxes by Gas Chromatography (D5442)–This method is similar to D2887 but supplies more detailed information about a wax by running the GC instrument at conditions that allow for the full resolution of paraffin wax individual peaks.

ASTM Test for Molecular Weight of Hydrocarbons by Thermoelectric Measurement of Vapor Pressures (D2503)–In this method, a small sample of wax is dissolved in a suitable solvent, and a droplet of the wax solution is placed on a thermistor in a closed chamber in close proximity to a suspended drop of the pure solvent on a second thermistor. The difference in vapor pressure between the two positions results in solvent transport and condensation onto the wax solution with a resultant change in temperature. Through suitable calibration, the observed effect can be expressed in terms of molecular weight of the wax specimen as a number average (Mn).

Regulatory Test for Food Grade Petroleum Wax: 21 Code of Federal Regulations 172.886–The polynuclear aromatic content of waxes can be estimated by the ultraviolet absorbance of an extract of the sample when scanned in the 280- to 400-nm range. The sample is first dissolved in isooctane and extracted using a dimethylsulfoxide-phosphoric acid solution to concentrate the aromatic material and remove it from the saturated hydrocarbons. The extract portion is tested for ultraviolet absorbance against specified maximum limits. These limits were established as being indicative of the maximum absorption peak areas associated with polynuclear aromatic hydrocarbons. If the wax sample under test falls within these limits, the sample is considered to pass, and no further continuation of the analysis is necessary. If, however, the results in this first phase of testing exceed any of the limits, there is a possibility that the high result may be interference due to single ring aromatic derivatives (these are highly absorptive, especially at lower wavelengths and would normally occur along with the polynuclear aromatics). The benzenoid structures are considered noncarcinogenic, and their interference in the test must be eliminated. This is done by continuing the sample extract into a second phase of analysis using a chromatographic separation procedure. The resulting concentrate of polynuclear aromatics is tested against the same specified limits for ultraviolet absorbance.

ASTM Test for Ultraviolet Absorbance and Absorptivity of Petroleum Products (D2008)–The procedure tests the product as a whole, without including any separation or fractionation steps to concentrate the absorptive fractions. When wax or petrolatum is tested in this procedure, the specimen is dissolved in isooctane, and the ultraviolet absorbance is measured at a specified wavelength such as 290 nm. The absorptivity is then calculated. This procedure, as such, is not a part of the federal specification.

ASTM Test for Carbonizable Substances in Paraffin Wax (D612)–Five milliliters of concentrated sulfuric acid are placed in a graduated test tube, and 5 mL of the melted wax is added. The sample is heated for 10 min at 70°C. During the last 5 min the tube is shaken periodically. The color of the acid layer is compared with the color of a standard reference solution. The wax sample passes if the color is not darker than the standard.

ASTM Test for Peroxide Number of Petroleum Wax (D1832)–In this test, a sample is dissolved in carbon tetrachloride, acidified with acetic acid, and a solution of potassium iodide is added. Any peroxide present will react with the potassium iodide to liberate iodine that is then titrated with sodium thiosulphate.

ASTM Standard Test Method for Measurement of Transition Temperature of Petroleum Waxes by Differential Scanning Calorimetry (DSC) (D4419)–This test method measures the transition temperatures of petroleum waxes, including microcrystalline waxes, by differential scanning calorimetry (DSC). These transitions may occur as a solid-solid transition or as a solid-liquid transition. The DSC is a very useful instrument for the characterization of waxes.

Functional Properties of Petroleum Wax

In the manufacture of wax-treated paper packaging, a variety of substrates are treated, and a variety of waxing methods and equipment are used. In some cases, commercial specimens of waxed packaging materials may be submitted to functional tests. Such testing serves to evaluate the total system, that is, the wax, paper, and coating condition. When it is desired to test the functional property of the wax itself or to compare it with other waxes apart from the variables of substrate and waxing process, a standard waxed specimen must be prepared under predetermined and controlled conditions. Several of the functional test methods include instructions for specimen waxing that usually employ standard paper substrate and standard coating conditions.

ASTM Test for Surface Wax on Waxed Paper or Paperboard (D2423)–This method determines the amount of wax present as a surface film on a substrate but not the absorbed wax. A waxed paper sample is cut to size and weighed; wax is carefully scraped from one surface with a razor blade and then the sample is reweighed to determine the amount of wax removed. The process is repeated on the reverse side if total surface wax is desired.

ASTM Test for Total Wax Loading of Corrugated Paperboard (D3344)–In this method, taking together all wax that may be present as impregnating or saturating or coating wax, the total wax loading of corrugated board is measured. The determination is made by extracting the waxed board specimen with warm solvent and evaporating to dryness to obtain extracted wax. The resulting residue will include any soluble additive materials associated with the wax, such as the polymeric additives normally used in the wax-based hot-melt coatings.

ASTM Test for Blocking and Picking Points of Petroleum Wax (D1465)–Two strips of wax coated paper, 1 in. wide, are placed face to face between two uncoated strips of paper on a calibrated temperature-gradient heating plate. The specimens are covered with foam rubber strips and steel bars to provide a moderate pressure loading and then subjected to the gradient heat of the blocking plate for 17 h. The specimens are removed, cooled, and peeled apart. The picking point is the lowest temperature at which the surface film shows disruption. The blocking point is the lowest temperature causing disruption of surface film over an area of 50 % of the strip width.

ASTM Test for Coefficient of Kinetic Friction for Wax Coatings (D2534)–A wax-coated paper is fastened to a horizontal plate attached to the lower, movable cross arm of an electronic load cell type tension tester. A second paper is taped to a 180-g sled that is placed on the first sample. The sled is attached to the load cell through a pulley. The kinetic coefficient of friction is calculated from the average force required to move the sled at 90 cm/min divided by the sled weight.

ASTM Test for Abrasion Resistance of Wax Coatings (D3234)–The abrasion resistance of a smooth wax coating on paper or paperboard is measured by determining how much change in gloss of the surface occurs under the abrading action of sand particles. A measured quantity of sand is allowed to impinge under specified conditions on the waxed surface, and the change in gloss is noted using ASTM Method D1834.

Bibliography

Annual Book of ASTM Standards, Vols. 05.01, 05.02, 05.03, ASTM International, West Conshohocken, PA.

Bennett, H., *Industrial Waxes*, Vol. I and II, Chemical Publishing Company, Inc., New York, 1963.

Petroleum Waxes: Characterization, Performance, and Additives, Technical Association of the Pulp and Paper Industry, Special Technical Association Publication No. 2, 1963.

Technical Evaluation of Petroleum Waxes, Technical Association of Pulp and Paper Industry, Special Technical Association Publication No. 6, 1969.

USP25-NF20, United States Pharmacopeial Convention, Inc. Rockville, MD.

Warth, A. H., *Chemistry and Technology of Waxes*, Reinhold Publishing Corp., New York, 1956.

Sequeira, A., Jr., *Lubricant Base Oil and Wax Processing*, Marcel Dekker, Inc. New York, 1994.

21

Methods for the Environmental Testing of Petroleum Products

Mark L. Hinman[1]

THE POTENTIAL FOR FUEL AND LUBRICANT MATE-rials to cause harmful effects in the environment needs to be evaluated. A reliable approach to evaluate the inherent hazard of a material to the environment is by testing the individual chemical properties in the environmental compartments of concern in separate, carefully controlled laboratory test systems:

- The media exchange rates of a material may be deduced from measurements of key physical properties.
- The degradation rates under specific conditions may be determined in tests for biodegradability, hydrolysis, photolysis, and photooxidation.
- The toxicity to or accumulation by organisms may be determined in single species tests using organisms of known number, age, health, and under constant environmental conditions.
- The specific test method to be used, the test organisms, and the data quality management of the testing are defined by the use of the data. In general, these test data are developed for submissions related to the registration of products to satisfy regulatory requirements, that allow the discharge of municipal and industrial wastes, or environmental (or ecological) risk assessment in response to perceived environmental contamination. This discussion provides a general overview of this testing and does not address specific requirements for individual test methods.

USES OF ENVIRONMENTAL CHARACTERISTICS

Registration of Products

PREMANUFACTURING/MARKETING NOTIFICATIONS

During the development of a chemical as an intermediate or a final product, it is necessary to consider the potential impact of that chemical on the environment. Corporate industrial decisions to proceed with the development or import of a material may include consideration of the potential effects of its manufacture, transport, use, and disposal on the environment. In the United States, government authorization for the manufacture of new chemicals is regulated under the Toxic Substances Control Act (TSCA), which requires a premanufacturing notification (PMN) be submitted for U.S. Environmental Protection Agency (EPA) review before full-scale manufacture of the material [1]. In Canada, the New Substances Provisions of the Canadian Environmental Protection Act [2] perform an equivalent role. In the European Union (EU), Council Directive 67/548/EEC–Classification, Packaging, and Labeling of Dangerous Substances (with its many amendments) drives the premarketing notification process [1].

The environmental characterization data required by each regulatory body are very specific in scope and test methods (Table 1). In order to formalize the international acceptance of data, the Organization for Economic Cooperation and Development (OECD) Decision Concerning the Mutual Acceptance of Data in the Assessment of Chemicals (i.e., the MAD decision) was adopted in 1981 and is legally binding on all member countries.[2] The MAD decision states that "Data generated in the testing of chemicals in an OECD member country in accordance with OECD Test Guidelines and OECD Principles of Good Laboratory Practices shall be accepted in the other Member countries for purposes of assessment and other uses relating to the protection of man and the environment" [3]. In an effort to meet both the requirements of the TSCA regulatory language and the MAD decision, the U.S. EPA has coordinated the original test methods supporting TSCA PMN submissions to increase the alignment with OECD test methods.

CLASSIFICATION AND LABELING

In the EU, Council Directive 67/548/EEC (sometimes known as the Dangerous Substances Directive) also requires that the party responsible for placing chemical substances on the EU market must classify them based on their intrinsic dangerous properties. Additionally, the substances must be labeled accordingly with required danger symbols, risk ("R"), and safety ("S") phrases indicting the risks involved and the safety measure to be taken when using the substance. This directive addresses materials that are composed of single chemical substances or complex chemicals that are composed of substances that are grouped together because they are normally only produced (or extracted) and used as mixtures (e.g., basestocks). In May 1999, the EU adopted Directive 1999/45/EC on the classification, packaging, and labeling of dangerous preparations that replaced the previous Dangerous Preparations Directive 88/379/EEC, as amended. The new directive introduced the classification "dangerous to the environment" for preparations and determined the criteria for evaluating the environmental properties of preparations, principally through conventional calculation methods. This classification was already applicable to substances [2].

[1] *Environmental Scientist, ExxonMobil Biomedical Sciences, Inc., Annandale, NJ 08801.*

[2] OECD member countries as of February 2003: Australia, Austria, Belgium, Canada, Czech Republic, Denmark, Finland, France, Germany, Greece, Hungary, Iceland, Ireland, Italy, Japan, Korea, Luxembourg, Mexico, Netherlands, New Zealand, Norway, Poland, Portugal, Slovak Republic, Spain, Sweden, Switzerland, Turkey, United Kingdom, and United States.

TABLE 1—Test Methods Commonly Used to Determine Relevant Environmental Characteristics of Materials in the PMN Process

Characteristic	USA-TSCA	European Union	Canada
Physicochemical			
Octanol/water partition coefficient	830.7550	OECD 107	OECD 107
	830.7560	OECD 117	OECD 117
	830.7570		
Water solubility	830.7840	OECD 105	OECD 105
	830.7860		
Vapor pressure	830.7950	OECD 104	OECD 104
Adsorption/desorption	. . .	OECD 106	OECD 106
Dissociation constant	830.7370	OECD 112	OECD 112
Persistence			
Hydrolysis	835.2110	OECD 111	OECD 111
	835.2130		
Photolysis	835.2210	. . .	. . .
	835.2310		
	835.5270		
Ready biodegradation	835.3110	OECD 301	OECD 301
		OECD 310	OECD 310
		OECD 311	OECD 311
Inherent biodegradation	835.3100	OECD 302	OECD 302
	835.3120	OECD 304	OECD 304
	835.3200		
	835.3400		
	835.5045		
Simulation biodegradation	. . .	OECD 303	. . .
		OECD 309	
Bioaccumulation in fish	850.1730	OECD 305	. . .
Environmental Toxicity			
Acute toxicity to fish	850.1075	OECD 203	EPS I/RM/9
	850.1085		OECD 203
Acute toxicity to *Daphnia*	850.1010	OECD 202	EPS I/RM/11
			OECD 202
Growth-inhibition test on algae	850.5400	OECD 201	EPS I/RM/25
Bacterial inhibition	850.6800	OECD 209	OECD 209
Daphnia reproduction (21 days)	850.1300	OECD 211	OECD 211
Test on higher plants	850.4100	OECD 208	OECD 208
	850.4150		
	850.4200		

TABLE 1—Test Methods Commonly Used to Determine Relevant Environmental Characteristics of Materials in the PMN Process (Continued)

Characteristic	USA-TSCA	European Union	Canada
	850.4225		
	850.4230		
	850.4250		
	850.4300		
	850.4400		
Test on earthworms	850.6200	OECD 207	OECD 207
Subacute toxicity on fish (14 days)	850.1400	OECD 204	OECD 204
Toxicity studies with birds	850.2100	OECD 205	OECD 205
	850.2200	OECD 206	OECD 206
	850.2300		

The classification "dangerous to the environment" required that specific environmental characteristics be determined. Generally, the testing that was reported in the PMN review is adequate to determine the risk phrases. As a result, the data for many PMNs are developed using the OECD test methods to allow the greatest amount of flexibility in data use.

Regulated Discharge to the Environment

The production of products results in some loss of the material into the environment. The environmental characteristics of the material define the environmental compartment(s) into which the material will move (soil, air, water, and sediment). The compartments most frequently considered when there is a discharge from a facility are the atmospheric and aquatic environments. This discussion will not address atmospheric discharges. Effluent content is regulated and compliance is monitored via permits that include regularly performing bioassays on the discharged effluent.

The regulatory body with jurisdiction over the facility specifies the test methods that will be used for the bioassays. In the United States, the U.S. EPA generally requires that the methods in 600/4-90/027F, 600/4-91/003, or 600/4-91/002 be used. In some cases, the state or regional body will have jurisdiction and require its own set of tests. Environment Canada has produced a series of test methods for effluent evaluation. In many regions of the world, the local test methods are based on either the U.S. EPA, Environment Canada, or OECD test methods (see box at end of chapter).

General Environmental Contamination

Toxicity testing with either standardized or indigenous species and site-specific conditions can be used to evaluate the potential environmental hazards and risks of an unanticipated, accidental release of a chemical into the environment. In these cases, the testing guidance is often less structured, and a variety of test methods or approaches are used. Common sources for test methods are ASTM and ISO (see box at end of chapter). Frequently, standards from one or both of these organizations will serve as the basis for a regional or local test method.

ENVIRONMENTAL CHARACTERISTICS

Physical Properties Important to Environmental Characterization

A number of physical properties are especially relevant to the environmental behavior of substances. Some of these characteristics are measured on a routine basis for products. However, complex mixtures often require more detailed information about the range of the property than pure substances. For example, with a product that is a mixture of components with greatly differing boiling points, the boiling range is typically reported as an average. When the boiling range is broad, there is also a broad range in vapor pressures of the components. This can lead to differences in the environmental compartmentalization of the components. The more volatile components will partition into the air phase to a greater extent and so will be present to a lesser extent in the soil and aquatic environments than the less volatile components. Consequently, for substances with multiple components, specific test methods need to be applied and reported that measure the range rather than the average of a property.

CHEMICAL STRUCTURE

The properties of chemicals depend upon their chemical structures. For fuels, this may involve a number of hydrocarbon structures having a range of molecular weights and chemical types, as well as a variety of organic and inorganic compounds included in additive treatments. For lubricants, mineral hydrocarbon basestocks, synthetic basestocks, or plant oil basestocks may be involved as well as a variety of additives.

OCTANOL/WATER PARTITION COEFFICIENT (K_{OW} OR P_{OW})

The octanol/water partition coefficient is the ratio of a chemical's solubility in *n*-octanol and water at steady state. Due to the wide range of measured values, K_{ow} is usually expressed as the logarithm (log K_{ow} or log P_{ow}). The K_{ow} is a characteristic of significant importance in environmental studies. It is a common surrogate for bioconcentration for classification determinations and is a key parameter in QSAR calculations. There are three basic approaches to the determination of K_{ow}:

- The original and reference method is the "shake flask method." An aqueous solution of the material is equilibrated with *n*-octanol. After equilibrium is reached, the water and the *n*-octanol phases are separated and analyzed for the material. The ratio of the material concentration in *n*-octanol to water is the K_{ow} (OECD 107, EPA 830.7550) [4].
- The HPLC (high-pressure liquid chromatography) method measures the retention time of the material on a hydrophobic column and compares it with the retention times of a number of standards with known K_{ow} values (ASTM E1147, OECD 117, EPA 830.7570). The method is based on the correlation of the retention time with the K_{ow}.
- The QSAR estimation method involves the calculation of K_{ow} based upon the known, empirical contribution of the various structural fragments of the molecule to the K_{OK} value. This approach generally gives good results for many hydrocarbons [5].

WATER SOLUBILITY

The persistence, biodegradation, and toxicity of a material in the environment are very dependent upon its water solubility. The measurement of this characteristic can be very straightforward:

- Increasing additions of the pure material are made to water (either directly or via a generator column) and the water analyzed until saturation is reached (ASTM E1148, OECD 105, EPA 830.7840, EPA 830.7860). The analysis must be specific to the material.
- A QSAR method is available for hydrocarbons that generally provide good water solubility estimates.

For complex mixtures, the individual components in the mixture have different water solubilities. Once the least soluble component saturates the water, an undissolved phase begins to form (typically at the water's surface). The remaining components will partition between the water and the undissolved phase and therefore never reach their individual water solubility limit. The composition of the water extract will be different from the composition of the complete mixture as it is also dependent on the ratio of material volume to the water volume. This solubility behavior is very complex. It should be remembered that the definition of solubility as it relates to pure substances does not readily apply to complex mixtures [6].

VAPOR PRESSURE

The vapor pressure of a substance is a measure of its tendency to enter the gas phase. This is important for determining both the rate of evaporation and the relative amount of the substance that will be in the air phase. For pure materials, direct pressure measurements may be used (ASTM E1194, OECD 104, EPA 830.7950). For materials with very low vapor pressure values or for a mixture, a vapor saturation method is used. The stationary medium in a generator column is coated with the material and air is cycled over the medium until it is saturated [7].

HENRY'S LAW CONSTANT

Henry's Law Constant (HLC or H) is the partition coefficient of a material between air and water. It is important in estimating the degree and rate of volatilization of a material from water into air. It may be measured directly by analytical means within a test system or taken as the ratio of the vapor pressure to the molar water solubility. The dimensions for this constant vary with the units used to define the concentration or pressure of the material in the gas phase and the concentration in the liquid phase. If the same units are used for both terms, the constant becomes a proportion with no dimensions. For complex mixtures, each component has a different water solubility and vapor pressure, so the HLC may cover a wide range.

ADSORPTION/DESORPTION

Adsorption is the taking up of a gas, vapor, or dissolved material on the surface of a solid. Desorption is the release of a gas, vapor, or dissolved material from a solid under the same conditions. The adsorption of organic chemicals from water into the organisms, sediment, or soil plays an important role in the fate and transport of these materials. For organic chemicals, the important determinant of adsorption is the organic matter content of the soil or sediment. As a result, the values for various soils or sediments are "normalized" by calculating the partitioning coefficient between water and soil organic matter (K_{oc}) (ASTM E1195, OECD 106, OECD 121). An estimate of the K_{oc} may be calculated directly from the K_{ow}.

DISSOCIATION CONSTANT

Some components of both fuels and lubricants dissociate into ionic form in an aquatic environment. Dissociation greatly influences all other environmental processes and properties such as sorption, bioconcentration, and toxicity [8]. Ions cannot partition into the air or nonpolar environments. Additionally, the ionic species may have very different aquatic toxicity from the un-ionized form as a result of different mechanisms to transport the molecule across cell membranes into the organism. Dissociation constants can be determined by OECD 112 or EPA 830.7370.

Persistence and Accumulation Properties Important to Environmental Characterization

Upon entering the environment, a material may be transported within and between environmental media (also called compartments or phases) and respond to ambient factors in the environment. The final equilibrium distribution of a material between the air, water, sediment, soil, and biota significantly influences the potential impact of the material within each media. The primary categories that affect the fate of materials are as follows:

- Transformation (biotic, abiotic)
- Accumulation (bioaccumulation, bioconcentration)

TRANSFORMATION PROCESSES

Standardized tests have been developed to allow some prediction of the degradability of materials in the environment. These are generally divided into chemical (abiotic) degradation and biodegradation tests. Standardized tests for determining the degradation rates are available from many standards developing organizations (e.g., ASTM, ISO, etc.), international consortia (e.g., OECD, CONCAWE, etc.), and regulatory bodies (e.g., U.S. EPA, Environment Canada, European Union, etc.).

Abiotic degradation of fuels and lubricants is primarily due to photolysis, hydrolysis, and photooxidation. These reactions convert the chemical into degradation products, but usually do not totally convert it to the inorganic forms (e.g., CO_2, H_2O). Transformation of a material may result in a change in toxicity and physical characteristics. Toxicity of the product(s) may change, but is not necessarily reduced. Thus, it is useful to understand the degradation products, if possible, to be able to evaluate their impact on the environment. There is much less known about the toxicity of these products.

Biodegradation is much more complicated than abiotic degradation. Microbial organisms may modify a material in a variety of ways, all of which can be considered biodegradation. Only certain microorganisms can accomplish it and not all microorganisms can degrade all materials in all environments. In some cases, a microorganism may be able to degrade only the products of a previous biodegradation process. The multitude of organisms and the highly variable genetic makeup of the microbial communities usually result in at least some microorganisms that can degrade an organic material to some extent. The outcome of this degradation may range from a simple chemical modification to a complete utilization by the organism for growth of new cells or for energy. The type and number of organisms are as important to the outcome as the structure of the material.

ABIOTIC DEGRADATION

- **Hydrolysis** Hydrolysis is a chemical transformation process in which an organic molecule, R-X, reacts with water to form a new carbon-oxygen bond and cleaving a carbon-X bond in the original molecule. It is a well-understood chemical process, and the types of chemicals that hydrolyze and the catalysis of the process have been studied extensively [8]. The standard tests for hydrolysis involve preparing aqueous solutions of the material at pH 4, 7, and 9, incubating at 50°C in the dark, and analyzing the remaining concentration at various time intervals (ASTM E895, OECD 111, EPA 835.2110, EPA 835.2130). Stability for more than a week indicates no significant hydrolysis has occurred.

  The bonds in most hydrocarbons do not tend to hydrolyze. In general, only those materials with ester bonds are expected to be affected by this process. Hydrolysis half-lives can range from days to years for these materials, with the rate of hydrolysis of organic compounds increasing with temperature in an exponential relationship [8]. Hydrolysis in not expected to be a major fate pathway for hydrocarbon-based fuels and lubricants, while synthetic and vegetable-oil–based lubricants may be more susceptible.
- **Photolysis** Photolysis is a light-initiated transformation reaction that is a function of the incident light energy (typically at wavelengths between 290 and 400 nm) and the structure of the material. Different classes of molecules absorb sunlight at different wavelengths and with different efficiencies. Molecules need to have a strong adsorption band in this region to undergo significant photolysis in the environment, so the UV/visible spectrum of chemicals is used to screen for the likelihood of photolysis (ASTM E896, EPA 835.2210, EPA 835.2310). It is not necessary that the material directly absorb the light to undergo photolysis. With indirect (or sensitized) photolysis, an intermediary chemical becomes energized and then energizes the material of interest (EPA 835.5270). Generally, phenols, furans, aromatic amines, sulfides, and nitro-aromatics are susceptible to indirect photolysis [8].

  Some polynuclear aromatic hydrocarbons (PAHs) can undergo a light-mediated increase in toxicity. Some intermediate weight PAHs, such as anthracene, fluoranthene, and pyrene, have demonstrated photo-induced toxicity (greatly enhanced toxicity when compared to the inherent toxicity of the material). When the PAH is present in the tissues of an organism, phototoxicity can occur when UV radiation is absorbed by the molecule.
- **Photooxidation** Atmospheric photooxidation is the degradation of a chemical in air due to reaction with ozone or hydroxyl radicals and is dependent upon the structure of the chemical, its concentration, and the ozone/hydroxyl radical concentration. Hydroxyl radical attack is the predominant mechanism for alkanes, olefins, alcohols, and simple aromatics, which are rapidly oxidized in air [8]. Hydroxyl radicals are produced through the interaction of sunlight with oxygen and other substances in the atmosphere. The volatility of hydrocarbons and the rapid oxidation in air through hydroxyl radical attack promotes photooxidation as a significant fate process for many hydrocarbons. There is a significant amount of test data on the rate of hydroxyl radical reaction for chemicals that has allowed for the development of effective QSAR models that calculate atmospheric half-lives [9].

BIODEGRADATION

As indicated by the name, biodegradation is the biologically mediated transformation of a material. The microorganisms that accomplish the biodegradation are very important to the nature of the result. Generally, the interest in studying biodegradation is to use the results to predict the environmental fate of a chemical. Thus, the microorganisms (the inoculum) used in laboratory tests are mixtures of species collected from particular environments. For a variety of reasons, sewage sludge has become the standard source of mixed inoculum for introduction into laboratory biodegradation test systems.

In the aquatic environment, bacterial numbers are lower than in sewage. Bacteria are also exposed to low concentrations of a variety of chemicals and nutrients. In the standardized tests, the organisms are constrained within a small volume and presented with a high concentration of the test material as the sole source of carbon. Those organisms capable of degrading the chemical will use it for new cells and energy. Preexposure to petroleum products can result in an increase of petroleum-degrading microorganisms from 1 to 10 % of the total population [10]. The rate of this biodegradation reflects the growth rate of those particular organisms under the conditions of the test. Although the kinetics are completely defined by the test and are not easily related to a rate in the environment, the test results are used to classify the persistence of the material and estimate its half-life in the environment for many regulatory programs.

Certain environmental conditions (e.g., availability of oxygen) greatly influence the types of organisms and their metabolic pathways. Standardized tests exist for aerobic and anaerobic conditions in the major environmental compartments (freshwater, marine water, sediments, and soils). In some cases, simulation tests are intended to replicate the natural conditions of the organisms, their environment, and their exposure to the chemical.

TABLE 2—Types and Characteristics of Biodegradation Tests

Attribute	Categories	Measurement	Properties
Extent of biodegradation	Ultimate	O_2	• Measures total conversion to inorganic forms (e.g., CO_2, etc.) • Mineralization
		CO_2	
		CH_3	
	Primary	Specific analysis	• Based on analysis of specific chemical (or chemicals) or chemical class • Abiotic losses controlled
	Removal	Specific analysis	• Primary biodegradation • Both abiotic and biodegradation
Rate of biodegradation	Ready	O_2	• Regulatory definition of "rapid" biodegradation. • Uses <30 mg/L nonadapted inoculum • Reaches minimum 60 % degradation to O_2 or CO_2 or 70 % removal of DOC in 28 days • Must go from 10 % to the pass level (60 or 70 %) in 10 days
		CO_2	
		DOC	
	Inherent	O_2	• Enhanced conditions to show possibility of eventually biodegrading • Generally a high biomass adapted inoculum
		CO_2	
		CH_3	
		DOC	
	Simulation	Specific analysis	• Reflects actual environmental behavior • Difficult to simulate most actual exposure situations

Certain standardized tests have been developed for regulatory purposes to evaluate both the rate and extent of biodegradation (Tables 2 and 3). The main terms used to describe biodegradation (*primary* and *ultimate*) distinguish between two extents of biodegradation. *Primary* refers to the initial transformation from the parent material and ultimate refers to mineralization of the material. Rate is described by the terms *inherent* and *Ready*. In general, to be classified as inherently biodegradable, there must be unequivocal evidence of biodegradation by any test method. The OECD methods require a 20 % degradation for this classification. "Ready" biodegradability is a regulatory classification originating in the European Union that has very specific criteria. The tests are characterized by low initial biomass (<30 mg/L) of sewage organisms that have not been previously exposed (adapted) to the material. The substance must meet the criterion of 60 % degradation to CO_2 or O_2 (or 70 % removal of dissolved organic carbon [DOC]) within 28 days. Further, the pass criteria must be met within 10 days after the biodegradation exceeds 10 % of the test material mass loaded (OECD 301, EPA 835.3110). The Ready biodegradation tests are considered so stringent that if a substance passes the test, it will also rapidly degrade under most environmental conditions. If a material does not pass a Ready test, it does not mean that it will not degrade in the environment.

A number of Ready biodegradability tests are available and differ mainly in the method of analysis (Table 3). The different assays often give different results. The analytical methods dictate details of the procedures and result in different applicabilities for each test. Generally, the disappearance of DOC is more rapid than the utilization of O_2 or the production of CO_2.

Comparing the results from different test methods for the same or similar materials can be problematic. The "percent age biodegradation" is very dependent upon what type of biodegradability was tested. A result of 100 % primary degradation in an inherent test cannot be compared to the results of a Ready test, where 100 % mineralization is not possible and where conditions are much more stringent. Even within similar test methods, the inoculum makes a great deal of difference in the outcome of the test. The number of competent organisms greatly affects the duration of the lag phases. If the number of these organisms in sewage sludge is low, different tests could show either good biodegradation or none at all. The microorganisms in the inoculum may vary significantly over time. The tests for biodegradation, even using a "standard" inoculum, will give much more variable results than the tests for physical properties or toxicity. As a result of this and various complications resulting from use of inappropriate tests for the physical characteristics of the material, the comparison of biodegradability between materials is only quantitative if they are tested in the same test systems using the same inoculum prepared at the same time. The major use of biodegradation tests is to provide an estimate of the potential of a material to degrade in the environment.

ACCUMULATION PROCESSES

Accumulation processes result in the transfer of a substance from an environmental medium (e.g., air, water, soil, and sediment) to a plant or animal. Materials that pose the greatest concern for accumulation potential enter the environment from diffuse sources, are poorly degraded in the environmental medium, exhibit a high affinity for lipids, and are resistant to the metabolic breakdown by plants and

TABLE 3—Examples and Descriptions of Relevant Biodegradation Test Methods

			Biodegradation			Suitability for Materials that Are:	
Test	**Standard/ Reference**	**Environment**	**Extent**	**Rate**	**Inoculum**	**Poorly Soluble**	**Volatile/ Gas**
DOC die-away	OECD 301 A ISO 7827	Aquatic	Ultimate primary	Ready	• WWTP[A] secondary effluent + activated sludge + surface water • Preexposure is option • 10^7 to 10^8 cells/L	No	No
CO_2 generation (Modified Sturm Method)	OECD 301 B ASTM D5864 ISO 9439 U.S. EPA 835.3110	Aquatic	Ultimate	Ready	• WWTP secondary effluent + activated good sludge + surface water • Preexposure is option • 10^5 to 10^6 cells/L	Good	No
CO_2 generation (Gledhill Method)	ASTM D6139	Aquatic	Ultimate	Ready	• Activated sludge (<25 % from industry) *or* soil *or* surface water • Preexposure is option • 10^7 to 10^8 cells/L	Good	No
MITI (I and II)	OECD 301C (I) OECD 302 C (II) U.S. EPA 835.3110	Aquatic	Ultimate	Ready	• Composite from 10 industrial effluents, WWTP secondary effluents, soils, etc. from different locations in Japan • No preexposure option • 10^7 to 10^8 cells/L	Good	Fair
Closed bottle	OECD 301 D ASTM El720 ISO 10707 U.S. EPA 835.3110	Aquatic	Ultimate	Ready	• WWTP secondary effluent that is primarily of domestic origin • Preexposure is option • 10^3 to 10^8 cells/L	Fair	Good
Two-phase closed bottle	ISO 10708	Aquatic	Ultimate	Ready	• WWTP secondary effluent *or* activated sludge *or* surface water • Preexposure is option • 10^7 to 10^{11} cells/L	Good	No
Modified screening	OECD 301 E ISO 7827	Aquatic	Ultimate	Ready	• WWTP secondary effluent *or* activated sludge *or* surface water • Preexposure is option • 10^4 to 10^5 cells/L	No	No
Respirometry	OECD 301 F ISO 9408 U.S. EPA 835.3110	Aquatic	Ultimate primary	Ready inherent	• Activated sludge (preferred) *or* WWTP secondary effluent *or* surface water • Preexposure is option • 10^7 to 10^8 cells/L	Excellent	Good
Seawater (shake flask method) (closed bottle method)	OECD 306	Aquatic marine	Ultimate	Inherent	• Indigenous microbial community present in media used for test (no inoculum) • 10^3 to 10^6 cells/L	(SF) Fair (CB) Fair	(SF) No (CB) Fair

(Continued)

TABLE 3—Examples and Descriptions of Relevant Biodegradation Test Methods (Continued)

			Biodegradation			Suitability for Materials that Are:	
Test	**Standard/ Reference**	**Environment**	**Extent**	**Rate**	**Inoculum**	**Poorly Soluble**	**Volatile/ Gas**
Zahn-Wellens	EMPA OECD 302 B ISO 9888 U.S. EPA 835.3200	Aquatic	Ultimate primary	Inherent	• Activated sludge • Preexposure is option • 10^{10} cells/L	No	No
Sludge semicontinuous (mod. SCAS)	OECD 302 A ASTM E1625 ISO 9887 U.S. EPA 835.3210 U.S. EPA 835.5045	Aquatic	Ultimate primary removal	Inherent	• Activated sludge • Preexposure is option • 10^9 to 10^{11} cells/L	Fair	No
Sludge-continuous (coupled units)	OECD 303 A ISO 11733	Aquatic	Ultimate simulation	Inherent	• Composite of WWTP secondary effluent + soil + surface water *or* activated sludge • 10^9 to 10^{11} cells/L	No	No
Soil	OECD 304 A ISO 14239 U.S. EPA 835.3300	Soil	Ultimate simulation	Inherent	• Indigenous microbial community present in media used for test (no inoculum)	Good	No → good (some variations)
Biogas generation—anaerobic	ECETOC TR 28 [20] ISO 11734	Aquatic	Ultimate primary	Inherent	• Anaerobic digester sludge inherent • Preexposure is option • 10^8 to 10^{12} cells/L	Good	Good
Two-stroke oil	CECL-33-A-94 [21]	Aquatic	Primary	Inherent	• Composite of primary and secondary WWTP effluents • No preexposure • 10^6 cells/L	Fair	No
CO_2 generation in sealed vessels	ISO 14593 CONCAWER Pt. No. 99/59 [22] Battersby 1997 [23] U.S. EPA 835.3120	Aquatic	Ultimate	Inherent	• Composite of WWTP secondary effluent + soil + surface water ultimate inherent • Preexposure is option • 10^7 to 10^8 cells/L	Good	Good
Assessing treatability or biodegradability of organic materials in porous pot test	ASTM E1798	Aquatic	Ultimate simulation	Inherent	• Activated sludge that is primarily of domestic origin • 10^7 to 10^8 cells/L	Fair	No
Shake flask die-away test	ASTM E1279 U.S. EPA 835.3170	Aquatic—surface water	Primary	Inherent	• 10^7 to 10^8 cells/L—indigenous microbial community present in media used for test	No	No

[A] Waste water treatment plant.

animals (ASTM E1022, ASTM E1688, ASTM E1676, OECD 305, EPA 850.1710, EPA 850.1730).

Past research has generally focused on accumulation in the aquatic environment. Terms commonly used to quantify accumulation potential by aquatic organisms are the bioconcentration factor (BCF) [11], the bioaccumulation factor (BAF) [12], and biota-sediment accumulation factor (BSAF) [13]. The BCF is a laboratory-derived measurement that simply defines the partition coefficient of the substance between aquatic organism tissue (primarily the lipid component) and the surrounding water. The BAF is generally a field-derived measurement determined from monitoring data. The BAF is calculated by dividing the observed field tissue concentration of the material by the corresponding concentration of the material in the water (soil, sediment, or food). This value can also be determined from microcosm and mesocosm experiments. The BAF differs from the BCF in that the BAF represents the accumulation via all possible exposure routes that may occur in the field (i.e., water, sediment, soil surface, and diet), whereas the BCF only reflects the accumulation due to aqueous exposure. The BSAF expresses the steady-state difference between the concentration of a bioaccumulating nonpolar organic chemical normalized to the organic carbon content of a sediment and the concentration measured in the total extractable lipids of an organism for which that sediment represents the source of the contamination in its habitat.

Although the BAF and BSAF values provide a more realistic appraisal of the accumulation potential of a material, there is a high degree of uncertainty associated with field data due to variability in the exposure values from various environmental sources and associated contaminants. Further, field-derived data are not available for most materials. Consequently, BCF provides a valuable surrogate parameter for assessing the relative bioaccumulation potential of materials under controlled laboratory conditions.

Based on the results of numerous BCF experiments for different substances, QSARs have been developed that relate the BCF value to the K_{ow}. In the absence of BCF data, the log K_{ow} values are used to predict the bioaccumulation potential for materials. These predictions are often used for hazard classification and risk assessment. However, this approach may be overly conservative for hydrocarbons.

Toxicological Properties Important to Environmental Characterization

STANDARDIZED TESTS

In standardized tests, biological variability is reduced by using organisms of selected species that are the same age range, similar in size, and in good health with no observable abnormalities. When the goal of the test is to measure inherent toxicity parameters such as mortality or effects on reproduction, a single species test is used. Concerns about different species sensitivities are addressed in environmental toxicology by testing representative or indicator species from various levels in the food chain (trophic levels). Considerable research and test method development has gone into the selection of test species that are representative of the more sensitive species (within the same grouping of organisms) in the environment. One advantage of using such representative species is that substances tested at different times and in different laboratories can be compared for relative toxicity. This has resulted in the development of QSARs for specific species and classes of organisms. These QSARs allow prediction of toxicity to aquatic organisms based on structural or physical properties of the substance (ASTM E1242). As with the biodegradation test methods, each specific test provides data that are representative of a specific aspect of the natural environment. The dominating factor relating the toxicity test to the environment is the mode of exposure.

EXPOSURES

As with the variability associated with the test organisms, each test material has unique characteristics that must be considered when conducting a standardized toxicity test. The volatility, water solubility, and complexity or variability of composition of the material can significantly affect the exposure of the test organism to the material and must be considered in the design of the test. Appropriate test designs allow for reproducibility of results within the same laboratory and comparison between laboratories for the same test material.

In order to relate effects to exposure, one needs to have a defined, quantified test material concentration for a specific period of time. Since the effect is related to the internal dose received by the organisms (a value not generally known), exposure duration must be adequately long to allow maximum uptake (or at least adequate uptake) of the material by the test organism. For low-solubility materials, such uptake may be quite slow. As a result of these considerations, the exposure duration of aquatic toxicity tests has been standardized. For acute toxicity tests, these periods range from 2 days for *Daphnia* tests to 4 days for most fish tests.

For stable, water soluble, pure materials, maintaining a constant exposure concentration is generally not difficult. Typically an arithmetic or geometric series of exposure concentrations is tested. The test method and preliminary testing usually determine the specific range and number of concentrations. This facilitates having a range in which the extremes provide no mortality and complete mortality. Test guidelines usually require that the actual concentration be analytically confirmed over the course of the exposure period. The resulting data allow the use of statistical techniques to develop a concentration-response curve and to find the mid-point or LC_{50} (ASTM E1847, ASTM E1023, ASTM E729, Env Can EPS I/RM/34). The same approach is used to establish other effect end points such as impact on growth or reproduction for chronic testing. For end points not involving lethality, EC_{50} is often used to indicate the "effect concentration" rather than the "lethal concentration." Discussing the concentration-response can be confusing. The smaller the concentration of the material that causes an effect, the greater is the toxicity of the material.

Typically, lubricant and fuel products are considered "difficult materials" for aquatic toxicity testing. Difficult materials may have low water solubility, be volatile, or be complex mixtures. Much of the historical information on the toxicity of these materials can be misleading or confusing. Often, the toxicity values are not reasonable (e.g., reported LC_{50} exceeds the water solubility of the material) and the value does not accurately represent the toxicity of the material. In this example, all that can be said is that the LC_{50} is greater than the water solubility. Generally, the material must be retested to develop scientifically defensible values. Designing, conducting, and interpreting aquatic toxicity studies on such materials involve special considerations [14].

CHARACTERISTICS OF "DIFFICULT MATERIALS"

Volatility

For volatile or unstable materials, it is difficult to maintain a consistent concentration for the exposure period. The traditional approach is to replace the exposure matrix concentrations with frequent renewals of the exposure medium from freshly prepared test solution (static renewal test) or even with a continuous flow of test solution through the exposure chambers (flow-through test). An approach that has been successfully used for volatile substances is to prevent loss by testing in tightly closed containers with no headspace. Precautions that organisms get sufficient oxygen must be taken. It should be re-emphasized that the purpose of these tests is to produce a quantitative concentration-response relationship, not to simulate the fate and effects of the substance in the environment. So the expected (nominal) concentration must be confirmed with periodic analyses, and, if necessary, the LC_{50} calculation may be based upon the measured concentrations [14].

Low Solubility

There are a number of issues related to low-solubility materials. The acute LC_{50} values for a homologous series of chemicals may sometimes seem confusing. In general, the members of the series with lower water solubility will have greater toxicity. However, as the number of carbon atoms in the series increases, the water solubility decreases. The decrease in solubility is usually greater than the increase in toxicity. So a point will be reached when no acute mortality is evident. This is sometimes referred as the "solubility cut-off." Larger members of the series are too insoluble to cause acute mortality. Although it is possible that the small amount of the larger members in solution may cause chronic toxicity and may contribute to the acute toxicity caused by other dissolved substances, they contribute little or no acute toxicity unto themselves. Further additions of a single material above its solubility limit will not increase the dissolved portion that is absorbable by the test organisms and therefore is not bioavailable. The conclusion of these tests should be "no acute toxicity at the maximum water solubility of the substance."

Sometimes tests are conducted as limit tests. These are tests in which the substance is added in an amount greater than its solubility to meet arbitrary toxicity limits (usually regulatory), such as 100 mg/L or 1,000 mg/L. In this example, if less than 50 % mortality is observed, the LC_{50} is reported as >100 mg/L or >1,000 mg/L, respectively, and is considered to have minimal toxicity. There are two potential sources of error inherent in this approach:

- The presence of minor components that are water soluble may affect the results. At additions above the water solubility of the main substance, the contaminants may continue to dissolve and may constitute a higher percentage of the water solution than the substance itself. In these cases, the toxicity observed is generally erroneously ascribed to the main substance.
- The occurrence of physical effects caused by the undissolved test material. Such effects may become confused with the inherent toxicity of the bioavailable portion of the material. Examples of such problems are the fouling of fish gills by oil droplets, which causes suffocation, or entrapment of *Daphnia* in a surface film of a material, causing immobility (an end point in the *Daphnia* acute toxicity test).

Complex Mixture Toxicity

The complexity of fuels and lubricants varies. Some products consist of isomers with a similar number of carbons that differ primarily in branching pattern. As a result, the toxicity of these materials is similar between individual components, since isomeric differences typically have little effect on aquatic toxicity. In contrast, other materials are composed of components that vary significantly in carbon number and chemical characteristics. Consequently, evaluating the toxicity of these materials is more complicated and involves both the amount and toxicity of individual components. Most hydrocarbons are believed to be toxic by the same mode of action and so are considered to have additive toxicity. See *Fundamentals of Aquatic Toxicology: Effects, Environmental Fate, and Risk Assessment* [15] for further discussion of the additivity of component toxicity.

For complex mixtures of poorly water-soluble substances, another challenge becomes apparent in testing and evaluating the results. Additions of a complex material to water at levels beyond the water solubility of the components results in a solution and a toxicity measurement that does not accurately reflect the composition or toxicity of the complete material. Some scientific and regulatory organizations have suggested using the "lethal loading" approach to overcome these difficulties [14]. This test methodology is specifically for poorly water-soluble complex mixtures and bases the toxicity measurement on the amount of substance added to water (not the concentration of the dissolved components that the organism is exposed to). The "amount of substance added" has been dubbed the "loading" as a simpler term. The common toxicity values derived from this testing approach is described as LL_{50} (EL_{50}) for "lethal loading, 50 %" ("effect loading, 50 %") to distinguish the results from tests on soluble, simple substances (LC_{50} or EC_{50}). For fuels and lubricants, the "lethal loading" approach is preferred for developing information on the quantity of released material that might be required to initiate significant ecological risk and classification or labeling schemes.

The lethal loading approach is not a reliable surrogate for quantified exposure data. An uninformed reader may confuse the loading values for the exposure values (which are usually much smaller than the loading values) and underestimate the inherent toxicity of a material. The information derived from the different exposure approaches is of great value if applied to appropriate environmental situations.

An additional reason for the use of a separate (LL_{50}) terminology is that, when conducting lethal loading tests, it is generally unknown whether the water phase is a true solution. The test substance is brought to equilibrium with water at each concentration (loading) to be tested. The portion of substance that separates after equilibration is generally removed and the aqueous phase referred to as the "water accommodated fraction" (WAF). The WAF may be a solution or an emulsion. Unlike previous methodologies, dilutions of the WAF are not tested, but rather WAFs are prepared at each loading tested. This allows the toxicity result to be related to the entire substance, giving a result that is relevant to realistic environmental situations. For example, if a product spill occurs, it will not be of a concentrated water extract of the substance but of the whole substance (generally) at rather low substance to water ratios. LL_{50} data allow a quantitative assessment of how much substance in a given volume of water is likely to result in toxicity [14,16].

TEST METHODS

Standardized Toxicity Test Procedures: Acute

The general approach to all toxicity tests is to expose the organisms to a series of concentrations of the test substance. The concentrations are chosen, often based on some preliminary tests, so that the biological effect (e.g., mortality) occurs over the range of concentrations. At the lowest concentration, little to no effect is observed, while at the highest concentration the effect is maximal. Table 4 compares acute and chronic test characteristics. Table 5 provides examples and descriptions of acute and chronic environmental toxicity test methods.

AQUATIC TOXICITY TESTS

Tests to determine the aquatic toxicity of substances have been in place longer than tests in other media. The intention is to provide testing at different levels within the aquatic food web. Algae, invertebrates (which feed on the algae), and fish (which feed on the invertebrates) are included in the usual aquatic testing scheme. A variety of species have been identified that are used as standard test organisms in these studies. The intent is to have representative or indicator species that have the following characteristics:

- Life histories are well understood and can be cultured in the laboratory
- Survive well in the laboratory
- Are relatively sensitive to a broad range of toxins
- Are well studied so that their responses may be readily interpreted

These species are intended to be a conservative representation of the entire aquatic environment because of the preference for selecting sensitive species and usually sensitive life stages for testing. The species frequently selected for freshwater toxicity tests are the fish–rainbow trout *(Oncorhynchus mykiss)* or fathead minnow *(Pimephales promelas)*, invertebrate–waterflea *(Daphnia magna* and *Ceriodaphnia dubia)*, and green algae *(Selenastrum capricornutum)*. (The scientific name of this green alga has recently been changed to *Pseudokirchneriella subcapitata* (Korshikov) Hindak [17]. In that this name is not yet in common use, the name *Selenastrum capricornutum* will be used in this document.) Data resulting from toxicity tests are generally used to calculate the concentration of material that will adversely affect 50 % of the test organisms (i.e., LC_{50}, LL_{50j}, and EC_{50}). The toxicity of these materials is generally grouped as follows: $LC_{50} \leq 1$ mg/L = very high (toxic); $1 < LC_{50} < 10$ mg/L = toxic (medium); $10 < LC_{50} < 100$ mg/L = harmful (low) [18].

TESTS WITH ALGAE

Cultures of selected algae in their exponential growth phase are exposed to various concentrations of a substance over several generations under defined conditions. Cell density is determined by microscopic counting of cells or by a spectrophotometric measurement of chlorophyll. The inhibition of growth in relation to a control (nonexposed) culture over a fixed period of time is determined. The algal cell density at each concentration is determined at 24, 48, and 72 h (and

TABLE 4—Comparison of Acute and Chronic Environmental Toxicology Tests

Test Characteristic	Acute	Chronic
Objective	• Determines test material concentration that causes effect on organism during short-term exposure • Evaluates relative toxicity of different materials to same organism • Evaluates relative sensitivity of different species to same material	• Determines the lowest observable effect and the no observable effect concentrations of a material
Onset of effect	• Sudden or delayed	• Delayed
Concentration of test material	• Relatively high	• Relatively low
Exposure frequency	• Static • Static renewal • Continuous (flow-through) • Pulsed	• Static renewal • Continuous (flow-through) • Pulsed
Test duration	• Typically <4 days • Exposure period small when life-cycle compared to total life-cycle	• Weeks to months to years • May include sensitive part of life-cycle, entire life-cycle, or multiple life-cycles
Effects typically measured	• Mortality • Immobilization • Changes in growth rates (algae)	• Organism development and growth • Reproductive success • Based on most sensitive endpoint
Common endpoints	• LL_0 (lethal load for 0 %) • LL_{50} (lethal load for 50 %) • EL_{50} (effect load for 50 %) • IL_{50} (inhibition load for 50 %) • LC_{50} (lethal cone for 50 %) • EC_{50} (effect cone for 50 %) • IC_{50} (inhibition cone for 50 %) • NEL (no effect level)	• IL_{20} (inhibition load for 20 %) • IL_{50} (inhibition load for 50 %) • NOEL (no observable effect level) • LOEL (lowest observable effect level)

TABLE 5—Examples and Descriptions of Acute and Chronic Environmental Toxicity Test Methods

Test	Duration	Measurements	Endpoints	Environment	Organisms	Scientific Names	Guidelines
Bacteria	5, 15, and 30 min	Luminescence inhibition	IC_{50}	Freshwater	Luminescent bacteria	*Photobacterium phosphoreum*	Env Can EPS I/RM/24
			IL_{50}	Marine water			
Algae	72–96 h	Growth	$E_{biomass}C_{50}$	Freshwater	Green algae	*Selenastrum capricomutum*	OECD 201
			$E_{growthrate}C_{50}$				
			$E_{biomass}L_{50}$				
			$E_{growthrate}L_{50}$				
		Growth rate	NOEC	Marine water		*Scenedesmus subspicatus*	U.S. EPA 850.5400
			LOEC			*Chlorella vulgaris*	
			EC_{50}			*Isocyrsis galbana* (most common species, but others can be used)	ASTM D3978
			EL_{50}				ASTM E1218
							ISO 8692
							ISO 10253
							Env Can EPS I/RM/25
	14 days	Fertilization survival	EC_{50}	Marine water	Seaweed	*Champia*	ASTM E1498
			EL_{50}			*Parvula*	
			NOEC			*Laminaria*	
			LOEC			*Saccharina*	
						Fucus sp.	
Plants	14 days	Growth	EC_{50}	Freshwater	Milfoil	*Myriophyllum sibiricum*	ASTM E1913
		Shoot length	EL_{50}				
		Root length	NOEC				
			LOEC				
	7–14 days	Reproduction	EC_{50}	Freshwater	Duckweed	*Lemna gibba, Lemna minor*	ASTM E1415, Env Can EPS, I/RM/37, U.S. EPA 850.4400
			EL_{50}				
			NOEC				
			LOEC				

	> 14 days	Growth	EC_{50}	Soil	Agricultural species		ASTM E1598
		Shoot length	EL_{50}		Native species		ASTM E1963
		Root length	NOEC				U.S. EPA 850.4100
		Germination emergence	LOEC				U.S. EPA 850.4150
			EC_{25}				U.S. EPA 850.4200
							U.S. EPA 850.4225
							U.S. EPA 850.4230
	14 days	Growth shoot length	EC_{50}	Freshwater	Emergent plants		ASTM E1841
		Root length	EL_{50}				
			NOEC				
			LOEC				
Acute invertebrate	48 h	Survival	LC_{50}	Freshwater	Daphnids	*Daphnia magna*	OECD 202
			LL_{50}			*Daphnia pulex, Daphnia sp.*	U.S. EPA 850.1010 ISO 6341
							Env Can EPS I/RM/11
	96 h	Survival	LC_{50}	Marine water	Mysids	*Americamysis bahia* (formerly *Mysidopsis bahia)*	U.S. EPA 850.1035
			LL_{50}		West Coast		ASTM E1463
					Mysids		
	14 days	Survival	LC_{50}	Soil	Earthworms	*Eisenia foetida, Lubricous terrestris*	OECD 207
			LL_{50}				ISO 11268-1
							U.S. EPA 850.6200
	3–10 days	Survival	LC_{50}	Marine and estuarine	Amphipods	*Amphipreia virginiana*	Env Can EPS I/RM/26
			LL_{50}	Sediment		*Corophium volutator, Eohaustorius estuarius, Eohaustorius*	ASTM E1367 U.S. EPA 850.1740
						Washingtonianus, Foxiphalus xiximeus, Leptocheirus pinguis, Rhepoxynius abronius	
	3–10 days	Survival	LC_{50}	Freshwater	Macroinvertebrates	*Hyalella azteca*	ASTM E729

(*Continued*)

TABLE 5—Examples and Descriptions of Acute and Chronic Environmental Toxicity Test Methods (Continued)

Test	Duration	Measurements	Endpoints	Environment	Organisms	Scientific Names	Guidelines
	≤ 14 days		LL_{50}	Sediments		*Chironomus sp.* *Hexagenia sp.* *Tubifex sp.* (most common species, but others can be used)	ASTM E1706 U.S. EPA 850.1735 U.S. EPA 850.1790
	48–96 h	Survival development	LC_{50}	Marine water	Bivalve molusks		ASTM E724 ASTM E1563 U.S. EPA 850.1055
			LL_{50}				
	24 h	Survival	LC_{50}	Freshwater	Rotifer	Brachionus sp.	ASTM E1440
			LL_{50}				
	10 days 96 h	Survival	LC_{50}	Marine and estuarine sediment	Polychaetous Annelid worms	*Neanthes arenacoedentata, Neanthes virens*	ASTM E1611 ASTM E1562
			LL_{50}				
	48–96 h	Survival	LC_{50}	Marine water	Copepods		ISO 14669
			LL_{50}				
Chronic invertebrate	≥ 14 days	Mortality	EC_{50}	Freshwater	Daphnids	*Daphnia magna*	OECD 202
	≤ 21 days	Reproduction	EL_{50}			*Daphnia pulex*	Env Can EPS I/RM/21
		Growth	NOEC			*Daphnia sp.*	ASTM E1192
			LOEC			*Ceriodaphnia dubia*	ASTM E1295
							ISO 10706
							U.S. EPA 850.1300
	28 days	Life-cycle	EC_{50}	Marine water	Mysids	*Americamysis bahia*	U.S. EPA 850.1350
			EL_{50}				
			NOEC				
			LOEC				
	7 days	Survival reproduction fertilization	EC_{50}	Freshwater	Daphnid Mysid Shrimp Sea Urchin	*Ceriodaphnia dubia, Americamysis bahia, Arbacia punctulata*	Env Can EPS I/RM/27 Env Can EPS I/RM/21
			EL_{50}	Marine water			
			NOEC				
			LOEC				
	Up to 3 months	Life-cycle	EC_{50}	Marine and estuarine sediment	Polychaetous annelid worms	*Neanthes arenacoedentata, Neanthes virens*	ASTM E1562
			EL_{50}				
			NOEC				
			LOEC				

	8 weeks 1 year	Reproduction	EC_{50}	Soil	Earthworms	*Eisenia foetida*	ISO 11268-2
			EL_{50}				ISO 11268-3
			NOEC				
			LOEC				
Acute fish	96 h	Survival	LC_{50}	Freshwater	Fathead Minnow Rainbow Trout Bluegill Sunfish Sheepshead Minnow Zebra Fish Common Carp Guppy Golden Orfe	*Pimephales promelas*	ASTM E729
	≤ 14 days		LL_{50}	Marine water		*Oncorhynchus mykiss*	OECD 203
						Lepomis macrochirus	Env Can EPS I/RM/13
						Cyprinodon variegatus	Env Can EPS I/RM/9
						Brachydanio rerio	Env Can EPS I/RM/22
						Cyprinus carpio	ISO 7346-1
						Poecia reticulata	ISO 7346-2
						Leuciscus idus	ISO 7346-3
							ISO 12890
							U.S. EPA 850.1075
Chronic fish	14 days	Mortality	EC_{50}	Freshwater	As for acute testing	As for acute testing	OECD 204
		Sublethal effects	EL_{50}	Marine water			
			NOEC				
			LOEC				
	28-60 days posthatch	Hatching mortality Length weight	EC_{50}	Freshwater	As for acute testing	As for acute testing	OECD 210
			EL_{50}	Marine water			U.S. EPA 850.1400
			NOEC				Env Can EPS I/RM/28
			LOEC				ASTM E1241
							ISO 10229
	7 days	Survival	EC_{50}	Freshwater	Sheepshead Minnow	*Cyprinodon variegatus*	Env Can EPS I/RM/22
		Length	EL_{50}	Marine water	Fathead Minnow	*Pimephales promelas*	
		Weight	NOEC				
			LOEC				

96 h for some test methods) after the start of the test. The mean cell density for each concentration is plotted against time to produce growth curves. The growth end point is determined by comparing the area under the growth curves for each concentration against the control. The average specific growth rate (the increase in cell density over time) is also evaluated (ASTM D3978, ASTM E1218, ISO 8692, ISO 10253, OECD 201, and EPA 850.5400). The test method required for Canadian regulatory submissions is similar in concept, but uses microliter volumes and only measures the algal cell density at 72 h (Env Can EPS I/RM/25).

TESTS WITH INVERTEBRATES

Acute toxicity tests with invertebrates generally expose organisms to the test substance for 2 to 4 days. This period of exposure encompasses at least one sensitive period of growth during the juvenile life-stage. In the acute immobilization test with *Daphnia,* effects on the swimming capability are investigated. The number of immobile organisms for each concentration is compared to the control to calculate the EC_{50} value.

TESTS WITH FISH

A variety of fish are routinely used for acute toxicity tests. Fish are exposed to the test substance for a 96-h period. Mortality and sublethal effect observations are performed daily, and the concentration that kills 50 % of the fish (LC_{50}) is calculated for each observation period.

Standardized Toxicity Test Procedures: Chronic and Subchronic

There are a variety of chronic toxicity test exposures:

- Short-term chronic, where organisms are exposed for a relatively short duration (often 7 days) during a critical life-stage
- Prolonged toxicity tests (usually 2 or more weeks)
- Full-life cycle test (usually two to three generations)

The chronic tests have in common the measurement of sublethal effects, such as growth (weight or length), development, reproduction (fecundity, number of offspring produced), or survival of offspring (Table 4). These end points are important to the survival of the population and are thus considered more sensitive end points than lethality. End points for chronic toxicity tests are usually expressed as a calculated concentration with a low level of toxicity (e.g., EC_{10}) or as a concentration that had an effect that is not statistically different than the control. The no observable effect concentration (NOEC) is the effect value corresponding to greatest concentration of the test substance that is not statistically different from the control, in contrast to the lowest observable effect concentration (LOEC), which is the effect value corresponding to the lowest concentration of the test substance that is statistically different from the control.

TESTS WITH INVERTEBRATES

A common chronic toxicity test with invertebrates is the *Daphnia* Reproduction Test in which less than 24-h-old *Daphnia* are exposed to a range of test substance concentrations for at least 14 days. Within this time period, each healthy daphnid can produce three or more broods of young. The number of broods and the number of young per brood are recorded. This test may be extended to 3 or 4 weeks, in which six to nine broods of young should be produced. This test is performed under semistatic or flow-through conditions. Both survival of the parent generation and the number of young produced are evaluated.

TESTS WITH FISH

There are two common chronic toxicity tests for fish. The first is a 14-day prolonged test, which is performed under conditions similar to the acute toxicity test. A representative weight and length measurement is taken for the fish at the start of the test. The test is performed using either semistatic or flow-through exposure conditions for 14 days. At termination, all surviving fish are weighed and measured. The NOEC and LOEC values are determined based on survival, weight, or length or other abnormal effects observed throughout the test. The second is an early life-stage test that begins exposure on newly fertilized eggs. This test encompasses the embryo, larval, and juvenile stages of various fish species. A flow-through test design is preferred to semistatic. This test design allows for evaluation of the following end points:

- Mortality during the embryonic stage, larval stage, and juvenile stage
- Days to hatch, and number hatched
- Length and weight data
- Any abnormalities observed

The exposure period for the early life-stage test varies for each species; for example, tests with the warm, marine water fish (Sheepshead Minnow) terminate after 28-days posthatch, whereas tests with the cold, freshwater fish (Rainbow Trout) terminate after 60-days posthatch.

NONAQUATIC TOXICITY TESTS

- **Sediment Toxicity Tests** Sediment tests are primarily performed on invertebrates that will either burrow into or ingest the sediment, using standard test protocols developed for freshwater amphipods *(Hyalella azteca),* midges *(Chironomus tentans, C. riparius),* mayflies (*Hexagenia* sp.), worms *(Tubifex tubifex),* saltwater amphipods (*Rhepoxynius abronius, Eohaustorius* sp.), and polychaetes *(Neanthes arenacoedentata, N. virens).* A control sediment can be spiked with a test chemical or a contaminated natural sediment can be tested. Exposure ranges from 10 to 30 days, with renewal of the overlying water during extended periods. End points evaluated during these short-term tests can consist of survival (LC_{50}), growth and development (EC_{50}), and reproduction (egg production and number of young produced).
- **Terrestrial Toxicity Tests** As with aquatic toxicity tests, a variety of species have been identified that are used as standard test organisms in terrestrial studies. These studies are focused on plants and animals that live in soil. For some applications, birds are tested [19]. The plant toxicity studies allow the use of a very long list of species, most of which are grains or vegetables of commercial or native value. The acute studies relate the concentration of the chemical to the mortality of the test organism or to germination (in the case of plants). For many tests with soil, the test material is dissolved or emulsified in deionized water and then thoroughly mixed with artificial or reference soil. Diluting a contaminated soil with an artificial or reference soil and then exposing the organisms can allow testing of a field sample.

- **Tests with Plants** Generally, a minimum of three species is exposed to the test chemical. One of the three species is usually a monocotyledon, and the other is a dicotyledon (a legume and a root crop). The test is terminated 14 days after 50 % of the control seedlings have emerged. Several end points are usually evaluated. The number of plants that emerge (LC_{50}) and the mean dry weight at termination (EC_{50}) as compared to the controls are common end points. Shoot or root length or mass may also be evaluated for this test.
- **Tests with Worms** There are many methods for testing material toxicity to earthworms, including spot application and immersion tests. Often, a simple paper contact toxicity test can be used as an initial screen test to identify those substances that should be tested in artificial soil. For this test, artificial soil is spiked with the test chemical, diluted into a concentration series, and added to the exposure chambers. Earthworms are rinsed and placed on top of the soil in each chamber. Since worms are negatively phototropic (they do not like light), the test is 182 performed under continuous lighting to promote borrowing and maximize exposure to the soil. Mortality observations are performed after 7- and 14-day exposures for subsequent LC_{50} determinations.

ASTM Test Methods to Test Petroleum and Lubricant Products

ASTM International[A]	Title
D3978	Standard Practice for Algal Growth Potential Testing with *Selenastrum capricomutum*
D5864	Standard Method for Determining Aerobic Aquatic Biodegradation of Lubricants or Their Components
D6006	Standard Guide to Assess the Biodegradability of Hydraulic Fluids
D6081	Standard Practice for Aquatic Toxicity Testing of Lubricants: Sample Preparation and Results Interpretation
D6139	Standard Test Method for Determining the Aerobic Aquatic Biodegradation of Lubricants or Their Components Using the Gledhill Shake Flask
D6384	Standard Terminology Relating To Biodegradability and Ecotoxity of Lubricants
D6731	Standard Test Method for Determining the Aerobic, Aquatic Biodegradability of Lubricants or Lubricant Components in a Closed Respirometer
E724	Standard Guide for Conducting Acute Toxicity Testing on Aqueous Effluents with Fishes, Macroinvertebrates, and Amphibians
E729	Standard Guide for Conducting Acute Toxicity Tests on Test Materials with Fishes, Macroinvertebrates, and Amphibians
E857	Standard Practice for Conducting Subacute Dietary Toxicity Tests with Avian Species
E895	Standard Practice for Determination of Hydrolysis Rate Constants of Organic Chemicals in Aqueous Solutions
E896	Solutions Standard Test Method for Conducting Aqueous Direct Photolysis Tests
E1022	Standard Guide for Conducting Bioconcentration Tests with Fishes and Saltwater Bivalve Mollusks
E1023	Standard Guide for Assessing the Hazard of a Material to Aquatic Organisms and Their Uses

ASTM International[A]	Title
E1147	Standard Test Method for Partition Coefficient (*N*-Octanol/Water) Estimation by Liquid Chromatography
E1148	Standard Test Method for Measurements of Aqueous Solubility
E1191	Standard Guide for Conducting Life-Cycle Toxicity Tests with Saltwater Mysids
E1192	Guide for Conducting Acute Toxicity Tests on Aqueous Effluents with Fishes, Macroinvertebrates, and Amphibians
E1193	Standard Guide for Conducting *Daphnia magna* Life-Cycle Toxicity Tests
E1194	Standard Test Method for Vapor Pressure
E1195	Standard Test Method for Determining a Sorption Constant (K_{oc}) for an Organic Chemical in Soil and Sediments
E1218	Standard Guide for Conducting Static 96-h Toxicity Tests with Microalgae
E1241	Standard Guide for Conducting Early Life-Stage Toxicity Tests with Fishes
E1242	Standard Practice for Using Octanol-Water Partition Coefficient to Estimate Median Lethal Concentrations for Fish Due to Narcosis
E1279	Standard Test Method for Biodegradation by a Shake-Flask Die-Away Method
E1295	Standard Guide for Conducting Three-Brood, Renewal Toxicity Tests with *Ceriodaphnia dubia*
E1367	Standard Guide for Conducting 10-Day Static Sediment Toxicity Tests with Marine and Estuarine Amphipods
E1415	Standard Guide for Conducting Static Toxicity Tests with *Lemna gibba* G3
E1439	Standard Guide for Conducting the Frog Embryo Teratogenesis *Assay-Xenopus* (Fetax)
E1440	Standard Guide for Acute Toxicity Test with the Rotifer *Brachionus*

ASTM International[A]	Title
E1463	Standard Guide for Conducting Static and Flow-Through Acute Toxicity Tests with Mysids from the West Coast of the United States
E1498	Standard Guide for Conducting Sexual Reproduction Tests with Seaweeds
E1562	Standard Guide for Conducting Acute, Chronic, and Life-Cycle Aquatic Toxicity Tests with Polychaetous Annelids
E1563	Standard Guide for Conducting Static Acute Toxicity Tests with Echinoid Embryos
E1598	Standard Practice for Conducting Early Seedling Growth Tests
E1611	Standard Guide for Conducting Sediment Toxicity Tests with Marine and Estuarine Polychaetous Annelids
E1625	Standard Test Method for Determining Biodegradability of Organic Chemicals in Semi-Continuous Activated Sludge (SCAS)
E1676	Standard Guide for Conducting Laboratory Soil Toxicity or Bioaccumulation Tests with the Lumbricid Earthworm *Eisenia fetida*
E1688	Standard Guide for Determination of the Bioaccumulation of Sediment-Associated Contaminants by Benthic Invertebrates
E1706	Test Method for Measuring the Toxicity of Sediment-Associated Contaminants with Fresh Water Invertebrates
E1720	Standard Test Method for Determining Ready, Ultimate, Biodegradability of Organic Chemicals in a Sealed Vessel CO_2 Production Test
E1798	Standard Test Method for Assessing Treatability or Biodegradability, or Both, of Organic Chemicals in Porous Pots
E1841	Standard Guide for Conducting Renewal Phytotoxicity Tests with Freshwater Emergent Macrophytes
E1913	Standard Guide for Conducting Static, Axenic, 14-Day Phytotoxicity Tests in Test Tubes with the Submersed Aquatic Macrophyte, *Myriophyllum sibiricum* Komarov
E1963	Standard Guide for Conducting Terrestrial Plant Toxicity Tests
E2170	Standard Test Method for Determining Anaerobic Biodegradation Potential of Organic Chemicals Under Methanogenic Conditions
Environment Canada[B]	**Title**
EPS I/RM/9	Biological Test Method: Acute Lethality Test Using Rainbow Trout
EPS I/RM/10	Biological Test Method: Acute Lethality Test Using Threespine Stickleback *(Gasterosteus aculeatus)*

Environment Canada[B]	Title
EPS I/RM/11	Biological Test Method: Acute Lethality Test Using Daphnia spp.
EPS I/RM/12	Guidance Document on Control of Toxicity Test Precision Using Reference Toxicants
EPS I/RM/13	Biological Test Method: Reference Method for Determining Acute Lethality of Effluents to Rainbow Trout
EPS I/RM/14	Biological Test Method: Reference Method for Determining Acute Lethality of Effluents to *Daphnia magna*
EPS I/RM/21	Biological Test Method: Test of Reproduction and Survival Using the Cladoceran *Ceriodaphnia dubia*
EPS I/RM/22	Biological Test Method: Test of Larval Growth and Survival Using Fathead Minnows
EPS I/RM/24	Biological Test Method: Toxicity Test Using Luminescent Bacteria *(Photobacterium phosphoreum)*
EPS I/RM/25	Biological Test Method: Growth Inhibition Test Using the Freshwater Alga *Selenastrum capricomutum*
EPS I/RM/26	Biological Test Method: Acute Test for Sediment Toxicity Using Marine or Estuarine Amphipods
EPS I/RM/27	Biological Test Method: Fertilization Assay Using Echinoids (Sea Urchins and Sand Dollars)
EPS I/RM/28	Biological Test Method: Toxicity Tests Using Early Life Stages of Salmonid Fish (Rainbow Trout)
EPS I/RM/30	Guidance Document on Measurement of Toxicity Test Precision Using Control Sediments Spiked with Reference Toxicants
EPS I/RM/32	Biological Test Method: Test for Survival and Growth in Sediment Using the Freshwater Midges *(Chironomus tentans* or *Chironomus riparius)*
EPS I/RM/33	Biological Test Method: Test for Survival and Growth in Sediment Using the Freshwater Amphipod *Hyalella azteca*
EPS I/RM/34	Guidance Document on Application and Interpretation of Single-Species Tests in Environmental Toxicology
EPS I/RM/35	Biological Test Method: Reference Method for Determining Acute Lethality of Sediment Toxicity to Marine or Estuarine Amphipods
EPS I/RM/37	Biological Test Method: Test for Measuring the Inhibition of Growth Using the Freshwater Macrophyte, *Lemna minor*
ISO[C]	
ISO 6341	Water quality—Determination of the inhibition of the mobility of *Daphnia magna* Straus (Cladocera, Crustacea)—Acute toxicity test

ISO[c]	
ISO 7346-1	Water quality—Determination of the acute lethal toxicity of substances to a freshwater fish *(Brachydanio rerio* Hamilton-Buchanan (Teleostei, Cyprinidae))—Part 1: Static method
ISO 7346-2	Water quality—Determination of the acute lethal toxicity of substances to a freshwater fish *(Brachydanio rerio* Hamilton-Buchanan (Teleostei, Cyprinidae))—Part 2: Semi-static method
ISO 7346-3	Water quality—Determination of the acute lethal toxicity of substances to a freshwater fish *(Brachydanio rerio* Hamilton-Buchanan (Teleostei, Cyprinidae))—Part 3: Flow-through method
ISO 7827	Water quality—Evaluation in an aqueous medium of the "ultimate" aerobic biodegradability of organic compounds—Method by analysis of dissolved organic carbon (DOC)
ISO 8192	Water quality—Test for inhibition of oxygen consumption by activated sludge
ISO 8692	Water quality—Fresh water algal growth inhibition test with *Scenedesmus* subspicatus and *Selenastrum capricornutum*
ISO 9408	Water quality—Evaluation of ultimate aerobic biodegradability of organic compounds in aqueous medium by determination of oxygen demand in a closed respirometer
ISO 9439	Water quality—Evaluation of ultimate aerobic biodegradability of organic compounds in aqueous medium—Carbon dioxide evolution test
ISO 9887	Water quality—Evaluation of the aerobic biodegradability of organic compounds in an aqueous medium—Semi-continuous activated sludge method (SCAS)
ISO 9888	Water quality—Evaluation of ultimate aerobic biodegradability of organic compounds in aqueous medium—Static test (Zahn-Wellens method)
ISO 10229	Water quality—Determination of the prolonged toxicity of substances to freshwater fish—Method for evaluating the effects of substances on the growth rate of rainbow trout (*Oncorhynchus mykiss* Walbaum (Teleostei, Salmonidae))
ISO 10253	Water quality—Marine algal growth inhibition test with *Skeletonema costatum* and *Phaeodactylum tricornutum*
ISO 10634	Water quality—Guidance for the preparation and treatment of poorly water-soluble organic compounds for the subsequent evaluation of their biodegradability in an aqueous medium
ISO 10706	Water quality—Determination of long term toxicity of substances to *Daphnia magna* Straus (Cladocera, Crustacea)
ISO 10707	Water quality—Evaluation in an aqueous medium of the "ultimate" aerobic biodegradability of organic compounds—Method by analysis of biochemical oxygen demand (closed bottle test)
ISO 10708	Water quality—Evaluation in an aqueous medium of the ultimate aerobic biodegradability of organic compounds—Determination of biochemical oxygen demand in a two-phase closed bottle test
ISO 11266	Soil quality—Guidance on laboratory testing for biodegradation of organic chemicals in soil under aerobic conditions
ISO 11268-1	Soil quality—Effects of pollutants on earthworms *(Eisenia fetida)*—Part 1: Determination of acute toxicity using artificial soil substrate
ISO 11268-2	Soil quality—Effects of pollutants on earthworms *(Eisenia fetida)*—Part 2: Determination of effects on reproduction
ISO 11268-3	Soil quality—Effects of pollutants on earthworms—Part 3: Guidance on the determination of effects in field situations
ISO 11269-1	Soil quality—Determination of the effects of pollutants on soil flora—Part 1: Method for the measurement of inhibition of root growth
ISO 11269-2	Soil quality—Determination of the effects of pollutants on soil flora—Part 2: Effects of chemicals on the emergence and growth of higher plants
ISO 11733	Water quality—Evaluation of the elimination and biodegradability of organic compounds in an aqueous medium—Activated sludge simulation test
ISO 11734	Water quality—Evaluation of the "ultimate" anaerobic biodegradability of organic compounds in digested sludge—Method by measurement of the biogas production
ISO 12890	Water quality—Determination of toxicity to embryos and larvae of freshwater fish—Semi-static method
ISO 14236	Soil quality—Laboratory incubation systems for measuring the mineralization of organic chemicals in soil under aerobic conditions
ISO 14442	Water quality—Guidelines for algal growth inhibition tests with poorly soluble materials, volatile compounds, metals and waste water
ISO 14593	Water quality—Evaluation of ultimate aerobic biodegradability of organic compounds in aqueous medium—Method by analysis of inorganic carbon in sealed vessels (CO_2 headspace test)
ISO 14669	Water quality—Determination of acute lethal toxicity to marine copepods (Copepoda, Crustacea)

ISO[C]	
ISO/TR 15462	Water quality—Selection of tests for biodegradability
ISO 15522	Water quality—Determination of the inhibitory effect of water constituents on the growth of activated sludge microorganisms
ISO 16221	Water quality—Guidance for determination of biodegradability in the marine environment
OECD[D]	
101	UV-VIS Absorption Spectra
102	Melting Point/Melting Range
103	Boiling Point
104	Vapor Pressure
105	Water Solubility
106	Absorption—Desorption Using a Batch Equilibrium Method
107	Partition Coefficient (*n*-octanol/water): Shake Flask Method
111	Hydrolysis as a Function of pH
112	Dissociation Constants in Water
117	Partition Coefficient (*n*-octanol/water), HPLC Method
121	Estimation of the Adsorption Coefficient (K_{oc}) on Soil and on Sewage Sludge using High Performance Liquid Chromatography (HPLC)
201	Alga, Growth Inhibition Test
202	*Daphnia* sp. Acute Immobilisation Test and Reproduction Test
203	Fish, Acute Toxicity Test
204	Fish, Prolonged Toxicity Test: 14-Day Study
205	Avian Dietary Toxicity Test
206	Avian Reproduction Test
207	Earthworm, Acute Toxicity Tests
208	Terrestrial Plants, Growth Test:
	A-Seedling Emergence and Seedling Growth
	B-Vegetative Vigor Test
209	Activated Sludge, Respiration Inhibition Test
210	Fish, Early-Life Stage Toxicity Test
211	*Daphnia magna* Reproduction Test
212	Fish, Short-Term Toxicity Test on Embryo and Sac-Fry Stages
215	Fish, Juvenile Growth Test
218	Draft—Sediment-Water Chironomid Toxicity Test Using Spiked Sediment

OECD[D]	
219	Draft—Sediment-Water Chironomid Toxicity Test Using Spiked Water
221	Draft—*Lemna* sp. Growth Inhibition Test
301	Ready Biodegradability:
	A—DOC Die-Away Test
	B—C02 Evolution Test
	C—Modified MITI Test (I)
	D—Closed Bottle Test
	E—Modified OECD Screening Test
	F—Manometric Respirometry Test
302	Inherent Biodegradability:
	A—Modified SCAS Test
	B—Zahn-Wellens/EMPA Test
	C—Modified MITI Test (II)
303	Simulation Test—Aerobic Sewage Treatment
304	Inherent Biodegradability in Soil
305	Bioconcentrations: Flow-Through Fish Test
306	Biodegradability in Seawater
307	Draft—Aerobic and Anaerobic Transformation in Soil
308	Draft—Aerobic and Anaerobic Transformation in Aquatic Sediment Systems
309	Draft—Aerobic Mineralisation in Surface Water-Simulation Biodegradation Test
310	Draft—Ready Biodegradability—CO_2 in Sealed Vessels (Headspace Test)
311	Draft—Ready Anaerobic Biodegradability: Gas Production from Diluted Anaerobic Sewage Sludge
U.S. EPA[E]	
830.7370	Dissociation constants in water
830.7550	Partition coefficient (*w*-octanol/water), shake flask method
830.7560	Partition coefficient (*w*-octanol/water), generator column method
830.7570	Partition coefficient (*w*-octanol/water), estimation by liquid chromatography
830.7840	Water solubility: Column elution method; shake flask method
830.7860	Water solubility, generator column method
830.7950	Vapor pressure
835.2110	Hydrolysis as a function of pH
835.2130	Hydrolysis as a function of pH and temperature

U.S. EPA[E]	
835.2210	Direct photolysis rate in water by sunlight
835.2310	Maximum direct photolysis rate in air from UV/visible spectroscopy
835.3100	Aerobic aquatic biodegradation
835.3110	Ready biodegradability
835.3120	Sealed-vessel carbon dioxide production test
835.3170	Shake flask die-away test
835.3180	Sediment/water microcosm biodegradation test
835.3200	Zahn-Wellens/EMPA test
835.3210	Modified SCAS test
835.3300	Soil biodegradation
835.3400	Anaerobic biodegradability of organic chemicals
835.5045	Modified SCAS test for insoluble and volatile chemicals
835.5270	Indirect photolysis screening test: Sunlight photolysis in waters containing dissolved humic substances
850.1010	Aquatic invertebrate acute toxicity, test, freshwater daphnids
850.1020	Gammarid acute toxicity test
850.1025	Oyster acute toxicity test (shell deposition)
850.1035	Mysid acute toxicity test
850.1045	Penaeid acute toxicity test
850.1055	Bivalve acute toxicity test (embryo larval)
850.1075	Fish acute toxicity test, freshwater and marine
850.1085	Fish acute toxicity mitigated by humic acid
850.1300	Daphnid chronic toxicity test
850.1350	Mysid chronic toxicity test
850.1400	Fish early-life stage toxicity test
850.1500	Fish life cycle toxicity
850.1710	Oyster BCF
850.1730	Fish BCF
850.1735	Whole sediment acute toxicity with invertebrates, freshwater

U.S. EPA[E]	
850.1740	Whole sediment acute toxicity with invertebrates, marine
850.1790	Chironomid sediment toxicity test
850.2100	Avian acute oral toxicity test
850.2200	Avian dietary toxicity test
850.2300	Avian reproduction test
850.4100	Terrestrial plant toxicity, Tier I (seedling emergence)
850.4150	Terrestrial plant toxicity, Tier I (vegetative vigor)
850.4200	Seed germination/root elongation toxicity test
850.4225	Seedling emergence, Tier II
850.4230	Early seedling growth toxicity test
850.4250	Vegetative vigor, Tier II
850.4300	Terrestrial plants field study, Tier III
850.4400	Aquatic plant toxicity test using *Lemna* spp. Tiers I and II
850.5400	Algal toxicity, Tiers I and II
850.6200	Earthworm subchronic toxicity test
850.6800	Modified activated sludge, respiration inhibition test for sparingly soluble chemicals
600/4-90/027F	Methods for Measuring the Acute Toxicity of Effluents and Receiving Waters to Freshwater and Marine Organisms
600/4-91/003	Short-Term Methods for Estimating the Chronic Toxicity of Effluents and Receiving Water to Marine and Estuarine Organisms
600/4-91/002	Short-Term Methods for Estimating the Chronic Toxicity of Effluents and Receiving Water to Freshwater Organisms

[A]May be purchased from ASTM International, 100 Barr Harbor Drive, West Conshohocken, PA 19428; www.ASTM.org.
[B]May be purchased from Environment Canada Environment Technology Centre, 3439 River Road South, Gloucester, Ontario KIA 0H3; www.ec.gc.ca/pands_e.html.
[C]May be purchased from ISO at www.iso.ch/iso/eri/prods-services/ISOstore/store.html.
[D]May be purchased from OECD Publications, Environmental Health and Safety Division, 2, rue Andre Pascal, 75775 Paris Cedex 16, France; www.OECD.org/ehs/test/testlist.htm.
[E]Available free of charge from U.S. Environmental Protection Agency; www.epa.gov/opptintr/oppts/.

References

[1] Rand, G. M., Wells, P. G., and McCarty, L. S., "Introduction to Aquatic Toxicology," *Fundamentals of Aquatic Toxicology: Effects, Environmental Fate, and Risk Assessment*, G. M. Rand, Ed., 1995, Taylor & Francis, Washington, DC, pp. 3–67.

[2] Government of Canada, Environment Canada, and Health Canada, *Guidelines for the Notification and Testing of New Substances: Chemicals and Polymers*, 1993, p. 217.

[3] Grandy, N. J., "Role of the OECD in Chemicals Control and International Harmonization of Testing Methods," *Fundamentals of Aquatic Toxicology: Effects, Environmental Fate, and Risk Assessment*, G. M. Rand, Ed., Taylor & Francis, Washington, DC, 1995, pp. 763–773.

[4] Lyman, W. J., "Octanol/Water Partition Coefficient," *Handbook of Chemical Property Estimation Methods: Environmental Behavior of Organic Compounds*, W. J. Lyman, W. F. Reehl,

and D. H. Rosenblatt, Eds., McGraw-Hill Book Co., New York City, 1982, pp. 1–54.

[5] Lipnick, R. L., "Structure-Activity Relationships," *Fundamentals of Aquatic Toxicology: Effects, Environmental Fate, and Risk Assessment*, G. M. Rand, Ed., Taylor & Francis, Washington, DC, pp. 609–656.

[6] Hinman, M. L., "Environmental Characteristics of Fuels and Lubricants," *Fuels and Lubricants Handbook: Technology, Properties, Performance, and Testing, MNL 37*, G. E. Totten, Ed., ASTM International, West Conshohocken, PA, 2003.

[7] Exxon Chemical Europe, I.: *A Practical Guide to Protecting Man and the Environment*, Exxon Chemical Europe, Inc., Machelen, 1999.

[8] Lyman, W. J., "Transport and Transformation Processes," *Fundamentals of Aquatic Toxicology: Effects, Environmental Fate, and Risk Assessment*, G. M. Rand, Ed., Taylor & Francis, Washington, DC, 1995, pp. 449–492.

[9] Atkinson, R., "Atmospheric Oxidation," *Handbook of Property Estimation Methods for Chemicals: Environmental and Health Sciences*, R. Boethling and D. Mackay, Eds., Lewis Publishers, Boca Raton, 2000, pp. 335–354.

[10] Wrabel, M. L., and Peckol, P., "Effects of Bioremediation on Toxicity and Chemical Composition of No. 2 Fuel Oil: Growth Responses of the Brown Alga Fucus vesiculosus," *Marine Pollution Bulletin*, Vol. 40, No. 2, 2000, pp. 135–139.

[11] Bysshe, S. E., "Bioconcentration Factor in Aquatic Organisms," *Handbook of Chemical Property Estimation Methods: Environmental Behavior of Organic Compounds*, W. J. Lyman, W. F. Reehl, and D. H. Rosenblatt, Eds., 1982, McGraw-Hill Book Co., New York, p. 30.

[12] Spacie, A., McCarty, L. S., and Rand, G. M., "Bioaccumulation and Bioavailability in Multiphase Systems," *Fundamentals of Aquatic Toxicology: Effects, Environmental Fate, and Risk Assessment*, G. M. Rand, Ed., Taylor & Francis, Washington, DC, pp. 493–521.

[13] McFarland, V. A., "Evaluation of Field-Generated Accumulation Factors for Predicting the Bioaccumulation Potential of Sediment-Associated PAH Compounds," U.S. Army Corps of Engineers, Washington, DC, 1995, p. 158.

[14] OECD, *Guidance Document on Aquatic Toxicity Testing of Difficult Substances and Mixtures*, OECD, Paris, 2000, p. 53.

[15] Rand, G. M., Ed., *Fundamentals of Aquatic Toxicology: Effects, Environmental Fate, and Risk Assessment*, 2nd ed., Taylor & Francis, Washington, DC, 1995, p. 1125.

[16] Singer, M. M., Aurand, D., Bragin, G. E., Clark, J. R., Coelho, G. M., Sowby, M. L., and Tjeerdema, R. S., "Standardization of the Preparation and Quantitation of Water-Accommodated Fractions of Petroleum for Toxicity Testing," *Marine Pollution Bulletin*, Vol. 40, No. 11, 2000, pp. 1007–1016.

[17] ISO, *Nomenclature for "Selenastrum capricornutum,"* ISO, Geneva, 1999, p. 4.

[18] European Union, *Commission Directive 93/21/EEC of 27 April 1993 Adapting to Technical Progress for the 18th Time Council Directive 67/548/EEC on the Approximation of the Laws, Regulations and Administrative Provisions Relating to the Classification, Packaging and Labelling of Dangerous Substances, Official Journal for the European Communities*, No. LI 10, 5 April 1993.

[19] OECD, *OECD Guideline for Testing of Chemicals: 205—Avian Dietary Toxicity Tests*, Organization for Economic Cooperation and Development, Paris, 1984, p. 9.

[20] ECETOC, *Evaluation of Anaerobic Biodegradation*, Technical Report No. 28, ECETOC, Brussels, 1988.

[21] CEC, "CEC L-33-A-93 Test Method," Coordinating European Council, Brussels, 1995, p. 18.

[22] CONCAWE, "A Test Method to Assess the 'Inherent' Biodegradability of Oil Products," 1999, CONCAWE, Brussels, p. 33.

[23] Battersby, N. S., "The ISO Headspace CO_2 Biodegradation Test," *Chemosphere*, Vol. 34, No. 8, 1997, pp. 1813–1822.

22

Determination of Inorganic Species in Petroleum Products and Lubricants

R. A. Kishore Nadkarni[1]

ALTHOUGH GENERALLY THE ANALYSIS OF organic components in petroleum products and lubricants receives the most attention in technology, the inorganic metals and nonmetals also constitute an important part of these products. In some products (e.g., crude oils) they occur naturally; in others (e.g., lubricants), they are purposely added to enhance performance of products. A majority of elements in the periodic table can be found in petroleum products. The major constituents are made from carbon-hydrogen-oxygen-sulfur (C-H-O-S) present at percent levels while metallic elements are present in levels of milligrams per kilogram or higher.

Generally, it is believed that the source of petroleum is of marine animal and vegetative origin deposited with sediment in the coastal waters in the prehistoric times. Over a period of millenniums, aided by heat and pressure, the material released various elemental and organic volatile compounds, leaving a mixture of hydrocarbons containing varying amounts of sulfur, oxygen, nitrogen, and trace amounts of metals and other elements as the end product.

Except for carbon and hydrogen, sulfur and nitrogen are the most abundant elements in the crude oils. Sulfur up to 6 mass percent and nitrogen over 1 mass percent have been reported in some crude oils. More than 40 other elements have been found in crude oils. Most common are vanadium (up to 1,000 mg/kg), nickel (up to 100 mg/kg), and iron (up to 30 mg/kg). In lesser amounts, other elements such as As, Ba, Be, Cd, Cr, Cu, Hg, Mn, Mo, Pb, Sb, Se, Sn, Te, etc. have been identified in the crude oils. These metals in crude oils are present as porphyrins, transition metal complexes, organometallic compounds, salts of carboxylic acids, and colloidal minerals.

Trace metals such as As, Fe, Na, Ni, Pb, and V present in the crude oils can have an adverse effect on the refining and processing operations, generally acting as catalyst poisons. Sulfur and mercury in crude oils are sources of atmospheric contamination by these harmful pollutants emitted to the environment during processing.

On the other hand, in formulating lubricating oils and additives, a number of organometallic compounds are purposely added to the products to enhance various oxidative and lubricating properties. Table 1 summarizes the various additive metals added to the lubricating oils to improve the engine performance.

USED OIL ANALYSIS

Trace metals in used oils, also called wear metals, originate from mechanical wear of the engine components as well as from contamination from air, fuel, and liquid coolant. Generally, the metals are present as particulate materials rather than as true solutions. The presence of specific metals in used lubricating oils is attributed to wear of specific metal parts of an engine. In a normally running engine, wear metal content of the oil slowly increases due to normal wear. However, a sudden increase in one or more metals in the in-service oils is indicative of excessive wear of some engine component. The analysis of in-service engine oil is used to estimate when the oil should be changed based on, among other measurements, the metal analysis. The analysis of used oils is also important for reprocessing the oils for environmental protection. Table 2 summarizes the common wear metals encountered and the engine wear they indicate.

Nonhomogeneous sampling due to the effect of particle size of wear metals is a great concern in used oil analysis. Severely worn engines produce large particles, and it is more important that these large rather than the small particles be included in such cases. To overcome this particles size effect, the so-called particle size–independent methods have been developed, which involve dissolution of oil in mineral acids followed by dilution with organic solvents and measurement by atomic absorption analysis (AAS) or inductively coupled plasma–atomic emission spectrometry (ICP-AES) [2–4].

Elemental analysis of used oils is generally performed using atomic spectroscopic methods, particularly ICP-AES, because of its multielement capability. However, the sampling system of ICP may be blocked by the wear metal particulates and it would not be able to "accurately" determine the elements because the particulates may also be settled and would not be fully decomposed with such a short residence time (about 20 ms). Rotrode may be a better method of choice for used oil analysis, and it has been used quite extensively for trend analysis. Rotating disk electrode emission spectrometry, either by a single spark technique or ashing rotrode technique, also allows for rapid multielement analysis of used oils. By analyzing two aliquots, one directly and one after acid dissolution, differentiation can be made between large and small wear particles. Rotrode system is perhaps more effective in sampling particulates than ICP-AES. Rotrode filter systems that collect particles on a porous electrode prior to analysis are generally considered a more effective sampling technique for sampling particulates compared to traditional ICP-AES technique.

ANALYTICAL METHODS FOR INORGANIC DETERMINATIONS

A vast body of literature exists detailing the elemental analysis of crude oils, petroleum products, additives, lubricating

[1] *Millennium Analytics, Inc., East Brunswick, NJ 08816.*

TABLE 1—Lubricant Additive Compounds and Their Performance Properties

Metal	Compounds	Performance Characteristics
Antimony	Dialkyldithiocarbamates, dialkylphosphorodithionates	Antiwear, extreme pressure, antioxidant
Barium	Sulfonates, phenates, diorganodiphosphates, phosphonates, thiophosphonates	Detergent inhibitors, corrosion inhibitors, detergents, rust inhibitors, automatic transmission fluids, greases
Boron	Borax and esters	Antiwear agents, antioxidant, deodorant cutting oils, greases, brake fluids
Cadmium	Dithiophosphates	Steam turbine oils
Calcium	Sulfonates, phenates	Detergent inhibitors, dispersants
Chromium	Salts	Grease additive
Lead	Naphthenate	Extreme pressure additive, greases, gear oils
Magnesium	Sulfonates, phenates	Detergent inhibitors
Mercury	Organic compounds	Bactericide in cutting oil emulsions
Molybdenum	Disulfide, phosphate, dibutyldithiocarbamate	Greases, extreme pressure additives
Nickel	Cyclopentadienyl complexes	Antiwear agents, carbon deposit reduction, improved lubrication and combustion
Phosphrous	Metaldialkyldithiophosphates	Detergents, antirusting agents
Selenium	Selenides	Oxidation and bearing corrosion inhibitors
Silicon	Silicone polymers	Foam inhibitors
Tin	Organocompounds	Antiscuffing additives, metal deactivators
Zinc	Dialkyldithiophosphates, dithiocarbamates, phenolates	Antioxidant, corrosion inhibitor, antiwear additive, detergents, extreme pressure additives, crankcase oils, hypoid gear lubricants, greases, aircraft piston engine oils, ATF, railroad diesel engine oils, brake lubricants

Excerpted from Reference [1].

oils, and used lubricating oils. At one time, before the advent of modern instrumental techniques, most of the elemental analysis was achieved by painstaking wet classical test methods. It is doubtful if there are any of today's commercial labs still routinely using such methods. A number of monographs are available on the subject of elemental analysis of petroleum products using instrumental methods [1,7,8].

All well-established analytical techniques are used in such analyses–AAS, graphite furnace AAS, hydride generation AAS, ICP-AES, ICP–mass spectrometry (ICP-MS), isotope dilution MS, wavelength and energy dispersive X-ray fluorescence (or XRF), ion chromatography, neutron activation analysis, oxidation/chemiluminescence or ultraviolet (UV) fluorescence, and microelemental analysis techniques. Many of the ASTM elemental analytical standards are based on most of these techniques. A list of ASTM elemental analysis standards available is given in Table 3. Several other standards are also under development in the D02.03 Subcommittee on Elemental Analysis. This table also contains cross-references to the Institute of Petroleum (IP) (now called Energy Institute) and International Organization for Standardization (ISO) technically equivalent standards.

Table 4 gives the categorization of these elemental analysis methods by the technique used for the final measurements.

SAMPLE PREPARATION

It is an old axiom that an analysis is only as good as the sampling used. Collection of a meaningful and representative sample is often the most critical step in an analytical procedure. In particular for trace element analysis, extreme care must be taken that the sample is not contaminated. Good lab practices (GLP) in this area include storage of the samples in a clean environment, use of clean glassware, and clean environment where the analytical operations and instruments are located.

Generally, ASTM Standard Practice D4057 or D4177 is used for manual or automatic sampling of petroleum products, respectively. Before taking an aliquot from a bulk material, the sample should be well mixed and homogeneous.

Once a representative aliquot is obtained, its further treatment will depend on the type of analysis that is to be done. Some test methods do not require any further sample preparation. Mostly these are XRF, NMR, C-H-N, and other miscellaneous test methods. Some methods require sample dilution with appropriate organic solvents. Other methods require ashing or sulfated ashing followed by mineral acid dissolution. Some methods require decomposition with other fusion agents or mineral acids. These three modes of preparation are used for AAS or ICP-AES measurements. Additionally, some methods use a pressure bomb or a lamp in an oxygen environment to effect dissolution of organic matrix. A number of methods use combustion followed by absorbent trains to isolate the species of interest that are then measured by a variety of means. See ASTM Standard Practice D7455, Sample Preparation of Petroleum Products

TABLE 2—Wear Metals in Used Lubricating Oils

Metal	Wear Indication
Aluminum	Piston and bearings wear, push rods, air cooler, pump housings, oil pumps, gear castings, box castings
Antimony	Crankshaft and camshaft bearings
Barium	Detergent additive, grease additive
Boron	Coolant leakage in system
Cadmium	Bearings
Calcium	Hard water, detergent additive, oxidation inhibitor, road salt
Chlorine	Antiwear additive, extreme pressure additive
Chromium	Ring wear, cooling system leakage, chromium plated parts in aircraft engines, cylinder liners, seal rings
Copper	Wear in bushings, injector shields, coolant core tubes, thrust washers, valve guides, connecting rods, piston rings, bearings, sleeves, bearing cages
Iron	Wear from engine block, cylinders, gears, cylinder liners, valve guides, wrist pins, rings, camshaft, oil pump, crankshaft, ball and roller bearings, rust
Lead	Bearings, fuel blowby, thrust bearings, bearing cages, bearing retainers
Magnesium	Cylinder liner, gear box housings in aircraft engines
Molybdenum	Wear in bearing alloys and in oil coolers, various Mo alloyed components in aircraft engines, piston rings
Nickel	Bearings, valves, gear platings
Phosphorus	Antiwear additive, extreme pressure additive
Potassium	Coolant additive leakage
Silicon	Dirt intrusion from improper air cleaner, seal materials, antifoamant leakage
Silver	Wrist pin bearings in railroad and auto engines, silver-plated spline lubricating pump
Sodium	Antifreeze leakage
Tin	Bearings and coatings of connecting rods and iron pistons
Titanium	Various titanium alloyed components in aircraft engines
Tungston	Bearings
Vanadium	Surface coatings on piston rings, turbine impeller beds, valves
Zinc	Neoprene seals, galvanized piping

Adapted in part from References [1,5,6].

and Lubricants for Elemental Analysis, for further discussion of this subject.

ASH–SASH

These two tests, ash (ASTM D482) and sulfated ash (SASH) (ASTM D874), are the two most primary and simple tests used in characterizing the metal content of petroleum products. Knowledge of the ash-forming materials present in the product can provide information whether or not the product is suitable for use in a particular application. Ash can result from oil or water-soluble metallic compounds or from extraneous contamination from dirt or rust. The tests simply involve combustion in the first case and combustion with concentrated sulfuric acid in the second case. At the end, the residue after combustion is measured in both the tests. Although, prima facie simple, these tests, particularly, the SASH test is fraught with interferences. In certain types of samples, all of the ash-forming metals are not retained quantitatively in the ash. For unused lubricating oils, SASH procedure should be used rather the ash procedure.

Detailed discussion about the interferences in the sulfated ash procedure is given by Nadkarni et al. [9]. There is evidence that magnesium does not react in the same way as other alkali metals in this test. If magnesium additives are present in the sample, the data should be used with caution. Samples containing molybdenum can give low results because the molybdenum compounds are not fully recovered at the temperature of ashing. When phosphorus is absent, Ba, Ca, Mg, Na, and K are converted to their sulfates, and tin and zinc to their oxides. When phosphorus is present along with metals, it remains partially or wholly in the sulfated ash as metal phosphates. Because zinc sulfate decomposes slowly to its oxide at the ignition temperature specified in the test method, samples containing zinc can give variable results unless the zinc sulfate is completely converted to zinc oxide.

Because of these interelement interferences, experimentally obtained sulfated ash values may differ from the sulfated ash values calculated from the elemental analysis. The formation of such nonsulfated species is dependent on the temperature of ashing, time ashed, and the composition of the metal compounds present in the oils. Hence, sulfated ash requirement generally should not be used in product specifications without a clear understanding between a buyer and a seller of the unreliability of the sulfated ash value as an indicator of the total metallic compound content of the oil.

CARBON-HYDROGEN-NITROGEN (C-H-N)

The determination of these matrix elements is useful in determining the complex nature of sample types covered in this test method. These results can be used to estimate the processing and refining potentials and yields in the petrochemicals industry. The concentration of nitrogen is typically a measure of the presence of nitrogen containing additives. This information can be used to predict performance. Some petroleum products also contain naturally occurring nitrogen. Knowledge of hydrogen content of the sample is helpful in addressing their performance characteristics. Hydrogen-to-carbon ratio is useful to assess the performance of upgrading processes.

Nadkarni et al. [10] demonstrated the use of such instrumental C-H-N determination of fossil fuels using the first commercially available instrument from Carlo Erba. Later, the D5291 standard was issued by ASTM, at that time the

TABLE 3—D02.SC 3 Elemental Analysis Test Methods

Designation-Year	Subject	IP	ISO
D129-00(05)	Sulfur by bomb method	61	
D482-07	Ash in petroleum products	4	6245
D808-05	Chlorine in petroleum products by bomb		
D874-07	Sulfated ash	163	3987
D1018-00(05)	Hydrogen in petroleum products		
D1091-00(05)	Phosphorus in lube oils and additives		
D1266-07	Sulfur by lamp method	107	
D1317-89(WD)	Chlorine in lubes by Na-alcoholate method	118	
D1318-00(05)	Sodium in fuel oils		
D1368-89(WD)	Trace lead in fuel standards		
D1548-92(WD)	Vanadium in heavy fuel oils		
D1552-07	Sulfur by high temperature method		
D2547	Lead volumetric titration	248	2083
D2599-87(WD)	Pb in gasoline by XRF	228	
D2622-08	Sulfur by wavelength dispersive XRF		
D2784-06	Sulfur in LPG		
D2785	Wickbold sulfur	243	4260
D3116-89(WD)	Trace lead in gasoline		
D3120-06e1	Trace sulfur by oxidative microcoulometry	16591	
D3227-04a	Mercaptan sulfur in gasoline	342	3012
D3228-05	Nitrogen in lubes by Kjeldahl method		
D3229-88(WD)	Low-level lead in gasoline		
D3230-08	Salts in crude oils		
D3231-07	Phosphorus in gasoline		
D3237-06e1	Lead by atomic absorthiumption spectrometry	428	
D3246-05	Sulfur by oxidative microcoulometry	373	
D3340-07	Lith and sodium in greases by flame photometry		
D3341-05	Lead in gasoline by iodine method		3830
D3348-07	Trace lead in gasoline		
D3431-87(WD)	Trace nitrogen by microcoulometry		
D3605-00(05)	Trace metals by atomic absorption and flame emission spectrometry	413	8691
D3701–01(06)	Hydrogen by low-resolution NMR	338	
D3831-01(06)	Manganese by atomic absorption spectrometry		
D4045-04	Sulfur by hydrogenolysis and rateometric colorimetry		
D4047-00(05)	Phosphorus in lubes by quinoline method	149	4265
D4294-08	Sulfur by energy dispersive XRF	336	8754
D4628-05	Additive elements by atomic absorption spectrometry	308	
D4629-08	Nitrogen by combustion–chemiluminescence	379	

(Continued)

TABLE 3—D02SC 3 Elemental Analysis Test Methods (Continued)

Designation-Year	Subject	IP	ISO
D4808-01(06)	Hydrogen by low-resolution NMR		
D4927-05	Additive elements by wavelength dispersive XRF spectrometry	407	
D4929-08	Organic chloride in crude oils		
D4951-06	Additive elements by ICPAES		
D4952-02(07)	Doctor test	30	5275
D5056-04	Trace metals in coke by atomic absorption spectrometry		
D5059-07	Lead by X-ray spectroscopy		
D5184-01(06)	Aluminum and silicon by ICPAES and AAS	377	10478
D5185-05	Additive elements by ICPAES		
D5291-02(07)	Carbon/hydrogen/nitrogen by instrumental techniques		
D5384-95(05)	Total chlorine in used oils by field kit		
D5453-08b	Sulfur by combustion–UV fluorescence		
D5600-04e1	Trace metals in coke by ICPAES		
D5622-95(05)	Oxygen by reductive pyrolysis		
D5708-05	Metals by ICPAES		
D5762-08	Nitrogen by combustion–chemiluminescence		
D5863-00a(05)	Metals in crude oils by atomic absorption spectrometry		
D6334-07	Sulfur by wavelength dispersive XRF		
D6443-04	Additive metals by wavelength dispersive XRF spectrometry		
D6444-99(WD)	Sulfur in gasoline by WD-XRF		
D6445-99(04)e1	Sulfur by energy dispersive XRF		
D6470-99(04)	Salt in crude oil		
D6481–99(04)	Additive metals by energy dispersive XRF		
D6595-00(05)	Wear metals in lubes by rotrode spectroscopy		
D6667-04	Sulfur in LPG by UV-fluorescence		
D6728-01(06)	Wear metals in fuels by rotrode spectroscopy		
D6732-04	Copper in jet fuels by GF-AAS		
D6920-07	Sulfur by EC-OCD		
D7039-07	Sulfur by MWD-XRF		
D7040-04	Phosphorus in GF 4 oils by ICP-AES		
D7041-04	Sulfur by GC-FID		
D7111-05	Trace elements in MD fuels by ICP-AES		
D7171-05	Hydrogen in MD fuels by NMR		
D7212-07	Sulfur by ED-XRF with low BG counter	531	
D7220-06	Sulfur by polarized ED-XRF	532	
D7260-06	Standard practice for ICP-AES		
D7303-06	Analysis of grease by ICP-AES		
D7318-07	Sulfate by potentiometric titration		
D7319-07	Sulfate by IC injection		

TABLE 3—D02SC 3 Elemental Analysis Test Methods (Continued)

Designation-Year	Subject	IP	ISO
D7328-07e1	Sulfate by aqueous IC		
D7343-07	Standard practice for XRF		
D7455-08	Standard practice for sample preparation		
D7482-08	Standard guide for storage of mercury samples		

WD, withdrawn.
Excerpted from Nadkarni [6].

only instrumental method available for simultaneous determination of these three elements in petroleum products. There are additional methods available for nitrogen and hydrogen determination; these are discussed later in this chapter. The ASTM D5291 method involves combusting the sample in a stream of oxygen at about 1800°C resulting in the production of CO_2, H_2O, N_2, N-oxides, and S-oxides. These oxides are removed by either absorption on various reagent columns or separated chromatographically, and at the end the species of interest are determined with thermal conductivity detector.

CHLORINE

The test methods available for this determination can be used to measure the level of chlorine containing compounds in petroleum products. This knowledge can be used to predict the performance or handling characteristics of the product. Chlorinated compounds can lead to corrosion of equipment and poisoning of the catalyst. They also present a health hazard when incompletely combusted. Chlorine content of petroleum products is determined prior to being recycled. Federal regulations mandate that often the chlorine content of used oil must be determined before recycling. These methods can also serve as a qualitative tool for the presence or nondetection of chlorine in petroleum products. In light of the industry efforts to prepare chlorine-free products, these test methods provide information regarding the chlorine levels, if any, in such products.

Also important is the determination of organic chloride. Organic chloride species are potentially damaging to refinery processes. Hydrochloric acid can be produced in hydrotreating or reforming reactors and the acid accumulates in condensing regions of the refinery. Unexpected concentrations of organic chlorides cannot be effectively neutralized and damage can result. Organic chlorides are not known to be naturally present in the crude oils and usually result from cleaning operations at producing sites, pipelines, or tanks. The three ASTM test methods used for this determination are:

D808	Chlorine in New and Used Petroleum Products (Bomb Method)
D4929	Organic Chloride Content in Crude Oil
D5384	Chlorine in Used Petroleum Products (Field Test Kit Method)

There are other methods based on XRF or ion chromatography, which are described later in this chapter.

HYDROGEN

Hydrogen content represents a fundamental quality of a petroleum product that has been correlated with many of the performance characteristics of that product. Hydrogen-to-carbon ratio is useful to assess the performance of upgrading processes.

The combustion quality of aviation turbine fuel has traditionally been controlled in specifications by such tests as smoke point (ASTM D1322), smoke volatility index, aromatic content, or luminometer number (ASTM D1740). Evidence is available that a better control of the quality may be obtained by limiting the minimum hydrogen content of the fuel.

Combustion properties of gas turbine fuels are primarily related to hydrogen content. As the hydrogen content of these fuels decreases, soot deposits, exhaust smoke, and thermal radiation increase. Soot deposits and thermal radiation can increase to the point that combustor liner burnout will occur. Hydrogen content is a procurement requirement of the following military fuels: JP-5 specified in MIL-DTL-5624U, JP-8 specified in MIL-DTL-83133E, and Naval Distillate specified in MIL-PRF-16884K.

The ASTM test methods available for the determination of hydrogen in petroleum products include:

D1018	Hydrogen in Petroleum Fractions
D3701	Hydrogen Content of Aviation Turbine Fuels by Low Resolution NMR
D4808	Hydrogen Content of Light Distillates, Middle Distillates, Gas Oils, and Residua by Low Resolution NMR Spectroscopy
D5291	Instrumental Determination of Carbon, Hydrogen, and Nitrogen in Petroleum Products and Lubricants
D7171	Hydrogen Content of Middle Distillate Petroleum Products by Low Resolution Pulsed NMR Spectroscopy

MERCURY

The presence of mercury in crude oil production, transport, and processing can be an environmental hazard and concern to industrial hygiene. Innumerable publications on this aspect are available; hence, this point will not be belabored here. ASTM Standard Practice D7482 describes the protocols to be followed for sampling, storage, and handling of hydrocarbon samples when intended for mercury determination. Two test methods are under development that involve sample combustion followed by gold amalgamation and cold vapor atomic absorption spectrometry measurements.

TABLE 4—Analytical Techniques Used for Elemental Analysis

ASTM	Analysis
Atomic Spectroscopy	
D1318	Sodium in fuel oils
D1368	Trace lead in fuel standards
D3116	Trace lead in gasoline
D3237	Lead by atomic absorption spectrometry
D3340	Lithium and sodium in greases by flame photometry
D3341	Lead in gasoline by iodine method
D3348	Trace lead in gasoline
D3605	Trace metals by atomic absorption and flame emission spectrometry
D3831	Manganese by atomic absorption spectrometry
D4628	Additive elements by atomic absorption spectrometry
D4951	Additive elements by ICP-AES
D5056	Trace metals in coke by AAS
D5184	Aluminum and silicon by ICP-AES and AAS
D5185	Additive elements by ICP-AES
D5600	Trace metals in coke by ICP-AES
D5708	Metals by ICP-AES
D5863	Metals in crude oils by AAS
D6595	Wear metals in lubes by rotrode spectroscopy
D6728	Wear metals in fuels by rotrode spectroscopy
D6732	Copper in jet fuels by GF-AAS
D7040	Phosphorus in GF 4 oils by ICP-AES
D7111	Trace elements in MD fuels by ICP-AES
D7260	Standard practice for ICP-AES
D7303	Analysis of greases by ICP-AES
D7482	Guide for storage, and handling of crude oils for mercury analysis
X-Ray Fluorescence Spectroscopy	
D2599	Lead in gasoline by XRF
D2622	Sulfur by WD-XRF
D2784	Sulfur in LPG
D4294	Sulfur by energy dispersive XRF
D4927	Additive elements by wavelength dispersive XRF
D5159	Lead by X-ray spectroscopy
D6334	Sulfur by wavelength dispersive XRF
D6443	Additive metals by wavelength dispersive XRF
D6444	Sulfur in gasoline by WD-XRF
X-Ray Fluorescence Spectroscopy	
D6445	Sulfur by ED-XRF
D6481	Additive metals by ED-XRF
D7039	Sulfur by MWD-XRF
D7212	Sulfur by ED-XRF with low background counter
D7220	Sulfur by polarized ED-XRF
D7343	Standard practice for XRF
Combustion–Elemental Detection	
D1552	Sulfur by high temperature method
D3120	Sulfur by oxidative microcoulometry
D3246	Sulfur by oxidative microcoulometry
D3431	Trace nitrogen by microcoulometry
D4045	Sulfur by hydrogenolysis and rateometric colorimetry
D4629	Nitrogen by combustion–chemiluminescence
D5291	Carbon, hydrogen, and nitrogen by instrumental techniques
D5453	Sulfur by combustion–UV fluorescence
D5622	Oxygen by reductive pyrolysis
D5762	Nitrogen by combustion–chemiluminescence
D6667	Sulfur in LPG by UV-fluorescence
D6920	Sulfur by EC-OCD
Wet Chemistry Methods	
D129	Sulfur by bomb method
D482	Ash
D808	Chlorine by bomb method
D874	Sulfated ash
D1091	Phosphorus in lube oils and additives
D1266	Sulfur by lamp method
D1317	Chlorine in lube oils by sodium-alcoholate method
D1548	Vanadium in heavy fuel oils
D2547	Lead by volumetric titration
D2785	Sulfur by Wickbold combustion
D3227	Mercaptan sulfur in gasoline
D3228	Kjeldahl nitrogen method for lubes
D3229	Low-level lead in gasoline
D3231	Phosphorus in gasoline
D4047	Phosphorus in lubes by quinoline method
D4929	Organic chloride in crude oils
D4952	Doctor test for sulfur
D5384	Trace chlorine in used oils by field kit

(Continued)

TABLE 4—Analytical Techniques Used for Elemental Analysis (Continued)

ASTM	Analysis
Miscellaneous Methods	
D1018	Hydrogen in petroleum products
D3230	Salts in crude oils
D3701	Hydrogen by low-resolution NMR
D4805	Hydrogen by low-resolution NMR
D6470	Salt in crude oil
D7041	Sulfur by GC-FID
D7171	Hydrogen in middle distillate fuels by NMR
D7318	Sulfate by potentiometric titration
D7319	Sulfate and chloride by ion chromatography injection
D7328	Sulfate and chloride by aqueous ion chromatography

NITROGEN

The concentration of nitrogen is a measure of the presence of nitrogen-containing additives. Knowledge of its concentration can be used to predict performance. Some process catalysts used in petroleum and chemical refining may be poisoned when even trace amounts of nitrogenous materials are contained in the feed stocks. Some nitrogen compounds tend to be the most difficult class of compounds to hydrogenate, so the nitrogen content remaining in the product of a hydrotreator is a measure of the effectiveness of the hydrotreating process. These tests can be used to determine bound nitrogen in process feeds and may also be used to control nitrogen compounds in finished products. Several ASTM test methods are available for the determination of nitrogen in lubricating oils and other petroleum products. These include:

D3228	Total Nitrogen in Lubricating Oils and Fuel Oils by Modified Kjeldahl Method
D3431	Trace Nitrogen by Microcoulometry
D4629	Trace Nitrogen in Liquid Petroleum Hydrocarbons by Syringe/Inlet Oxidative Combustion and Chemiluminescence Detection
D5291	Instrumental Determination of Carbon, Hydrogen, and Nitrogen in Petroleum Products and Lubricants
D5762	Nitrogen in Petroleum and Petroleum Products by Boat-Inlet Chemiluminescence

The classical Kjeldahl method has been in use for over 100 years. In this case, a sample is digested with a hot catalyzed mixture of concentrated sulfuric acid and potassium sulfate that converts the nitrogen compounds to ammonium salts. These salts are then decomposed in a hot alkaline solution, releasing ammonia, which is then distilled into either standard sulfuric or boric acid solution, and finally determined by alkalimetric or acidimetric titration. However, the method is not applicable to the compounds containing an N-N or N-O linkage, since nitrogen in such compounds may be converted to oxygen-nitrogen compounds (i.e., nitrates). Thus, often Kjeldahl method tends to give lower biased results compared with instrumental methods described here. For an estimation of true total nitrogen, only instrumental nitrogen methods will give the correct results.

In addition to the C-H-N method, ASTM D5291, mentioned, there are at least two other instrumental methods standardized just for nitrogen determination at high and trace amount levels. These are based on similar principles of combustion, chemical absorption, and separation followed by chemiluminescence methods of detection. In the latter mode of detection, a sample is introduced into a quartz combustion tube by syringe, by boat inlet, or by solid sample inlet with a programmable furnace. The sample is heated to 950° to 1,100°C in an oxygen-rich atmosphere. The sample is pyrolyzed and combusted with all nitrogen being selectively converted to NO. The gaseous products are conducted through a drier and subsequently reacted in another chamber with ozone to form metastable NO^*_2, which almost instantaneously degrades to its ground state with radiation emission at 650 to 900 nm. The intensity of emission (chemiluminescence) is measured with a photomultiplier tube and quantitated using a nitrogen calibration standard.

The nitrogen results from chemiluminescence methods tend to be lower than the nitrogen results from C-H-N analyzer technique. This is because not all nitrogen species produced from the combustion would react with ozone; thus, the nitrogen results are lower than what they should be.

OXYGEN

Determination of total oxygen in gasoline and methanol fuels by reductive pyrolysis is described in ASTM Test Method D5622. This test method complements ASTM Test Method D4815, which covers the determination of several specific oxygen-containing compounds in gasoline. The presence of oxygen-containing compounds in gasoline can promote more complete combustion, which reduces carbon monoxide emissions. The Clean Air Act (1992) requires the gasoline sold within certain specified geographic areas to contain a minimum percentage of oxygen by mass (presently 2.7 mass percent) during certain periods of the year. This requirement can be met by blending compounds such as methyl *tert-butyl* ether, ethyl *tert-butyl* ether, and ethanol into the gasoline. This test method covers the quantitative determination of total oxygen, which is the regulated parameter.

PHOSPHORUS

Knowledge of the phosphorus content, and thus the phosphorus containing additives, in lubricating oil or additive can be used to predict performance characteristics. Phosphorus in gasoline can damage catalytic convertors used in automotive emission control systems, and its level therefore is kept low. GF 4 and later grades of engine oils marketed in the years 2004 onward have a maximum phosphorus concentration limit of 500 to 800 mg/kg. These limits are required to minimize poisoning of automotive emission control by volatile phosphorus species. It is anticipated that the later grades of oil may have even lower phosphorus levels allowed in them. In addition to the test methods designed only for phosphorus determination (a), there are several other ASTM test methods using ICP-AES

(b) or XRF (c) technique that can accurately determine in a variety of matrices. These include:

D1091	Phosphorus in Lubricating Oils and Additives (a)
D3231	Phosphorus in Gasoline (a)
D4047	Phosphorus in Lubricating Oils and Additives by Quinoline Phosphomolybdate Method (a)
D4927	Elemental Analysis of Lubricant and Additive Components by Wavelength Dispersive X-Ray Fluorescence Spectroscopy (c)
D4951	Determination of Additive Elements in Lubricating Oils by ICP-AES (b)
D5185	Determination of Additive Elements, Wear Metals, and Contaminants in Used Lubricating Oils and Determination of Selected Elements in Base Oils by ICP-AES (b)
D5600	Trace Metals in Petroleum Coke by ICP-AES (b)
D6443	Determination of Ca, Cl, Cu, Mg, P, S, and Zn in Unused Lubricating Oils and Additives by WD-XRF Spectrometry (Mathematical Correction Procedure) (c)
D7040	Determination of Low Levels of Phosphorus in ILSAC GF 4 and Similar Grade Engine Oils by ICP-AES (a, b)
D7111	Determination of Trace Elements in Middle Distillate Fuels by ICP-AES (b)
a, Method for phosphorus only; b, method using ICP-AES technique for multielement analysis; c, method using XRF technique for multielement analysis.	

SALTS

Excessive chloride left in crude oil frequently results in higher corrosion rates in refining units and has detrimental effects on the catalysts used in these units. It is important to know the approximate content of chloride in crude oils to decide whether the crude oil needs desalting. The efficiency of the process desalter can also be evaluated. This information is useful for crude oil processors. Two ASTM test methods available for such analysis are:

D3230	Salts in Crude Oils (Electrometric Method)
D6470	Salt in Crude Oils (Potentiometric Method)

SULFUR

A very large amount of literature exists on the significance of presence of sulfur in various petroleum products, particularly gasoline and diesel. The U.S. Environmental Protection Agency (EPA) and similar regulatory bodies throughout the world regulate the sulfur content of these fuel products with a view to reduce the atmospheric emissions from their combustion, mainly from automobiles and power generators. The quality of many petroleum products is related to the amount of sulfur present. Knowledge of sulfur concentration is necessary for processing purposes. In distillate fuels and solvents, sulfur can be present as mercaptans or as hydrogen sulfide. These can attack many metallic and nonmetallic materials in fuel and other distribution systems. A negative Doctor Test (ASTM D4952) ensures that the concentration of these compounds is insufficient to cause such problems in normal use. In liquefied petroleum gases, presence of sulfur can result in corrosion of metal surfaces. Sulfur can be poisonous to catalysts in subsequent processing.

A large number of test methods are also available for this analysis. The principles and precision capability of these techniques have been evaluated elsewhere [11–13]. A list of methods available for sulfur determination is given later. Generally, the methods are based on XRF spectroscopy or combustion-detection by various physical means. Of these, the U.S. EPA has approved ASTM D2622 WD-XRF, D3120 microcoulometry, D5453 UV-fluorescence, and D7039 MWD-XRF as acceptable methods for analyzing ultra low sulfur diesels.

D1266	Sulfur in Petroleum Products (Lamp Method)
D1552	Sulfur in Petroleum Products (High Temperature Method)
D2784	Sulfur in LPG by Oxy-Hydrogen Burner or Lamp
D3120	Trace Quantities of Sulfur in Light Liquid Petroleum Hydrocarbons by Oxidative Microcoulometry
D3227	(Thio Mercaptan) Sulfur in Gasoline, Kerosine, Aviation Turbine, and Distillate Fuels by Potentiometric Method
D3246	Sulfur in Petroleum Gas by Oxidative Microcoulometry
D4045	Sulfur in Petroleum products by Hydrogeneolysis and Rateometric Colorimetry
D4294	Sulfur in Petroleum and Petroleum Products by ED-XRF Spectrometry
D4952	Qualitative Analysis for Active Sulfur Species in Fuels and Solvents (Doctor Test)
D5453	Total Sulfur in Light Hydrocarbons, Spark Ignition Engine Fuel, Diesel Engine Fuel, and Engine Oil by UV Fluorescence
D6334	Sulfur in Gasoline by WD-XRF
D6445	Sulfur in Gasoline by ED-XRF Spectrometry
D6667	Total Volatile Sulfur in Gaseous Hydrocarbons and Liquefied Petroleum Gases by UV Fluorescence
D6920	Total Sulfur in Naphthas, Distillates, Reformulated Gasolines, Diesels, Biodiesels, and Motor Fuels by Oxidative Combustion and Electrochemical Detection
D7039	Sulfur in Gasoline and Diesel Fuel by Monochromatic WD-XRF Spectrometry
D7041	Total Sulfur in Light Hydrocarbons, Motor Fuels, and Oils by On-Line Gas Chromatography with Flame Photometric Detection
D7212	Low Sulfur in Automotive Fuels by ED-XRF Spectrometry Using a Low Background Proportional Counter
D7220	Sulfur in Automotive Fuels by Polarization XRF Spectrometry

Additionally, some methods have been specifically designed for measuring sulfate and chloride in gasolines. The reason for the interest in these species is because of the use of ethanol as a blending agent in gasoline. Sulfates and chlorides are indicated in the filter plugging deposits and fuel injector deposits. When fuel ethanol is burned, sulfates may contribute to sulfuric acid emissions. Ethanol acceptability for use depends on the sulfate content.

D7318	Total Inorganic Sulfate in Ethanol by Potentiometric Titration
D7319	Total and Potential Sulfate and Inorganic Chloride in Fuel Ethanol by Direct Injection Suppressed Ion Chromatography
D7328	Total and Potential Inorganic Sulfate and Total Inorganic Chloride in Fuel Ethanol by Ion Chromatography Using Aqueous Sample Injection

METALS

The determination of metals in petroleum products is a thriving area with a number of techniques and standard test methods available for this work. All stages of petroleum products from crude oils to finished products are analyzed for metals. At one time, lead was an important metal analyzed in gasoline; however, with lead compounds being banned from motor fuels, this importance has waned. Some specific reasons for metal analysis of various petroleum products and lubricants are:

- Some oils are formulated with metal-containing additives that act as detergents, antioxidants, antiwear agents, etc. (see Table 1). Some of these additives contain one or more of these metals: barium, calcium, copper, magnesium, molybdenum, zinc. Additive packages are blends of individual additives, whose specifications are based in part on elemental composition. Lubricating oils are typically blends of additive packages, and their specifications are also determined, in part, by elemental composition.
- Catalyst fines in fuel oils can cause abnormal engine wear. The major constituents of these catalysts are aluminum and silicon, and it is necessary to estimate the amount of these metals being transported into the fuel oils.
- When fuels are combusted, vanadium present in the fuel can form corrosive compounds. The value of crude oil can be determined, in part, by the concentration of nickel, vanadium, and iron. Nickel and vanadium present at trace levels in petroleum fractions can deactivate catalysts during processing.
- Operating experience of gas turbines and diesel engines has shown that some of the ash forming substances present in a fuel can lead to high temperature corrosion, ash deposition, and fouling of fuel systems. Ash-forming materials may be in a fuel as oil-soluble metallo-organic compounds, as water-soluble salts, or as solid foreign contamination. Their presence and concentration vary with the geographic source of a crude oil, and they are concentrated in the residual fractions during the refining process. Although the distillate fuel oils are typically contaminant free, ash-forming materials may be introduced later in the form of salt-bearing water or by contact with other petroleum products during transportation and storage. Specifications for such contamination and presence of trace metals are given in ASTM Specifications D2880 and D975 for gas turbine and diesel engine fuels, respectively, and in U.S. military specification MIL-F-1688-J. Preconditioning of the fuel before it reaches the gas turbine or diesel engine has become a prerequisite for installations that use heavy petroleum fuel and for sites that use light distillate fuel oils. On-site fuel analysis to determine the extent of contamination is an integral part of a fuel quality management program. Fuel contamination specifications vary among the different gas turbine manufacturers. However, without exception, each requires that the contaminants must be as low as possible. In most power generation installations, it is the owner who has the responsibility of verifying fuel cleanliness in compliance with the turbine manufacturer's warranty specifications. This leads to an on-site analytical instrument performance requirement of below 1.0 mg/kg for several elements.
- There are several sources of multielement contamination of naval distillate fuel. Sea water is pumped into the diesel fuel tanks (as ballasts) to trim ships. In addition, some of the oilers (fuel supply ships) can have multiuse or dirty tanks. Corrosion products can also come from unlined tanks, piping, pumps, and heat exchangers.
- The concentrations of wear metals in used lubricating oils can be indicative of abnormal wear if there are baseline concentration data for comparison. A marked increase in the elemental levels can also indicate contamination from other fluids in the engine or contamination due to dirt, etc. (see Table 2).
- The presence and concentration of various metallic elements in petroleum coke are major factors in determining the suitability of the coke for various end uses. The standardized analysis methods provide a procedure for use by a buyer and a seller in the commercial transfer of petroleum coke to determine whether the material meets the specifications of the purchasing party.

A brief description of each of the major techniques used for metal analysis follows.

ATOMIC ABSORPTION SPECTROSCOPY

In the past 50 years, AAS has truly become a "cookbook" type of analysis. Most modern AAS instruments are fully automated. In this technique, a sample solution is vaporized in a nebulizer, which enters the flame and the compounds become atomized. The element concentration is determined based on the attenuation or absorption by the analyte atoms of a characteristic wavelength emitted from a light source. This is typically a hollow cathode lamp, which consists of a tungsten anode and a cylindrical cathode enclosed in a gas-tight chamber. The cathode is made up of the analyte metal. Usually separate lamps are necessary for each metal, but multielement lamps also are now widely available. The detector is usually a photomultplier tube. A monochromator is used to separate the element spectral line, and the light source is modulated to reduce the amount of unwanted radiation reaching the detector. Usually the AAS instruments use a flame as an atomization system. The fuel/oxidant ratio and fuel composition determine the flame temperature. Refractory oxide forming elements usually need a stronger oxidizing flame such as nitrous oxide + acetylene (2,955°C) instead of more commonly used air + acetylene flame (2,300°C). Figure 1 shows a schematic diagram of an atomic absorption spectrometer.

Over 70 elements can be determined by AAS. The precision of the results is of the order of about 1 to 5 %. Once the sample is in a suitable solution, an analysis can be completed in just a few minutes. Although usually free of spectral interferences, there could be ionization, chemical, and matrix interferences. See Nadkarni [8, Chapter 2] for further discussion of this technique in the oil industry.

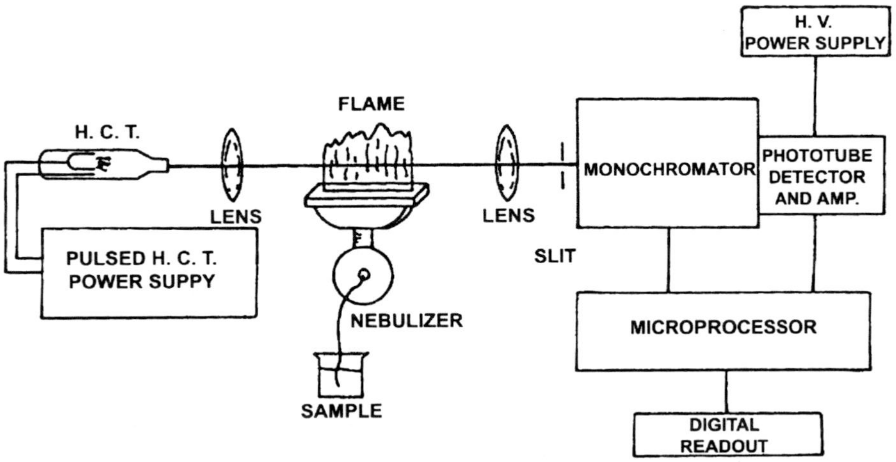

Fig. 1—Schematic representation of atomic absorption spectrometer.

ASTM	Analysis
D3237	Lead by Atomic Absorption Spectrometry
D3605	Trace Metals by Atomic Absorption and Flame Emission Spectrometry
D3831	Manganese by Atomic Absorption Spectrometry
D4628	Additive Elements by Atomic Absorption Spectrometry
D5056	Trace Metals in Coke by AAS
D5184	Aluminum and Silicon by ICP-AES and AAS
D5863	Metals in Crude Oils by AAS

Some specific elements determined due to special interest in petroleum products are described in the following sections.

LEAD

Toxic nature of lead in environment and to human health is well known. Test methods for determination of lead in gasoline were primarily designed for measuring compliance of trace lead levels as required by federal regulation for lead-free gasoline (40 CFR part 80). In addition to the ICP-AES methods, specific ASTM test methods for lead determination include:

ASTM	Scope	Matrix	Measurement
D3237	2.5–25 mg/L	Gasoline	AAS
D3341	0.026–1.3 g/L	Gasoline	EDTA titration
D3348	2.64–26.4 g/L	Field test	Colorimetric
D5059	0.0026–1.32 g/L	Gasoline	XRF

MANGANESE

ASTM Test Method D3831 can be used for AAS determination of manganese in gasoline in concentrations from 0.25 to 40 mg/L. The method was developed and tested specifically for the determination of methylcyclopentadienyl manganese tricarbonyl (MMT) added to gasoline, reformulated gasolines, and gasohols. These organomanganese compounds were added to gasolines as antiknock agents. Because generally the use of MMT has been discontinued in recent years, the widespread use of this method has declined.

VANADIUM

Vanadium in fuels can form low-melting compounds that are severely corrosive to metal parts. ASTM Test Method D1548 is used for such analysis using sulfuric acid decomposition of the fuel oil followed by formation and measurement of phosphotungstovanadic acid by colorimetry.

SODIUM

Excessive amounts of sodium can indicate the presence of materials that cause high wear of burner pumps and valves and contribute to deposits of boiler heating surfaces. Two test methods, ASTM D1318 and D3340, were specifically designated for the determination of sodium in addition to several other methods that determine sodium along with several other metals using ICP-AES. In both test methods, the sample is ashed, the residue is dissolved in dilute acid, and sodium in the resultant solution is measured with a flame photometer. As may be inferred from the discussion earlier on sodium determination, all of these metals can also be easily determined by the ICP-AES methods described later.

GRAPHITE FURNACE AAS

Substituting flame with a graphite furnace (also called as carbon rod or electrothermal) increases the sensitivity of AAS by several-fold. It is probably the most sensitive atomic spectroscopic method, rivaling mass spectrometry. Its strength lies at the ability to determine elements present in samples at sub-parts per million levels. The heated graphite furnace consists of a hollow graphite cylinder placed in the light path of an AAS instrument so that the light beam passes through it. A small hole on the top in the middle of the cylinder allows about 100 μL of sample solution to be injected in the furnace. As the furnace is gradually heated to higher temperature, it goes through the cycles of drying, charring, and atomizing. As the metal is atomized, the transient absorption signal is measured. All advantages and drawbacks for GF-AAS technique are the same as those found in flame AAS. See Tittarelli in Nadkarni [8, Chapter 3] for further discussion of this technique in the oil industry.

An example of the use of the GF-AAS technique is ASTM Standard Test Method D6732, in which copper is determined in jet fuels. The scope of the method is 5 to 100 μg/kg. The specific interest in copper determination is because at high temperatures aviation turbine fuels can oxidize and produce insoluble deposits that are detrimental to aircraft propulsion systems. Very low copper concentrations (in excess of 50 μg/kg) can significantly accelerate this thermal instability of aviation turbine fuel. Naval shipboard aviation fuel delivery systems that contain copper-nickel piping can increase copper levels in the fuel. Thus, this test method can be used for checking the copper content of aviation turbine fuels taken on shipboard, in refineries, and at fuel storage depots.

HYDRIDE GENERATION–AAS

This technique is used in special cases where high sensitivity is required but traditional techniques such as AAS or GF-AAS are not applicable. In this technique, the elemental forms in the sample solution are reacted with in situ produced hydrogen, thus converting the elements into their hydrides (e.g., selenium, bismuth, arsenic) or elemental vapors (e.g., mercury). The vapor is fed into the flame and, in the usual AAS fashion, is detected and determined by the instrument. Although at present there are no ASTM methods published for such analysis in oil industry, an example of such usage can be found in Nadkarni [14].

ATOMIC FLUORESCENCE SPECTROMETRY

In this technique, the hollow cathode lamp light source is positioned so that it does not shine directly into the instrument monochromator. The sample is aspirated into the flame, and the analyte element excited by the light source emits fluorescent radiation. This technique has better detection limits and has potential for multielement analysis. It has been demonstrated to determine mercury, sulfur, and other elements, but its use is not as widespread as that of AAS, GF-AAS, or ICP-AES. See Stockwell in Nadkarni [8, Chapter 12] for further discussion of this technique used in the oil industry.

INDUCTIVELY COUPLED PLASMA–ATOMIC EMISSION SPECTROMETRY

The most significant development in the analytical atomic spectroscopy area in the past 30 years has been the development of the technique of ICP-AES. This instrument is widely used in the oil industry laboratories for determination of metals, and in many cases it has essentially replaced the use of AAS or GF-AAS. Its main advantages are wide dynamic range, simultaneous multielement capability, and extremely low detection limits. Argon is the most commonly used gas to produce a plasma, which is an electrically neutral, highly ionized gas consisting of ions, electrons, and neutral particles. However, in the ICP-AES technique, the plasma is ionized by a strong electric field by radiofrequency. The radiofrequency discharge at 27.12 or 40.68 MHz frequency causes the flowing argon gas to become electrically conductive and reach a temperature up to 15,000°C. The sample solution is sprayed into this plasma through a nebulizer where the elements of interest are vaporized, atomized, or ionized and electronically excited. While returning to their ground states, the electrons emit photons at wavelengths characteristic of an element.

This emitted light is measured by a polychromator in a simultaneous spectrometer or is measured by a single photomultplier in a scanning or sequential spectrometer. A schematic diagram of an ICP-AES system is shown in Fig. 2.

Botto in Nadkarni [8, Chapter 4] provides further discussion of this technique and its use in the oil industry. ASTM standards using ICP-AES for oil analysis are listed next; these have been summarized by Nadkarni [15].

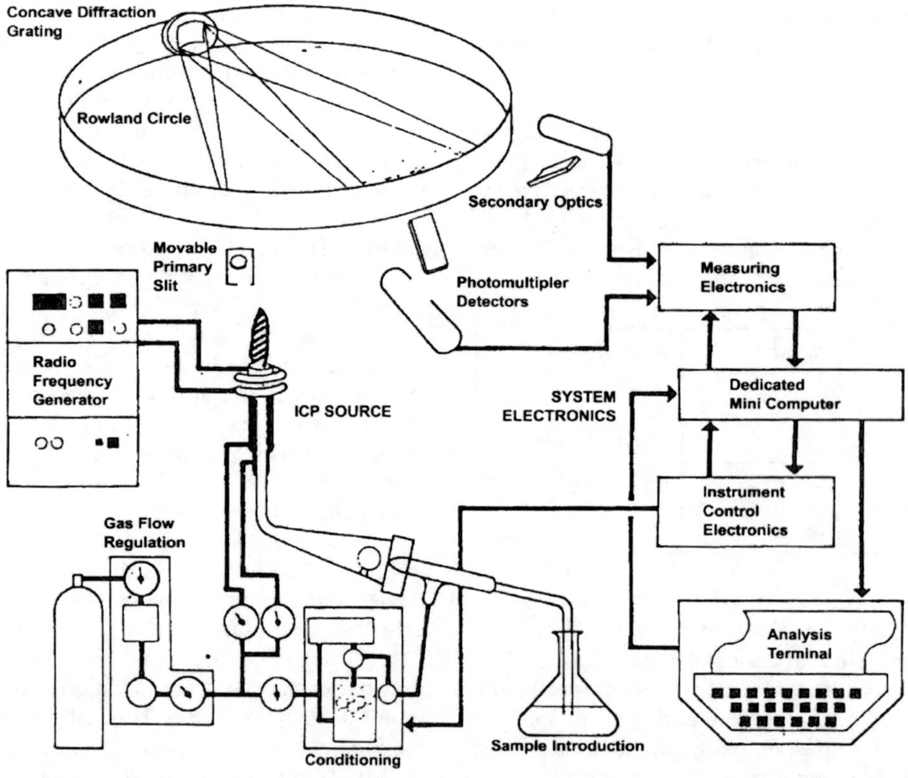

Fig. 2—Schematic representation of inductively coupled plasma–atomic emission spectrometer.

D4951	Additive Elements by ICP-AES
D5184	Aluminum and Silicon by ICP-AES and AAS
D5185	Additive Elements by ICP-AES
D5600	Trace Metals in Coke by ICP-AES
D5708	Metals by ICP-AES
D7040	Phosphorus in GF 4 Oils by ICP-AES
D7111	Trace Elements in MDFuels by ICP-AES
D7260	Standard Practice for ICP-AES
D7303	Analysis of Greases by ICP-AES

X-RAY FLUORESCENCE SPECTROMETRY

XRF technique rivals atomic spectroscopic techniques in its applicability, sensitivity, ease of operation, nondestructive nature, and applicability to various matrices without having to change the physical state of the sample. There are two principal historic modes of XRF analysis: wavelength dispersive (WD) and energy dispersive (ED). More recently, monochromatic wavelength dispersive (MWD) instruments have achieved wide acceptability in the oil industry, particularly for sulfur determination.

When a sample is bombarded with X-rays, electrons are ejected from the various energy levels of the target atoms. In the process, X-rays of discrete characteristic energy are emitted. This is called *X-ray fluorescence* (XRF). Every element has a unique electron distribution; thus, every element produces a unique secondary X-ray spectrum whose intensity is proportional to the elemental concentration in the sample. XRF can be applied to all elements with atomic numbers greater than 10 (i.e., neon). However, modern XRF instruments with improved detector technology have the capability of determining lower atomic number elements such as boron or magnesium. The basic components of an XRF system are a source of excitation consisting of a high-energy X-ray tube, an optical system for selecting characteristic C-rays, and integrating or counting circuits for measurements (Fig. 3). The common X-ray tubes used are made of W, Mo, Cr, Rh, or Sc, which have high-energy output, spectral purity, and long-term stability.

Due to numerous interelement matrix effects, standards identical in composition to the samples need to be used. Particle shape and size are also important; however, this may not be an important factor in liquid hydrocarbon samples. Many of these interferences and matrix effects can be compensated by corrective techniques such as standard addition, internal standard addition, matrix dilution, thin film method, and mathematical correction methods.

In the WD-XRF technique, an X-ray impinges on the sample and generates fluorescent X-rays. These are then diffracted on a crystal. A goniometer selects the geometry between the crystal and the detector. The geometry controls the detection of X-rays from the element of interest. Different crystals have different sensitivities. ASTM methods available for WD-XRF analysis of petroleum products include:

D2622	Sulfur by WD-XRF
D4927	Additive Elements by Wavelength Dispersive XRF
D5059	Lead by X-Ray Spectroscopy
D6334	Sulfur by Wavelength Dispersive XRF
D6443	Additive Metals by Wavelength Dispersive XRF
D6444	Sulfur in Gasoline by WD-XRF
D7039	Sulfur by MWD-XRF
D7343	Standard Practice for XRF

The ED-XRF technique uses a much less energetic X-ray tube. The emitted X-rays from the sample impinge directly on a detector, typically an Si(Li), which generates pulses on an incident beam, and these pulses are sorted and counted by a multichannel analyzer. Simultaneous determination of elements with atomic numbers greater than magnesium is possible. Optimization for specific elements is accomplished through the use of secondary targets and filters. Radioisotope sources can be used in place of X-ray tubes in instruments designed for limited element applications. ASTM methods available for ED-XRF analysis of petroleum products include:

D4294	Sulfur by Energy Dispersive X-Ray Fluorescence
D6445	Sulfur by ED-XRF
D6481	Additive Metals by ED-XRF
D7212	Sulfur by ED-XRF with Low Background Counter
D7220	Sulfur by Polarized ED-XRF
D7343	Standard Practice for XRF

In the MWD-XRF technique, a monochromatic X-ray beam with a wavelength suitable to excite the K-shell electrons of an element is focused onto a test specimen contained in a sample cell. The fluorescent K alpha radiation emitted by sulfur is collected by fixed monochromator. The intensity of the sulfur X-rays is measured using a suitable

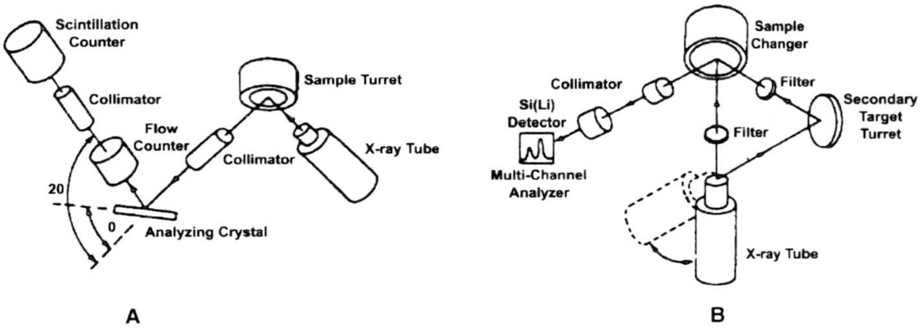

Fig. 3—Schematic representation of an X-ray fluorescence spectrometer. *A*, Wavelength dispersive XRF. *B*, Energy dispersive XRF.

detector. Excitation by monochromatic X-rays reduces background, simplifies matrix correction, and increases the signal-to-background ratio compared to polychromatic excitation used in the conventional WD-XRF instruments. So far this technique has only been used for the determination of sulfur (ASTM D7039) and chlorine in gasoline and diesel samples.

For further detailed discussions on these techniques and their applications in the oil industry, see Chapter 8 by Vrbos and Glose on WD-XRF, Chapter 9 by Pohl and Petiot on ED-XRF, and Chapter 10 by Chen on MWD-XRF in Nadkarni [8].

SPECIFIC METAL ANALYSIS

In addition to the plethora of AAS, ICP-AES, and XRF methods listed earlier for the determination of a large number of metals in petroleum products and lubricants, there are several other methods available for these determination but they are specifically meant for individual elements. These include:

ASTM Standard	Metal Determined	Technique Used
D6732	Copper in Jet Fuels	GF-AAS
D1318	Sodium in Residual Fuel Oils	Flame Photometry
D3340	Lithium and Sodium in Lubricating Greases	Flame Photometry
D1548	Vanadium in Navy Fuels	Colorimetry
D3237	Lead in Gasoline	AAS
D3341	Lead in Gasoline	Iodine Monochloride
D3348	Lead in Gasoline	Colorimetric (Field Test Method)
D5059	Lead in Gasoline	XRF
D3831	Manganese in Gasoline	AAS

ION CHROMATOGRAPHY

Until the advent of this technique in 1975, all anions in the any matrices had to be determined by wet chemistry techniques such as gravimetry or titrimetry. Ion chromatography (IC) is a combination of ion exchange chromatography, eleuent suppression, and conductometric detection. For anion analysis, a low-capacity anion exchange resin is used in the separator column and a strong cation exchange resin in the H^+ form is used in the suppressor column. A dilute mixture of sodium carbonate and bicarbonate is used as the eluent, because carbonate and bicarbonate are conveniently neutralized to low conductivity species and the different combinations of carbonate-bicarbonate give variable buffered pH values. This permits the ions of interest in a large range of affinity to be separated. The anions are eluted through the separating column in the background of carbonate-bicarbonate and conveniently detected based on electrical conductivity. See Fig. 4 for a schematic diagram of a basic ion chromatographic system.

In practice, a dilute aqueous sample is injected at the head of the separator column. The anion exchange resin selectively causes the various sample anions of different types to migrate through the bed at different respective rates, thus, effecting the separation. The effluent from the separator column then passes to the suppressor column

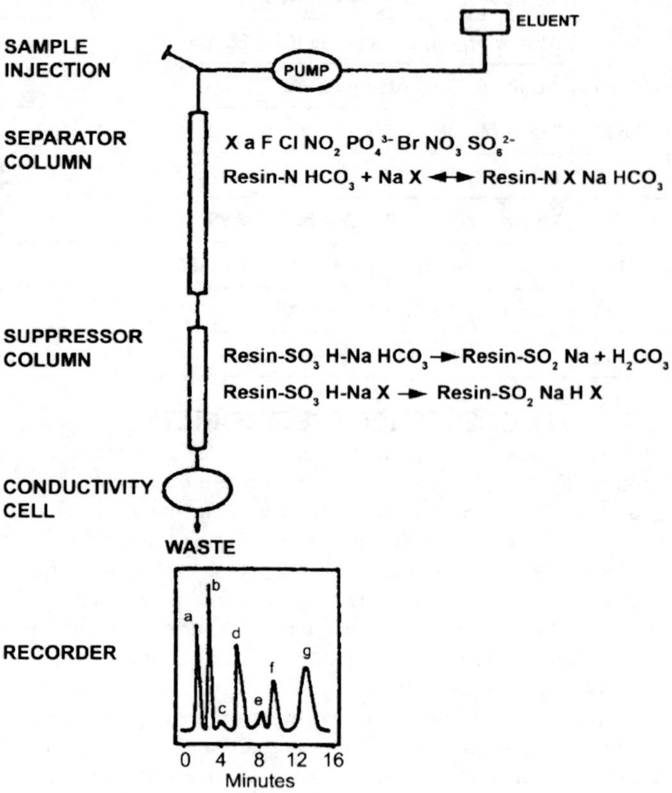

Fig. 4—Schematic representation of an ion chromatograph.

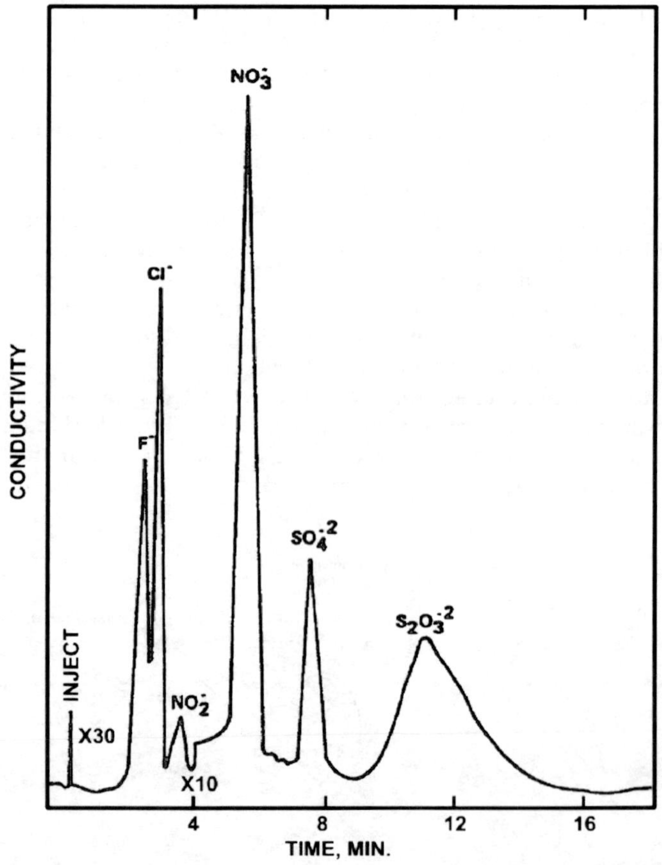

Fig. 5—Example of a multianion ion chromatogram.

where the H^+ form cation exchange resin absorbs the cations in the eluent stream. Finally, the suppressor column effluent passes through a conductivity cell. The highly conductive anions in a low background conductance of carbonic acid are detected at high sensitivity by the conductivity detector. The nonspecific nature of the conductometric detection allows several ions to be sequentially determined in the same sample. At the same time, differential elution rates from the column make the conductometric detection highly specific and relatively free from interferences. Different stable valence states of the same element can be determined. See Fig. 5 for an example of a multianion ion chromatogram.

Because of the nonspecific nature of the conductivity detector, the chromatograph peaks are identified by their retention times, using a standard anion mixture to calibrate the retention time and the peak strength. Sometimes two ions having the same or close retention times will be detected as one broad peak, giving erroneous results. Also, if one ion is present in very large amounts compared to its neighbor ion in the chromatograph, the broad peak of the first major ion will change the baseline and the shape of the second weaker ion. See Nadkarni and colleagues [16,17] for examples of ion chromatographic applications in the fossil fuel industry.

In addition to the determination of anions, other IC methods have been developed for the determination of cations using different exchangers and reagent-based eluents or solvents. However, their usefulness, at least in the oil industry, is lagging because of the wide availability of other instrumental techniques for cation analysis, such as AAS, ICP-AES, XRF, etc.

See Chassional in Nadkarni [8, Chapter 15] for further detailed discussion of this technique and its application to oil chemistry.

QUALITY ASSURANCE OF RESULTS

Virtually all ASTM standard test methods issued by D02.SC 3 Committee on Elemental Analysis include mandatory sections on quality control and assurance of results, because we consider it critical in elemental analysis just as in other forms of analysis. Thus, each test method contains an Appendix that summarizes the expected quality control/quality assurance protocols to be used. A greater discussion of quality control/quality assurance systems to be used in petroleum products and lubricants testing laboratories can be found in ASTM Standard Practice D6792 and in Nadkarni [8, Chapter 22].

FUTURE PROSPECTS

As is evident from this discussion, there are an abundant number of test methods available for the determination of inorganic constituents of the petroleum products and lubricants. Virtually all known inorganic analytical techniques have been used in the oil industry laboratories. Most modern test methods have adequate detection limits and appropriate precision for use in production and commerce. Yet no one method will deliver all requirements of an ideal test method.

TABLE 5—Comparison of Elemental Analysis Spectroscopic Methods

Parameter	AAS	GF-AAS	ICP-AES	ICP-MS	WD-XRF	ED-XRF
Instrument cost[A]	\$25 K	\$50 K	\$100 K	\$200+ K	\$100–200 K	\$50 K
Maintenance	Minimal	Minimal	Moderate	High	Minimal	Minimal
Operation	Easy	Easy	High	High	Easy	Easy
Sample preparation	Dissolution or dilution	Dissolution or dilution	Dissolution or dilution	Dissolution or dilution	None	None
Analysis time	Few minutes	>AAS	Few minutes	Few minutes	Few minutes	Few minutes
Elemental capability	Single	Single	Multi	Multi	Multi	Multi
Dynamic range (orders of magnitude)	1–3	1–2	5–6	5–8	3–6	3–6
Detection limits	mg/kg	Sub-mg/kg	Sub-mg/kg	µg/kg–ng/kg	mg/kg	g/kg
Interferences	Chemical; ionization	Chemical; matrix; physical; molecular absorption	Spectral; matrix	Spectral; chemical	Matrix	Matrix
Sample state	Liquid	Liquid	Liquid	Liquid	Liquid or solid	Liquid or solid
Sample volume	Few mL	µL	Few mL	Few mL	10–15 mL	10–15 mL
Analysis precision, %RSD	~ 1	1–3	0.5–2	0.5–3	0.2–2	0.2–2
Applications	Extensive	Limited	Extensive	Limited	Extensive	Extensive
ASTM D02 STMs	7	1	9	0	7	6

[A]Additional accessories will increase the cost.

It is up to an individual laboratory to choose what is the best method for their needs, weighing the costs versus the benefits of these techniques. Table 5 provides a comparison between the most widely used spectroscopic techniques for elemental analysis.

Developmental research is still going on in oil industry and instrument vendor laboratories to improve on the existing test methods and prepare more sensitive and precise techniques. In particular, work on mercury in crude oil, sulfur in gasoline and diesel, and analysis of biofuels is at the forefront of such research. We should soon be seeing the results of this work in formalization of more ASTM standard test methods in this area.

References

[1] Nadkarni, R. A., *Modern Instrumental Methods of Elemental Analysis of Petroleum Products and Lubricants,* STP 1109, ASTM International, West Conshohocken, PA, 1991.

[2] Saba, C. S., and Eisentraut, K. J., "Determination of Molybdenum Wear Metal in Lubricating Oils by Atomic Absorption Spectrophotometry with a Particle Size Independent Method," *Anal. Chem.*, Vol. 51, No. 1927, 1979.

[3] Brown, J. R., Saba, C. S., Rhine, W. E., and Eisentraut, K. J., *Anal. Chem.*, Vol. 52, No. 2365, 1980.

[4] Kauffman, R. E., Saba, C. S., Rhine, W. E., and Eisentraut, K. J., "Quantitative Multielement Determination of Metallic Wear Species in Lubricating Oils and Hydraulic Fluids," *Anal. Chem.*, Vol. 54, No. 975, 1982.

[5] Nadkarni, R. A., "Advances in Elemental Analysis of Hydrocarbon Products," C. S. Hsu, Ed., *Analytical Advances for Hydrocarbon Research,* Kluwer/Academic/Plenum Publishers, New York, 2003, Chapter 2, pp. 27–56.

[6] Nadkarni, R. A., "Elemental Analysis," in G. E. Totten, Ed., *Fuels and Lubricants Handbook,* ASTM International, West Conshohocken, PA, 2003, Chapter 26, pp. 707–716.

[7] Nadkarni, R. A., *Elemental Analysis of Fuels and Lubricants: Recent Advances and Future Prospects,* ASTM International, West Conshohocken, PA, STP 1468, 2005.

[8] Nadkarni, R. A., *Spectroscopic Analysis of Petroleum Products and Lubricants,* Monograph 9, ASTM International, West Conshohocken, PA, 2009, in press.

[9] Nadkarni, R. A., Ledesma, R. R., and Via, G. H., "Sulfated Ash Test Method: Limitations of Reliability and Reproducibility," SAE Technical Paper No. 952548, 1995.

[10] Nadkarni, R. A., Cornett, R. B., and Bredweg, R. L., "Evaluation of an Elemental Analyzer," *Amer. Lab.*, Vol. 17, No. 2, 1985, pp. 69–81.

[11] Nadkarni, R. A., "Determination of Sulfur in Petroleum Products and Lubricants: A Critical Review of Test Performance," *Amer. Lab.*, Vol. 32, No. 22, 2000, pp. 16–25.

[12] Nadkarni, R. A., "The Challenge of Sulfur Analysis in the Fuels of the Future," *ASTM Standardization News,* Vol. 32, No. 6, 2004, pp. 32–35.

[13] Nadkarni, R. A., "Trace Levels of Sulfur in the Fuels of the Future: An Analytical Perspective," *J. ASTM International,* Vol. 2, No. 2, 2005.

[14] Nadkarni, R. A., "Applications of Hydride Generation Atomic Absorption Spectrometry for Coal Analysis," *Anal. Chim. Acta.*, Vol. 135, No. 135, 1982, 363–368,

[15] Nadkarni, R. A., "Use of ICP-AES for Metal Analysis in the Oil Industry," *ICP Information Newsletter,* Vol. 30, No. 10, 2005, pp. 1059–1061,

[16] Nadkarni, R. A., and Pond, D. M., "Applications of Ion Chromatography for Determination of Selected Elements in Coal and Oil Shale," *Anal. Chim. Acta.*, Vol. 146, 1983, pp. 261–266.

[17] Nadkarni, R. A., and Brewer, J. M., "Applications of Ion Chromatography in Environmental Analysis," *Amer. Lab.*, Vol. 19, No. 2, 1987, pp. 50–59.

23

Standard Test Method Data Quality Assurance

Alex T. C. Lau[1]

FOR THE PETROLEUM INDUSTRY, STANDARD TEST methods, specifications, and practices developed under the jurisdiction of ASTM D02 Committee on Petroleum and Petroleum Products or ISO TC28 Technical Committee on Petroleum Products are globally accepted. A conscious decision has been made to maintain technical harmony between standards produced by both organizations on similar subject matters. The purpose of this chapter is to familiarize readers with statistics-based data quality metrics for ASTM and ISO test methods that have been adopted by both organizations, and the relevant ASTM standard practices to quantify, monitor, maintain, and apply these metrics.

TEST METHOD DATA QUALITY METRICS

Precision

A standard test method (STM) comprises a series of interconnected work processes and equipment specifications required to produce a final test result. Outcome of work processes is subject to random variation from sources that are inherent to the process itself. In a perfect world, if the same material is repeatedly tested using the same STM, whether by the same operator in the same laboratory using the same apparatus or by different laboratories or operators using different apparatus meeting the specified requirements, all test results generated will be numerically identical.

However, we do not live in a perfect world. Repeated execution of the same STM on the same material, whether by the same operator in the same laboratory using the same apparatus, or by different operators in different laboratories using apparatus of similar design, will not always yield numerically identical results. Minor differences will exist between test results due to variation in factors that are inherent (embedded) within the STM. Variation attributable to factors considered inherent to the STM is generally referred as common cause variation and can be thought of conceptually as the inherent "imperfections" associated with the STM. In statistical nomenclature, this inherent "imperfection" is known as precision, although some might prefer to think of it as imprecision.

Intuitively, the precision (variation) associated with multiple test results, obtained by the same operator on the same test material, using the same apparatus and STM, and, within a very short time, should be less than the variation associated with an equivalent number of test results, where each test result is produced by a different laboratory on the same test material using the same STM. This intuitive expectation comes from recognizing that, in the former scenario, the number of factors that can affect the precision of the test results is less than the latter. Therefore, precision must be associated with specific conditions.

Outlined below are the ASTM E456, Standard Terminology Relating to Quality and Statistics, definitions of precision, and the three sets of conditions under which precision is measured, representing the most controlled to the least controlled (Table 1 and Fig. 1). ASTM E456 is under the jurisdiction of ASTM Main Committee E11 on Quality and Statistics. While other ASTM and ISO main committees can and do use definitions that are not identical in wording to those in E456 to better serve the industry they represent, the technical intent of the E456 definitions is preserved.

- *Precision*—The closeness of agreement between independent test results obtained under stipulated conditions.
- *Repeatability conditions*—Conditions where independent test results are obtained with the same method on identical test items in the same laboratory by the same operator using the same equipment within short intervals of time.
- *Intermediate conditions*—Conditions under which test results are obtained with the same test method using test units or test specimens taken at random from a single quantity of material that is as nearly homogeneous as possible, and with changing conditions such as operator, measuring equipment, location within the laboratory, and time.
- *Reproducibility conditions*—Conditions where test results are obtained with the same method on identical test items in different laboratories with different operators using different equipment.

All STMs developed by ASTM are required to have published estimates of precision estimated under both repeatability and reproducibility conditions. These estimates must be calculated from data collected in accordance with interlaboratory study protocols meeting the technical requirements prescribed in E691, Standard Practice for Conducting an Interlaboratory Study to Determine the Precision of a Test Method, or by an interlaboratory test program that yields equivalent information, such as a standard practice developed by an ASTM technical committee. For a majority of test methods developed by ASTM D02 Committee on Petroleum and Petroleum Products, precision and bias are estimated in accordance with ASTM D6300, Standard Practice for Determination of Precision and Bias Data for Use in Test Methods for Petroleum Products and Lubricants, or its predecessor, Manual on Determining Precision Data for ASTM Test Methods on Petroleum Products and Lubricants (RR: D02-1007), published in December 1972. RR: D02-1007 significantly predates E691 and has been used by D02 until the

[1] *Process Analytics and Blending Specialist, Engineering Services Canada, Toronto, Ontario, Canada.*

TABLE 1—Standardized Conditions for Precision

	Repeatability Condition	Intermediate Precision Condition	Reproducibility Condition
Laboratory	Same	Same	Different
Operator	Same	Different	Different
Apparatus	Same	Same[A]	Different
Test material	Same	Same	Same
Time between test	Short[B]	Multiple days	Not relevant

[A]Can be different instruments meeting the same design requirement.
[B]STM dependent, typically does not exceed 1 day.

development of D6300. For ISO TC28 test methods, the corresponding standard to ASTM D6300 is ISO 4259. The technical committees responsible for these standards (ASTM D02.CS94, ISO TC28 WG2) have agreed to maintain technical harmony between D6300 and ISO 4259.

An additional term associated with intermediate precision is defined in ASTM D6299, Standard Practice for Applying Statistical Quality Assurance Techniques to Evaluate Measurement System Performance, as follows:

- Site precision (R′), *n*–the value below which the absolute difference between two individual test results obtained under site precision conditions may be expected to occur with a probability of approximately 0.95 (95 %). It is defined as 2.77 times the standard deviation of results obtained under site precision conditions.
- Site precision conditions, *n*–conditions under which test results are obtained by one or more operators in a single site location practicing the same test method on a single measurement system that may comprise multiple instruments, using test specimens taken at random from the same sample of material, over an extended period of time spanning at least a 15-day interval.

While repeatability and reproducibility conditions represent "most controlled" and "least controlled" testing environments, site precision conditions represent the typical environment under which test results are produced at a single laboratory, and hence is the most relevant set of conditions for the conduct of test method quality assurance activities within a single laboratory or site.

It should be noted that site precision conditions should include all sources of variation that are typically encountered during normal, long-term operation of the measurement system. Thus, all operators who are involved in the routine use of the measurement system should contribute results to the site precision determination. Since the objective is to estimate long-term random variation, use of an excessive number of results obtained within a 24-h period to calculate R' is discouraged as this may cause R' to be more representative of shorter-term variation than the long-term effect.

Given the different conditions under which these precision metrics are estimated, it is a statistical expectation that $r < R' < R$. Furthermore, since different sites may differ in how tightly their site conditions are controlled (i.e., numbers of operators, instruments, and general competency), R' is expected to vary from one site to another.

R, r, and R' can be interpreted nonstatistically as the "recommended maximum difference between two results obtained under specified conditions that can be attributed to the test method precision." Hence, these values can be used as decision limits to support or challenge the validity of the assumption that both test results have been produced on the same material in a correct manner under the associated specific conditions.

A more rigorous statistical interpretation of r, R', and R is that these represent numerical limits that are expected to be exceeded about 5 % of the time by the absolute (sign-less) difference between any two results obtained under the prescribed conditions. It should be noted that r, R, and R' may not necessarily remain constant throughout the measurement range or material types included in the scope of the STM. These limits are related to (but are not) the standard deviations of test results obtained on the same material under the prescribed conditions.

Three common scenarios under which R, r, and R' can be used are described below:

- R (reproducibility)–if the difference between two results produced under reproducibility conditions exceeds R, it is reasonable to conclude that one or both laboratories' test method work processes are not in statistical control, or, not executed with acceptable performance proficiency (discussed later in this section), or, that the materials tested by the labs are not the same.
- r (repeatability)–if the difference between two results obtained under repeatability conditions exceeds r, it is reasonable to conclude that the test method has not been executed properly for at least one of the test results, or, there are other assignable causes (e.g., equipment or reagents issues) that adversely affected the data quality.
- R' (site precision)–use to decide if the difference between two results produced by the same laboratory over some period of time on the same material is reasonably representative of a test method work process that is in statistical control. R' can also be compared to R by forming the ratio R/R' to validate that the site is executing the test method proficiently. This ratio is defined in ASTM D6792, Standard Practice for Quality System in Petroleum Products and Lubricants Testing Laboratories, as test performance index, and is expected to be greater than 1. See ASTM D6792 for detailed discussions.

Bias

Bias is a term that describes a systemic difference between the results from a test method and an accepted benchmark or reference value for property of the material tested. Bias is not a random effect in that this effect (systemic difference) is associated with all test results produced. The following are definitions for bias and two key related terms quoted from ASTM E456 and D3244:

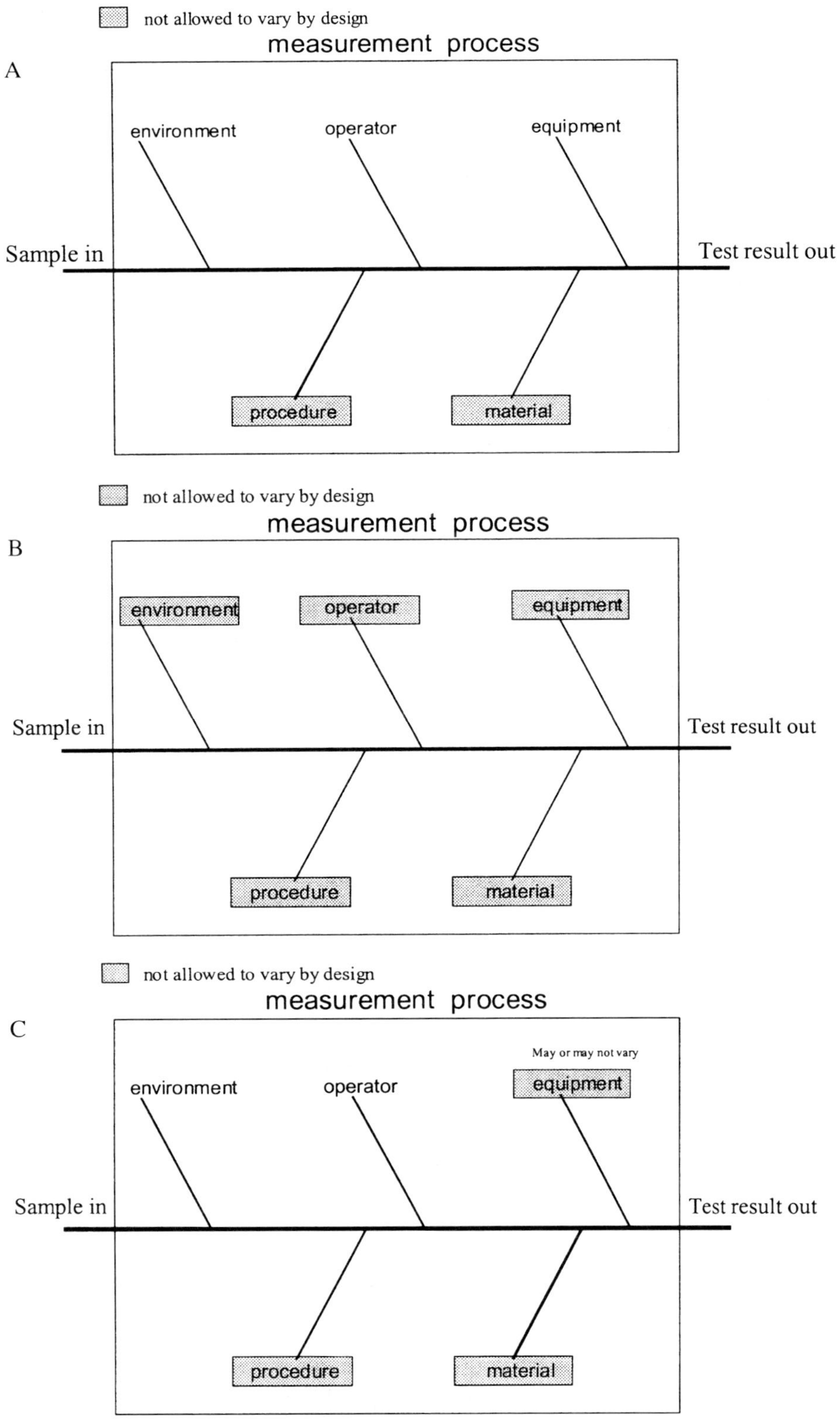

Fig. 1—*A*, Conditions under which ASTM reproducibility (*R*) is estimated. *B*, Conditions under which ASTM repeatability (*r*) is estimated. *C*, Conditions under which site precision (*R'*) is estimated.

- Bias–The difference between the expectation of the test results and an accepted reference value. (E456)
- Accepted reference value (ARV)–A value that serves as an agreed-upon reference for comparison and that is derived as (1) a theoretical or established value, based on scientific principles; (2) an assigned value, based on experimental work of some national or international organization such as the U.S. National Institute of Standards and Technology (NIST); or (3) a consensus value, based on collaborative experimental work under the auspices of a scientific or engineering group. (E456)

- True value–for practical purposes, the value toward which the average of single results obtained by *N* laboratories converges, when *N* becomes very large. Consequently, such a true value is associated with the particular test method used. (D3244)

The term *consensus value*, as described in the earlier definition of *accepted reference value*, can be practically interpreted as the *true value* as defined by D3244, provided *N* is sufficiently large. This is the most common type of ARV for ASTM D02 Committee on Petroleum and Petroleum Products test methods.

All STMs developed by ASTM are required to include bias statements. For empirical properties that are defined by the test method (e.g., octane, cetane, cloud, pour, flash), a bias statement for the property measured is not possible because the test method defines the property.

The practical interpretation of "bias" can best be illustrated by the following example:

> If a single laboratory obtained *N* test results for the same material from a single test method under site precision conditions, *AND*, if the material has an ARV, then, for large values of *N* (> 30), a formal statistical test of significance can be performed on the difference between the average of the *N* results versus the ARV of the material. The statistical test estimates the probability that a difference as large as the one observed can be obtained purely by chance under the assumption that there is no bias. If this probability is less than a user prespecified value (typically 5 %), then the test is said to be "statistically significant." The conclusion drawn is that the data produced by this test method from the laboratory have a "statistically significant" bias, with no more than about a 5 % chance that this conclusion is made in error.

Detailed discussion on statistical tests of significance is beyond the intended scope of this chapter. However, it should be noted a detected "statistically significant" bias may not be of practical significance since the magnitude of bias detectable decreases as the number of results (*N*) increases. Hence, it is possible to detect a "trivially small" bias that is of little practical consequence with a large *N*.

Accuracy

Accuracy is defined as the closeness of agreement between a test result and an accepted reference value (ASTM E456).

The term *accuracy*, as defined by ASTM E456, describes a combined performance attribute associated with a test method that includes the random (common cause) and systemic effects from all factors that influence the final test method work process outcome (test result). Hence, the combined effects of both precision and bias affect the accuracy. The term *accuracy*, as defined in ASTM E456, is intended to be a qualitative term that can be interpreted as the overall trustworthiness of any test result produced by the test method. Accuracy is not a term associated with a quantitative measure.

Bias between Standard Test Methods (Relative Bias)

ASTM D02.CS94 Coordinating Subcommittee on Quality Assurance and Statistics developed D6708, Standard Practice for Statistical Assessment and Improvement of Expected Agreement Between Two Test Methods that Purport to Measure the Same Property of a Material, to assess and quantify biases (systemic differences) between test methods that claim to measure the same property of a material. This practice defines statistical methodology to assess and decide whether a simple linear bias correction can further improve the expected agreement between the test methods. It is intended for use with results collected from an interlaboratory study meeting the requirement of Practice D6300 or its equivalent (for example, ISO 4259). The interlaboratory study must be conducted on at least ten materials that span the intersecting scopes of the test methods, and results must be obtained from at least six laboratories using each method. The statistical methodology is based on the premise that a bias correction will not be needed. In the absence of strong statistical evidence that a bias correction would result in better agreement between the two methods, a bias correction is not made. If a bias correction is required, the "parsimony principle" is followed whereby a simple correction is to be favored over a more complex one.

DATA QUALITY ASSURANCE MONITORING

In today's quality-conscious and cost-competitive environment, the ability to demonstrate "statistical control" of the test method used for product certification is one key mutual expectation between suppliers and customers. Being "in statistical control" implies the test method work process is operated with a mean (location) and precision (variation from common causes) that are stable, quantifiable, and reasonable compared to other similar work processes that are operated in accordance with an industry agreed-on standard work process such as an ASTM or an ISO STM. Statistical control status can be assessed using control chart techniques to analyze repetitive results obtained over time on standard reference materials with industry-accepted reference values, secondary working standards (typically prepared from a batch of production material), or both. A favorable outcome from the control chart assessment can be extrapolated to infer acceptable test method data quality in routine testing. In order for this extrapolation to be technically appropriate, standard reference materials and secondary working standards should be compositionally representative of material being routinely tested. As well, control chart related testing should preferably be done under "blind" conditions, that is, without revealing to the tester that these materials are standards. A nonfavorable outcome from the control chart assessment should cast doubt on the data quality integrity and should trigger a follow-up investigation.

Acquisition of standard reference materials that are compositionally representative of the product routinely tested can be a time-consuming and expensive undertaking for ASTM test methods, especially for test methods where only the equipment design, setup, and operating procedures are standardized (e.g., test methods that measure empirical bulk properties such as octane, cloud, melt/flow ratio). This is because the industry-accepted reference value for these types of material has to be established by an interlaboratory testing protocol involving multiple test method work processes from different testing organizations. For data quality assurance of these types of test methods, a cost-effective and reasonably timely alternative to standard reference material testing is to participate in regularly conducted interlaboratory exchange testing programs where typical production samples are tested by multiple production test facilities. Participants can control chart their deviations from the consensus results established by exchange testing program to determine if their testing process is or is not biased relative to the testing community. (ASTM Subcommittee D02.CS92 conducts regular exchange testing programs specifically for this purpose.)

ASTM D6299 provides detailed guidance on the relevant statistical techniques to monitor test method precision and bias performance using production materials, reference standards, and results from interlaboratory exchange testing programs.

USE OF STANDARD TEST METHOD DATA QUALITY METRICS TO DETERMINE CONFORMANCE WITH PRODUCT QUALITY SPECIFICATION

Product quality specifications quantitatively define the desired value and limits of acceptance for specific product qualities. These specifications typically are referenced in business contractual agreements established between supplier and recipient. For the refining industry, test methods to be used for specification conformance assessment are an integral part of the product specification. Test methods developed by international standards development bodies such as ASTM or ISO are usually referenced for this purpose. Due to the precision and bias attributes associated with the test methods described earlier, application of the same test method to the same material by different parties rarely yields the same numerical results. Hence, when a product release or acceptance decision has to be made using test results that fall on either side of the specification limit (conforming and nonconforming), the testing (im)precision needs to be appropriately accounted for.

Listed next are two industry standards that provide relevant discussion and guidance on how to account for test method–related differences when interpreting multiple test results to determine product quality conformance to specification:

- ASTM D3244, Standard Practice for Utilization of Test Data to Determine Conformance with Specifications
- ISO 4259, Petroleum Products: Determination and Application of Precision Data in Relation to Methods of Test

A detailed discussion of these practices is beyond the intended scope of this chapter. Interested readers are encouraged to study these practices in detail. It should be noted that ASTM D3244 is a standard practice specifically intended to provide guidance in using test results to determine specification conformance based on the criticality of the specification. Therefore, the degree of criticality needs to be agreed on a priori between supplier and customer prior to the execution of ASTM D3244, whereas ISO 4259 discusses the relevant issues only as a topic related to test method precision study.

SUMMARY

In conclusion, this chapter has described the STM performance metrics adopted by ASTM D02 Committee on Petroleum and Petroleum Products and the standard practices available to quantify and monitor these metrics using statistical quality assurance techniques. Knowledge of these performance metrics can drive test method improvements in both technical revisions and work practices, resulting in improved confidence of the product quality information represented by test method results, thus facilitating efficient and effective commerce between supplier and consumer.

ASTM Standards

ASTM	Title
E691	Standard Practice for Conducting an Interlaboratory Study to Determine the Precision of a Test Method
E456	Standard Terminology for Relating to Quality and Statistics
D6300	Standard Practice for Determination of Precision and Bias Data for Use in Test Methods for Petroleum Products and Lubricants
D6299	Standard Practice for Applying Statistical Quality Assurance Techniques to Evaluate Analytical Measurement System Performance
D3244	Standard Practice for Utilization of Test Data to Determine Conformance with Specifications
D6792	Standard Practice for Quality System in Petroleum Products and Lubricants Testing Laboratories
D6708	Standard Practice for Statistical Assessment and Improvement of the Expected Agreement Between Two Test Methods that Purport to Measure the Same Property of a Material

24

Synthetic Liquid Fuels

Lelani Collier,[1] Carl Viljoen, Mirriam Ajam, Mazwi Ndlovu, Debby Yoell, Paul Gravett, and Nico Esterhuyse[2]

AS THE WORLD SEEKS ALTERNATIVE AND SUPERIOR ways to meet the demands for cleaner and reliable energy supplies, Sasol's commercial experience in the production of synthetic fuels by using coal-to-liquids (CTL) and gas-to-liquids (GTL) technology is increasingly being recognized and sought after worldwide. Sasol plays a vital role in fueling South Africa and supplies, through the retail and wholesale sector, between 30 and 40 % of the country's liquid fuel from coal and crude oil, including sulfur-free diesel, gasoline, jet fuel, illuminating paraffin (IP) and liquefied petroleum gas (LPG). The transportation fuels produced from Sasol's High-Temperature Fischer-Tropsch (HTFT) technology are fit for purpose and have been used in modern technology vehicles for many years [1]. Sasol's first international joint venture GTL plant using Sasol's proprietary Low-Temperature Fischer-Tropsch (LTFT) technology in Ras Laffan, Qatar, was commissioned in 2006 and the first products were produced in 2007. Synthetic fuel products produced by Sasol such as LPG, final gasoline, naphtha blending components, middle distillates (IP, diesel, and aviation fuel), and heavy fuel oil have successfully been analyzed and certified for commercial use utilizing American Society of Testing and Materials (ASTM) methods over the past 50 years. The Sasol production processes are also controlled and managed using data obtained with mostly ASTM methods, thus enabling the conversion of coal and gas to liquid synthetic fuels.

FISCHER-TROPSCH TECHNOLOGY

Synthetic fuels are new-generation transport fuels that are produced via a three-stage process based on the Fischer-Tropsch (FT) technology using three main feedstocks–namely, (1) coal (CTL), (2) natural gas (GTL), or (3) the greener alternative biomass (biomass to liquids [BTL]) [2]. Of the three options, GTL and CTL are the most advanced commercially, with plants such as Shell's Bintulu plant in Malaysia, which is to be the basis for a much larger Shell GTL plant in Qatar, to be commercialized in 2010, and the already fully operational Sasol Qatar Petroleum plant, Oryx GTL. The only other commercial GTL plant, PetroSA in South Africa, is an HTFT plant converting natural gas to transportation fuels, using Sasol Synthol technology. The largest commercial CTL operation is the Sasol Synfuels plant in Secunda, which uses HTFT for the production of mainly chemicals and gasoline (approximately 160,000 barrels per day) [1]. Other companies that pursue the FT technology, through either research, pilot plants, or planned large-scale plants, to produce synthetic fuels from varying feedstock include the Royal Dutch-Shell group, ExxonMobil, Syntroleum, Rentech, Texaco, and ARCO. Syntroleum produces synthetic fuels from a wide variety of feedstock including natural gas to fats, oils, and greases. Syntroleum is also using its Bio-Synfining™ technology to produce renewable synthetic fuels with the same superior qualities as coal and natural gas-based FT fuels. Similarly, Rentech has successfully applied the Rentech® Process in facilities that range in size from pilot scale to 300 barrels per day of synthetic fuels and chemicals production [3].

A diagrammatic representation of a generic FT synthesis process is shown in Fig. 1.

A generic FT process comprises three main process steps, namely:

1. Synthesis gas (syngas) preparation, which involves either reforming of natural gas or gasification of coal or biomass to produce syngas ($CO + H_2$), which are the building blocks for synthesis, by gasification.
2. Conversion of syngas into FT synthetic crude (syncrude) by using FT technology via a chain growth process according to the equation written as follows:

$$CO + 2H_2 \rightarrow -[CH2]- + H_2O$$

where $-[CH_2]-$ = the basic building block [4].

Four types of reactor systems for FT technology are currently in use, namely:

- Circulating fluidized bed reactor
- Fluidized bed reactor
- Tubular fixed bed reactor
- Slurry phase reactor

Fluidized bed reactors operate at temperatures approximately 100°C higher than the tubular or slurry phase reactors–hence, the terms HTFT and LTFT [4].

3. Upgrading of the syncrude to fuels via different refinery processes. Refinery design reflects the prevailing fuel specifications, and the refinery configuration used will depend on the feed to be processed (feed composition is dependent on the nature of the FT catalyst and FT operating conditions) and the desired split between transportation fuels and chemicals [4].

With regard to the refinery flow schemes, the HTFT facilities are more complex than the LTFT facilities. LTFT flow schemes are simple, consisting mainly of hydroprocessing and fractionation steps producing naphtha and paraffinic diesel fuel oil. HTFT flow schemes, on the other hand, require specific types of refining such as oligomerization, hydrocracking, aromatization, and hydrogenation. Other well-known conventional crude refining technologies like catalytic and thermal cracking, catalytic reforming, alkylation, etherification, and coking can also be included depending on the specific final product requirements. Final fuels are produced

[1] *Principal Scientist*
[2] *All of the authors are from Sasol Limited, Johannesburg, South Africa.*

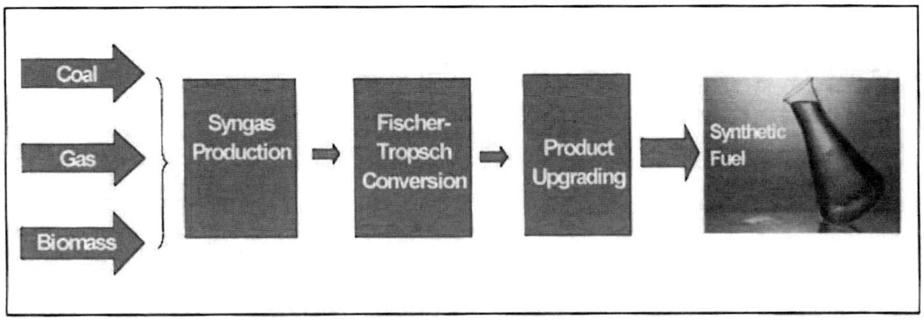

Fig. 1—Diagrammatic representation of synthetic fuels production [2].

with HTFT technology by blending different hydrocarbon components to produce LPG, gasoline, middle distillates (diesel fuel oil and kerosene), and fuel oil. In addition to the transportation fuels, HTFT produces ethylene and propylene as well as co-monomers that can be used in the polymer industry [4]. The typical yield structures of the current product slate derived from LTFT and HTFT refining is presented in Table 1.

The difference in the above product slates and yields is primarily attributed to the FT technology itself and to a lesser extent to the differences in the plant configurations, operating conditions, and catalysts selected for LTFT and HTFT refining. A summary of the main differences between HTFT and LTFT is presented in Table 2.

The synthetic fuel products of the FT process are characterized by their very consistent quality. These fuels contain extremely low levels of sulfur (<5 ppm), nitrogen-containing compounds, and poly-nuclear aromatics and, in the case of diesel, comprise mostly linear hydrocarbons with a low degree of branching.

HTFT Product Slate

The Sasol Secunda HTFT refinery produces hydrocarbon components from different streams, which are blended according to the prevailing fuel specifications to produce the final desired LPG, gasoline, middle distillates (diesel fuel oil and kerosene), and fuel oil [1].

Only a limited amount of C_5 and lighter material can be blended into fuel, and therefore processing large quantities of HTFT material lighter than C_5 is less desirable. Most of the propylene produced during HTFT is recovered for chemical use and the unused propene-rich C_3 and butene-rich C_4 material is beneficiated via alkylation, aromatization, and oligomerization [4].

The unrefined HTFT fraction heavier than C_{22} has low levels of polynuclear aromatics and can be blended into diesel, the direct inclusion being limited by the T_{90} (and T_{95}) boiling point specification [6]. This heavy fraction is suited to hydrocracking as a means of shifting the carbon number distribution to lighter products [1].

LTFT Product Slate

A typical LTFT plant would produce two major products: a light fraction (hydrocarbon condensate), which is usually a liquid at room temperature, and a heavy fraction (wax), which is usually a solid at room temperature [4]. These products are upgraded in the refinery to produce a range of FT fuels. The use of diesel as a transportation fuel is increasing, FT synthesis is well positioned. LTFT processes are ideally suited to the production of distillate with naphtha as the main co-product [4]. The simplest option to upgrade LTFT syncrude to commercially viable products is by hydroprocessing to yield LPG, naphtha, kerosene, and diesel. The option also exists to obtain certain specialty products such as lubricant base oils and chemicals (derived from the oxygenated products dissolved in the reaction water) or to produce gasoline [4]. In the absence of hydrocracking, specialty wax products can also be derived. The diesel produced from hydrocracking of LTFT syncrude is clean burning and sulfur free and is ideally suited for use as a blend component to enhance the properties of petroleum derived diesel. Naphtha produced from this FT process can be converted into light olefins using steam cracking, while the kerosene is suitable as a jet fuel component.

ANALYSIS OF SASOL SYNTHETIC PRODUCTS

Final synthetic fuels produced by Sasol are certified at the Sasol laboratories. The methods used for certifying the products are predominantly ASTM test methods and the Energy Institute IP test methods. Specifications for final fuel such as gasoline and diesel are regulated via the South African National Standards (SANS). These specifications are generally aligned with European specifications, although there is a current lag between South African and European specifications of approximately 10 years. SANS also coordinates South

TABLE 1—Typical Ranges of Product Yields from an LTFT and an HTFT Refining Process [5]

Product	LTFT	HTFT
Diesel	30–80 %	15–20 %
Naphtha	15–25 %	. . .
LPG	5 %	5 %
Gasoline	. . .	30–40 %

TABLE 2—Summary of the Key Differences Between HTFT and LTFT

Designation	HTFT	LTFT
Product slate	LPG, gasoline, kerosene (jet fuel), diesel, and chemicals; products boiling in the gasoline and distillate boiling range containing synthetic aromatics	LPG, naphtha, kerosene, diesel, base oils, and waxes; products are paraffinic in nature

Africa's input to the development of international standards by the International Organization for Standardization (ISO).

All fuel companies in South Africa, as well as some of the third party certifying laboratories, are enrolled in the Measuring Assurance Program governed by the South African Bureau of Standards to ensure consistent results when analyzing fuel samples by ASTM and IP methods. The products that are part of this program are gasoline, diesel, and jet fuel. Samples of fuels produced and marketed in South Africa are submitted as part of this program and both Sasol and PetroSA submit synthetic (FT-derived) final fuel samples. The results from the different laboratories have delivered data indicating that ASTM and IP methods are fit for use with synthetic fuels.

Liquid Petroleum Gas

LPG is produced in both the LTFT and HTFT processes. LPG sold in South Africa needs to adhere to SANS 1774:2007 specification. This specification, as well as the ASTM D1835-05 specification, employs various ASTM methods to determine the specific properties of LPGs. Table 3 lists the ASTM methods used by Sasol to analyze synthetic LPG. A discussion on selected LPG properties and the use of ASTM methods to analyze synthetic LPG follows this table.

Composition

Depending on the application of the product, LPG can be a single-component product like propane or a two-component mixture, with the composition determined by ASTM Method D2163.

TABLE 3—ASTM Methods Used by Sasol to Analyze Synthetic LPG

Property	Units	ASTM Test Methods Used for Synthetic LPG
Total acetylenes	Mole percent	D2163
Total C_2 hydrocarbons	Mole percent	D2163
Total ethylene	Mole percent	D2163
Total dienes	Mole percent	D2163
Total C_5 and higher hydrocarbons (calculated as *n*-pentane)	Mole percent	D2163
Vapor pressure at 37.8°C	kPa	D1267, D2598^A
Copper corrosion (1 h at 37.8°C)		D1838
Residue:		
On evaporation	mL	D2158
Oil stain (ring after 2 min)		
Total sulfur	Mass percent	D2784, D5453
Stenching agent as ethyl mercaptan	Mass percent	D5305
Density 20°C	kg/L	D1657, D2598

^{A}In the event of a dispute regarding the vapor pressure determination, ASTM D1267 method is used.

Volatility

The volatility of LPG is determined by the vapor pressure using ASTM Method D1267 (gauge vapor pressure). ASTM Method D2598 is a calculation method based on the composition of the gas to obtain properties such as motor octane number, relative density, and vapor pressure and therefore can also being used to determine the vapor pressure of synthetic LPG. ASTM Method D1267 is currently being used to certify LPG's vapor pressure.

Copper Corrosion

Copper corrosion and residue are successfully determined by ASTM Methods D1838 (copper strip corrosion method) and D2158, respectively. Although synthetic LPG is extremely low in sulfur content, the copper corrosion determination and limitations are still set to ensure that copper or copper alloy fittings in transportation, storage, or utility equipment are not deteriorated due to the presence of possible hydrogen sulfide (H_2S). This limit will also ensure that H_2S is not present in quantities that may present health or safety hazards.

Sulfur Content

Sulfur content and the amount of stenching agent added are analyzed using ASTM Method D2784 (oxyhydro burner/lamp). Synthetic LPG's sulfur content can also be determined by using ASTM D5453 (ultraviolet fluorescence). The stenching agent added to synthetic LPG is ethyl mercaptan and ASTM Method D5305 is used to determine the amount of ethyl mercaptan in the vapor phase of LPG systems.

Density

Although density determination of LPG has little significance, it is determined to ensure transportation and storage requirements are met. Density of LPG can be measured by ASTM D1657 (pressure hydrometer to determine density of light hydrocarbons), although the calculation method in D2598 is currently being used.

Gasoline

Currently SANS Specification 1598:2006 for unleaded gasoline uses various ASTM methods to determine the specified gasoline properties. Table 4 lists the ASTM methods used by Sasol to analyze synthetic gasoline according to the requirements as set out by SANS. A discussion on selected HTFT gasoline properties and the use of ASTM methods to analyze this gasoline follows.

Composition

The hydrocarbon types present in synthetic HTFT gasolines are determined by ASTM D1319. Compositionally, HTFT gasoline is low in sulfur, nitrogen, and aromatic content; contains olefins; and is high in paraffins [1]. As with crude derived gasoline, synthetic fuel also contains benzene and monoaromatics (although low levels of total aromatics). The benzene and aromatic content of HTFT gasoline is specified in line with international trends due to the health risks and toxic emissions associated with these molecules. Some of the benzene in gasoline is emitted from vehicles as unburned fuel, while benzene is also formed as a partial combustion product of larger aromatic components. It may also be emitted as a combustion product of naphthenics such as cyclohexane, which are products of subsequent reactions of combustion products such as acetylene and 1,3-butadiene

TABLE 4—ASTM Methods Used by Sasol to Analyze Synthetic (HTFT) Gasoline

Property	Units	ASTM Test Method Used for Synthetic Gasoline
Olefins	Volume percent	D1319
Aromatics	Volume percent	D1319, D5580, D5443[A]
Benzene content	Volume percent	D5580, D5443, D4815, D3606, D6730[A]
Density	kg/L	D1298, D4052
Distillation corrected to 101.325 kPa		D86
Initial boiling point	°C	
10 % evap (v/v)	°C	
50 % evap (v/v)	°C	
90 % evap (v/v)	°C	
Final boiling point	°C	
Residue	Volume percent	
Reid vapor pressure at 37.8°C	kPa	D5191 (no alcohol), D4953 (containing oxygenates)
Octane rating		
Research octane number (RON)		D2699
Motor octane number (MON)		D2700
Existent gum	mg/100 mL	D381
Potential gum (2.5 h at 100°C)	mg/100 mL	D873
Induction period	Minutes	D525
Total sulfur	Mass percent	D5453, D2622[B]
Doctor test OR mercaptan sulfur	Mass percent	D4952 or D3227
Copper corrosion (3 h at 50°C)		D130
Oxygen content	Mass percent	D4815
Ether content	Volume percent	D4815

[A]In the event of a dispute regarding the determination of benzene and aromatics, the ASTM D5443 method is used.
[B]In the event of a dispute regarding the determination of total sulfur, the ASTM D5453 method is used.

[7]. The benzene and monoaromatic content of synthetic gasoline is determined using ASTM D1319 and D5443, respectively. The ASTM Methods D4815, D3606, D5580, and D6730 can also be used successfully to determine benzene. The olefin content is also specified due to the contribution of olefins to evaporative emissions and ozone formation. The combustion of olefins can also form toxic dienes such as 1,3-butadiene [8].

Density

Density (in conjunction with other properties) characterizes both the heavy and the light components in petroleum fuel. A density range for a given fuel is required for the satisfactory operation of the vehicle and is taken into consideration when designing and developing fuel delivery systems to improve combustion and to optimize fuel economy [9]. The density of synthetic gasoline is currently determined by using ASTM D1298 and ASTM D4052 (digital density meter) methods. ASTM Method D1250 (petroleum measurement tables) is also successfully used for correlation of densities between 15° and 20°C.

Volatility

Some aspects of gasoline performance are correlated with ranges of the distillation profile.

Front end volatility determines aspects such as:

- Ease of cold or hot start
- Freedom from vapor lock
- Reduced evaporative and running loss emissions

Midrange volatility provides:

- Rapid warm up and smooth running of the vehicle
- Good fuel economy over a short trip
- Good power acceleration
- Protection against carburetor icing and hot stalling

Tail-end volatility influences:

- Fuel economy once the engine has warmed up
- Reduced engine deposits
- Minimal fuel dilution of the crankcase oil
- Minimal volatile organic carbons (VOCs) for exhaust [9]

The most common method to determine the distillation profile of fuels is ASTM D86, and this method has been used by Sasol since synthetic gasoline and diesel fuel were first produced in South Africa in the late 1950s. Volatility of a fuel is also determined by measuring the vapor pressure at 37.8°C by using ASTM Methods D4953 (modified method from Reid method ASTM D323, developed for alcohol/oxygenate containing fuel), D5190 (automated method), and D5191 (mini method).

Antiknock Rating

Antiknock rating and gasoline volatility are two of the most important characteristics of gasoline. The definition of antiknock rating of automotive spark-ignition engine fuels is set out in detail in Appendix X1 of ASTM Specification D4814. In short, the octane number is a measure of a gasoline's ability to avoid knocking as it burns in a combustion chamber. The knocking phenomenon is caused by the uncontrolled autoignition of unburnt fuel (called end-gas) brought into contact with elevated temperatures and pressures [10].

The methods listed to determine the antiknock rating of gasoline in single-cylinder engines are ASTM D2699 (Research Octane Number [RON]) and D2700 (Motor Octane Number [MON]), and these methods are successfully used to determine the antiknock characteristics of the synthetic gasoline. ASTM D2885 (on-line analyzer) is also being used successfully for antiknock ratings of synthetic fuel, although final fuel produced in South Africa may not be certified using on-line analyzers as fuel needs to be certified batchwise.

Stability

The degradation of fuel results from oxidation, which can occur during both storage and usage and can lead to the formation of gums. Further, exposing a less stable fuel to elevated temperatures and pressures, such as during combustion, can also lead to undesired deposit formation. Readily oxidizable species are also referred to as gum precursors. They are responsible for gum formation and include, but are not limited to, dienes, olefins (branched and cyclic), and polyaromatic hydrocarbons [11]. The SANS specification for gasoline sold in South Africa specifies that gasoline must still meet the specification after storage of 6 months. ASTM D381 is used to determine the existing gum content of synthetic gasoline, while ASTM D873 (although not designed for gasoline) is used to determine the tendency of the gasoline to form gums and deposits under severe aging conditions. Determining the induction period of gasoline is also a measurement to determine the tendency to form gums during storage under accelerated oxidation conditions. ASTM D525 is currently being used for synthetic fuels.

Sulfur

Although synthetic fuels have extremely low sulfur content, the sulfur content is still determined due to the requirement to report it. The sulfur content of gasoline can be determined by various ASTM methods. Sasol utilizes ASTM Methods D2622 and D5453 to determine the total sulfur content of synthetic gasoline, while the mercaptan content in the gasoline is determined by ASTM Methods D3227 and D4952. ASTM D130 is also used to detect possible corrosion due to sulfur species present in the gasoline.

Oxygenates and Ethers

Oxygenates and ethers can be used as blending components to improve the octane rating of the gasoline, and therefore these parameters in gasoline are also specified in the SANS standard. The total alcohol content method specified for South African gasoline is a simple method to merely identify the presence of any alcohol in the gasoline. However, ASTM D4815 (gas chromatography) is successfully being used for synthetic gasoline to determine the specific alcohols and ethers concentrations.

Synthetic alcohol is produced by Sasol during the HTFT process. Some of the synthetic ethanol, produced as a co-product from the FT process (currently up to 2 volume percent) is being added into the synthetic gasoline to improve the octane rating of the fuel but also as part of Sasol's drive for the production of cleaner fuels. The synthetic ethanol adheres to the SANS standard 465:2005 based on ASTM D4806-06.

Water Tolerance

Currently, the water tolerance method for gasoline–alcohol mixtures measures the ability of fuel to retain water in solution, at a specific temperature, to which the gasoline is likely to be exposed when in use. Sasol uses ASTM D6422 to measures the temperature at which a gasoline–alcohol blend separates into two distinct phases and therefore is not being used for determining the water tolerance of synthetic gasoline. Testing synthetic fuel using ASTM Method D6422 is, however, possible and has been used successfully.

Aviation Fuels

Over a period spanning the past 17 years, Sasol has followed an extensive and comprehensive process for the approval and qualification of first semisynthetic jet fuel and later fully synthetic jet fuel, engaging all relevant international stakeholders [12].

Historical Development of Synthetic Aviation Fuels

Air traffic to the Republic of South Africa (RSA) has increased significantly and is projected to continue to increase especially up to 2010. There are only two sources of jet fuel that could meet the demand at the busiest airport in Africa, OR Tambo International Airport (ORTIA)–the inland Natref refinery and the rail-car service transporting Jet A-1 from two coastal crude oil refineries in Durban. The maximum production capacity at Natref had, however, already been reached, while the coastal refineries could not easily meet the extra demand due to serious logistical constraints [12].

As a solution, Sasol proposed to increase the supply from the Natref refinery by blending in a hydrocarbon stream from its refinery at Secunda. The result is a blend of 50 % of a synthetic kerosene stream with 50 % of the conventional petroleum-derived kerosene to yield a "semisynthetic" Jet A-1 fuel. This was the first known instance of commercial jet fuel containing hydrocarbons not refined from petroleum crude oil. There was thus no accepted approval or qualification process by which this new fuel could gain formal acceptance by the airframe manufacturers, regulatory bodies, and original engine manufacturers (OEMs). In 1996, Southwest Research Institute (SwRI®) in San Antonio, Texas, was contracted by Sasol to lead the process of qualifying the Semi-Synthetic Jet Fuel (SSJF) for use as a commercial jet fuel [12].

Qualification of Semisynthetic Jet Fuel for Use as Commercial Aviation Turbine Fuel

The U.K. Aviation Fuel Committee Strategy Group, AFCSG, concluded that there was insufficient knowledge available about the characteristics of SSJFs. Therefore, a temporary modification was made to Defence Standard (DEF STAN) 91-91 to preclude the use of hydrocarbons from nonpetroleum sources until certain issues and concerns were addressed to the satisfaction of the British Ministry of Defence [12]. All of the issues and concerns were identified, investigated, and discussed in the final research report, published by SwRI in November 1997 [13]. Over the following 2 years, all the data could be assimilated and the concerns of this very conservative industry could be addressed to the satisfaction of all the parties (representatives of the airline, engine, fuel, and ground handling communities), leading to the U.K. Ministry of Defence finally giving its approval to SSJF by incorporating it into DEF STAN 91-91, Issue 3, in November 1998 [14]. SSJF was incorporated into the ASTM D1655 Specification for Aviation Turbine Fuel in 1999.

Almost all jet fuel batches dispatched from Natref to ORTIA since July 1999 contained synthetic paraffinic kerosene (SPK) material from Secunda (0 to 48 volume percent). Semisynthetic Jet A-1 is fungible with crude oil–derived Jet A-1 and therefore loses its synthetic tag once it enters the logistical distribution chain. Sasol also demonstrated the ability to successfully incorporate hydrotreated and hydrocracked synthetic kerosenes into the production of SSJF, and the use of these blending streams was approved in DEF STAN 91-91, Issue 4, in November 2002 [12].

Qualification of Fully Synthetic Jet Fuel for Commercial Use as Jet A-1

In 2000, Sasol again commissioned SwRI to qualify 100 % synthetic jet fuel as a commercial Jet A-1, getting buy-in and approval from the various stakeholders. In December 2003, a report [15] on the evaluation of the properties and characteristics prescribed in Annex D [13]. DEF STAN 91-91 was submitted to the Aviation Fuel Committee of the Ministry of Defence in the United Kingdom. Four samples of synthetic kerosenes were blended by Sasol to represent the practical range of jet fuels that could be made from four of the five synthetic refinery streams at the Sasol refinery in Secunda. These four synthetic kerosene blends met all of the property requirements of Table 1 of DEF STAN 91-91 and had properties and characteristics typical of conventional petroleum-derived jet fuel. After the successful completion of additional engine and hardware testing requested by the turbine engine manufacturers, it was concluded that jet fuels blended from Sasol synthetic refinery streams were fit for use as aviation kerosene [16].

Sasol's CTL jet fuel has become the first fully synthetic fuel to be approved for use by commercial aviation as of April 8, 2008, when its approval was published in the British Ministry of Defence's (UK MoD) DEF STAN 91-91, Issue 6. The inclusion of Sasol 100 % synthetic jet fuel is currently being written into ASTM D1655, which will complete the approval process.

Analyzing Semisynthetic and Fully Synthetic Jet Fuel

Table 5 presents the ASTM methods used to certify Sasol's semisynthetic and fully synthetic jet fuel according to ASTM D1655 and the British Ministry of Defence (DEF STAN 91-91, Issue 6). A discussion on selected fuel properties and the use of the ASTM methods follows.

TABLE 5—ASTM Methods Employed by Sasol for Semisynthetic and Fully Synthetic Jet Fuel

Fuel Property	Units	ASTM Test Methods Used for Semisynthetic and Fully Synthetic Jet Fuel
Color	. . .	D156
Particulate contaminants	mg/L	D5452
Acidity	mg KOH/g	D3242
Aromatics	% v/v	D1319/D6379[A]
Total sulfur	% m/m	D4294/D5453
Mercaptan sulfur	% m/m	D3227
Copper corrosion	Class	D130
Boiling point distribution	°C, volume percent	D86
Flash point	°C	D3828
Density at 20°C	kg/m^3	D1298/D4052
Freezing point	°C	D5972/D2386/D5901[B]
Viscosity at −20°C	mm^2/s	D445
Specific energy	MJ/kg	D4809/D3338/D4529[B]
Smoke point	mm	D1322
Naphthalenes	% v/v	D3338
Thermal stability	Class mm Hg	D3241
Existent gum	mg/100 mL	D381
Water content	mg/kg	D1744
Water separation characteristics	. . .	D3948
Conductivity	pS/m	D2624
Lubricity, wear scar diameter (WSD)	mm	D5001

[A] D1319, not suitable for synthetic jet fuel component; aromatics less than 5 volume percent.
[B] In the event of a dispute regarding the determination of specific energy, ASTM D4809 is used.

Composition

Accurate and detailed characterization of jet fuel by quantification of the hydrocarbon classes is essential in understanding how the physical properties of jet fuel are influenced by its chemical composition. Jet fuel composition can vary significantly depending on the process and refining conditions used in the production of the jet fuel. Currently at Sasol, jet fuel composition is characterized using two-dimensional gas chromatography (GC × GC)-MS to classify different classes of compounds quantitatively and qualitatively. This is not a widely recognized method for analyzing the composition of jet fuel. The more recognized ASTM method used to determine types of hydrocarbons in middle distillates, with an average carbon number of C_{10} to C_{18}, is ASTM D2425. ASTM D2425 makes use of a mass spectrometer (MS) to analyze saturated hydrocarbon and aromatic fractions that have been separated using ASTM D2549 (elution chromatography). The reproducibility of the saturated hydrocarbon fraction at a 40 to 50 mass percent is ±4.0 % and the reproducibility error would increase for higher amounts of paraffins. This would therefore not be an ideal method for synthetic jet fuel components containing predominantly paraffins. The aviation fuels community is currently actively exploring alternative analytical methods for jet fuels (including synthetic components), which will have better precision statements.

Aromatics

Aromatics are considered important constituents in jet fuel since they can cause various elastomers to swell, resulting in a better seal. For this reason, a minimum requirement of 8 volume percent aromatics is stipulated for fully and semisynthetic jet fuel in DEF STAN 91-91, Issue 6, as a safeguard for meeting elastomer compatibility concerns. When the synthetic jet fuel component is analyzed by fluorescent indicator adsorption (FIA) according to ASTM D1319, aromatics in the range of 0.1 to 2 % are sometimes detected. Subsequent analysis by ASTM D5186 (supercritical fluid chromatography) showed that this was due to the tail end of the isoparaffins and that in fact very few aromatics were present. For this reason, the FIA method (ASTM D1319) is not considered accurate for fuels with aromatic contents of less than 5 volume percent.

Distillation Profile

Volatility of jet fuel is its tendency to vaporize and is characterized by the flash point and distillation profile. A minimum distillation gradient specification was included for synthetic jet fuel to ensure that it has a distillation slope matching that of conventional crude oil derived jet fuel. For FSJF, the boiling point distribution should have a minimum distillation gradient defined by T_{50} to T_{10} $\geq$20°C and T_{90} to T_{10} $\geq$40°C when measured by ASTM D86.

Freezing Point

Jet fuel is exposed to very low temperatures at altitude and must be able to retain its fluidity and pumpability at these extreme temperatures. The freezing point characteristics of synthetic jet fuel were investigated using technically equivalent methods to correlate results to referee methods. Freezing point analyses were performed on five Sasol semisynthetic jet fuel blends using ASTM D2386 (manual method), ASTM D5901 (automated method), and ASTM D5972 (Phase Technology's instrument). The results obtained from the three methods were repeatable and reproducible and compared well with one another. The Sasol Fuels Laboratory makes use of the new Phase Technology unit, ASTM D5972, to measure freezing point. It was also conclusively proved that the above analytical methods were as effective on synthetic fuels as on conventional fuels. Sasol synthetic jet fuels have a higher concentration of isoparaffins than most jet fuels, explaining the better low-temperature characteristics, and will not cause any low-temperature operability problems.

Specific Energy

Space is an important consideration on an aircraft, and thus the amount of energy contained in a given quantity of fuel is important. The energy content of jet fuel is a measure of the heat released when a known quantity of fuel is burned under specific conditions. Jet fuel composition has a direct effect on energy content, and this effect is normally predicted by the fuel density. To determine specific energy, the referee method (ASTM D4809) makes use of a bomb calorimeter. In addition to this, there are two other methods for calculating specific energy: ASTM D3338 and ASTM D4529. ASTM D3338 calculates specific energy from a correlation with density, aromatic content, and boiling point distribution, with a correction for sulfur content. ASTM D4529 calculates specific energy from a correlation with density and aniline point on a sulfur-free basis. The results obtained from these three methods correlated well with one another and provided proof that these correlations could also be applied successfully to synthetic jet fuels.

Thermal Sability

The thermal stability of jet fuel is one of the most important properties because the fuel serves as a coolant in aircraft engines and airframes. Synthetic jet fuel derived from the FT process exhibits exceedingly high thermal and oxidative stability due to its negligible quantities of heteroatomic species (sulfur-, nitrogen-, and oxygen-containing species) and aromatics. The thermal stability of synthetic jet fuel is measured by the Jet Fuel Thermal Oxidation Tester (JFTOT) according to ASTM D3241. Sasol's synthetic jet fuels have excellent thermal stability and exhibit breakpoints higher than the industry average.

Conductivity

The DEF STAN 91-91 fuel specification requires a fuel's conductivity to be in the range of 50 to 450 pS/m to prevent accidents due to static discharge during fueling operations. Static dissipator additive (SDA) increases the fuel conductivity and therefore aids in dissipating electrostatic charge. Without an SDA, synthetic fuels do not meet this requirement. In the presence of SDA, the electrical characteristics of Sasol's synthetic jet fuel are acceptable, as measured by ASTM D2624.

Lubricity

Lubricity is defined as the ability to reduce friction between moving solid surfaces. Jet engines rely on the fuel to lubricate some of the moving parts and should therefore possess a certain degree of lubricity. Jet fuels with low aromatic and sulfur content usually exhibit poor lubricity properties. Synthetic jet fuel component when blended with crude oil–derived jet fuel has lubricity, meeting the lubricity specification limit of 0.85 mm maximum, when measured with the Ball on Cylinder Evaluator (BOCLE), ASTM D5001.

Diesel Fuel Oil

ASTM D975 is the standard specification for diesel fuel oils (DFOs). It covers various grades of petroleum-derived diesel fuel oils and prescribes the required properties of the diesel fuels at delivery to a customer. This specification uses a number of ASTM methods to quantify specified property of the diesel fuel. Currently, the South African National specification SANS 342:2006 for automotive diesel fuel uses several ASTM methods to determine the specified diesel properties. Table 6 lists the ASTM methods used by Sasol to analyze both HTFT and LTFT diesel successfully.

A discussion on selected diesel fuel properties and the use of ASTM methods to analyze synthetic diesel fuels follows.

Cetane Index and Number

Currently, three ASTM methods can be used to quantify the ignition and combustion characteristics of diesel fuels: D613, D6890, and D7170. The determination of the cetane number via ASTM D613 poses a practical challenge for its application to LTFT diesel fuels. The proposed ASTM D613 method is the original Cooperative Fuel Research (CFR) engine test method, which is currently used as a reference method for quantifying the ignition quality of a diesel fuel. The cetane number scale covers the range from 0 to 100; however, in practice, testing is done in the range of 30 to 65 cetane number units due to the low reproducibility (ranges from ±6.8 to ±7.8) at high cetane ratings (72 to 80). LTFT DFOs are characterized by a high cetane number (>70), due to the high *n*-paraffin content of LTFT diesel fuels, and therefore ASTM D613 is used to determine the cetane number of LTFT diesel.

HTFT DFOs have a lower cetane number (<65) than LTFT DFOs (typically >80) and have successfully been analyzed and certified using ASTM D613 for the past 50 years.

The ASTM D6890 method, also known as the Ignition Quality Tester™ (IQT) method, describes the quantitative determination of the ignition characteristics of conventional (petroleum-based) diesel fuels, oil-sands–based fuels, and blends of fuel containing biodiesel and diesel fuel oils containing cetane number improvers. This method uses a mathematical equation that correlates an ignition delay determination to the cetane number determined by ASTM D613 test method, thus giving a derived cetane number (DCN). The alternative ASTM D6890 method is preferred for the determination of the cetane number of LTFT synthetic diesel fuel oils because of its ease of use and higher accuracy [17].

The cetane index determination of synthetic fuels, by ASTM D976 or by D4737, is not being used at the Sasol FT

TABLE 6—ASTM Methods Employed by Sasol to Analyze Synthetic Diesel Fuel Oils

Property	Units	ASTM Test Methods Used for Synthetic Diesel
Cetane index		Not analyzed
Cetane number	Rating	D613/D6890/ D7170[A]
Paraffinic hydrocarbons	Mass percent	D2425
Polynuclear aromatic hydrocarbons	Mass percent	D5186
Sulfur	mg/g	D5453
Copper strip corrosion (3 h at minimum 50°C)	Rating	D130
Flash point	°C	D93
Ash content	Mass percent	D482
Carbon residue (on 10 % distillation residue)	Mass percent	D524
Water and sediment	Mass percent	D6304 water, D473 sediment
Oxidation stability, 16 h at 95°C	mg/100 mL	D2274
HT thermal stability, 90 min at 150°C	% Filter pad rating	D6468
Lubricity, corrected WSD at 60°C	mm	D6097
Viscosity @ 40°C	mm^2/s	D445
Distillation, T90°C	Volume percent	D86/D2887[B]
Cloud point	°C	D2500
Density		D1298, D4052[C]

[A]ASTM D613, referee method for synthetic HTFT diesel oil; ASTM D6890 preferred method for LTFT diesel fuel oils.
[B]In the event of a dispute regarding the determination of T90 distillation point, ASTM D2887 is used.
[C]In the event of a dispute regarding the determination of density, ASTM D4052 is used.

refinery. The method for determining cetane index is based on crude derived fuel and is therefore not applicable to synthetic diesel fuel oils.

Paraffinic Hydrocarbons

Synthetic LTFT diesel fuels are characterized by a high paraffinic content, which, as mentioned earlier, gives rise to the high inherent cetane number of the fuels. The ASTM D2425 test method is an MS method developed to determine types of hydrocarbons present in middle distillates with a boiling range of 204 to 343°C; corresponding to 5 to 95 % (v/v), as determined by ASTM D86. The ASTM D2425 test method covers analytical samples containing paraffins with average carbon number values ranging from C_{10} to C_{18}. The reproducibility of this method on paraffin concentrations from 40 to 50 mass percent (saturated fraction) is 4.0. The precision at higher paraffin concentrations than those mentioned earlier is therefore higher. Synthetic diesel fuels contain greater than 75 % paraffins and therefore other techniques such as GC, GC-MS and GC × GC are also used in conjunction with the prescribed ASTM D2425 method.

Aromatic and Polynuclear Aromatic Hydrocarbons

Aromatics have a negative effect on the ignition quality (i.e., cetane number) of diesel fuel, while polynuclear aromatics have a negative effect on the levels of particulate matter emissions from a diesel engine. In contrast to the paraffin content, the aromatics content in synthetic diesel fuels is lower and consists mainly of monoaromatics. ASTM D5186 is the standard test method that is used for the determination of the monoaromatic and polynuclear aromatic hydrocarbon contents in diesel fuels and aviation turbine fuels. This method uses supercritical fluid chromatography (SFC) to determine aromatics concentration in the range from 1 to 75 mass percent and polynuclear aromatic hydrocarbon concentrations in the range from 0.5 to 50 mass percent. This method is therefore applicable to synthetic diesel fuels with a minimum polynuclear aromatic concentration of 0.5 mass percent. Due to the low aromatics content of LTFT diesel, Sasol uses the UOP 495 test method to determine aromatics in LTFT DFOs, which can detect aromatics in the range of 3 ppm to 5 % (m/m).

Sulfur

Besides the detrimental effect that sulfur in diesel has on the environment, modern exhaust after treatment devices also requires low levels of sulfur. Synthetic DFOs sulfur content is very low (less than 5 ppm) and is well below current Euro-4 specifications. ASTM Test Method D5453 is used to determine the total sulfur content in liquid hydrocarbons, which boil in the range of 25 to 400°C and have viscosities between 0.2 and 20 cSt (mm^2/S) at room temperature. Based on three separate interlaboratory studies on precision and three other investigations by ASTM, it was established that this test method is applicable to naphthas, distillates, engine oil, ethanol, fatty acid methyl ester (FAME), gasoline fuels, diesel, biodiesel, diesel/biodiesel blends, and jet fuel. The method is capable of analyzing samples containing 1.0 to 8,000 mg/kg total sulfur and therefore is successfully being used to determine the sulfur content of synthetic DFOs. Sulfur species can also cause copper corrosion. ASTM D130 is used to detect possible copper corrosion due to sulfur species present in the DFOs.

Ash Content

Ash content (noncombustible material) in DFOs can be due to the presence of solid material such as dirt, rust, or oil- or water-soluble metallic compounds. The ash content in synthetic DFOs is determined by ASTM D482, where the fuel is burned until only ash and carbon remains. The carbonaceous residue is then reduced to ash by heating it in a furnace (775°C).

Stability

The stability of DFOs can be categorized into three types: oxidative, thermal, and storage stability. The SANS specification for DFOs sold in South Africa specifies that the DFOs after conventional storage of 12 months must still comply to the specifications as set out in the standard. Synthetic DFOs derived from the FT process exhibit excellent oxidative and thermal stability due to the negligible quantities of heteroatomic species, aromatics, and olefins in LTFT diesel. The oxidative and thermal stability of synthetic DFOs is determined by ASTM D2274 (accelerated method for storage stability under oxidizing conditions) and D6468 (thermal oxidative stability).

Lubricity

Diesel fuel injection equipment relies on the lubricating properties of the diesel fuel. Thus, if the diesel fuel lacks lubricating ability, there is a likelihood that it would result in a shortened life of engine components, such as fuel injection pumps and injectors. FT fuels tend to have poor lubricity due to the absence of polar compounds (such as oxygenates, sulfur and nitrogen species, and aromatics). Lubricity additives are thus needed for synthetic diesel fuels (which have a typical WSD of 500 to 600 μm) to minimize wear between rubbing surfaces [18]. The addition of lubricity additives to synthetic DFOs has been proved in field trials and market experience. The interlaboratory round robin study (ASTM D6079) used diesel fuels with WSDs in the range 143 to 772 μm, which accommodates the lubricity values of synthetic diesel fuels.

Viscosity

The ASTM D445 standard test method determines the kinematic viscosity (ν) of transparent and opaque liquid petroleum products and covers a wide range of kinematic viscosities (0.2 to 300,000 mm^2/s) at different temperatures. The viscosities of synthetic diesel fuels fall within the range that is prescribed by the ASTM D445 test method.

Distillation

ASTM D86 and D2887 are the two methods that can be used to measure the distillation characteristics of a diesel fuel. It should, however, be noted that the ASTM D86 standard test method is the referee method for the determination of the distillation profile of diesel fuel and, as mentioned during the gasoline discussion, has been used by Sasol for the past 50 years. The distillation curve of a diesel fuel is characterized by the lower boiling (lighter ends) components and higher boiling (heavier ends) components. For classes of compounds with the same carbon number, the order of increasing boiling point by class is isoparaffins, *n*-paraffins, naphthenes, and aromatics [18].

The specifications that are considered to be the most important to meet are the distillation T_{90} and flash point, as these define the distillation curve. The distillation T_{90} value is used to set the bottom/heavier end of the diesel, while the flash point defines the top/lighter end of the diesel cut. A higher T_{90} is usually indicative of higher particulate matter emissions, and therefore the distillation range should be as low as possible without affecting other properties such as flash point, viscosity, and combustion quality.

Density

Density is an important physical property that can be used, together with other properties, to characterize petroleum products. It is important to accurately determine the density of petroleum products for the conversion of measured volumes to volumes at the standard temperature of 15°C.

The performance of a diesel engine is determined by the fuel's density (power and fuel consumption). The density of synthetic paraffinic diesel fuels is lower than that of petroleum-derived diesel fuels. The use of lower-density, synthetic diesel fuels in compression ignition (CI) engines has been shown to result in improved performance such as lower ignition delay, easier cold start, reduction in exhaust emissions, noise reduction, and improved engine durability [19].

The density of synthetic paraffinic diesel fuels falls within the range covered by the scope of ASTM D4052 and D1298 methods.

Cloud Point

As with all diesel fuels, synthetic diesel forms wax crystals at low temperatures, and as the temperature decreases, these wax crystals can precipitate out of the fuel as solid wax. If sufficient wax crystals precipitate, fuel flow to an engine can be restricted. The temperature at which the first wax crystals precipitate (or are large enough to be visible or impart a cloudy appearance to the fuel) is known as the cloud point of the diesel fuel. The cloud point of synthetic DFOs is successfully determined by ASTM Method D2500.

Heavy Fuel Oil

Synthetically derived fuel oils are widely used in South Africa in low- to high-temperature processes for steam and air heating in the manufacturing sector. Synthetic heavy fuel oils are certified using ASTM methods according to the specifications as set out in the CKS 142-1989 specification for heavy fuel oil by the South African Bureau of Standards (SABS). Table 7 specifies the ASTM methods used to analyze synthetic heavy fuel oils [10]. All the above mentioned ASTM methods that are used to test and certify crude-derived heavy fuel oils are valid for the testing and certification of synthetically derived heavy fuel oils.

Viscosity

Synthetic fuel oils do not behave as Newtonian liquids at ambient temperatures, due to the waxy nature of the fuels. The kinematic viscosity of synthetic heavy fuel oils is determined at 100°C to ensure homogeneous sample. This is in contrast to the determination of the viscosity of crude-derived heavy fuel oils, which is determined at 40 or 50°C using ASTM D445.

Pour Point

Synthetic heavy fuel oil could contain long carbon chain *n*-paraffinic components, which form wax crystals at low

TABLE 7—ASTM Methods Used by Sasol to Analyze Synthetic Heavy Fuel Oil

Property	Units	ASTM Test Method Used for Synthetic Heavy Fuel Oil
Viscosity at 100°C	cSt	D445
Density at 20°C	kg/L	D1298
Flash point	°C	D93
Pour point	°C	D97
Sulfur content	Mass percent	D129
Ash content	Mass percent	D482
Water content	Volume percent	D95

temperatures and blocks filters and leads to pump ability and combustion problems. Pour point is the lowest temperature at which oil will start flowing. The pour point of synthetic fuel oil is being determined by ASTM Method D97.

Ash Content

Ash content is usually inorganic noncombustible material or metals present in fuel oils. Synthetic heavy fuel oil contains very low concentrations of inorganic material or metals. Inorganic material or metal residue is normally found after the fuel oil is burned in a furnace or boiler. High concentration of these noncombustible materials will cause nozzle blockages, with deposits of solids in the boiler or furnace resulting in damage in the furnace or boiler. Ash content of synthetic heavy fuel oil is successfully determined by ASTM Method D482.

Water Content

The presence of any appreciable water will cause the discontinuance of flames. Although synthetic heavy fuel oils contain a low concentration of water, ASTM Method D95 is successfully being used to determine the water content.

CONCLUSION

The Sasol production processes are controlled and managed using data obtained mostly with ASTM methods, thus enabling the conversion of coal and gas to liquid synthetic fuels serving Sasol's synthetic industry. ASTM methods have also been used successfully over the past 50 years to analyze and certify synthetic oil products such as LPG, final gasoline, naphtha blending components, middle distillates (IP, diesel, and aviation fuel), and heavy fuel oil, with excellent results for Sasol and its long list of customers who use these products.

Applicable ASTM Methods and Standards for Synthetic Fuels

ASTM	Title
D86	Test Method for Distillation of Petroleum Products at Atmospheric Pressure
D93	Test Method for Flash Point for Pensky Martens Closed-Cup Tester
D95	Test Method for Water in Petroleum Products and Bituminous Materials by Distillation
D97	Test Method for Pour Point of Petroleum Products
D130	Test Method for Corrosiveness to Copper from Petroleum Products by Copper Strip Test
D156	Test Method for Saybolt Color of Petroleum Products (Saybolt Chromometer Method)
D323	Test Method for Vapor Pressure of Petroleum Products (Reid Method)
D381	Test Method for Gum Content in Fuels by Jet Evaporation
D445	Test Method for Kinematic Viscosity of Transparent and Opaque Liquids (and Calculation of Dynamic Viscosity)
D473	Test Method for Sediment in Crude Oils and Fuel Oils by Extraction Method
D482	Test Method for Ash from Petroleum Products
D524	Test Method for Ramsbottom Carbon Residue of Petroleum Products
D525	Test Method for Oxidation Stability of Gasoline (Induction Period Method)
D613	Test Method for Cetane Number of Diesel Fuel Oil
D873	Test Method for Oxidation Stability of Aviation Fuels (Potential Residue Method)
D1267	Test Method for Vapor Pressure of Liquefied Petroleum (LP) Gases (LP-Gas Method)
D1298	Practice for Density, Relative Density (Specific Gravity) or API Gravity of Crude Petroleum an Liquid Petroleum Products by Hydrometer Method
D1319	Test Method for Hydrocarbon Types in Liquid Hydrocarbon Products by Fluorescent Indicator Adsorption
D1322	Test Method for Smoke Point of Kerosene and Aviation Turbine Fuels
D1552	Test Method for Sulfur in Petroleum Products (High Temperature Method)
D1657	Standard Test Method for Density or Relative Density of Light Hydrocarbons by Pressure Hydrometer
D1744	Standard Test Method for Determination of Water in Liquid Petroleum Products by Karl Fischer Reagent
D1838	Test Method for Copper Strip Corrosion by Liquefied Petroleum (LP) Gases
D2015	Standard Test Method for Gross Calorific Value of Coal and Coke by Adiabatic Bomb Calorimeter
D2158	Test Method for Residues in Liquefied Petroleum (LP) Gases
D2163	Analysis of Liquefied Petroleum (LP) Gases and Propane Concentrates by Gas Chromatography
D2274	Test Method for Oxidation Stability of Distillate Fuel Oil (Accelerated Method)
D2386	Test Method for Freezing Point of Aviation Fuels
D2425	Test Method for Hydrocarbon Types in Middle Distillates by Mass Spectrometry
D2500	Test Method for Cloud Point of Petroleum Products
D2598	Calculation of Certain Physical Properties of Liquefied Petroleum (LP) Gases from Compositional Analysis
D2622	Test Method for Sulfur in Petroleum Products by Wavelength Dispersive Fluorescence Spectrometry
D2624	Test Method for Electrical Conductivity of Aviation and Distillate Fuel
D2699	Test Method for Research Octane Number of Spark Ignition Engine Fuel
D2700	Test Method for Motor Octane Number of Spark Ignition Engine Fuel
D2784	Test Method for Sulfur in Liquefied Petroleum Gases

ASTM	Title
D2887	Test Method for Boiling Range Distribution of Petroleum Fractions by Gas Chromatography
D3227	Test Method for Thiol (Mercaptan) Sulfur in Gasoline, Kerosene, Aviation Turbine and Distillate Fuels
D3241	Test Method for Thermal Oxidation Stability of Aviation Turbine Fuels (JFTOT Procedure)
D3242	Test Method for Acidity in Aviation Turbine Fuel
D3338	Standard Test Method for Estimation of Net Heat of Combustion of Aviation Fuels
D3828	Test Method for Flashpoint by Small Scale Closed Tester
D3948	Test Method for Determining Water Separation Characteristics of Aviation Turbine Fuels by Portable Separometer
D4052	Test Method for Density and Relative Density of Liquids Digital Density Meter
D4294	Test Method for Sulfur in Petroleum Products by Energy Dispersive X-Ray Fluorescence Spectroscopy
D4529	Test Method for Estimation of Net Heat of Combustion of Aviation Fuels
D4809	Test Method of Heat of Combustion of Liquid Hydrocarbons by Bomb Calorimeter (Precision Method)
D4815	Test Method of Determination of MTBE, ETBE, TAME, DIPE, *Tertiary*-Amyl Alcohol and C_1 to C_4 Alcohols in Gasoline by Gas Chromatography
D4952	Test Method for Qualitative Analysis for Active Sulfur Species in Fuels and Solvents (Doctor Test)
D4953	Test Method for Vapor Pressure of Gasoline and Gasoline-Oxygenate Blend (Dry-Method)
D5001	Test Method for Measuring Lubricity of Aviation Turbine Fuels by Ball-on-Cylinder Lubricity Evaluator (BOCLE)
D5186	Standard Test Method for Determination of Aromatic Content and Polynuclear Aromatic Content of Diesel Fuels and Aviation Turbine Fuels

ASTM	Title
D5191	Test Method for Vapor Pressure of Petroleum Products (Mini-Method)
D5305	Standard Test Method for Determination of Ethyl Mercaptan in LP-Gas Vapor
D5443	Standard Test Method for Paraffin, Naphthene, and Aromatic Hydrocarbon Type Analysis in Petroleum Distillates Through 200 Degrees
D5452	Standard Test Method for Particulate Contamination in Aviation Fuels by Laboratory Filtration
D5453	Test Method for Determination of Total Sulfur in Light Hydrocarbons, Motor Fuels and Oil by Ultraviolet Fluorescence
D5580	Test Method for Determination of Benzene, Toluene, Ethylbenzene, *p/m*-Xylene, *o*-Xylene, C_9 and Heavier Aromatics, Total Aromatics in Finished Gasoline by Gas Chromatography
D5901	Standard Test Method for Freezing Point of Aviation Fuels (Automated Optical Method)
D5972	Standard Test Method for Freezing Point of Aviation Fuels (Automatic Phase Transition Method)
D6304	Standard Test Method for Determination of Water in Petroleum Products, Lubricating Oils, and Additives by Coulometric Karl Fischer Titration
D6379	Test Method for Determination of Vapor Pressure (VP_X) of Petroleum Products, Hydrocarbons, and Hydrocarbon-Oxygenate Mixtures (Triple Expansion Method)
D6468	Standard Test Method for High Temperature Stability of Middle Distillate Fuels
D6890	Standard Test Method for Determination of Ignition Delay and Derived Cetane Number (DCN) of Diesel Fuel Oils by Combustion in a Constant Volume Chamber
D7170	Standard Test Method for Determination of Derived Cetane Number (DCN) of Diesel Fuel Oils—Fixed Range Injection Period, Constant Volume Combustion Chamber Method

References

[1] Kamara, B. I. and Coetzee, J., "An Overview of High Temperature Fischer-Tropsch Gasoline and Diesel Quality," *Energy and Fuels*, Vol. 23, 2009, pp. 2242–2247.

[2] Alliance for Synthetic Fuels in Europe, ASFE Website, November 2007, http://www.synthetic-fuels.org/index_en.php

[3] Matthew, J., Overett, R., Hill, O., and Moss, J. R., "Organometallic chemistry and surface science: Mechanistic models for the Fischer-Tropsch synthesis," *Coordination Chem. Reviews*, 2000, pp. 206–207, 581–605.

[4] Steynberg, A. P. and Dry, M. E., "Fischer-Tropsch Reactors," in A. P. Steynberg and M. E. Dry, Eds., *Studies in Surface Science and Catalysis 152: Fischer-Tropsch Technology: 64-183.* Netherlands: Elsevier, 2004.

[5] Kamara, B. I. and Coetzee, J., "An Overview of High Temperature Fischer-Tropsch (HTFT) Gasoline and Diesel Quality," *SAE International Powertrains, Fuels and Lubricants Congress*, Shanghai, China, 2008.

[6] Steyn, C., "The role of Mossgas in Southern Africa," *Presented at the 2nd Sub-Saharan Africa. Catal. Symp.*, Swakopmund, Namibia, 2001.

[7] Kaiser, E. W., Siegl, W. O., Cotton, D. F., and Anderson, R. W., "Effect of Fuel Structure on Emissions from a Spark-Ignite Engine. 2. Naphthene and Aromatic Fuels," *Environ. Sci. Technol*, Vol. 26, 1992, pp. 1581–1586,

[8] SAPIA, "Petrol and Diesel in South Africa and the Impact of Air Quality," *A Comprehensive Reference Guide*, SAPIA Publication,-South Africa 2008, http://www.sapia.co.za/pdf/2008-Sapia-AR.pdf (accessed January 2010).

[9] Anon. *Motor Gasolines Technical Review (FTR-1)*, Chevron Products Company, 1996, http://www.chevron.com/prodserv/fuels/bulletin/motorgas/Motor_Gas.pdf (accessed January 2010).

[10] Gravett, P. S., Roets, P. N. J., and Strauss, H. F., "Sasol Technology Fuels Research," *The Use of Environmental Acceptable Synthetic Fuel Oils and Blends thereof with Crude Derived Heavy Fuel Oil*, 9th International Conference on Stability and Use of Liquid Fuels, Spain, 2005.

[11] O'Rear, D. J., Bacha, J. D., and Tiedemann, A. N., *Energy Fuels*, Vol. 18, No. 682, 2004.

[12] Viljoen, C. L., Moses, C. A., and Morgan, P. M., "The Sasol Synthetic Jet Fuel Story," *2nd SAIAS Symposium*, Stellenbosch, 14–16 September 2008.

[13] Moses, C. A., Stavinoha, L. L., and Roets, P. N. J., "Qualification of SASOL Semi-Synthetic Jet A-1 as Commercial Jet Fuel," *Final Report*, Southwest Research Institute, San Antonio, TX, 1997, http://www.crcao.com/reports/recentstudies2007/AV-02-04/AV-2-04%20Final%20Report%20-%20FINAL2%20March%202-2008.pdf (accessed January 2010).

[14] UK Ministry of Defence Standard 91-91, Issue 3, 1999.

[15] Moses, C. A., Wilson, III, G. and Roets, P. N. J., *Evaluation of Sasol Synthetic Kerosene for Suitability as Jet Fuel*, San Antonio, TX: Southwest Research Institute, Report no. 08-04438, 2003.

[16] Moses, C. A., *Evaluation of Sasol Synthetic Kerosene for Suitability as Jet Fuel: Phase II: Engine and Combustor Tests*,- Southwest Research Institute, Report No. 08-04438-2, 2007, http://www.crcao.com/reports/recentstudies2007/AV-02-04/AV-2-04%20Final%20Report%20-%20FINAL2%20March%202-2008.pdf (accessed January 2010).

[17] Yates, D. B., Viljoen, C., and Metcalf, O., "An Accurate Determination of the Cetane Number Value of GTL Diesel, SAE Paper Series No. 2007-01-0026, SAE Transactions," *J. Fuels and Lubricants*, Vol. 116, 2007, pp. 18–27.

[18] Technical Review (FTR-2), "Diesel Fuels," *J. Diesel Fuel Refining and Chemistry*, Chevron Products Company, 1998.

[19] ACEA, AAM, EMA, and JAMA, "Technical Background for Harmonised Fuel Recommendations: Diesel Fuels," *Worldwide Fuel Charter*, 4th ed., 2006, http://oica.net/wp-content/uploads/2007/06/wwfc-fourth-edition-sep-2006.pdf (accessed January 2010).

Index

B

P

Q

R

T

X

Y